High Resolution X-ray Spectroscopy of Cosmic Plasmas

International Astronomical Union
Union Astronomique Internationale

The following Colloquia of the International Astronomical Union are published for the Union by Cambridge University Press.

82. Cepheids. *Edited by Barry F. Madore.* 0 521 30091 6. 1985

91. History of Oriental Astronomy. *Edited by G. Swarup, A.K. Bag and K.S.Shukla.* 0 521 34659 2. 1987

92. Physics of Be Stars. *Edited by A. Slettebak and T.P. Snow.* 0 521 33078 5. 1987

101. Supernova Remnants and the Interstellar Medium. *Edited by R.S. Roger and T.L. Landecker.* 0 521 35062 X. 1988

105. The Teaching of Astronomy. *Edited by Jay M. Pasachoff and John R. Percy.* 0 521 35331 9. 1990

106. Evolution of Peculiar Red Giant Stars. *Edited by Hollis Johnson and Ben Zuckerman* 0 521 36617 8. 1989

111. The Use of Pulsating Stars in Fundamental Problems of Astronomy. *Edited by Edward G. Schmidt* 0 521 37023 X. 1989

115. High Resolution X-ray Spectroscopy of Cosmic Plasmas. *Edited by Paul Gorenstein and Martin Zombeck* 0 521 37018 3. 1990

International Astronomical Union
Union Astronomique Internationale

High Resolution X-ray Spectroscopy of Cosmic Plasmas

The proceedings of International Astronomical Union 115th colloquium

Edited by

PAUL GORENSTEIN AND MARTIN ZOMBECK

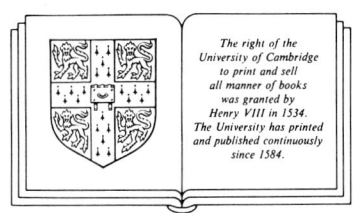

CAMBRIDGE UNIVERSITY PRESS
Cambridge
New York   Port Chester
Melbourne   Sydney

Published by the Press Syndicate of the University of Cambridge
The Pitt Building, Trumpington Street, Cambridge CB2 1RP
40 West 20th Street, New York, NY 10011, USA
10 Stamford Road, Oakleigh, Melbourne 3166, Australia

© Cambridge University Press 1990

First published 1990

Printed in Great Britain at the University Press, Cambridge

British Library cataloguing in publication data

Library of Congress cataloguing in publication data

ISBN 0 521 37018 3

CONTENTS

Organizing Committees	viii
Acknowledgements	viii
Preface	ix
Colloquium Photograph	x
List of Participants	xi

1. X-rays from a Hot Plasma

Emission Lines From Hot Astrophysical Plasmas
    J.C. Raymond   1
Spectroscopic Diagnostics for Ions Observed in Solar and Cosmic Plasmas
    H. Mason   11
Recent Advances in Atomic Modelling
    W. Goldstein   21
Relativistic Free-free Gaunt Factor of the Dense High Temperature Stellar Plasma
    N. Itoh, M. Nakagawa and Y. Kohyama   32
The Differential Emission Measure of λ And
    M. Landini, B.C. Monsignori Fossi and R. Pallavicini   36
X-ray Lines From Mg VIII and Si X Ions and Their Diagnostic Use
    B.N. Dwivedi   40
Thermal Instability in a Hot Plasma
    S.A. Balbus and N. Soker   44
X-ray Line Formation in Astrophysical Environments: The L-shell, Spectra of Ionized Iron
    D.A. Liedahl, S.M. Kahn, A.L. Osterheld and W.H. Goldstein   49
Measurement of Carbon Ion Photoabsorption Cross Sections Using Laser Plasmas
    B.J. Wargelin, S.M. Kahn, W.W. Craig and R. London   53
Tokomak Plasmas: A Paradigm for Coronal Equilibrium and Disequilibrium
    R. Petrasso   57

2. Magnetic Effects

Low Energy Lines in Spectra of Gamma Ray Bursts
    G. Bisnovatyi-Kogan and A.F. Illarionov   63
The Effects of Magnetic Fields Upon Cosmic X-ray Spectra
    P. Mészáros   70
Effect of Vacuum Polarization in a Strong Magnetic Field and Spectral Features of X-Ray Source Emission
    Y. Gnedin   78
Merging of Iron Lines in Spectra of X-ray Burst Sources
    J. Madej   85
Role of Magnetospheric Plasma Physics for Understanding Cosmic Phenomena
    I.M.L. Das   90

3. Stellar Coronae

Goals for the Application of High Resolution X-ray Spectroscopy to the Diagnosis of Stellar Corona Plasmas
    J.L. Linsky   94
X-ray Spectroscopy across the HR Diagram
    J.H.M.M. Schmitt   110
Spectroscopy of Stellar Coronal Sources with the Medium Energy Experiment on EXOSAT
    R. Pallavicini, L. Pasquini, J.H.M.M. Schmitt and G. Tagliaferri   122
X-ray Spectral Synthesis in Hydrodynamic Flare Models
    S. Serio, E. Antonucci, M.A. Dodero, G. Peres and F. Reale   126
X-ray Spectral Diagnostics for Coronal Loops in the Active K Dwarf AB Doradus
    O. Vilhu and A.C. Cameron   132

4. Supernova Remnants, Soft X-ray Background

High Resolution Spectroscopy and Plasma Diagnostics of Supernova Remnants
    C. Canizares   136

Features in the Soft X-ray Background
    R. Rothenflug   146

Non-equilibrium Ionisation in Tycho's Supernova Remnant
    W. Brinkmann   156

Fe VIII, IX, and X Line Emission From the Soft X-ray Background: Previous Limits and a Future Measurement
    J.J. Bloch, W.C. Priedhorsky and B.W. Smith   160

X-ray emission from Reverse-Shocked Ejecta in Supernova Remnants
    D.F. Cioffi and C.F. McKee   164

Integrated X-ray Surface Brightnesses of Supernova Remnants and Comparison with Radio and Infrared Values
    J.R. Dickel, L.K. Norton and P.J. Gensheimer   168

A Comparison of Three Regions of Puppis A
    M.F. Fischbach, L.M. Bateman, C.R. Canizares, T.H. Markert and P.J. Saez   172

5. Compact Binaries

An EXOSAT Observation of Spectral Variability From RS CVn Binary AR LAC
    N.E. White, R.A. Schafer, A.N. Parmer, K. Horne and J.L. Culhane   176

Theory of Accretion Disk Coronae
    T. Kallman   187

Lin Structures in the X-ray Spectra of Cygnus X-2 Observed With EXOSAT
    P.E. Freeman, S.M. Kahn, L. Chiapetti, A. Ciapi, L. Maraschi, E.G. Tanzi, A. Treves, E.G. Branduardi-Raymont and E.N. Ercan   197

X-ray Absorption by Ionized Gas in EXOSAT Spectra from the Binary System 4U1700-37/HD153919
    F. Haberl, T.R. Kallman and N.E. White   201

X-ray Spectroscopy of the Ultra-soft Transient 4U1543-47
    H. van der Woerd, N.E. White and S.M. Kahn   205

6. Clusters of Galaxies, Cooling Flows

X-ray Spectra of Clusters of Galaxies
    C. Sarazin   209

Signatures of Cooling Flows
    A. Fabian   219

Alternatives to the Existence of Large Cooling Flows
    W. Tucker   232

Evolution of the Coronae in Early Type Galaxies
    L. David, W. Forman and C. Jones   240

Implications of Abundance Measurements of the Intracluster Medium
    W. Forman, C. Jones and L. David   245

Clusters of Galaxies and the Hot Intracluster Medium
    C. Jones, W. Forman and L. David   249

The Evolution of Cooling Flow and the Mass Deposition Process
    M. Hattori and A. Habe   255

A Cooling Flow Cluster at Redshift, Z=0.2
    A. Wolter, I.M. Gioia, T. Maccacaro, S.L. Morris, R. Nesci, G.C. Perola and R.E. Schild   259

7. Active Galactic Nuclei, Cosmic X-ray Background

High Resolution X-ray Spectroscopy of Active Galactic Nuclei
    J. Krolik   263

Comsic X-ray Background from Early Active Galactic Nuclei
    A. Zdziarski   274

Cosmic X-ray Background from Self-absorbed Low Luminosity AGNs
    J.E. Grindlay and M. Luke   276

8. Future X-ray Observatories

Comments on the Future Observatories and Their X-ray Spectroscopy Capability
    J.L. Culhane   281
The ROSAT Mission
    J. Trümper   291
The SPEKTROSAT Mission
    P. Predehl   295
The X-ray Astronomy Satellite SAX
    R.C. Butler   302
XSPECT: A Telescope/Spectrometer System on SPECTRUM-RÖNTGEN-GAMMA
    H. Schnopper   307
JET-X: A Joint European X-ray Telescope for SPECTRUM-X
    A. Wells, D.H. Lumb, K..A. Pounds, G.C. Stewart, B. Aschenbach, H. Brauninger, G. Hasinger,
    J. Trümper, O. Citterio, L. Scarsi, A. Peacock and B. Taylor   318

9. Future X-ray Observatories, Detectors and Instrumentation

A High Resolution Echelle Spectrometer for Soft X-ray and EUV Astronomy
    J. Green and S. Bowyer   331
Expected Scientific Performance of the Three Spectrometers on the Extreme Ultraviolet Explorer
    J. Vallerga, P. Jelinsky, P.W. Vedder and R. Malina   335
Dispersive Spectroscopy on AXAF
    T. Markert   339
Thermal Detectors for X-ray Astronomy
    S. Holt   346
Innovative Techniques for X-ray Calorimetry
    S. Labov, E. Silver, D. Landis, N. Madden, F. Goulding, J. Beeman, E. Haller, J. Rutledge,
    G. Bernstein and P. Timbie   357
Observing Soft X-ray Line Emission from the Interstellar Medium with X-ray Calorimeter on a
Sounding Rocket
    J. Zhang, B. Edwards, M. Juda, R. Kelley, G. Madejski, D. McCammon, H. Moseley, M. Skinner,
    R. Schoelkopf and A. Szymkowiak   361
High Throughput Soft X-ray Spectroscopy With Reflection Gratings
    S. Khan   365
The SHEAL Diffuse X-ray Spectrometer Experiment
    W.T. Sanders, S.L. Snowden and R.J. Edgar   376
Stigmatic Spectroscopic Instruments for the Wavelength Range 30–300 Å With High Angular and
Spectral Resolution Using Multi-layer Mirrors
    E.N. Ragozin   380
Subject Index   384
Object Index   389

IAU Colloquium 115

High Resolution X-ray Spectroscopy of Cosmic Plasmas

Scientific Organizing Committee	Local Organizing Committee
Paul Gorenstein, Chairman J. Bleeker C. Canizares L. Culhane A. Fabian S. Kahn J. Linsky M. Oda J. Raymond C. Sarazin H. Schnopper R. Sunyaev A. Treves J. Trümper	Martin Zombeck, Chairman Paul Gorenstein Kathleen Lestition Thomas Markert

Acknowledgements

We would like to thank the following organizations for their generous financial contributions in support of IAU Colloquium 115. These grants helped us to defray standard meeting costs such as facility and equipment rental, daily coffee breaks, and colloquium receptions, and they have allowed us to provide travel grants to participants.

Ball Aerospace
Bendix-Allied Signal
Eastman-Kodak
Lockheed Missiles and Space
Perkin-Elmer
TRW
NASA
Smithsonian Institute
International Astronomical Union

PREFACE

The maturing of X-ray astronomy as an observational science is reflected in the increasing emphasis being placed upon high resolution spectroscopy. Gas at temperatures from 10^6 to 10^8 K is found in cosmic environments ranging in scale from the coronae of stars to the medium between members of a rich cluster of galaxies. It is also found in accretion disks of compact objects, supernova remnants, and active galactic nuclei. Thermal X-ray emission characterised by lines and edges is a prominent feature of these objects. Hence, a great deal of effort is being devoted to improving theoretical models for the interpretation of spectroscopic observations. As the power of high resolution X-ray spectroscopy to elucidate astrophysical conditions becomes more widely appreciated experimenters are making significant improvements to the sensitivity and resolution of instruments.

X-ray spectroscopy is an important objective of virtually every new major national and international mission in X-ray astronomy that is now being planned or under discussion. These initiatives include ASTRO-D (Japan), *SAX* (Italy) *Spectrum-X-Gamma* (USSR), AXAF (USA),. XMM (European Space Agency), and a spectroscopic successor to ROSAT (Germany). *In toto*, X-ray astronomy missions account for a very large fraction of the total resources that the world will devote to astronomy from space over the next two decades.

IAU Colloquium 115 was the first major international meeting to provide a comprehensive forum with a cosmic focus for all aspects of high resolution X-ray spectroscopy. The sequence of papers ranges from detailed theoretical models of X-ray emission, to observational results, to instrument concepts for the future. This collection should serve as an introduction, as well as a comprehensive review of the current status of high resolution spectroscopy in a cosmic setting.

The sessions were held August 22–25, 1988 at the Harvard University Science Center in Cambridge Massachusetts. Special thanks are due to the other members of the local organizing committee, namely Kathy Lestition and Tom Markert for their efforts in making the sessions function smoothly and organizing social programs. We are specially grateful to Sharlene Ford whose services as chief secretary were very valuable through all phases of IAU Colloquium 115 beginning with the organization of the meeting to the delivery of the manuscripts to Cambridge University Press.

Paul Gorenstein
Martin Zombeck

List of Participants

E. Antonucci	Istituto Di Fisica, Torino, Italy
M. Arnaud	Service d'Astrophysique, CEN Saclay, Cedex, France
S. A. Balbus	University of Virginia, Charlottesville, Virginia, USA
M. Bautz	MIT, Cambridge, Massachusetts, USA
D. Bhattacharya	Raman Research Institute, Bangalore, India
G. Belvedere	Istituto Di Astronomia, Citta Universitaria, Catania, Italy
R. Berlot	Los Angeles, California 90027
G. Bisnovatyi-Kogan	Institute for Space Research, Moscow, USSR
J. Bixler	LLNL, Livermore, California
J. Bloch	LANL, Los Alamos, New Mexico
H. V. Bradt	MIT, Cambridge, Massachusetts, USA
U. G. Briel	MPE, Garching bei München, FRG
A. C. Brinkman	Laboratory for Space Research, Utrecht, The Netherlands
W. Brinkmann	MPE, FRG
C. Budtz-Jorgensen	Danish Space Research Institute, Lyngby, Denmark
R. C. Butler	PSN/CNR, Roma
C. Canizares	MIT, Cambridge, Massachusetts, USA
W. C. Cash	CASA, University of Colorado, Boulder, Colorado, USA
J. Chappell	Harvard/Smithsonian CfA, Cambridge, Massachusetts, USA
F. E. Christensen	Danish Space Research Institute, Lyngby, Denmark
D. F. Cioffi	NASA/GSFC, Greenbelt, Maryland, USA
G. W. Clark	MIT, Cambridge, Massachusetts, USA
T. J-L. Courvoisier	Geneva Observatory
J. L. Culhane	MSSL, Surrey, England, UK
I. M. L. Das	University of Allahabad, Allahabad, India
L. P. David	Harvard/Smithsonian CfA, Cambridge, MA 02138
P. De Korte	SRON, Utrecht, The Netherlands
D. Dewey	MIT, Cambridge, Massachusetts, USA
J. R. Dickel	University of Illinois, Urbana, Illinois, USA
B. N. Dwivedi	Banaras Hindu University, Varanasi, India
R. J. Edgar	University of Wisconsin, Madison, Wisconsin, USA
A. Fabian	Institute of Astronomy, Cambridge, UK
W. Forman	Harvard/Smithsonian CfA, Cambridge, Massachusetts, USA
G. W. Fraser	University of Leicester, Leicester, UK
P. E. Freeman	University of California, Berkeley, California
M. R. Garcia	Harvard/Smithsonian CfA, Cambridge, Massachusetts, USA
D. Gergal	Observertoire de Paris-Meudon, Cedex, France
I. M. Gioia	Harvard/Smithsonian CfA, Cambridge, Massachusetts, USA

Y. Gnedin	Astronomical Observatory, Pulkovo, Leningrad, USSR
W. Goldstein	LLNL, Livermore, California, USA
P. Gorenstein	Harvard/Smithsonian CfA, Cambridge, Massachusetts, USA
J. Green	University of California, Berkeley, California, USA
J. Grindlay	Harvard/Smithsonian CfA, Cambridge, Massachusetts, USA
J. M. Grunsfeld	University of Chicago, Chicago, Illinois, USA
H. Gursky	NRL, Washington, DC, USA
F. Haberl	EXOSAT Observatory, SSD, ESTEC, The Netherlands
A. Hamilton	JILA, University of Colorado, Boulder, Colorado, USA
J. A. Hoffman	NASA/JSC, Houston, Texas
S. Holt	NASA/GSFC, Greenbelt, MD 20771
A. Hornstrup	Danish Space Research Institute, Lyngby, Denmark
J. P. Hughes	Harvard/Smithsonian CfA, Cambridge, Massachusetts, USA
N. Itoh	Sophia University, Tokyo, Japan
F. A. Jansen	Lab. for Space Research, Leiden, The Netherlands
C. Jones	Harvard/Smithsonian CfA, Cambridge, Massachusetts, USA
S. M. Kahn	University of California, Berkeley, California, USA
T. R. Kallman	NASA/GSFC, Greenbelt, Maryland, USA
E. Kellogg	Harvard/Smithsonian CfA, Cambridge, Massachusetts, USA
K. Koyama	Nagoya University, Nagoya, Japan
J. H. Krolik	Johns Hopkins University, Baltimore, Maryland, USA
S. Labov	LLNL, Livermore, California
M. Landini	Arcetri, Naples, Italy
K. Lestition	Harvard/Smithsonian CfA, Cambridge, Massachusetts, USA
D. Liedahl	University of California, Berkeley, Oakland, California, USA
J. Linsky	JILA, University of Colorado, Boulder, Colorado, USA
K. S.K. Lum	MIT, Cambridge, Massachusetts, USA
T. Maccacaro	Harvard/Smithsonian CfA, Cambridge, Massachusetts, USA
J. Madej	Astronomical Observatory of Warsaw University, Warszawa, Poland
T. Markert	MIT, Center for Space Research, Cambridge, Massachusetts, USA
H. Mason	Cambridge University, Cambridge, UK
D. Moses	Am. Sci. & Eng., Fort Washington, Cambridge, MA 02139
J. C. McDowell	Harvard/Smithsonian CfA, Cambridge, Massachusetts, USA
P. Meszaros	Pennsylvania State University, University Park, Pennsylvania, USA
B. Monsignori-Fossi	Osservatorio di Arcetri, Florence, Italy
S. M. Murray	Harvard/Smithsonian CfA, Cambridge, Massachusetts, USA
S. Naranan	Tata Institute of Fundamental Research, India
J. E. Neff	NASA/GSFC, Greenbelt, Maryland, USA
R. Pallavicini	Osservatorio di Arcetri, Florence, Italy

R. Petrasso	MIT, Cambridge, Massachusetts
P. Predehl	MPE, Garching bei Muenchen, FRG
F. A. Primini	Harvard/Smithsonian CfA, Cambridge, Massachusetts, USA
E. N. Ragozin	P.N. Lebedev Physics Institute, Moscow, USSR
G. L. Rawley	NASA/GSFC, Greenbelt, Maryland, USA
J. Raymond	Harvard/Smithsonian CfA, Cambridge, Massachusetts, USA
G. R. Ricker	MIT, Cambridge, Massachusetts, USA
R. R. Ross	MIT, Cambridge, Massachusetts
R. Rothenflug	Centre d'Etude Nucleaire de Saclay, Cedex, France
R. Sadat	Observertoire de Paris-Meudon, Cedex, France
W. T. Sanders	University of Wisconsin, Madison Wisconsin, USA
C. Sarazin	University of Virginia, Charlottesville, Virginia, USA
J. Schmitt	MPE, Garching Bei München FRG
H. W. Schnopper	Danish Space Research Institute, Lyngby, Denmark
D. Schwartz	Harvard/Smithsonian CfA, Cambridge, Massachusetts, USA
S. Serio	Istituto Di Fisica, Torino, Italy
F. D. Seward	Harvard/Smithsonian CfA, Cambridge, Massachusetts, USA
E. Silver	LLNL, Livermore, California
B. W. Smith	LANL, Los Alamos, New Mexico, USA
G. Trinchieri	Harvard/Smithsonian CfA, Cambridge, Massachusetts, USA
J. Trümper	MPE, Garching bei Muenchen, FRG
K. Tucker	Harvard/Smithsonian CfA, Cambridge, Massachusetts. USA
W. Tucker	Harvard/Smithsonian CfA, Cambridge, Massachusetts, USA
J. Vallerga	SSL,University of California, Berkeley, California, USA
H. van der Woerd	EXOSAT Observatory, ESTEC-SAE, The Netherlands
L. Van Speybroeck	Harvard/Smithsonian CfA, Cambridge, Massachusetts, USA
O. Vilhu	University of Helsinki, Helsinki, Finland
S. D. Vrtilek	NASA/GSFC, Greenbelt, Maryland
S. Wagh	Harvard/Smithsonian CfA, Cambridge, Massachusetts, USA
B. Wargelin	University of California, Berkeley, California, USA
A. Wells	University of Leicester, UK
N. E. White	EXOSAT Observatory, SSD, ESTEC, Noordwijk, The Netherlands
B. L. Whitten	Colorado College, Boulder, Colorado, USA
B. J. Wilkes	Harvard/Smithsonian CfA, Cambridge, Massachusetts, USA
F. Winkler	Middlebury College, Middlebury Vermont, USA
A. Wolter	Harvard/Smithsonian CfA, Cambridge, Massachusetts, USA
D. M. Worrall	Harvard/Smithsonian CfA, Cambridge, Massachusetts, USA
A. Zdziarski	Space Telescope Science Institute, Baltimore, Maryland, USA
J. Zhang	University of Wisconsin, Madison, Wisconsin, USA
M. V. Zombeck	Harvard/Smithsonian CfA, Cambridge, Massachusetts, USA

1. X-rays from a Hot Plasma

EMISSION LINES FROM HOT ASTROPHYSICAL PLASMAS

John C. Raymond

Harvard-Smithsonian Center for Astrophysics, Cambridge, MA, USA

ABSTRACT. The spectral lines which dominate the X-ray emission of hot, optically thin astrophysical plasmas reflect the elemental abundances, temperature distribution, and other physical parameters of the emitting gas. The accuracy and level of detail with which these parameters can be inferred are limited by the measurement uncertainties and uncertainties in atomic rates used to compute the model spectrum. This paper discusses the relative importance and the likely uncertainties in the various atomic rates and the likely uncertainties in the overall ionization balance and spectral line emissivities predicted by the computer codes currently used to fit X-ray spectral data.

1. INTRODUCTION

High resolution X-ray spectroscopy gives us access to the huge amount of information contained in X-ray spectral lines. The *Einstein* and EXOSAT satellites have provided a taste of detailed spectroscopy of cosmic plasmas, and the next generation of X-ray satellites should give us spectra comparable to those obtained from solar flares. The interpretation of spectral lines from cosmic plasmas, which so far has been limited mostly by the spectral resolution or statistical quality of the data, may then be limited by the accuracy of theoretical models of the X-ray emission. The accuracy of our inferences about the elemental composition, the density, and the temperature structure of the emitting plasma is no better than the accuracy of the atomic rates and other assumptions which go into the model.

Here we'll concentrate on the atomic rates and their reliability. To keep the discussion simple, the temperature of the emitting gas is assumed constant as a function of time. Density effects are ignored and diffusion due to steep temperature and density gradients is taken to be negligible. The electron velocity distribution is taken to be Maxwellian, and electric and magnetic fields are assumed to be weak. We'll also assume that photoionization is insignificant and that line and continuum optical depths are small. The limits of validity of these assumptions and some of the consequences of their violation are discussed in Raymond (1988).

The following sections discuss the processes which determine the ionization state of the plasma and the intensities of the spectral lines. Comparisons are then made between various spectral calculations and between a model computation and the X-ray spectrum of a solar flare.

2. IONIZATION BALANCE

Under the above assumptions, the ionization state of each element in the plasma is determined by a set of equations

$$0 = \frac{1}{n_e}\frac{dn_i}{dt} = q_{i-1}n_{i-1} - (q_i+\alpha_i)n_i + \alpha_{i+1}n_{i+1} \qquad (1)$$

for each ion of the element, where q_i is the ionization rate coefficient from ion i and α_i is the recombination rate coefficient from ion i.

The accuracy of the available ionization and recombination rate coefficients varies widely. Ionization cross sections have been measured for a number of low and moderate ions by crossed beam experiments to 10 - 20% accuracy (e.g. Gregory *et al.* 1987). For many ions, cross sections computed in the Distorted Wave approximation are likely to be good to 20 - 30% (Younger 1981), but these are not yet available for all the astrophysically important ions. Auger ionization following innershell excitation makes an important contribution for some iso-electronic sequences. The best compilation of ionization rates is given by Arnaud and Rothenflug (1985). At fairly high densities, such as solar flares, corrections due to the population of metastable levels may increase the ionization rates somewhat.

The recombination coefficient is made up of radiative recombination and dielectronic recombination. The former,

$$A^{+j} + e^- \rightarrow A^{+j-1} + h\nu \qquad (2)$$

is the inverse process of photoionization, so the detailed balance relation is used with the photoionization cross section to compute the rate, and the rate is just as accurate as the photoionization cross sections. For hydrogenic ions, the rates are accurate to a few percent. For more complex ions, the recombination to excited levels is computed by scaling hydrogenic rates, and the rate of recombination to the ground state is computed from the ground state photoionization cross section (e.g. Reilman and Manson 1979). Overall, radiative recombination the rates are generally good to around 30%. Radiative recombination rates to ions of the Li-like and Na-like sequences, which are quite a lot like hydrogenic ions, are probably more accurate. Radiative recombination rates for low ionization states of iron and nickel, which have very complicated structures, are worse. More accurate photoionization cross sections for both ground and excited levels should be available before long as a result of the Opacity Project (see Pradhan 1987).

In the dielectronic recombination process, an electron encounters an ion and excites it, but finds itself with negative energy, so that it is bound in some high *nl* state of the recombined ion. For example, a Li-like ion has a ground 2s level and excited 2p level. The process

$$2s + \varepsilon l \rightarrow 2pnl \tag{3}$$

has a very large cross section at energies ε just below the threshold for 2s - 2p excitation. The ion in the 2p*nl* state can either autoionize (reversing the process which formed it) or emit a photon, usually a 2s - 2p photon, to reach the bound, highly excited 2s*nl* state of the Be-like ion. The contributions of a large number of *nl* levels must be summed to find the total dielectronic recombination rate. Burgess (1965) fit the dielectronic recombination rates of many ions to a simple formula which is still widely used. It is probably good to its advertised accuracy of about 30% in the cases for which it was intended; elements up to calcium, and temperatures near the peak of the ionic abundance in coronal equilibrium. The Burgess formula requires modification at lower temperatures (Nussbaumer and Storey 1983), at high densities (Summers 1974), in strong electromagnetic fields (Müller *et al.* 1987), and in cases in which the doubly excited level can autoionize to levels other than the ground state (Jacobs *et al.* 1978). At present there is no no compliation of dielectronic recombination rates which adequately includes all these effects. Even sophisticated calculations of dielectronic recombination in complex ions can disagree by factors of two, and the only measurements of dielectronic recombination rates are dominated by field effects. Hahn (1985) reviews the theory, and recent calculations for some ions are given by Roszman (1987), Smith *et al.* (1985), and McLaughlin and Hahn (1984). Figure 1 compares several computed dielectronic recombination rates for the Li-like ion Fe XXIV. All the calculations agree at low temperatures, where recombination by way of 2s - 2p dominates, but they disagree badly at high temperatures, where excitation to the *n* = 3 levels becomes important. Unfortunately, the Fe XXIV emission lines are formed in the high temperature regime.

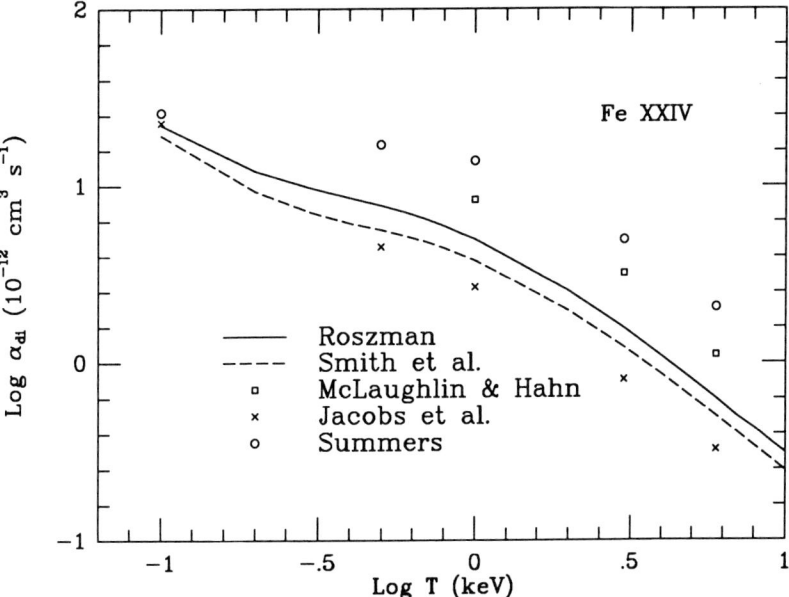

Figure 1. Dielectronic recombination of Fe XXIV.

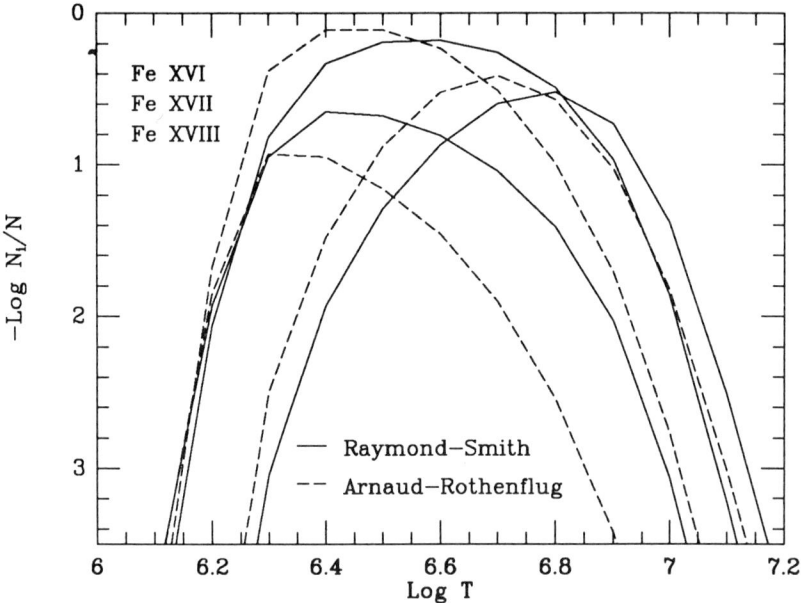

Figure 2. Concentrations of Fe XVI, Fe XVII and Fe XVIII.

Putting all the ionization and recombination rates together, we arrive at a computed ionization balance by solving the set of equations (1) for each element. Figure 2 compares the ionization balances computed for a few ions of iron by Arnaud and Rothenflug (1985) and the current version of the Raymond and Smith (1977) X-ray emission code. The ionic abundance curves in the different calculations look quite similar, but the peak temperatures are shifted by around 0.1 in log T. Figure 2 also shows an example of a more extreme difference. The Fe XVI abundance, which is controlled by dielectronic recombination of the complicated Ne-like ion Fe XVII, differs by more than a factor of two even at its peak.

3. RADIATION

The continuum emission includes bremsstrahlung, which has been computed to a few percent accuracy up to temperatures around 10^8 K, where relativistic corrections and electron-electron bremsstrahlung begin to be significant. The radiative recombination continuum is computed along with the radiative recombination rates, and is equally accurate. Radiative recombination continua of H- and He-like ions are generally by far the most important, and accurate cross sections are available. The two-photon continua of H- and He-like ions (arising from the metastable 2s and 1s2s ^{1}S levels respectively) make significant contributions to the X-ray continuum, and reliable cross sections for the excitation rates to these levels are available.

At temperatures below about 10^7 K, and in some energy bands at higher temperatures, the X-ray radiation is largely due to bound-bound transitions of the more abundant elements, especially O, Si, and Fe. The emissivity in a given line depends on the elemental abundance, the ionization balance and the excitation rate of the line. The reliability of the excitation rates varies widely. For excitations to $n = 2$ states of H- and He-like ions, extensive theoretical calculations in Close-Coupling and Distorted Wave methods are available, and for these simple ions the excitation rates are probably good to 10-20%. Thus the O VII and O VIII, Si XIII and Si XIV, or the Fe 6.7 keV features are not only the strongest lines in most X-ray spectra, they are also the most reliably interpreted. Higher lying levels of these ions, as well as most of the more complex ions, have not been as extensively treated. Distorted Wave cross sections are available for most of the $n = 2$ to $n = 3$ lines of Fe XVII through Fe XXIV which dominate the 1 keV complex of lines, but neither cascades nor resonances in the excitation cross sections, which significantly enhance some of the lines of Fe XVII (Smith *et al.* 1985) have been included for most of the ions. Some indication of trouble with the theoretical excitation cross sections comes from studies of the relative intensities of the $n = 2$ to $n = 3$ lines of Li- and Be-like ions in laboratory and a solar flare plasmas, in which most relative line intensities agree to the expected 20% level, but a few lines discrepant by a factor of two (Huang *et al.* 1988; Mackenzie *et al.* 1985).

A correction which must be included for some lines is the contribution of dielectronic recombination satellites. The decay of the doubly excited level in the dielectronic recombination process produces a photon near the wavelength of the resonance transition. Satellites to lines of He- and Ne-like ions tend to be most important, and the satellite contributions are largest for highly charged ions. The satellites are strongest relative to direct excitation of the resonance lines at low temperatures, so the cool side of the temperature range of an He- or Ne-like ion gives the largest satellite contribution. The satellite lines to the He-like resonance lines have been extensively studied for their diagnostic uses, and their predicted intensities are probably good to 20% (see review by Bely-Dubau 1988).

Another correction is the contribution of recombination to emission line intensities, following radiative recombination to excited levels or during the cascade of the recombining electron following dielectronic recombination. These terms are not generally included except for H- and He-like ions, where they increase the intensities of some lines by 20 - 30% in collisional ionization equilibrium (Mewe and Schrijver 1978). In a rapidly cooling plasma, or in a photoionized gas such as the accretion disk corona of an X-ray binary, this contribution can be much larger.

4. COMPARISON AMONG THEORIES

Several computer codes put together ionization balance calculations with excitation rates to predict X-ray emission spectra as a function of temperature for use in interpreting X-ray observations (Mewe, Gronenschild, and van den Oord 1985; Raymond and Smith, updated version of 1977 code; Gaetz and Salpeter 1983; Hamilton and Sarazin 1984; Shull 1981; Landini *et al.* 1985). Doschek and Cowan (1985) give a line emissivity list for the 10 - 200 Å range based on observed solar X-ray spectra rather than theoretical calculations. Comparison of these calculations gives some idea of their accuracy. Figure 3 shows the emissivity of the O VII $\lambda 21.6$

line from several of these models. The uncertainties in both the ionization balance and the excitation rate for this line are fairly small, and the four calculations give quite similar results. The Fe XVII line at λ15.01 is shown in Figure 4. Here the

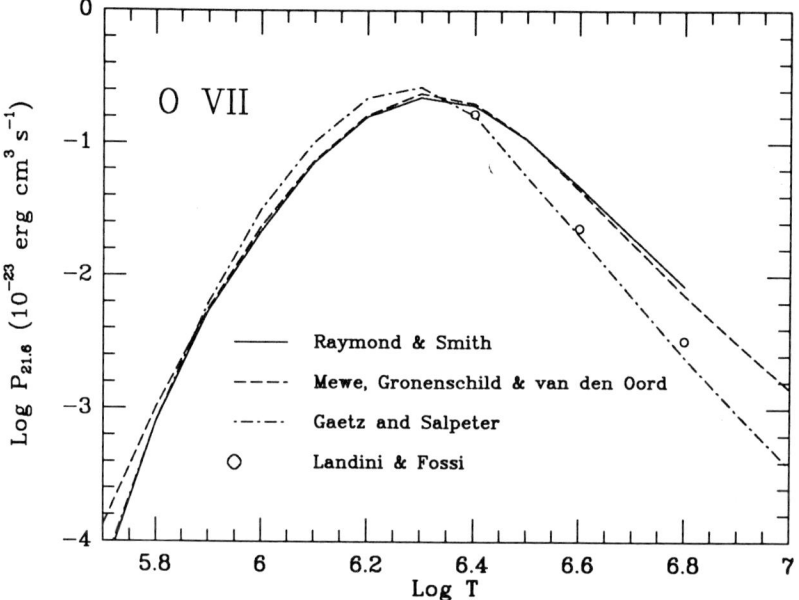

Figure 3. The O VII resonance line predicted by different models.

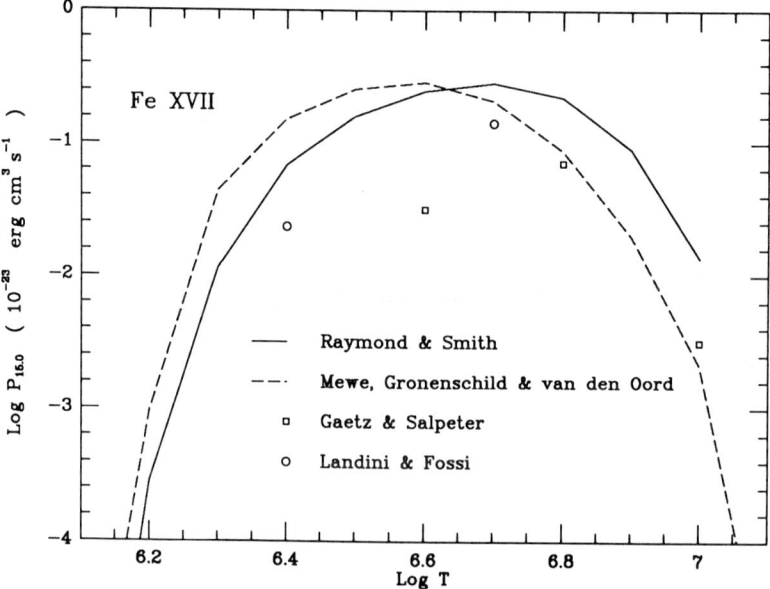

Figure 4. The Fe XVII resonance line.

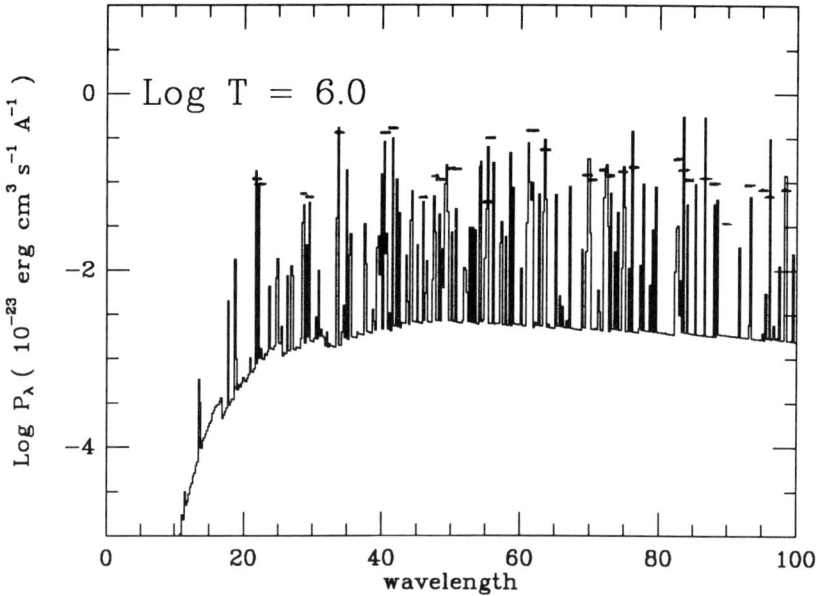

Figure 5. Comparison of X-ray emission codes.

disagreement among the calculations is much worse, mostly due to differing ionization balances used, though the excitation rates also differ somewhat.

Figure 5 shows the overall spectrum emitted by 10^6 K gas with solar abundances and 0.2 Å resolution as computed with the current version of the Raymond and Smith (1977) code. The emissivities of the strong lines given by Mewe, Gronenschild and van den Oord (1985) are indicated by the horizontal bars. It can be seen that the lines of H- and He-line C, N, and O agree quite well, as expected from the reliable rate coefficients for these simple ions. Many of the longer wavelength lines disagree by a factor of two, however. Much of this is due to differing ionization balance calculations, so that a line which peaks at log T = 6.0 in one model may peak at log T = 6.1 in the other. In general, low resolution spectra fit with different thermal X-ray emission models yield the same temperature, but somewhat different emission measures (Schmitt *et al.* 1985). With higher resolution data, the inferred plasma parameters will depend on the particular lines used. For the usual case in which a fairly broad range of temperatures is present, the shift in ionization balance of about 0.1 in log T among the various calcualtions will probably lead to a similar difference in inferred temperatures.

5. COMPARISON WITH OBSERVATIONS

Testing the models against observed spectra is obviously necessary. It is obviously necessary to have many more independent line intensities than model parameters. Only solar X-ray spectra provide adequate spectral resolution, statistical quality and wavelength coverage.

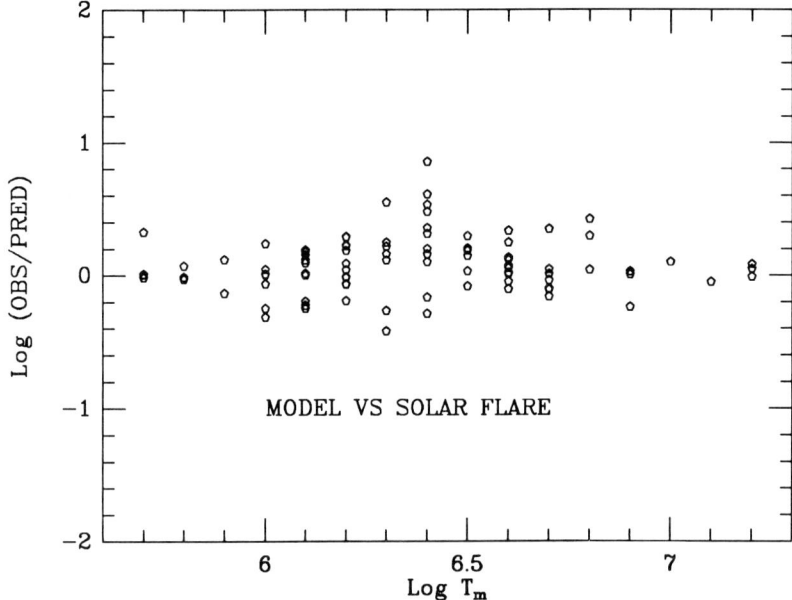

Figure 6. Comparison of model with solar flare observations.

A composite solar flare spectrum can be constructed by joining the 8-22 Å spectrum obtained with the SOLEX experiment on the P78-1 satellite (McKenzie *et al.* 1980) to the 15-100 Å flare spectrum obtained during a rocket flight (Acton *et al.* 1985) by scaling to the same O VII λ21.6 intensity. Some error results from this merging, but the long wavelength range contains mostly low temperature lines, while the short wavelength lines are formed at high temperatures. One then chooses a distribution of emission measure with temperature to match the intensities of some of the "most reliable" lines and iterates to improve the overall fit. The electron density is chosen to match the relative intensities of the O VII lines. Standard abundances were assumed, except that Mg was increased by a factor of 1.6, and Ca and Ni were doubled. Figure 6 shows a plot of the ratio of observed intensity to the intensity predicted by the model for 95 lines. The average absolute value of the log of the ratio of observed to predicted intensities is 0.17, indicating a factor of 1.5 typical error, with the worst line being stronger than predicted by a factor of 5. Based on the uncertainties quoted in the observational papers, it is likely that half the error is observational and half in the model. Further tweaking of the model parameters and further work on deconvolving line blends could reduce the discrepancy somewhat. It is not clear whether the remaining errors come from inadequate atomic physics or from a breakdown in some assumption such as ionization equilibrium or Maxwellian velocity distribution. Better model calculations, starting with improved collisional excitation rates for the complex ions, will be needed for the interpretation of AXAF and XMM X-ray spectra.

REFERENCES

Acton, L.W., Bruner, M.E., Brown, W.A., Fawcett, B.C., Schweizer, W., and Speer, R.J. 1985, *Ap. J.*, **291**, 865.
Arnaud, M., and Rothenflug, R. 1985, *Astr. Ap. Suppl.*, **60**, 425.
Bely-Dubau, F. 1988, in *Hot Thin Plasmas in Astrophysics*, Ed. R. Pallavicini (Kluwer: Dordrecht), p. 21.
Burgess, A. 1965, *Ap. J.*, **141**, 1588.
Doschek, G.A., and Cowan, R.D. 1985, *Ap. J. Suppl.*, **56**, 67.
Gaetz, T.E., and Salpeter, E.E. 1983, *Ap. J. Suppl.*, **52**, 55.
Gregory, D.C., Wang, L.J., Meyer, F.W., and Rinn, K. 1987, *Phys. Rev. A*, **35**, 3256.
Hahn, Y. 1985, *Adv. Atomic and Mol. Phys.*, **21**, 123.
Hamilton, A.J.S., and Sarazin, C.L. 1984, *Ap. J.*, **284**, 601.
Huang, L.K., Lippmann, S., Stratton, B.C., Moos, H.W., and Finkenthal, M. 1988, *Phys. Rev. A.*, **37**, 3927.
Jacobs, V.L., Davis, J., Kepple, P.C., and Blaha, M. 1977, *Ap. J.*, **211**, 605.
Landini, M., Monsigniori Fossi, B.C., Paresce, F., and Stern, R.A. 1985, *Ap. J.*, **289**, 709.
McKenzie, D.L., Landecker, P.B., Broussard, R.M., Rugge, H.R., Young, R.M., Feldman, U., and Doschek, G.A. 1980, *Ap. J.*, **241**, 409.
McKenzie, D.L., Landecker, P.B., Feldman, U., and Doschek, G.A. 1985, *Ap. J.*, **289**, 849.
McLaughlin, D.J., and Hahn, Y. 1984, *Phys. Rev. A.*, **29**, 712.
Mewe, R., and Schrijver, J. 1978, *Astr. Ap.*, **65**, 99.
Mewe, R., Gronenschild, E.H.B.M., and van den Oord, G.H.J. 1985, *Astr. Ap. Suppl.*, **62**, 197.
Müller, A., Belic, D.S., DePaola, B.D., Djuric, N., Dunn, G.H., Mueller, D.W., and Timmer, C. 1987, *Phys. Rev. A*, **36**, 599.
Nussbaumer, H., and Storey, P.J. 1983, *Astr. Ap.*, **126**, 75.
Pradhan, A.K. 1987, *Physica Scripta*, **35**, 840.
Raymond, J.C. 1988, in *Hot Thin Plasmas in Astrophysics*, R. Pallavicini, ed. (Dordrecht: Kluwer), p. 3.
Raymond, J.C., and Smith, B.W. 1977, *Ap. J., Suppl.*, **35**, 419.
Reilman, R.F., and Manson, S.T. 1979, *Ap. J. Suppl.*, **40**, 815.
Roszman, L.J. 1987, *Phys. Rev. A.*, **35**, 3368.
Schmitt, J.H.M.M., Harnden, F.R., Peres, G., Rosner, R., and Serio, S. 1985, *Ap. J.*, **288**, 751.
Shull, M.J. 1981, *Ap. J., Suppl.*, **46**, 27.
Smith, B.W., Raymond, J.C., Mann, J.B., Jr., and Cowan, R.D. 1985, *Ap. J.*, **298**, 898.
Summers, H.P. 1974, *Culham Laboratory Internal Memo* IM-367, Culham Laboratory, Ditton Park, Slough, England.
Younger, S.E. 1981, *JQSRT*, **26**, 329.

DISCUSSION-J. Raymond

S. Kahn: At what density are dielectronic recombination rates affected, say from Fe XVII?

J. Raymond: A highly charged ion with no δ n=0 transitions to the ground state is only affected at extremely high densities: around 10^{20} cm^{-3} for Fe XVII.

Y. Gnedin: You made comparison of predictions and observations mainly for solar flares. What is a situation with supernova remnants?

J. Raymond: Supernova remnants have more model parameters – abundances and non-equilibrium ionization – as well as fewer observed lines. One can choose the best fit model, but not really test the model beyond that.

J. Schmitt: You performed a comparison of line emission [ratios] between the Raymond & Smith and Mewe & Gronenschild code at Log T = 6.0. Could you comment on what this comparison looks like at Log T = 7.0 especially longward of 50Å ?

J. Raymond: The shift of around 0.1 in Log T should occur there, as well as at lower temperatures. Mewe & Gronenschild treat some multiplets as individual lines which Raymond & Smith lump together.

Dickel: Individual lines can be off a lot, but if you have many lines of one or more atoms and ions, how well can you get the temperature and density?

J. Raymond: Many atoms and ions still leave the problem of the uncertainty of ~ 0.1 in Log T of the ionization balance calculations. Temperature sensitive line ratios of individual ions give a check on this. In the case of Fe XVII, though a similar scatter among several ratios is seen.

J. Canizares: In anticipation of the future X-ray spectroscopy missions, it would be very useful to have "empirically corrected" emission measures for some of the stronger lines that tend to be used for plasma diagnostics (for example the Fe XVII lines). In other words, it would be useful for observers to know which lines you have the most confidence in.

J. Raymond: The Doschek and Cowan line list determined from solar spectra is the thing to use.

H. Gursky: Can you comment on the effect of the physics of plasmas on the calculations you present?

J. Raymond: The major plasma effect to worry about is a non-Maxwellian electron distribution, since the only astrophysical plasma we can measure, the solar wind, shows non-Maxwellian tails. If such a tail is present at energies like 10kT, it can drastically affect the ionization.

SPECTROSCOPIC DIAGNOSTICS FOR IONS OBSERVED IN SOLAR AND COSMIC PLASMAS

Helen E Mason

Department of Applied Mathematics and Theoretical
Physics, Silver Street, Cambridge, CB3 9EW, U. K.

ABSTRACT. The X-ray wavelength region (1-200Å) is rich in spectral lines from highly ionised systems. Spectra from the solar atmosphere have been studied extensively with various instruments covering different wavelength regions. In this paper, we discuss the solar spectral line emission with particular reference to iron ions and helium like ions observed during solar flares. The atomic processes involved in the calculation of theoretical intensities for low density plasmas are outlined together with the diagnostic properties of the emission lines. Comparisons are made with available cosmic X-ray spectra and predicted spectra for future projects, such as AXAF.

1. SOLAR ATMOSPHERE

The solar corona has a characteristic temperature of around 2×10^6K and a number density of about 10^8 cm^{-3}. This is in sharp contrast to the visible surface of the sun (photosphere) which has a temperature of 6000°K and a number density of 10^{17}cm^{-3}. There is a steep transition region between the chromosphere and corona as illustrated in figure 1. X-ray and UV images show that the distribution of material in the solar atmosphere is far from homogeneous - indeed it is determined by the magnetic field configurations. Regions of open magnetic fields (coronal holes), which are believed to be the source of the solar wind, have very little coronal emission, whereas active regions show complex loop structure of closed magnetic fields. Small bipolar regions, bright points of X-ray emission, are also evident. These phenomena were studied extensively by instruments on Skylab - Apollo Telescope Mount (ATM)(1972). When the magnetic structures in active regions re-configure, vast amounts of energy are released (solar flares). The temperature in these explosive events rises to over 10^7K and the electron number density exceeds 10^{12}cm^{-3}. The Solar Maximum Mission (SMM), launched in 1980, was designed specifically to study solar flares.

The higher the spatial resolution, the more complex are the structures which are observed. The HRTS instrument flown on Spacelab II (1985) obtained a spectral resolution better than 1". This instrument obtained spectra along a stigmatic slit from 1000-1800Å. Observations of transition region ions (eg CIV) indicate spatial inhomogeneities, variations in line width and shifts which dispel all notions of a homogeneous spherically symmetric transition region.

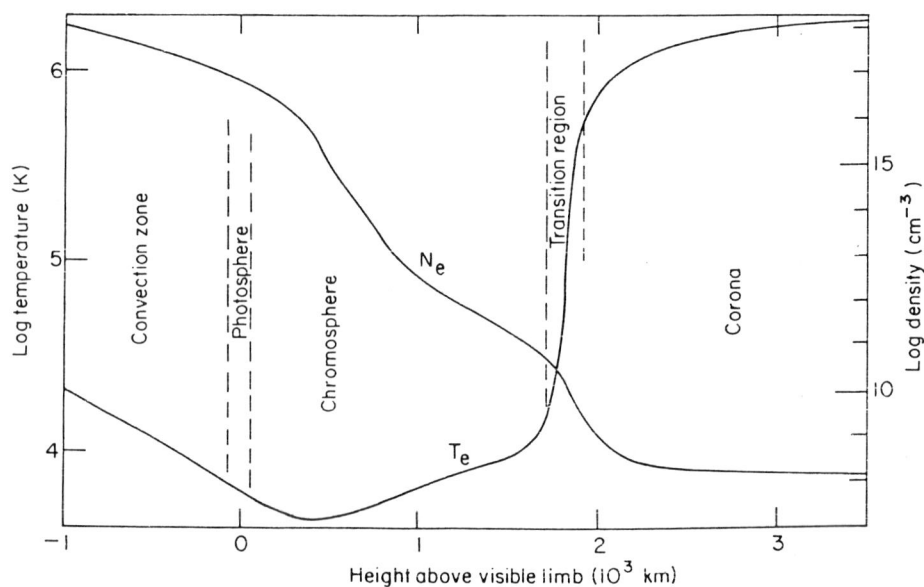

Figure 1   Schematic representation of the variation with height of the mean values of temperature and density in the outer layers of the sun (Gabriel and Mason, 1982).

2. SPECTRAL LINE EMISSIVITY

The line emissivity (per unit volume, per unit time) is

$$\epsilon(\lambda_{ij}) = N_j A_{ji} \frac{hc}{\lambda_{ij}} \qquad (j>i)$$

where A_{ji} - spontaneous radiative transition probability
 h - Planck's constant
 c - velocity of light
 λ_{ij} - wavelength for transition i-j

N_j is the number density of level j

$$N_j(X^{+m}) = \frac{N_j(X^{+m})}{N(X^{+m})} \frac{N(X^{+m})}{N(X)} \frac{N(X)}{N(H)} \frac{N(H)}{N_e} N_e$$

$\frac{N_j(X^{+m})}{N(X^{+m})}$ - population of level j relative to the total number density of the ion X^{+m}. This is a function of electron density and temperature.

$\dfrac{N(X^{+m})}{N(X)}$ — ionisation ratio of the ion X^{+m} relative to the total number density of element X. This is a function of temperature.

$\dfrac{N(X)}{N(H)}$ — element abundance. This is usually assumed to be constant but recent analyses of solar spectra indicate that there may be variations in the solar abundances in different features.

$\dfrac{N(H)}{N_e}$ — hydrogen abundance.

N_e — electron number density

In most solar analyses we assume that the ionisation/recombination processes can be de-coupled from the level population calculations for individual ions. The intensity of a spectral line is then given by the integration of the line emissivity over the volume.

$$I(\lambda ij) = \int_V \epsilon(\lambda ij)\, dV$$

For the simple two level ion in coronal equilibrium, the population of the upper level (j) is negligible compared to that of the ground level (i) ($N_i(X^{+m})/N(X^{+m}) \approx 1$). The upper level is excited by electron collisional excitation and de-excited by radiative decay.

$$N_i N_e C_{ij} = N_j A_{ji}$$

where C_{ij} is the electron excitation rate coefficient, which is the integration of the electron excitation cross-section over a Maxwellian velocity distribution, and A_{ji} is the radiative transition probability.

Thus
$$\epsilon(\lambda_{ij}) = \dfrac{N_j(X^{+m})}{N(X^{+m})} \dfrac{N(X^{+m})}{N(X)} \dfrac{N(X)}{N(H)} C_{ij} \dfrac{hc}{\lambda_{ij}} N_e^2$$

$$\epsilon(\lambda_{ij}) = G(T) N_e^2$$

$$I(\lambda ij) = \int_V G(T) N_e^2\, dV$$

This can be re-written in the form

$$I(\lambda ij) = \int_T G(T)\, \phi(T)\, dT$$

where ⌀(T) is called the differential emission measure which indicates the distribution of material as a function of temperature.

If the radiative decay rate is small, eg for forbidden or inter-system transitions, then other de-populating processes (eg electron collisions) compete and the ion departs from coronal equilibrium. The population of the upper ("metastable") level can become comparable to that of the ground level. At high densities the intensity of the line emission from the metastable level is proportional to N_e rather than N_e^2. The ratio of this spectral line relative to an allowed line can be used as a density diagnostic (see figure 2). When the population of the metastable level is significant other levels can be excited from it as well as from the ground level. This again produces an electron density diagnostic line ratio. Once departures from coronal equilibrium occur, it is necessary to solve the level population equations including all relevant excitation and de-excitation processes to determine theoretical line intensities as a function of temperature and density.

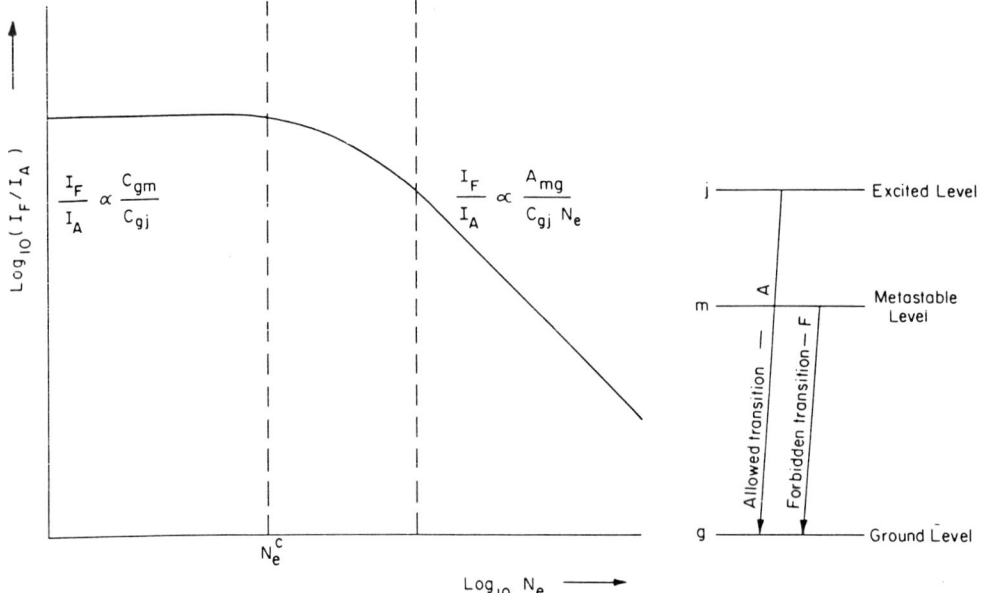

Figure 2 The variation with density of the intensity ratio of a forbidden to an allowed transition (Gabriel and Mason, 1982).

Spectroscopic diagnostics involve the analyses of one or more spectral lines to determine the characteristics of the emitting plasma. From a single line one can measure the line shape, width and shift or asymmetry. These parameters are determined by the ion temperature, any micro-turbulence, flows, waves etc. The absolute line intensity is a function of the electron temperature, density and the element abundance. The relative intensities of two or more lines can be used to determine the electron density, temperature and equilibrium state of the plasma. The lines used must be carefully selected. There have been numerous reviews of spectroscopic diagnostics for the solar atmosphere eg Dere and Mason (1981), Gabriel and Mason (1982), Doschek (1985), Mason (1988). In this review paper we discuss the wavelength region 1 - 200Å giving a few specific examples.

3. HIGHLY IONISED IRON IONS

During solar flares, we see a whole sequence of spectral lines from highly ionised iron ions (FeXVIII - FeXXVI). These fall into different wavelength regions. Take for example FeXXI ($T_e(max) \approx 10^7 K$), the transitions between $1s^2 2s^2 2p^2$ and $1s^2 2s 2p^3$ fall between 90 and 150 Å, between $1s^2 2s^2 2p^2$ and $1s^2 2s^2 2p 3d$ fall around 12Å, and between $1s^2 2s^2 2p^2$ and $1s 2s^2 2p^3$ fall around 1.9Å. Theoretical calculations for FeXXI were published by Mason et al (1979).

Only one set of flare spectra have been obtained in the EUV wavelength region. These data were obtained by Kastner et al (1974) with the OSO-5 satellite (figure 3). This spectral range which has great diagnostic potential (see Mason et al, 1984) has been sadly neglected. In figure 4 is shown a spectrum of Capella obtained by the transmission grating spectrometer on EXOSAT. Some of the highly ionised iron spectral features are clearly resolved.

Figure 5 shows a solar flare spectrum in the X-ray wavelength region which was obtained by Neupert et al (1973). This can be compared with the X-ray spectrum of the Puppis A (figure 6) measured with the Focal Plane Crystal Spectrometer on the Einstein Observatory. Several higher resolution solar flare spectra are now available, the best being that obtained by Phillips et al (1982) with the SMM XRP Flat Crystal Spectrometer.

The Kα transitions around 1.9Å, together with the satellite lines of FeXIV, have been extensively observed by SMM-XRP, SOLFLEX, and HINATORI instruments. Sample spectra are shown in figure 7.

Many of these spectral lines can be used as diagnostics for electron density or temperature (see for example Gabriel and Mason, 1982).

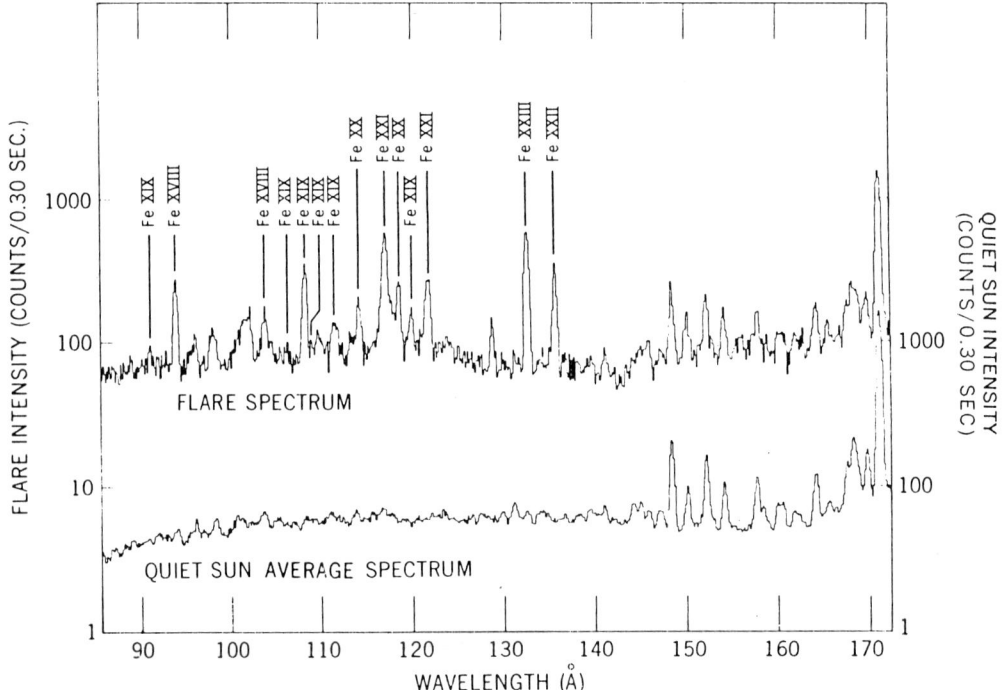

Figure 3 OSO-5 solar flare spectrum of emission lines from highly ionised iron ions (Kastner *et al*, 1974).

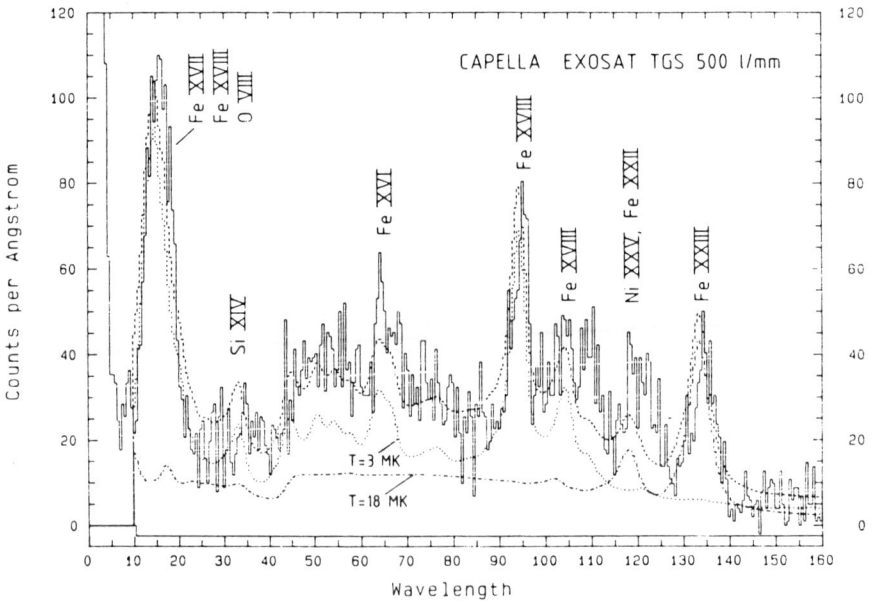

Figure 4 EXOSAT (500 l/mm) Transmission Grating spectrum of Capella (Brinkman *et al*, 1987).

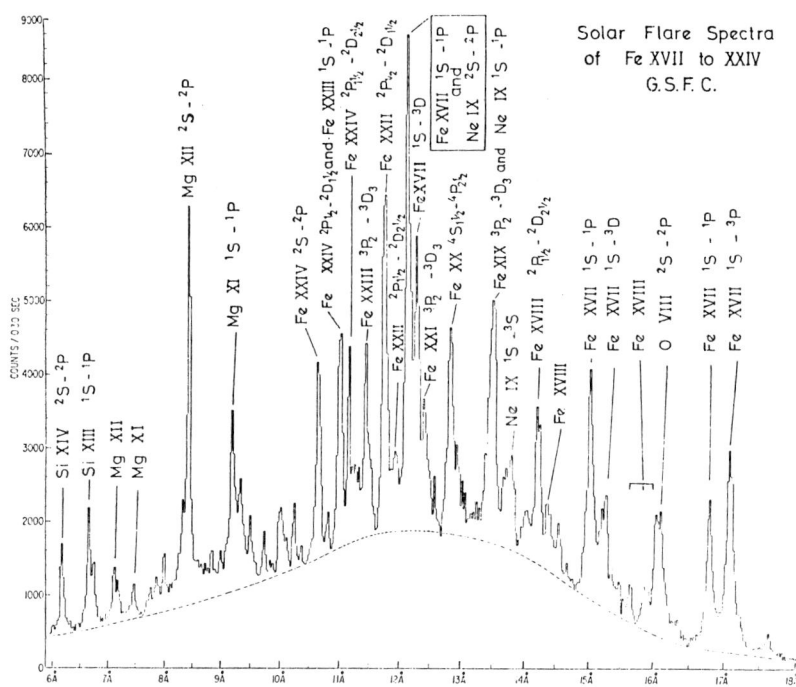

Figure 5 Solar flare spectrum from OSO-5 (Neupert et al, 1973).

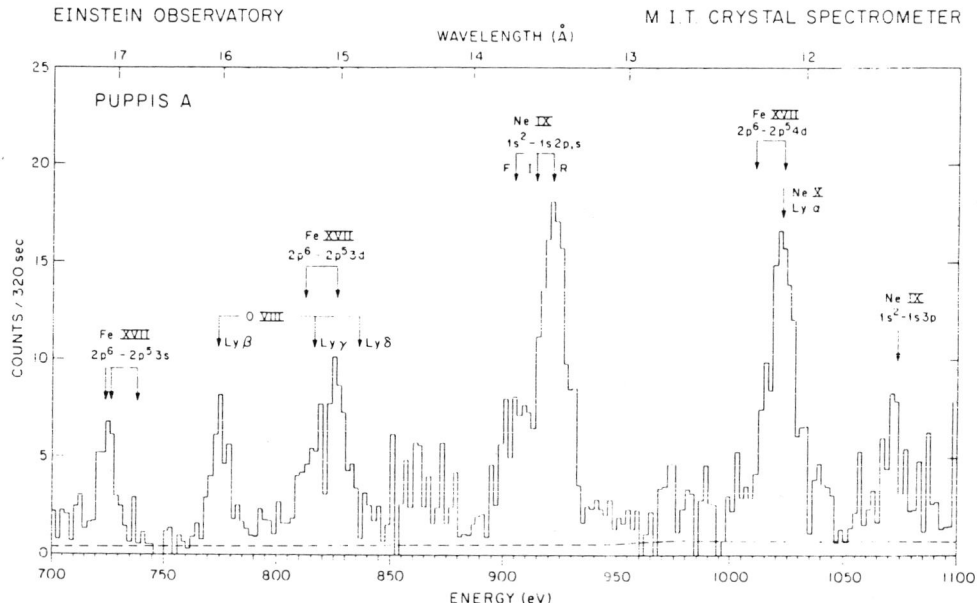

Figure 6. X-ray spectrum of Puppis A measured with the Focal Plane Crystal Spectrometer on the Einstein Observatory (Winkler et al, 1981).

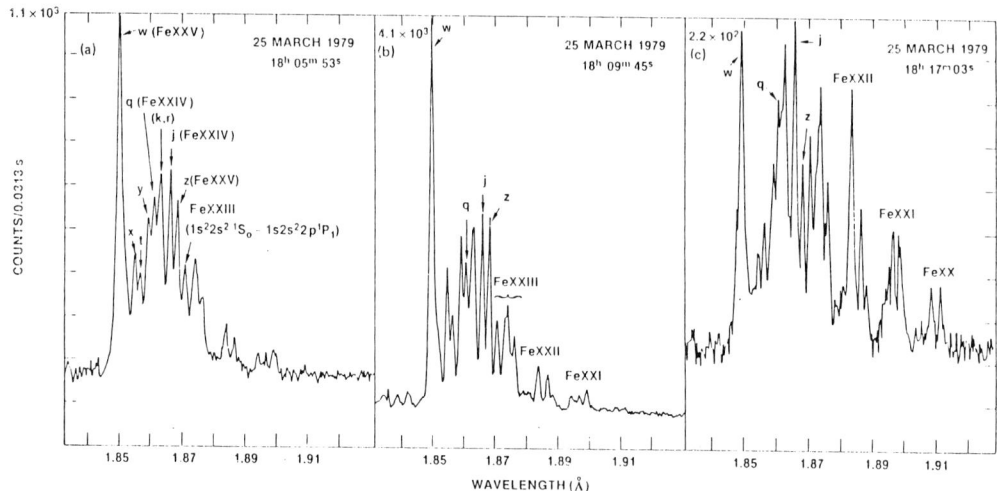

Figure 7 Flare spectra of iron line emission near 1.86Å obtained with the Naval Research Laboratory's X-ray crystal spectrometer SOLFLEX (Doschek et al, 1979).

4. HELIUM LIKE IONS

The ratio R of the forbidden to the intercombination line from helium like ions can be used to determine electron density (see Gabriel and Mason, 1982). The lines from OVII were studied using the SOLEX instrument and electron densities were derived for solar flares (McKenzie et al, 1980). The NeIX lines observed with SMM-XRP have also been used to obtain electron density variations for solar flares (Wolfson et al, 1983). Unfortunately the NeIX lines which lie around 13Å are contaminated by FeXIX lines. However, it is possible to separate the two sets of lines using theoretical intensity ratios (Bhatia et al, 1989). Several helium like ions were studied by Brown et al (1986) from solar flare spectra obtained with a grazing incidence spectrograph (10-95Å) aboard a sounding rocket.

5. THEORETICAL CALCULATIONS

The theoretical intensities of spectral lines depend on many atomic parameters. It is necessary to solve the atomic structure and electron scattering problems to obtain radiative data and collision rates. Many such calculations have been carried out by the author and colleagues. A programme of work is in progress in Dr A Burgess' group at

Cambridge to calculate and assess available atomic data for astrophysical applications (Burgess et al, 1988, 1989). An atomic data bank is being compiled for input into analysis programs such as time dependant ionisation calculations and differential emission measure analyses. These are of special relevance to future solar and astrophysical missions, particularly those discussed in this conference.

REFERENCES

Bhatia, A. K., Fawcett, B. C., Lemen, J. R., Mason, H. E., and Phillips, K. J. H., 1989, *M.N.R.A.S.*, submitted.
Brinkman, A. C., VanRooijen, J. J., Bleeker, J. A. M., Dijkstru, J. H., Heise, J., deKorte, P. A. J., Mewe, R., and Paerels, F., 1987, *Astro. Lett. and Commun.*, **26**, 73.
Brown, W. A., Bruner, M. E., Acton, L. W., and Mason H. E., 1986, *Ap. J.*, **301**, 981.
Burgess, A., Mason, H. E., and Tully, J. A., 1988, *J de Phys.*, **49**, C1-107.
Burgess, A., Mason, H. E., and Tully, J. A., 1989, *Astr. Ap.*, in press.
Dere, K. P., and Mason, H. E., 1981, in *Solar Active Regions*, ed F. Orrall (Boulder:Colorado Associated University Press) chap. 6.
Doschek, G. A., 1985, *Autoionisation*, ed A. Tenkin, chap. 6.
Doschek, G. A., Kreplin, R. W., and Feldman, U., 1979, *Ap. J. Lett.*, **233**, 457.
Gabriel, A. H., and Mason, H. E., 1982, in *Applied Atomic Collision Physics*, Vol. 1, ed. H. S. W. Massey, E. W. McDaniel, and B. Bederson (New York:Academic), p. 345.
Mason, H. E., 1988, *J. de Phys.*, **49**, C1-13.
Mason, H. E., Bhatia, A. K., Kastner, S. O., Neupert, W. M., and Swartz, M., 1984, *Solar Phys.*, **92**, 199.
Mason, H. E., Doschek, G. A., Feldman, U., and Bhatia, A. K., 1979, *Astr. Ap.*, **73**, 74.
Kastner, S. O., Neupert, W. M., Swartz, M., 1974, *Ap. J.*, **191**, 26.
McKenzie, D. L., Broussard, R. M., Landecker, P. B., Rugge, H. R., Young, R. M., Doschek. G. A., and Feldman, U., 1980, *Ap. J.*, **238**, L43.
Neupert, W. M., Swartz, M., and Kastner, S. O., 1973, *Solar Phys.*, **31**, 171.
Phillips, K. J. H., Leibacher, J. W., Wolfson, C. J., Parkinson, J. H., Fawcett, B. C., Kent, B. J., Mason, H. E., Acton, L. W., Culhane, J. L., and Gabriel, A. H., 1982, *Ap. J.*, **256**, 774.
Winkler, P. F., Canizares, C. R., Clark, G. W., Markert, T. H., Kalata, K., and Schnopper, H. W., 1981, *Ap. J. Lett.*, **246**, L27.
Wolfson, C. J., Doyle, J. G., Leibacher, J. W., and Phillips, K. J. H., 1983, *Ap. J.*, **269**, 319.

DISCUSSION-H. Mason

Y. Gnedin: You showed you last slide with fine structure of a C IV region. Could you explain the spatial resolution of this region?

H. Mason: The spatial resolution achieved by the HRTS instrument is about 1″. The structures seen are close to the spatial resolution of the instrument. Detailed studies of these structures have been carried out by Brueckner, Cook and Derel at the Naval Research Laboratory.

O. Vilhu: Have you ever observed any thermal or non-thermal velocities (asymmetries, red and blue shifts) during solar X-ray flares?

H. Mason: Non-thermal velocities have been observed during solar flares. In the early stages, the spectral lines are broad with blue shifted components. A lot of work has been done on this phenomena with the Ca XIX lines (~ 3Å) studied by SMM-XRP. The origin could be chromospheric evaporation from the foot points of the flare loops. Blueshifts in the Fe XXI (1354.1Å) line have also been studied with SMM-UVSP (Mason et al, 1986, *Ap.J.* **309**, 435).

R. Pallavicini: As you know, XMM will try to measure densities using density sensitive line ratios in the spectral region ~ 5 to 50 Å . Besides OVIII, do you think there are appropriate lines – especially <u>higher</u> <u>temperature</u> lines – in that region which are suitable for deriving electron densities?

H. Mason: There are several density sensitive ratios in this wavelength region – A review is given in Brown et al (1986, *Ap.J.* **301**, 981).

H. Gursky: Could you comment on the question I asked Dr. Raymond; namely, the effect of purely plasma physics effects on the calculations that you present?

H. Mason: In "coronal equilibrium" it is usually assumed that the ionization/recombination and excitation/ de-excitation processes can be treated separately. If the timescale for changes in the plasmas parameters are faster than the timescale for atomic processes, it is necessary to couple the ionization/recombination and level population calculations. The collisional excitation and ionization rates etc. are usually calculated assuming a Maxwellian velocity distribution. This is not necessarily a good assumption for all plasmas. Strong magnetic fields can also affect atomic processes.

P. DeKorte: What spectra resolution is needed to carry out diagnostics on He-like ions like Ne IX in conjunction with the presence of Fe XIX lines?

H. Mason: About 005-0.1 Å . The Ne IX lines are contaminated by Fe XIX lines. Even the resolution achieved by SMM-XRP was not sufficient to separate these blends. However, the amount of FeXIX contributing to the NeIX lines could be estimated theoretically if a spectral resolution ~ 0.05-0.1 Å is achieved. This would resolve the strongest FeXIX lines 135. 04-13.520 Å from the NeIX resonance line (13.446 Å), (Bhatia et al 1989 *M.N.R.A.S*, submitted).

RECENT ADVANCES IN ATOMIC MODELING

W. H. Goldstein[1]

[1]High Temperature Physics Division, Lawrence Livermore National Laboratory, Livermore, CA 94550 USA

ABSTRACT. Precision spectroscopy of solar plasmas has historically been the goad for advances in calculating the atomic physics and dynamics of highly ionized atoms. Recent efforts to understand the laboratory plasmas associated with magnetic and inertial confinement fusion, and with X-ray laser research, have played a similar role. Developments spurred by laboratory plasma research are applicable to the modeling of high-resolution spectra from both solar and cosmic X-ray sources, such as the photo-ionized plasmas associated with accretion disks. Three of these developments in large scale atomic modeling are reviewed: a new method for calculating large arrays of collisional excitation rates, a sum rule based method for extending collisional-radiative models and modeling the effects of autoionizing resonances, and a detailed level accounting calculation of resonant excitation rates in FeXVII.

1. INTRODUCTION

The possibility of creating hot, dense, earthbound plasmas using powerful lasers and pulsed power sources has opened new avenues of research into the atomic physics of highly ionized atoms and their behavior under a broad range of plasma conditions. High resolution spectroscopy of these sources has posed significant new challenges to modelers of the atomic kinetics of non-equilibrium (non-LTE) plasmas. In particular, neither coronal nor Saha equilibrium typically obtains, and the application of detailed collisional-radiative physics is essential.

The collisional-radiative regime yields little simplification in atomic kinetics. States and processes that could be neglected in coronal equilibrium cannot be ignored. Excited atomic levels may be populated by recombination, ionization and photo-driven processes, in addition to simple collisional excitation. Ionization equilibrium cannot be assumed. Line formation mechanisms in astrophysical plasmas also run a wide gamut. The standard collisional models, appropriate for solar coronal plasmas, are not applicable to recombining, photo-ionized sources like accretion disks. As high resolution spectral data from cosmic plasmas becomes available, the need will arise for more suitable modeling tools.

In this paper I'll review three advances in collisional-radiative modeling: an improvement in computational efficiency; an improvement in the approximate treatment of kinetics; and, the converse, improvements in the level of detail. My discussion of the first two topics will be limited to theoretical notes and brief comments on phenomenological applicability. Under the third heading I'll present results of a new calculation, in a detailed level accounting scheme, of resonant excitation rates in FeXVII.

2. THE PAINLESS WAY TO ACCURATE, RELATIVISTIC, MULTICONFIG- URATIONAL, DISTORTED WAVE COLLISIONAL EXCITATION RATES

Electron densities well above 10^{19} cm^{-3} are easily achieved in vacuum spark and laser-produced plasmas. Though still far from LTE (the high density limit where atomic dynamics become irrelevant), these conditions cannot be treated in the coronal approximation. Rather, it is necessary to account for multiple collisional excitations, and collisional, as well as radiative, deexcitations.

For a perspective on this complication, consider a model for the neon-like ion. From its $1s^22s^22p^6$ ground state, we can construct the 36 lowest lying excited states of this ion by promoting an electron out of the n=2 shell and into the n=3 shell. In the coronal limit, a kinetics model of this ion would require the calculation of 36 collisional excitation rates (in addition to the cascade matrix). Because of the low density, an excited state, once created by a collision, decays radiatively long before the ion can undergo a second collision. However, at high density, excited states are collisionally coupled to each other, as well as to the ground state, so that (37×36)/2=666 excitation rates are required.

The problem is that accurate collisional excitation rates are computationally expensive -- until recently it took anywhere from one half to several hours (K. Reed, private communication; Goldstein and Reed 1986) of Cray CPU time to obtain just the 36 rates needed for the coronal model in the relativistic distorted wave scheme. (To my knowledge, no benchmark existed for the 666 rate high density case.) An elegant solution to this problem was recently discovered by computational atomic physicists at Hebrew University (Bar-Shalom, Klapisch and Oreg 1988). Their solution is based on a powerful factorization theorem, combined with an empirically motivated interpolation approximation.

The factorization theorem is an exact formula for separating a collisional excitation cross section into an angular recoupling coefficient and a radial part.(Oreg 1971, 1975; Bar-Shalom et al. 1988) The former subsumes the details of the target state angular couplings, and is entirely independent of the dynamics of the interaction and the kinematical variables used for its description. The radial factor is a sum over partial waves, and depends on the bound orbitals involved in the transition as well as the continuum electron's energy and wave function, but is independent of the specific atomic states. Thus, it is formally common to many level-to-level transitions, and need not be recalculated for each one. (This result may sound suspiciously like the Wigner-Eckart theorem, but note that we're talking about factorizing cross sections -- including interference and exchange terms and a partial wave sum -- not just matrix elements.)

Factorization is not the whole story, though, since the radial components for each transition are not completely independent of the detailed levels involved. The dependence is indirect, arising from the change in energy of the continuum electron, i.e., the threshold energy, ΔE. But this dependence is smooth, and for sufficiently small changes in threshold energy, the dependence can be treated, to arbitrary accuracy, by linear interpolation. In fact, by performing many calculations, the Hebrew University group determined empirically that, when the energy of the outgoing electron is held fixed, the dependence of the radial factor on ΔE (or, in the case of dipole-allowed transitions, $\log \Delta E$) is linear over a quite large range (Bar-Shalom et al. 1988).

Since the radial integrals represent the largest expense in calculating cross sections, the factorization-interpolation solution yields tremendous savings in computational effort. For example, far from taking on the order of an hour for the 36 rate neon-like problem, this method requires about 4 minutes in Cray CPU time to obtain the 666 rates needed under non-coronal conditions, with no significant loss in accuracy (Goldstein and Reed 1986). Figure 1 should help explain whence this savings comes, as well as demonstrate an application to the fluorine-like ion. This charge state has 111 low-lying excited levels of the form $1s^2 2s^1 2p^6$ and $1s^2(2s2p)^6 3l$, l=s,p,d, and even the coronal problem of generating all 111 ground state excitation rates is daunting. As shown in Figure 1, eight of these rates describe excitation from the $1s^2 2s^2 2p^5(J=3/2)$ ground level to the levels belonging to the $1s^2 2s^2 (2p_{1/2})(2p_{3/2})(3d_{5/2})$ multiplet. (I've adopted relativistic notation for the sub-shell structure: nl_j.) According to the factorization theorem, the cross section for each of these transitions has the form

$$\sigma(2p_{1/2} \to 3d_{5/2}, E_{out}, \Delta E; J=3/2 - J_f) = A(p_{1/2} \to d_{5/2}; J=3/2 - J_f) R(2p_{1/2} \to 3d_{5/2}, E_{out}, \Delta E), \quad (1)$$

where R is common to all eight transitions. If we compute R at $\Delta E = \Delta E_{min}$ and at ΔE_{max}, and interpolate for the remaining transitions, the eight original calculations are reduced to two. (The angular coefficients are computationally trivial combinations of 3j and 6j symbols). It takes only a moment's reflection to realize the possible savings when this method is applied to more complicated situations, for example, to transitions between a pair of eight level multiplets. As a rough rule of thumb, the computational effort scales with the square of the number of sub-shells in the model, rather than the square of the number of levels.

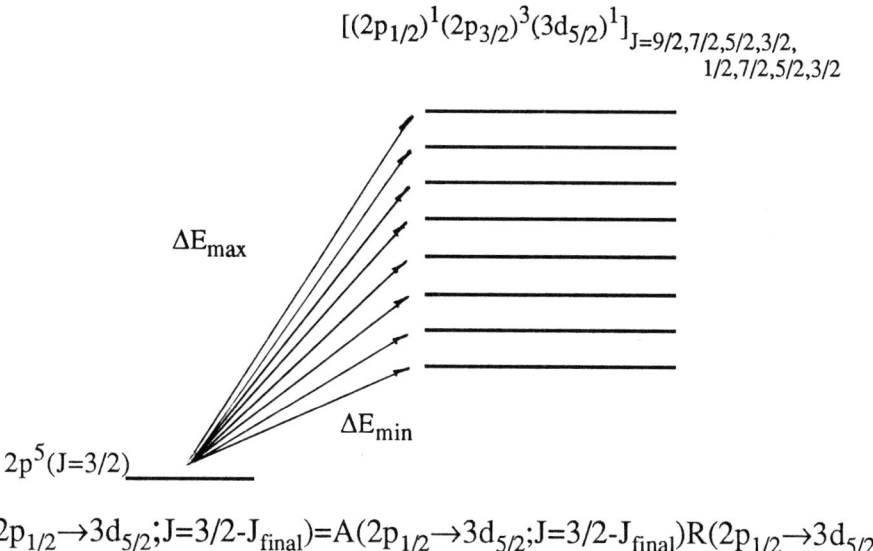

Figure 1 -- Representation of the eight collisional cross sections for exciting the $1s^2 2s^2 2p^5(J=3/2)$ ground level of the fluorine-like ion to levels of the $1s^2 2s^2 (2p_{1/2})(2p_{3/2})(3d_{5/2})$ multiplet. Factorization reduces these to an angular coefficient times a radial factor that is common to all eight transitions. By interpolating the value of this factor, all eight cross sections can be obtained from a small subset.

The fluorine-like example should suggest that the factorization-interpolation method is a boon to low density, coronal modeling as well as to the high density regime. With it, ground state excitation rates for complicated ions -- like fluorine- or oxygen-like -- can be obtained, for a range of temperatures, at minimal cost.

3. LEVEL-CONFIGURATION SUM RULES FOR COLLISIONAL-RADIATIVE PROCESSES.

Even the largest models for population kinetics in plasma take into account a very small subset of the accessible atomic levels. Various assumptions and approximations are invoked to justify neglecting high-lying Rydberg levels, autoionizing manifolds and, in fact, entire charge states. But in most cases, the effects of circumscribing the kinetics are not well understood or controlled.

An obvious solution to this problem is to treat a limited set of levels in detail, while adopting a hydrogenic description for a larger set of surrounding configurations. Since it can replace high multiplicity configurations with single effective levels, this approach overcomes the hardship of calculating a host of new rate constants when a manifold is added to the model, and limits the growth of the system of rate equations. Unfortunately, there is not a consistent way to couple hydrogenic "effective" levels to detailed atomic states, and this leads typically to spurious behavior at the boundary between the hydrogenic and detailed descriptions. In particular, not only must one assume that the levels represented by the hydrogenic configuration are statistically populated, but to couple them to detailed levels, statistical assumptions must be made for their population as well, and for the behavior of branching ratios. Furthermore, hydrogenic approximations for rate constants are often very inadequate.

The factorization-interpolation method described in Section 1 underlies an alternative approach to expanding collisional-radiative models that avoids the inconsistencies of the hydrogenic approximation. This model is based on sum rules that express the total rate or cross section for a transition between a detailed level and a configuration. For collisional excitation, the sum rule follows trivially from Eq. (1):

$$\sum_{j_{final}} \sigma(nl_j \to n'l'_{j'}, \Delta E; J_{initial} \to J_{final}) \approx R(nl_j \to n'l'_{j'}, \Delta E_{avg}) \sum_{j_{final}} A(l_j \to l'_{j'}; J_{initial} \to J_{final}) \quad (2)$$

The summation on the right hand side of Eq. (2) can be performed analytically, while the radial factor can be computed in any atomic structure model (hydrogenic, Hartree- or Dirac-Fock, parametric potential, etc.). The only approximation involved in Eq. (2) is the necessity of using an average transition energy, ΔE_{avg}, but this is common to all effective level methods. Note that deriving the sum rule depends crucially on being able to factor out of the cross section all dependence on continuum electrons. Analogous factorization theorems have been obtained for collisional ionization, autoionization (Oreg, Goldstein, Bar-Shalom and Klapisch 1988), and radiative transitions (Condon and Shortley 1935, Bauche, Bauche-Arnoult and Klapisch 1988).

The sum rules represent exact, analytic expressions for the rate of a process summed over final states, and, conversely, for the inverse process averaged over initial states. Together with a statistical ansatz for the distribution of population within a summed configuration, they provide a consistent and efficient -- not to mention elegant -- way to collapse high multiplicity configurations down to effective levels, and to couple these levels to detailed atomic structure.

Sum rules fit neatly and unobtrusively at any point in a collisional-radiative model. They are as computationally efficient as the hydrogenic approach, but far more accurate and require no assumptions about branching ratios into detailed levels. They have specific applications to treating Rydberg series (note that the analytic factor in (2) is independent of n), and to calculating branching ratios, that, after all, are precisely sums over final states. When applied to autoionizing configurations, we obtain a novel configuration average approach to evaluating multi-step, resonant contributions to recombination, ionization and excitation processes, i.e., dielectronic recombination, excitation or ionization followed by autoionization and resonant excitation (Oreg et al. 1988).

4. A DETAILED CALCULATION OF RESONANT EXCITATION IN FeXVII

Access to super-computers and the development of powerful, efficient new computer codes has facilitated a deeper level of detail in the atomic modeling we can apply to the collisional-radiative problem. Single configuration LS coupling atomic structure calculations have given way to multiconfigurational, intermediate coupling. Calculations that involved averaging or lumping together the levels of a configuration can now be done using detailed level accounting. This development is most evident in the modeling of resonant processes involving the branching ratios of autoionizing levels. M. H. Chen's recent calculations of dielectronic recombination rates (Chen 1986) and the work of Reed, Chen and Hazi (1988) on resonant excitation in oxygen-like selenium are examples of this trend. The former work, in particular, replaced earlier, less accurate, configuration average, LS coupling results (LaGattuta and Hahn 1983; Roszman 1979; Griffin, Pindzola and Bottcher 1985).

In photo-ionized astrophysical plasmas, ionization of inner-shell electrons can lead to the emission of one or more additional Auger electrons, and this process is governed, also, by branching ratios for autoionizing levels. Calculations of the rate at which this process populates excited levels of neon- and fluorine-like ions of iron will be necessary to properly analyze high resolution spectra of accretion disk sources.

Even in the case of the simpler solar coronal plasmas, the importance of resonant excitation and dielectronic recombination for understanding FeXVII spectra has been recognized. In a comprehensive analysis of this problem, Smith et al. (1985) calculated resonant excitation rates and the dielectronic satellites of neon-like iron. There is the suggestion as well that dielectronic recombination from FeXVIII is important in the formation of FeXVII lines in the corona (Liedahl et al. 1988).

Resonant excitation of neon-like ions, particularly of the $2p^53s$ levels whose direct excitation cross sections are small, is of interest in the laboratory plasma community as well. Determining the population of these levels is a challenge to modelers because under almost any plasma conditions, it is populated both directly and indirectly, by both collisional and cascade processes. And the record in accurately modeling the intensity of the $2p^6$-$2p^53s$ lines is spotty at best. Relative intensities in the 2p-3s lines far exceeding what would be expected based on comparative oscillator strength have been observed in both laser-produced and vacuum spark plasmas (Goldstein et al. 1987; Finkenthal et al.

1988) as demonstrated in Figure 2. One hypothesis for these anomalies is that they represent contributions from a low temperature phase where recombination and/or resonant excitation is important.

Figure 3 shows a SOLEX spectrum (Rugge and McKenzie 1985). The wavelength region covered is the analog in iron of the titanium spectrum in 2(b). Here the M2 line to the long wavelength side of G appears owing to the much lower electron density. Otherwise, there is much in common between the laboratory and solar spectra, including the signal strength of the 2p-3s lines. As an example of an application of detailed level accounting, we have performed a new calculation of resonant excitation rates for the $2p^5 3s (J=0,1,1,2)$ levels in FeXVII. Our goal is to improve on the original estimates of Smith et al. for this astrophysically important process, and to help gauge its impact on laboratory plasma analysis. (A recent calculation by Omar and Hahn, 1988, uses a non-relativistic, LS coupling, term average method that is unlikely to be adequate.)

Figure 4 is a reminder of how resonant excitation works. An electron with the correct energy colliding with a neon-like ion in its ground state can excite a resonance that is essentially a doubly excited sodium-like level. Once a free electron is captured this way, it can de-excite in several ways, depending on the total energy of the doubly excited state. If the energy is below that of the first excited neon-like level (the $2p^5 3s, J=2$), as depicted in figure 4(a), the state either autoionizes back to $2p^6$ plus a free electron -- a contribution to elastic scattering, or it radiatively relaxes, forming a stable sodium-like state, a process called dielectronic recombination. At higher incident electron energies, as shown in figure 4(b),resonances are excited that, in addition to the deexcitation pathways enumerated in 4(a), have the option of autoionizing to excited neon-like levels.

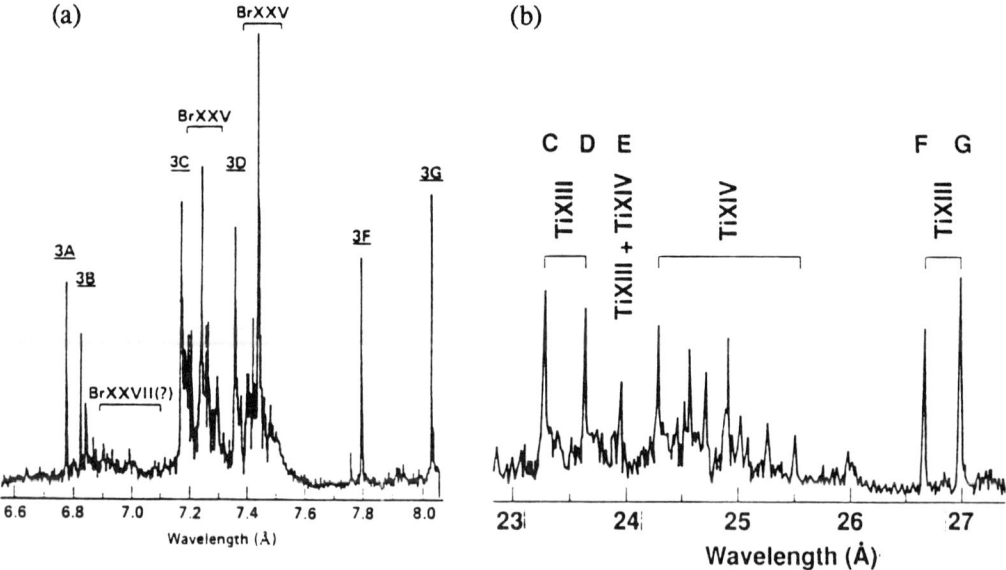

Figure 2 -- (a)Time-integrated spectrum of neon-like bromine produced by 3×10^{13} W/cm^2 from .53µm laser. N=3-n=2 resonance lines are labeled 3A,B,C,D,F,G. The latter two are anomalously strong 2p-3s transitions (Goldstein et al. 1987). (b) Time-integrated spectrum of neon-like titanium produced in a vacuum spark (Finkenthal et al. 1988).

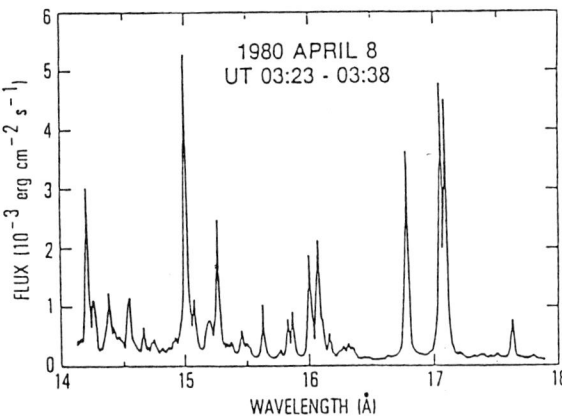

Figure 3 -- Solar flair spectrum from Rugge and McKenzie (1985) showing the neon-like iron 2p-3d resonance lines at 15.0, 15.3 and 15.5 Å, and the 2p-3s lines at 16.8 and 17.0 Å. The direct excitation rates for the latter are smaller by a factor of 30 than that of the 15Å line. The extra line at 17.1 Å is the $2p^6$-$2p^53s(J=2)$ E2 transition. This line is absent in the high density laser-produced and vacuum spark plasmas shown in Fig. 2.

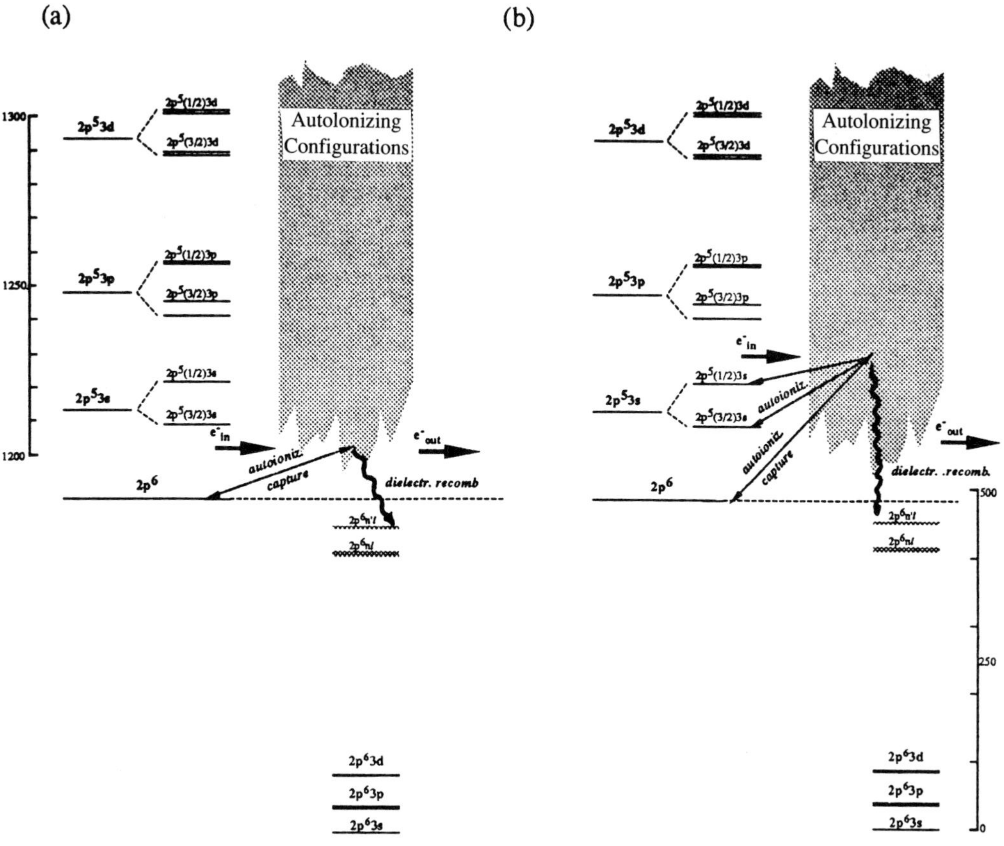

Figure 4 -- (a) Free electrons with energy below threshold for excitation of the neon-like $2p^53s$ level undergo either elastic scattering or dielectronic recombination. (b) Electrons with energy above threshold have additional deexcitation channels available: resonant excitation of neon-like excited states. (Energy scales are in eV.)

This sequence of events is a "resonant" contribution to collisional excitation since it leaves a free electron and an excited neon-like level in the final state. In the isolated resonance approximation, the rate coefficient for this process is given by

$$R^{res.exc.}(\text{final}) = \sum_{C_{auto}} \sum_{level \in C_{auto}} R^{capt.}(2p6 \to level) \frac{A^{autoioniz.}(\text{level} \to \text{final})}{\sum_j [A^{autoioniz.}(\text{level} \to j) + A^{rad.}(\text{level} \to j)]}$$

(3)

where "final" is a neon-like level and the sodium-like "level" belongs to the autoionizing configuration C_{auto}. The capture rate is obtained by detailed balancing the autoioization rate, and the last factor is the branching ratio for "level." The sum in the denominator can generally be truncated to include only the dominant radiative decay branch.

The importance of accounting explicitly for each level (both "final" and "level") in Eq. (3) can be seen by noting that the branching ratio is very sensitive to the opening of new autoionization channels with increasing energy. The intricate interleaving of sodium-like doubly-excited and neon-like levels makes it essential to retain the full detail of the level structure to obtain accurate branching ratios. A good example of this effect is found in the

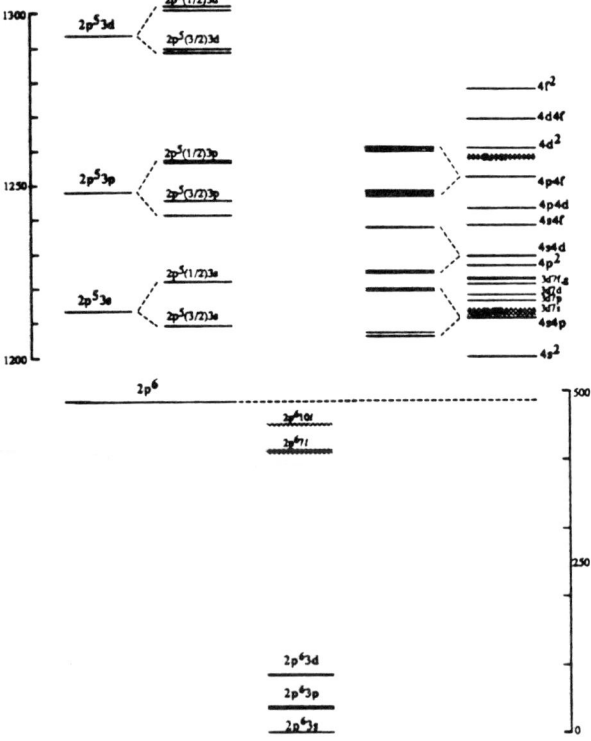

Figure 5 -- Schematic energy level diagram for neon-like and sodium-like iron. The (broken) energy scale is in eV. On the left are the shown the $2p^5 3l$ configurations of the neon-like ion, broken down as far as their relativistic sub-shell structure. The branching into individual levels is not shown. On the right are the configurations of the form $2p^5 4l\,4l'$. To convey an idea of how these manifolds interleave with the neon-like, several have been broken into sub-configurations. Note that the actual detailed level structure used in the calculation is not shown here.

$2p^5 4l4l'$ configurations, which, as shown in Figure 5, are spread over many autoionizing channels. Furthermore, detailed level accounting is necessary to avoid having to make statistical assumptions about population distribution at some point in the calculation. Such assumptions are always hazardous in calculations involving fluorescent yields where selection rules play an important role (Chen, Crasseman and Matthews 1975).

The autoionizing configurations included in the outer summation in Eq. (3), and the radiative and autoionizing channels included in the branching ratios for the present calculations are listed in Table 1. Besides applying detailed level accounting for all the configurations listed, this calculation differs from its predecessors in including the $2p^53p$ and $2p^53d$ autoionization channels for $C_{auto}=2p^54l4l'$. As shown by figure (5), these channels cannot be neglected for these configurations, and they have the effect of reducing the partial contribution to the rate of excitation of the $2p^53s$ by more than a factor of two. Atomic structure, including energy levels, autoionization rates and Einstein coefficients, were obtained in the relativistic parametric potential model (Klapisch et al. 1977).

In light of our significant disagreement with previous results, a comment is in order on consistency checks of this calculation. The same set of data was used to calculate partial dielectronic recombination rates for the $2p^53l3l'$ and $2p^54l4l'$ configurations. The results fell within 20% of the calculation of Chen (1986). In addition, an independent calculation was solicited from Chen (1988) of the $2p^54l4l'$ contribution to resonant excitation of the $2p^53s$ levels. Using his multiconfigurational Dirac-Fock code (MCDF) and the detailed level accounting scheme, he obtained results that agreed with the present calculation to better than 5%. Finally, the convergence of the series in n and l were verified; no extrapolation beyond n=14 was performed for the $3dnl'$ and $3pnl'$ series. The latter series had not quite converged, and we estimate that our results in this case may undershoot the converged value by up to 15%.

The results of this calculation, at an electron temperature of 200 eV, are presented in Table 2, along with the results quoted by Smith et al. (1985), broken down by final state and autoionizing series. The importance of detailed level accounting is evident in the reduction found in rates by factors ranging from 3 to 5. (The factor of 10 between the $4l4l'$ contributions is reduced to 5 when the a correction is made for the $2p^53p$ channel not included in the earlier calculation.)

C_{auto}	C_{Ne} (autoion. channels)	C_{Na} (rad. channels)
$2p^53dnl$, $l \leq 4$, $6 \leq n \leq 14$	$2p^6$, $2p^53s$, $2p^53p$	$2p^6nl(E1), 2p^63d(E1)$
$2p^53pnl$, $l \leq 4$, $8 \leq n \leq 14$	$2p^6$, $2p^53s$	$2p^63p(E1), 2p^6nl(E1,E2),$ $2p^53snl(E1)$
$2p^54l4l'$, $l \leq 3$	$2p^6$, $2p^53s$, $2p^53p, 2p^53d$	$2p^64l(E1), 2p^53l4l'(E1)$

Table 1 -- The autoionizing configurations included in the present calculation, with their Auger and radiative decay branches. Configuration interaction (CI) among autoionizing levels was included to the extent that for each n, the energy was diagonalized on the space of all l($\leq$4). CI between the $2p^53pnl$ and $2p^53snl$ was included to allow for E1 decays to the $2p^5nl$. All levels of the $2p^54l4l'$ were included in a single structure calculation.

$2p^53s$ state	AUTOIONIZING SERIES						TOTAL RATE $\times 10^{-13} cm^3 sec^{-1}$	
	$2p^53pnl$		$2p^53dnl$		$2p^54l4l'$			
	A	B	A	B	A	B	A	B
J=2	5.72	1.51	12.2	7.14	31.0	2.94	49.0	11.6
J=1	13.4	4.40	10.7	4.61	20.7	2.06	44.8	11.1
J=0	0.86	0.18	3.12	1.11	6.70	1.08	10.7	2.37
J=1	<u>8.65</u>	<u>4.27</u>	<u>8.17</u>	<u>4.43</u>	<u>18.3</u>	<u>2.25</u>	<u>35.1</u>	<u>11.0</u>
All J	28.6	10.4	34.2	17.3	76.7	8.33	140.	36.1

Table 2 -- Partial and total resonant excitation rate coefficients (10^{-13} cm^3 sec^{-1}) for the $2p^53s$ levels of FeXVII. Values for this calculation (B) are at an electron temperature of 200 eV and for n≤14, while those of Smith et al. (1985) (A) are at 217 eV and include an extrapolation of n to 100.

ACKNOWLEDGEMENT

I'd like to acknowledge my co-workers, J. Oreg, A. Bar-Shalom, M. Klapisch and A. Osterheld, and many very helpful consultations with M. H. Chen. Work performed under the auspices of the U.S. Department of Energy by Lawrence Livermore National Laboratory under contract #W-7405-Eng-48.

REFERENCES

Bar-Shalom, A., Klapisch, M. and Oreg, J. 1988, *Phys. Rev. A*, **38**, 1773.
Bauche, J., Bauche-Arnoult, C. and Klapisch, M. 1988, *Transition Arrays in the Spectra of Ionized Atoms*, to appear in Advances in Atomic and Molecular Physics.
Chen, M. H., Crasseman, B., Matthews, D. L. 1975, *Phys. Rev. Lett.*, **34**, 1309.
Chen, M. H. 1986, *Phys. Rev. A*, **34**, 1073.
Chen, M. H. 1988, private communication.
Condon, E. U. and Shortley, G. H. 1935, Theory of Atomic Spectra (London : Cambridge University Press)
Finkenthal, M., Stutman, D., Mandelbaum, P., Osterheld, A. L., Goldstein, W. H. and Chen, M. H. 1988, "Electron Density Measurement of a Pre-pinched Vacuum Spark Plasma Using Soft X-ray Titanium and Vanadium Emission," UCRL-99351, to appear in *J. Phys. B: At. Mol. Phys.*
Goldstein, W. H., Walling, Bailey, J., Chen, M. H., Fortner, R., Klapisch, M., Phillips, T. and Stewart, R. E. 1987, *Phys. Rev. Lett.*, **58**, 2300.
Goldstein , W. H. and Reed, K. J. 1986, in Conference on Atomic Processes in Hot Dense Plasma, Jerusalem (unpublished).
Griffin, D. C., Pindzola, M. S. and Bottcher, C. 1985, *Phys. Rev. A*, **31**, 568.
Klapisch, M., Schwob, J. L., Fraenkel, B. S. and Oreg, J. 1977, *J. Opt. Soc. Am.*, **61**, 148.
LaGattuta, K. and Hahn, Y. 1983, *Phys. Rev. A*, **27**, 1675.
Liedahl, D., Kahn, S., Osterheld, A. L. and Goldstein, W. H., in preparation.
Omar, G. and Hahn, Y. 1988, *Phys. Rev. A*, **37**, 1983.
Oreg, J. 1971, *J. Math. Phys.*, **12**, 1018.
Oreg, J. 1975, Ph.D. Thesis (Hebrew University).
Oreg, J., Goldstein, W.H., Bar-Shalom, A. and Klapisch, M. 1988, UCRL-98456, "SumRules for the Collisional Radiative Model," submitted. to *Phys. Rev. A*.
Reed, K. J., Chen, M. H. and Hazi, A. U. 1988, "Autoionizing Resonances in Electron Impact Excitation of Oxygen-like Selenium," UCRL-98374, submitted to Phys. Rev. A.
Roszman, L. J. 1979, *Phys. Rev. A*, **20**, 673.
Rugge, H. R. and McKenzie, D. L. 1985, *Ap. J.*, **297**, 338.
Smith, B. W., Raymond, J. C., Mann, J. B. and Cowan, R. D. 1985, *Ap. J.*, **298**, 898.

DISCUSSION-W. Goldstein

B. Smith: You began your talk by describing some shortcuts you took in computing atomic rates. I am not sure these represent an advance. It is certainly not safe to compute populations of levels within configurations according to statistical weights.

I would also like to correct one or two errors you made in describing our work. We did include configuration interaction in all detailed calculations. Quantum defect methods were used only to obtain the n-dependence of autoionization rates.

W. Goldstein: In the case of excitation cross-sections, the factorization is exact, whereas the interpolation in threshold energy that this factorization allows, can be handled with any degree of precision required. Since this technique opens the door to modeling systems that were not that were not tractable previously – those in which, say hundreds, or more excitation rates are relevant – it certainly represents an advance.

The sum rules are also exact, given the necessity of using configuration average transition energies. This would be the only approximation involved, then, in using the sum rules in the calculation of branching ratios (sums over final state). I certainly agree that it is dangerous to use them , if the intention is to distribute this rate statistically amongst individual levels. The accuracy of such a procedure must be considered on a case-by-case basis. (The accuracy will depend on electron density, ionic charge, principal quantum number, etc).

With regard to your calculation of resonant excitation, the absence of any reference to configuration interaction led me to the assumption that it was not included. Also, the paper seemed to suggest that, for all but a handful of doubly-excited configurations, a quantum-defect estimate was adopted for the total autoionization rate from the group of levels with an "n" electron above a neon-like core, to each neon-like level. This rate, I assumed, was distributed statistically amongst the "l" values of the Rydberg configuration.

RELATIVISTIC FREE-FREE GAUNT FACTOR OF THE DENSE HIGH-TEMPERATURE STELLAR PLASMA

Naoki Itoh, Masayuki Nakagawa, and Yasuharu Kohyama

Department of Physics, Sophia University, Tokyo, Japan

ABSTRACT The free-free Gaunt factor of the dense high-temperature stellar plasma is calculated by using the accurate relativistic cross section, and is compared with the Gaunt factor derived by using Sommerfeld's exact nonrelativistic cross section. A wide range of electron degeneracy is accurately taken into account. Significant deviations from the nonrelativistic relsults are found for high-temperature cases.

1. INTRODUCTION

The present authors (Itoh, Nakagawa, and Kohyama 1985) have investigated the relativistic free-free Gaunt factor for a high-temperature stellar plasma. They have used the relativistic Bethe-Heitler cross section (Bethe and Heitler 1934) corrected by the Elwert factor (Elwert 1939). Itoh, Nakagawa, and Kohyama (1985) have found that the relativistic Gaunt factor deviates significantly from the nonrelativistic one at high temperatures.

The exact nonrelativistic bremsstrahlung cross section for the pure Coulomb field is given by Sommerfeld's formula (Sommerfeld 1953). For practical calculations we use the series expantion method of Karzas and Latter (1961) by taking into account the electron degeneracy accurately. Green (1960) has also given numerical tables of the exact nonrelativistic free-free Gaunt factor taking into account the electron degeneracy accurately. Berger (1956, 1957) has examined the accuracy of the Elwert approximation. However, his calculation is limited to the case of nonrelativistic nondegenerate electrons. Elwert and Haug (1969) and Pratt and Tseng(1975) has confirmed that the Bethe-Heitler cross section corrected by the Elwert factor gives excellent results for ions with small atomic number.

2. FORMULATION

The detailed derivation of the relativistic free-free Gaunt factor with the use of the Elwert approximation has been presented in Iton, Nakagawa, and Kohyama (1985).

When the electrons are nonrelativistic, it is customary to express the thermally averaged inverse bremsstrahlung cross section in terms of the thermally averaged Kramers cross section per electron,

$$\sigma_K = \frac{2}{3}\left(\frac{2mc^2}{3\pi kT}\right)^{1/2} \frac{(2\pi c)^3}{\omega^3} \alpha r_0^2 n_j Z_j^2$$

$$= \frac{16\sqrt{2}}{3\sqrt{3}} \pi^{5/2} \alpha r_0^2 \frac{m^{1/2} h^3 c^4}{(kT)^{7/2} u^3} n_j Z_j^2 , \qquad (1)$$

where m is the electron mass, ω is the angular frequency of the absorbed photon, α is the fine-structure constant, r_0 is the classical electorn radious, n_j is the number density of ions with charge $Z_j e$, and $u = \hbar\omega/kT$. We define the relativistic free-free Gaunt factor by

$$\langle g_{Z_j} \rangle = \frac{n_- \langle \sigma^- \rangle + n_+ \langle \sigma^+ \rangle}{n_j Z_j \langle \sigma_K \rangle} , \qquad (2)$$

where n_- and n_+ are the number densities of electrons and positrons, and $\langle \sigma^- \rangle$ and $\langle \sigma^+ \rangle$ are the thermally averaged inverse bremsstrahlung cross sections for the electrons and the positrons. The thermal averages of the these cross sections are carried out relativistically.

Green (1960) calculated the thermally averaged exact nonrelativistic free-free Gaunt factor for the case of degenerate electrons by using the energy-dependent Gaunt factor of Karzas and Latter (1961). In using their data, he had to make interpolations with respect to $u = \hbar\omega/kT$. Nakagawa, Kohyama, and Itoh (1987) calculated the exact nonrelativistic energy-dependent free-free Gaunt factor, and then used this for the calculation of the thermally averaged Gaunt factor without making any interpolation.

3. NUMERICAL RESULTS

We have carried out the numerical calculations of the thermally averaged nonrelativistic and relativistic free-free Gaunt factors. The temperature parameter

$$\gamma^2 = \frac{Z_j^2 Ry}{kT} = Z_j^2 \frac{1.579 \times 10^5 K}{T} \qquad (3)$$

is adopted following Karzas and Latter (1961) and Green (1960). For the nonrelativistic case, the thermally averaged free-free Gaunt factor does not depend on Z_j and T separately, but on the combination Z_j^2/T. Examples of the calculations are shown in Figures 1 and 2. Agreement of the results of the calculations with different methods proves the accuracy of the present calculations. The overall accuracy of the present calculations is about 0.2 %.

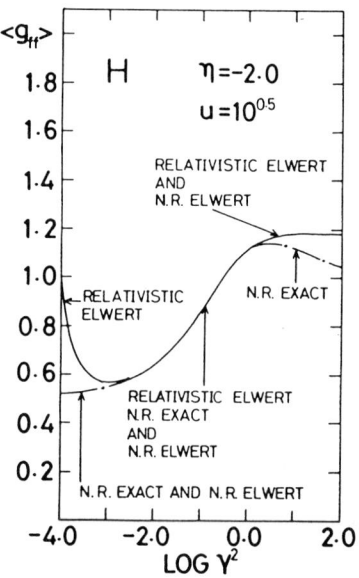

FIG. 1. Comparison of the various thermally averaged free-free Gaunt factors for the case of pure hydrogen plasma with $\eta = -2.0$ and $u = 10^{0.5}$.

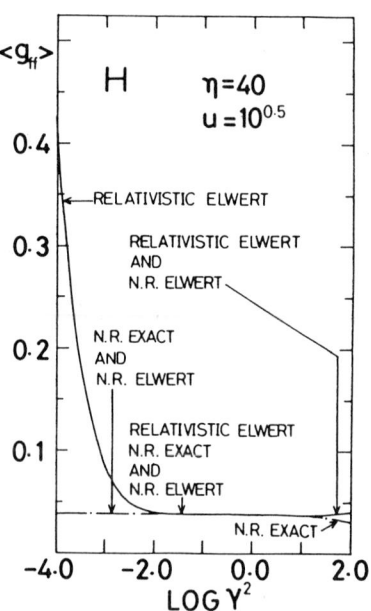

FIG. 2. Same as FIG. 1, but for $\eta = 40$.

REFERENCES

Itoh,N., Nakagawa,M., and Kohyama,Y. 1985, Ap.J., 294, 17.
Bethe,H.A., and Heitler,W. 1934, Proc.Roy.Soc.London, A146, 83.
Elwert,G. 1939, Ann.Phys., 34, 178.
Sommerfeld,A. 1953, Atombau und Spektrallinien (Braunschweig:Vieweg), Vol. 2, Chap. 7.
Karzas,W.J., and Latter,R. 1961, Ap.J.,Suppl., 6, 167.
Green,J.M. 1960, Fermi-Dirac Averages of the Free-Free Hydrogenic Gaunt Factor (RAND Corporation Rept. RM-2580-AEC).
Berger,J.M. 1956, Ap.J., 124, 550.
Berger,J.M. 1957, Phys.Rev., 105, 35.
Elwert,G., and Haug,E. 1969, Phys.Rev., 183, 90.
Pratt,R.H., and Tseng,H.K. 1975, Phys.Rev., A11, 1797.
Nakagawa,M., Kohyama,Y., and Itoh,N. 1987, Ap.J., Suppl., 63, 661.

THE DIFFERENTIAL EMISSION MEASURE OF λ And

M.Landini[1], B.C.Monsignori Fossi[2] and R.Pallavicini[2]
[1](Department of Physics,University of Naples, Italy)
[2](Arcetri Astrophysical Observatory,Florence, Italy)

ABSTRACT. Simultaneous X-ray and UV observations of λ And from EXOSAT and IUE are analyzed and the differential emission measure as a function of temperature is derived.

1 - INTRODUCTION

λ And is a well known single line spectroscopic binary with an orbital period of 20.5^d. Its very low mass function (0.006 $M_\odot$) and no evidence of eclipses indicate that the system is seen almost pole on. A photometric optical variability with a period of 54^d has been shown to be produced by a system of two large spots on the primary star which, therefore, does not rotate synchronously with the orbital motion. The primary is a G 7-8 III-IV star which has long been known to have strong chromospheric activity in the Ca II (Gratton, 1950) and Mg II lines (Baliunas and Dupree, 1982). Its chromospheric and coronal activity is also confirmed by microwave (Bath and Wallestein,1976), X-ray (Walter and Bowyer,1980; Vaiana et al.,1981) and UV emissions(Linsky et al. 1978).These observations indicate that λ And is a RS CVn star (Hall, 1981) and one of the brightest members of this class.

Observations of λ And with the EXOSAT satellite were included in a large program devoted to the study of activity in late type stars (Pallavicini et al., 1988). On November 11, 1985 simultaneous observations were obtained with EXOSAT LE and ME experiments and with IUE . For the purpose of the present analysis, the ME data have been binned in 3 bands (table 1).

Table 1
Observations of λ And on Nov 11,1985

EXOSAT ME (keV)	$c\,s^{-1}$		EXOSAT LE	$c\,s^{-1}$		IUE	$c\,s^{-1}$	
1.17-2.14	7.54	10^{-1}	3 Lex	4.815	10^{-1}	N V	2.5	10^{-2}
2.14-3.40	8.58	10^{-1}	4 lex	3.957	10^{-1}	C II	1.23	10^{-1}
3.40-5.26	4.30	10^{-1}	Al/Pa	1.279	10^{-1}	Si IV	1.1	10^{-1}
			Boron	3.253	10^{-2}	C IV	3.11	10^{-1}

The aim of this paper is to analyze these observations in order to deduce the differential emission measure as a function of temperature. Uncertainties are assumed to be ±10% for the EXOSAT data and ±30% for the IUE data.

2 - THE DIFFERENTIAL EMISSION MEASURE DISTRIBUTION.

It is not possible to reproduce both the UV and X-ray observations with a single temperature model; this is well known in the solar case and is clearly shown by fig. 1, where for each temperature we plot the emission measure that would be needed to reproduce the observed emissions. The same figure suggests that most of the emission in the ME and LE bands must come from a " quasi isothermal " plasma with a temperature higher than $20\,10^6\,°K$, while no isothermal solution appears to be possible for the UV lines.

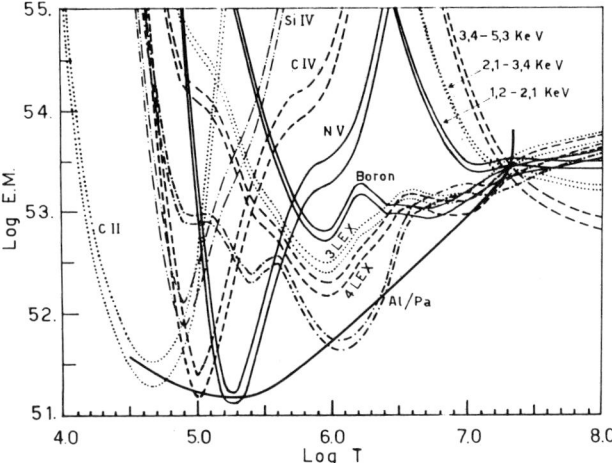

FIG. 1 - *Log of the emission measure versus temperature allowed by the observed fluxes (including errors) in binned ME bands (1.17-2.14 KeV, 2.14-3.4 KeV, 3.4-5.26 KeV), LE filters (Boron, 3Lexan, 4 Lexan, Al/Pa) and UV lines (C II, C IV, N V, Si IV).*

The envelope of curves shown in fig.1 suggests the shape of the differential emission measure (d.e.m.) as a function of the temperature. We assume the following "parameterized" form of d.e.m.:

$$n^2 (dV/ dT) = C\, e^{T_0 / T} (T / T_c)^\beta [1- (T / T_c)^\gamma]^{-1/2}$$

Apart from the exponential term, which modulates the low temperature region (UV emission), this form is suggested by a theoretical investigation of the energy balance (Landini and Monsignori Fossi, 1975,1981) and has been proved to be useful also for solar active regions (Monsignori Fossi and Landini, 1988).

The plasma emissivity for each spectral region of interest has been integrated over the assumed d.e.m. and a best χ^2 fit procedure has been used to evaluate the free parameters T_0, T_c, b, g and C; in the present case the best fit procedure has been applied to all data in table 1 except the C II line which may be optically think and the Al/Pa filter, which proved to give an unusually low flux (see later).

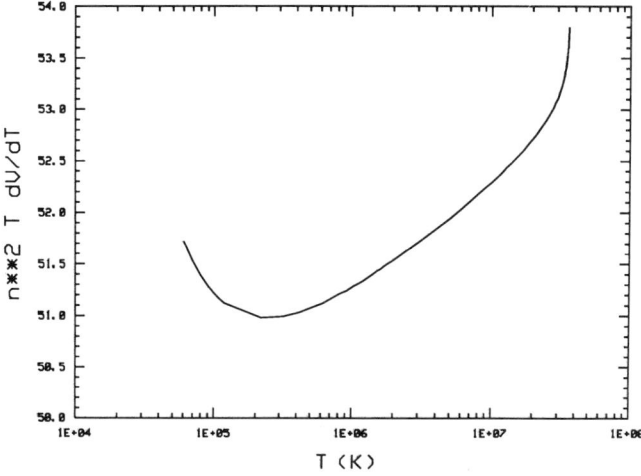

FIG. 2 - *The differential emission measure distribution that allows the best χ^2 fit between predicted and observed data.*

The following values give a reasonable good fit (reduced χ^2 = 1.2): log T_c =7.56; log T_0 =5.4; γ=.64; β =.015 ; C =1.5 10^{45} cm^{-3}K^{-1}.

The differential emission measure is plotted in fig.2 and is used to generate a synthetic spectrum (Landini and Monsignori Fossi, 1984) of the radiation emitted by the portion of the loop at temperatures larger than 6 10^4 °K.

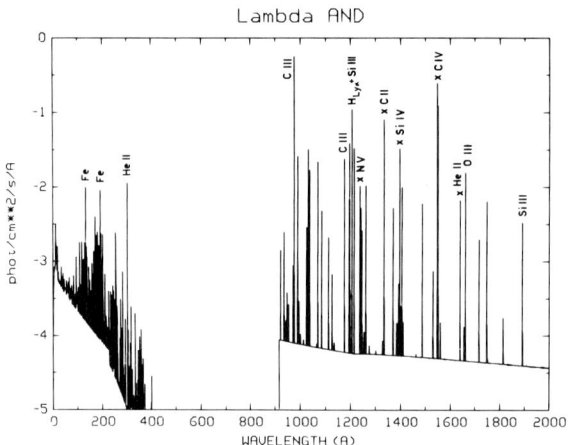

FIG. 3 - *The synthetic spectrum from 1 to 2000 A obtained for λ And assuming the parameterized differential emission measure distribution shown in fig.2. The interstellar absorption, evaluated with a column density N_H = 5.2 10^{18} cm^{-2} , affects the region from 200 to 980 A.*

The spectral distribution over the region 1-2000 A is shown in fig. 3: the most prominent spectral lines are indicated and crosses give the observed fluxes for the IUE lines. A few lines which have not been used in the χ^2 fit are present in the synthetic spectrum and need comments: Si IV (1397 A) is too low by a factor 3 than predicted: this fact was already noticed and is attributed to the density dependence of ion abundance (Jordan et al., 1987). He II (1640 A) in λ And is much higher than predicted: this fact can be attributed to an overpopulation of the upper level due to recombination of He nuclei following photoionization by photons with λ< 227 A. These photons are much more numerous in λ And than in the Sun, due to the higher coronal temperature of this star (Linsky et al., 1978). Photoionization modifies the ionization balance of the He ions and this fact may also explain the disagreement of the Al/Pa signal which is be affected by the strong He II 303 A line visible in the synthetic spectrum. Also a sligthly different value for the interstellar absorption may change the expected Al/Pa signal.

In the Sun, transition region and coronal emission originates from loop-like structures which to first approximation can be considered of costant pressure and costant cross section. We can make the same assumption for λ And. Two large spots, covering 12% of the surface of the primary star was inferred by Bopp and Noha (1980), at about 30^0 in latitude and separated by 160^0 in longitudine. If we assume that they indicate the presence of two "preferential longitudines" of activity we may postulate the existence of a large loop connecting them.

For a star radius is 5-6 $R_\odot$, the two spots are about 10^{12} cm apart and this value can also be assumed as a reasonable estimate of the half length L of the loop. If this loop is similar to solar ones, an aspect ratio of 0.1 may be assumed and the loop cross-section is given by:

$$S = \pi\ 10^{-2}\ L^2 \simeq 3\ 10^{22}\ cm^2$$

Since the projected disk area of the star is about 4.5 10^{23} cm^2, the loop cros-section is somewhat less than 10% of the stellar disk, very near the value obtained from the optical observations.

Assuming a constant pressure and a constant cross-section, the semilength of the loop is given by:

$$L/2 = \int dh = C \ (4 k^2/p^2 S) \int_{T_1}^{T_c} T^2 \ e^{T_0/T} (T/T_c)^\beta [1 - (T/T_c)^\gamma]^{-1/2} dT.$$

With $T_1 = 2 \ 10^4$ °K and the best-fit parameters given above, one derives $p = 15$ dyn cm^{-2}, using the mentioned values of S and L.

3. CONCLUSIONS

The analysis of X-ray and UV emission of λ And observed with EXOSAT (ME and LE) and IUE shows that:

- a continous distribution of d.e.m. with temperature up to a maximum temperature $\simeq 35 \ 10^6$ °K can reproduce (reduced $\chi^2 \simeq 1.2$) all available observations. The worst fit occurs for the low energy part of the ME spectrum, which requires a somewhat larger d.e.m. around $5 \ 10^6$ °K, and for the Al/Pa filter, which is lower than theoretical predictions (a larger interstellar column density ?);
- the d.e.m. [T n^2 (dV/dT)] presents a minimum around $2 \ 10^5$ °K and increases at higher temperatures;
- the derived d.e.m. is compatible with a loop model having: half length = 10^{12} cm, cross-section = $3 \ 10^{22}$ cm^2, and pressure =15 dyn cm^{-2}. In contrast to the model of Rosner et al.(1978) the heating deposition along this loop turns out to be non uniform.

REFERENCES

Baliunas S.L. and Dupree A.K.: 1982, Ap.J. **252**, 668
Bath G.T. and Wallestein G., Publ.Astr.Soc.Pac.:1976, **88**, 759
Bopp B.W. and Noha P.V., Publ.Soc.Pac.:1980, **92**, 717,
Gratton L.:1950, Ap.J., **111**, 31
Hall D.S. :1981, in 'Solar Phenomena in Star and Stellar Systems' R. Bonnet and A. Dupree eds.,p.431
Jordan C. et al.:1987,Mon.Not.R.astr.Soc.,**225**,903
Landini M. and Monsignori Fossi B.C.:1975, Astron. Astrophys.,**42**,213
Landini M. and Monsignori Fossi B.C.:1981, Astron. Astrophys.,**102**,391
Landini M. and Monsignori Fossi B.C.:1984, Physica Scripta,T7,53
Linsky J.L. et al.,Nature : 1978 , **275**,389
Monsignori Fossi B.C. and Landini M.:1988 in "Activity in Cool Star Envelopes" O.Havens et al. eds.,237
Pallavicini R., Monsignori Fossi B.C., Landini M. and Schmitt J.H.M.M. : 1988, A.Ap. **191**, 109
Rosner R.,Tucker W.H. and Vaiana G.S.,1978, Ap.J., **220**,643
Vaiana G.S.et al.: 1981, Ap.J.,**245**,163
Walter F.and Bowyer S.:,1981, Ap.J.**245**, 671

X-RAY LINES FROM Mg VIII AND Si X IONS AND THEIR DIAGNOSTIC USE

Bhola N. Dwivedi

Department of Applied Physics, Institute of Technology,
Banaras Hindu University, Varanasi-221 005, India

ABSTRACT. The solar X-ray emission lines from Mg VIII and Si X ions have been studied. The variation of theoretical line intensity ratio $I(\lambda 75.03)/I(\lambda 74.86)$ from Mg VIII and $I(\lambda 50.69)/I(\lambda 50.52)$ from Si X as a function of electron density is found to be good density monitors of the emitting regions of solar plasma. The computed values of line intensity from these ions based on Kopp and Orrall model have been used to derive electron density of the quiet Sun and coronal holes. The electron densities of 10^9 cm^{-3} and 4.6×10^8 cm^{-3} are estimated at the electron temperatures of 8×10^5 K and 1.6×10^6 K for the quiet Sun whereas the respective values of 5.4×10^8 cm^{-3} and 1.7×10^8 cm^{-3} are obtained for the coronal holes. The line intensity ratios studied here are independent of temperature variation and are therefore excellent candidates for electron density diagnostics. However, observational data with improved spectral resolution is needed for using X-ray line pairs studied for their diagnostic use.

INTRODUCTION

Ions in the boron sequence have rich emission line spectra in the X- and EUV region. Observations of the relative strengths of EUV lines from these ions especially Mg VIII and Si X have been extensively used to probe the solar plasma (Elwert and Raju 1975; Flower and Nussbaumer 1975a,b; Vernazza and Mason 1978; Dwivedi and Raju 1980; Saha and Trefftz 1983; Dwivedi 1988). However, little attention has been paid so far to use X-ray lines from these ions for electron density diagnostics of solar and astrophysical plasmas mainly due to lack of observational data. Recently Brown et al. (1986) have studied the electron density diagnostics in the 10-100 Å interval for a solar flare with considerable emphasis on line intensity ratios from helium-like ions. In the present investigation X-ray lines from Mg VIII and Si X have been considered for electron density diagnostics of cosmic plasmas. According to the ionization equilibrium calculations of Jordan (1969), Mg VIII has maximum relative ion abundance at 8×10^5 K and Si X at 1.6×10^6 K. Because of its maximum relative ion abundance around 8×10^5 K, lines emitted from Mg VIII ion are most suitable to probe chromosphere-corona transition region and coronal holes.

METHOD

If the computed line intensity ratio is found to vary with electron density, one finds it useful for density diagnostics. Moreover, the observed line intensity ratio should fall well within the density-sensitive portion of the curve in order to reliably determine the electron density. However, the observed lines with calibrated intensities have to be distinctly free from any ambiguity with regard to the blending, masking or other observational problems. Therefore, one finds very few observed lines with calibrated intensities to exploit them for density determinations. One of the other possible ways to look at the problem is to use a model atmosphere to compute the corresponding line intensities to find out the theoretical values of N_e. Such a study is also useful in identifying close lines arising from different ions prevalent in the atmosphere and provides first hand information to predict lines with observable intensity, which have hitherto not been observed, for future observations. This study also acts as a test or constraint on the model atmosphere when compared with the observational data. We have, therefore, used a model atmosphere of Kopp and Orrall (1976) for the quiet Sun and coronal holes to estimate the line intensities in order to use them for density determinations. The necessary steady state equilibrium equations for various levels accounting for different physical processes as well as the atomic data used have been described by Dwivedi and Raju (1980).

RESULTS AND DISCUSSION

Density sensitivity of line intensity ratios considered in this study arises because of the existence of the metastable levels.

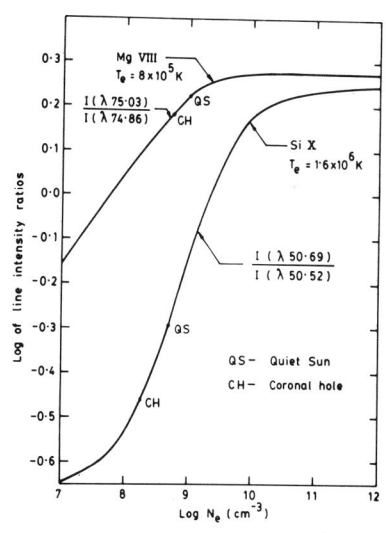

FIG.1. Electron density dependence of Mg VIII and Si X density-sensitive X-ray line intensity ratios. Dots correspond to the computed intensity ratios based on the model atmosphere of Kopp and Orrall (1976).

Figure 1 shows the variation of the intensity ratios of X-ray lines from Mg VIII and Si X ions as a function of electron density. The values of temperature indicated in this figure are those at which the relative ion abundances of the elements are reported to be the maximum. In order to study the temperature dependence of line intensity ratios, we have carried out the computation at two different temperatures on either side of the temperature at which the relative ion abundance of the element is maximum. We find that the line intensity ratios discussed here are rather insensitive of temperature variation.

The intensity ratio of the $(\frac{3}{2} - \frac{5}{2})$ and $(\frac{1}{2} - \frac{3}{2})$ lines of the multiplet $(2s^2\, 2p\, ^2P^o - 2s^2\, 3d\, ^2D)$ is an electron density diagnostic. The $(\frac{3}{2} - \frac{5}{2})$ transition is usually blended with the $(\frac{3}{2} - \frac{3}{2})$ transition. X-ray line pair $\lambda 75.03 - \lambda 74.86$ from Mg VIII seems to be an excellent candidate for electron density diagnostic for the quiet Sun and coronal holes. Similarly the X-ray line pair $\lambda 50.69 - \lambda 50.52$ from Si X could be equally useful for probing the solar plasma. The X-ray lines from Mg VIII and Si X corresponding to the $(\frac{3}{2} - \frac{1}{2})$ of $(2s^2\, 2p\, ^2P^o - 2s^2\, 3s\, ^2S)$ transition have observable intensity and could also be useful for diagnostic studies. Unfortunately, there are not many observed X-ray lines from these ions with calibrated intensities suitable for density determinations. In order to determine the electron density from the quiet Sun and coronal hole regions, the computed line intensities based on Kopp and Orrall model have been used. The line intensity ratios thus obtained are shown by dots in Figure 1 for the quiet Sun and coronal holes. The electron density thus derived are listed in Table 1 using X-ray lines from Mg VIII and Si X ions which seem to be quite reasonable and await observational support by future solar missions. We have also listed the line intensity ratios based on the model atmosphere of Elzner (1976) for the quiet Sun for the sake of comparison. Some of the observed lines by Malinovsky and Heroux (1973) are also indicated. Computed line intensities using models are meant only to assert the potentiality of the line intensity ratios studied here for density diagnostics. However, the variation of the line intensity ratio as a function of electron density is independent of any model parameters and could be used to derive electron density from observed values of line fluxes that future missions might provide.

CONCLUSION

The X-ray lines from Mg VIII and Si X are exellent candidates for probing cosmic plasmas. A model atmosphere of Sun has been used to ensure the utility of the line intensity ratios for diagnostic studies. Notwithstanding the derived electron densities being model dependent, the line intensity ratio curves as a function of electron density are independent of any model parameters-elemental abundances and relative ion abundances etc. It is, therefore, concluded that the X-ray lines studied could potentially be used as density indicators

TABLE 1. Computed line intensity ratios and derived electron densities using Kopp and Orrall model

Line ratio	Quiet Sun		Coronal hole	
	Intensity ratio	N_e (cm^{-3})	Intensity ratio	N_e (cm^{-3})
Mg VIII-ion				
$\frac{I(\lambda 75.03)}{I(\lambda 74.86)}$	1.67 1.56 1.47*	10^9 6.8×10^8 4.6×10^8	1.52	5.4×10^8
$\frac{I(\lambda 82.82)}{I(\lambda 74.86)}$	0.53 0.52+	10^9 6.3×10^8	0.51	5.8×10^8
Si X-ion				
$\frac{I(\lambda 50.69)}{I(\lambda 50.52)}$	0.50 0.37+	4.6×10^8 2.2×10^8	0.34	1.7×10^8
$\frac{I(\lambda 55.01)}{I(\lambda 50.52)}$	0.32 0.30+	4.6×10^8 2.7×10^8	0.29	1.5×10^8

+Computed line intensity ratio using Elzner model (1976)
*Observed line intensity ratio from Malinovsky and Heroux (1973).

of the solar and astrophysical plasmas. None the less, the observational data with improved spectral resolution is awaited by future solar missions in order to use these lines for density diagnostics in a realistic fashion.

ACKNOWLEDGEMENT. I am grateful to Dr. Paul Gorenstein for providing me local hospitality during IAU Colloquium 115.

REFERENCES

Brown, W.A., Bruner, M.E., Acton, L.W., and Mason, H.E. 1986, Ap. J., **301**, 981.
Dwivedi, B.N. and Raju, P.K., 1980, Solar Phys., **68**, 111.
Dwivedi, B.N. 1988, Solar Phys. Lett., in press.
Elwert, G. and Raju, P.K. 1975, Ap. Space Sci., **38**, 369.
Elzner, L.R. 1976, Astr. Ap., **47**, 9.
Flower, D.R. and Nussbaumer, H. 1975a, Astr. Ap., **45**, 145.
Flower, D.R. and Nussbaumer, H. 1975b, Astr. Ap., **45**, 349.
Jordan, C. 1969, M.N.R.A.S., **142**, 501.
Kopp, R.A. and Orrall, F.Q. 1976, Astr. Ap., **53**, 363.
Malinovsky, M. and Heroux, L. 1973, Ap. J., **181**, 1009.
Saha, H.P. and Trefftz, E. 1983, Solar Phys., **87**, 233.
Vernazza, J.E. and Mason, H.E. 1978, Ap. J., **226**, 720.

THERMAL INSTABILITY IN A HOT PLASMA

Steve A. Balbus and Noam Soker.
Department of Astronomy
University of Virginia
P. O. Box 3818 University Station
Charlottesville, VA 22903 U. S. A.

ABSTRACT. The nature of local thermal instability in static and dynamic radiating plasmas described by an equilibrium cooling function has been reexamined. Several new results have been found. In a plasma in both thermal and hydrostatic equilibrium, if the cooling function is not an explicit function of position, and does not display isentropic thermal instability (i.e. sound waves are thermally stable), then isobaric thermal instability by the Field criterion is present if and only if convective instability is present by the Schwarzschild criterion. In this case, thermal overstability does not occur. For the case of a dynamical plasma we present a very general Lagrangian equation for the development of nonradial thermal instability. In the limit of large cooling time to free-fall time ratio, the equation is solved analytically by WKBJ techniques. Results are directly applicable to cluster X-ray cooling flows. Such flows are surprisingly *stable* except for perturbation wavenumbers that are very nearly radial. We believe that the origin of cooling flow optical filaments is not to be found in linear thermal instability.

1. INTRODUCTION

That a diffuse hot plasma can be thermally unstable is well-known (Field 1965, Mathews and Bregman 1978). We have reexamined the nature of thermal instability in hot plasmas. Using standard fluid techniques, we have found some surprising results. They include: (a) In a static plasma characterized by a mass-specific radiative loss function $\mathcal{L}$ that depends upon density ρ and temperature T (but not position r), thermal instability by the Field criterion generally occurs if and only if convective instability by the Schwarzschild criterion is present. If the explicit spatial gradient $\partial \mathcal{L}/\partial r$ is sufficiently large and pointed opposite to the direction of gravity, then thermal instability or convective instability will *necessarily* be present. (b) Nonradial thermal perturbations ("blobs") in cooling flows are stable throughout most of the flow. Radial instabilities are present, but mode-coupling may severely restrict their nonlinear development. At cool temperatures ($< 10^7 K$) and small radii, flow convergence can make buoyant oscillations significantly overstable. This may lead to nonlinear clumping in the accreting gas at the center of cooling flows. We present here an explanation of these finding and speculate on their implications.

2. THERMAL INSTABILITY OF A STATIC PLASMA

A one-dimensional static plasma is described by the loss function equation $\mathcal{L}(\rho, T, r) = 0$, or upon differentiation:

$$\frac{d\mathcal{L}}{dr} = \frac{\partial \mathcal{L}}{\partial r} + \frac{d\rho}{dr}\left(\frac{\partial \mathcal{L}}{\partial \rho}\right)_T + \frac{dT}{dr}\left(\frac{\partial \mathcal{L}}{\partial T}\right)_\rho = 0. \tag{2.1}$$

Using standard transformations, one can rewrite the thermodynamic partial derivatives of $\mathcal{L}$ in terms of temperature derivatives at constant pressure P and constant entropy S (Balbus and Soker 1988):

$$\frac{\partial \mathcal{L}}{\partial r} + T\left(\frac{\partial \mathcal{L}}{\partial T}\right)_P \left(\frac{3}{5}\frac{d\ln P}{dr} - \frac{d\ln \rho}{dr}\right) + \frac{2}{5}T\left(\frac{\partial \mathcal{L}}{\partial T}\right)_S \frac{d\ln P}{dr} = 0. \tag{2.2}$$

The product of the isobaric temperature derivative of $\mathcal{L}$ and the spatial entropy gradient has the same sign as the isentropic derivative of $\mathcal{L}$, unless the explicit spatial gradient of $\mathcal{L}$ is sufficiently large, in which case the product has the opposite sign. Since the isentropic derivative of $\mathcal{L}$ is generally positive for any standard astrophysical cooling function, we may conclude that *if $\mathcal{L}$ is independent of r, a medium is thermally unstable by the Field criterion if and only if it is convectively unstable by the Schwarzschild criterion.* This suggests that static models of gaseous galactic haloes will quite generally display convective instability, and that thermal instabilities will form in a dynamically active background. Equation (2.2) applied to a static cooling flow model heated by a $1/r^2$ source (say relativistic particles from an AGN) suggests that regardless of the form of $\mathcal{L}$, either thermal or convective instability *must* be present (but not both).

3. THERMAL INSTABILITY OF A DYNAMICAL PLASMA

The equilibrium fluid is considered to be a flowing, spherically symmetric, time-independent, optically thin plasma subject to bulk heating/cooling processes. Self-gravity is assumed to be negligible, but an external gravitational potential is present. Quasi-hydrostatic equilibrium need not prevail. We consider the local stability of this flow to general spheroidal perturbations. We introduce the quantity a, which measures the radial separation of two close points as a function of time t. The spheroidal perturbations of the fluid displacement vector $\boldsymbol{\xi}$ have associated radial wave number k, $(kr \gg 1)$, and spherical harmonic index l. Neglecting thermal conduction the evolutionary equation for the radial displacement amplitude ξ of a comoving fluid element is found to be:

$$\left(\frac{d}{dt} + \frac{2}{5}T\Theta_{T,P}\right)\frac{1}{ag}\left(\frac{d}{dt}a^2\frac{d}{dt}\frac{\xi}{a} + \frac{d}{dt}\frac{k^2 r^2}{l(l+1)}\frac{d}{dt}\frac{\xi}{a}\right)$$
$$+ a\left(\frac{3}{5}\frac{\partial \ln P}{\partial r} - \frac{\partial \ln \rho}{\partial r}\right)\frac{d}{dt}\frac{\xi}{a} = 0. \tag{3.1}$$

where

$$\Theta_{T,P} = \left[\frac{\partial(\mathcal{L}/T)}{\partial T}\right]_P, \qquad q^2 = \frac{k^2}{a^2} + \frac{l(l+1)}{r^2}, \qquad g = -\frac{1}{\rho}\frac{\partial P}{\partial r} \tag{3.2}$$

and μ is the mean mass per particle and k_B is the Boltzmann constant. The quantity ξ enters into the equation only in the ratio ξ/a since a radial displacement that is "frozen" into the flow would scale proportional to a with no physical consequences.

The applications of eq. (3.1) to cooling flows are ideally suited to the use of WKBJ techniques because the cooling time is long compared to sound crossing time. We define

$$\beta^2 \equiv \left(1 + \frac{k^2 r^2}{l(l+1)a^2}\right)^{-1}, \quad \omega_{BV}^2 \equiv g\left(\frac{3}{5}\frac{\partial \ln P}{\partial r} - \frac{\partial \ln \rho}{\partial r}\right), \quad (3.3)$$

where ω_{BV} is the effective Brunt-Väisälä frequency. The WKBJ solution to eq. (3.1) is

$$\frac{\xi}{\beta} = \left[\frac{e^{\pm i \int^t \beta \omega_{BV} dt'}}{(\beta \omega_{BV})^{1/2}}\right] \exp\left[-\int^t \frac{\frac{2}{15} T \Theta_{T,S} \frac{\partial \ln P}{\partial r} + \frac{1}{3} \frac{\partial \Theta}{\partial r}}{\frac{5}{3}\frac{\partial \ln \rho}{\partial r} - \frac{\partial \ln P}{\partial r}} dt'\right]. \quad (3.4)$$

Equation (3.4) may be interpreted as follows. The term $1/\beta$ on the left-hand-side of the equation is a geometric factor which converts ξ to $|\xi|$ for large l. The first grouping of terms on the left-hand-side is simply the WKBJ expression for comoving Brunt-Väisälä oscillations. In the following group, there are two thermal terms. If the isentropic condition $\Theta_{T,S} < 0$ holds, then the nearly adiabatic oscillations are overstable, pumped by buoyancy forces that are aided by radiative losses. The second thermal term involves the explicit spatial gradient of the loss rate, and is generally present if there is a central heating source in the flow, as in a Compton driven wind (Begelman et al. 1983). In this case, overstability becomes possible if the heating increases on a downward displacement of the fluid element, and decreases on an upward displacement.

The physically important quantity $\delta\rho/\rho$ (relative Eulerian density perturbation) for WKBJ solutions is:

$$\frac{\delta\rho}{\rho} = -\left(\frac{3}{5}\frac{\partial \ln P}{\partial r} - \frac{\partial \ln \rho}{\partial r}\right)\xi \quad (3.5)$$

In other words, the relative Eulerian density amplitude is simply the radial displacement divided by the entropy scale height of the flow. Note that the true test of instability is the behavior of $\delta\rho/\rho$, since it is ultimately Eulerian perturbations that measure physical changes in flow quantities. Equations (3.4) and (3.5) can be used to give the rough scaling of $\delta\rho/\rho$ as a function of equilibrium Eulerian position r of an accreting fluid element, assuming power-like behavior for background flow variable. For oscillatory perturbations with $\beta \sim$ unity, on the most unstable part of the equilibrium cooling curve where $\pounds/T \sim T^{-3/2}$ (Raymond, Cox, and Smith 1976), $\Theta_{T,S}$ is very small. Then $\delta\rho/\rho \sim r^{-1/2}$ for an isothermal sphere potential, and $\sim r^{-1/4}$ for a central point mass. Under these conditions, oscillatory instabilities are generally mild away from $r = 0$. As shown in the two figures, they all but disappear in more detailed treatments of cooling flows. Deep in the core radius of the cluster potential, if $g \sim r$ then $\delta\rho/\rho \sim r^{-1}$ and an overstability of an essentially adiabatic character becomes important. Clumpy gas at the center

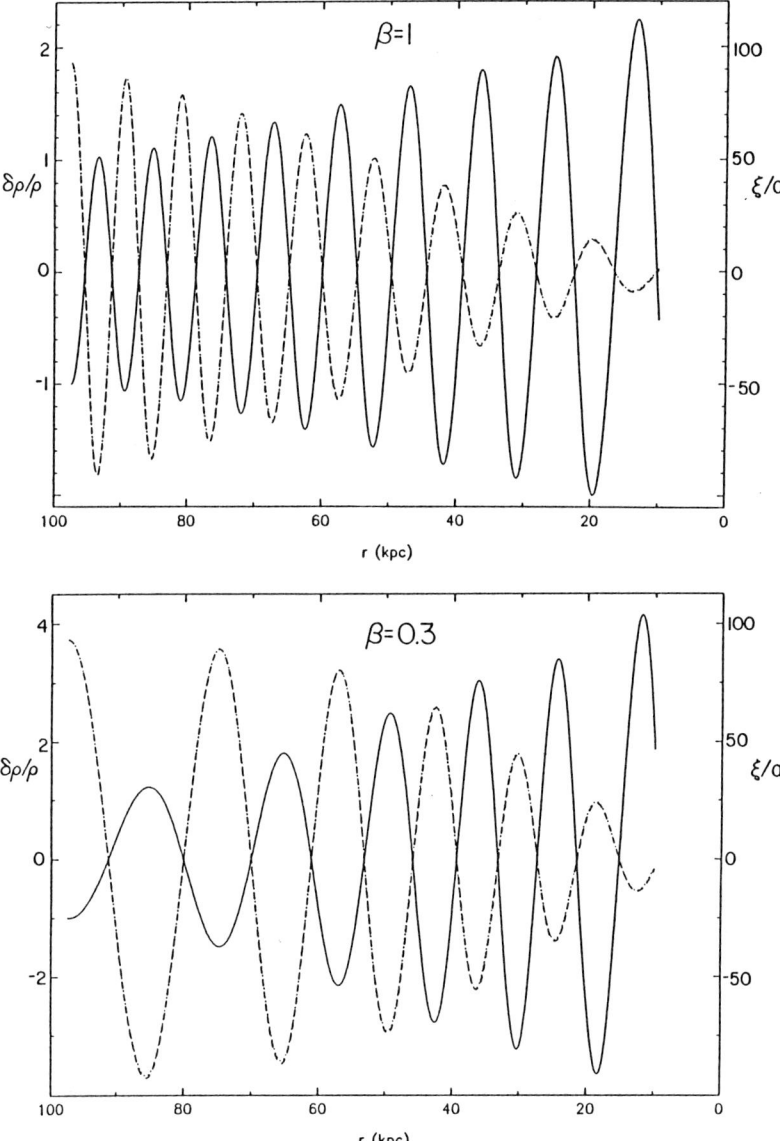

Figure 1. Evolution of two perturbations in typical cluster cooling flow: $\beta = 1$ corresponds to radially displaced perturbation ($kr \ll l(l+1)$), $\beta = 0.3$ to a more nearly azimuthal displacment ($kr \sim 3 \times i(i+1)$). Background flow is "standard, no star-formation" model of White and Sarazin (19897). Dashed line is numerical solution to eq. (3.1), dotted line is WKBJ solution eq. (3.4). (They are indistinguishable.) Solid line is $\delta\rho/\rho$; little growth is evident down to 10 kpc. Field type instability would obtain as $\beta \to 0$.

of cooling flows is in fact seen or inferred in millimeter, optical and x-ray studies (Lazareff et al. 1988, Hu et al. 1985, Canizares et al. 1987).

The importance of our findings is that if significant amounts of matter are cooling and dropping out of x-ray accretion flows at large radii, highly nonlinear (and nonacoustical) disturbances are necessary *ab initio*. To both grow and avoid detection (there is no direct evidence of cooling gas at large distances from cluster or galactic center [Hu et al. 1985]), several constraints on the matter are necessary (e.g. Nulsen 1986, Thomas 1987): no buoyant oscillations, small blob sizes, no dynamical or conductive assimilation into diffuse flow, $\sim 100\%$ efficiency of dark matter formation, etc. Alternatively, if one adopts the straightforward implication of this work that rapidly cooling, unstable blobs are not present at large flow radii, nonsteady accretion is needed to explain the x-ray observations. This conclusion is also indepndently supported by the recent discovery of large amounts of accreting molecular gas in the inner 5 kpc of NGC1275 (Lazareff et al. 1988).

5. CONCLUSIONS

[1] In gravitationally bound hot plasmas, there is an important connection between thermal instability by the Field criterion and convective instability by the Schwarzschild criterion. If the radiative loss function $\mathcal{L}$ is independent of position and is of any standard astrophysical form, the two occur simultaneously. If thermal instability is present under these conditions, it will form in a convectively unstable background. A large spatial gradient (opposite to gravity) in $\mathcal{L}$ forces either one or the other of the instabilities to be present.

[2] In slowly settling cooling flows, nonradial perturbations are essentially stable at large radii, and potentially overstable at very small radii. It is only when the oscillation frequency approaches the cooling time that Field-type thermal instability becomes important. In particular, cooling blobs must be highly nonlinear to grow and drop out of accretion flows. The alternative is young or transient cooling flows.

REFERENCES

Balbus, S.A., and Soker, N. 1988, *Ap.J.*, in press.
Begelman, M.C., McKee, C.F., and Shields, G.A. 1983, *Ap. J.*, **271**, 70.
Canizares, C.R., Markert, T.H., and Donahue, M. 1987, in *Cooling Flows in Clusters and Galaxies*, ed. A.C. Fabian, (Kluwer: Dordrecht), p. 73.
Field, G.B. 1965, *Ap. J.*, **142**, 531.
Hu, E.M., Cowie, L.L., and Wang, Z. 1985, *Ap. J. Suppl.*, **59**, 447.
Lazareff, B., Castets, A., Kim, D.-W., and Jura, M. 1988, preprint.
Mathews, W.G., and Bregman, J.N. 1978, *Ap. J.*, **224**, 308.
Nulsen, P.E.J. 1986, *M.N.R.A.S.*, **221**, 337.
Raymond, J.C., Cox, D.P., and Smith, B.W. 1976, *Ap. J.*, **204**, 290.
Sarazin, C.L. 1986, *REv. Mod. Phys.*, **58**, 1.
Thomas, P.A. 1987, in *Cooling Flows in Clusters and Galaxies*, ed. A.C. Fabian, (Kluwer: Dordrech), p. 361.
White, R.E. III, and Sarazin, C.L. 1987, *Ap. J.*, **318**, 612.

X-RAY LINE FORMATION IN ASTROPHYSICAL ENVIRONMENTS: THE L-SHELL SPECTRA OF IONIZED IRON

Duane A. Liedahl[1], Steven M. Kahn[1], Albert L. Osterheld[2], William H. Goldstein[2]

[1]Department of Physics and Space Sciences Laboratory
University of California, Berkeley, CA, 94720, USA

[2]High Temperature Physics Division
Lawrence Livermore National Laboratory, Livermore, CA, 94550, USA

ABSTRACT. We have initiated an extensive atomic modeling effort applicable to X-ray line emission in high-temperature astrophysical plasmas. The emphasis of our program is on the detailed accounting of the mechanisms which populate excited states of highly ionized atoms over a wide range of electron temperatures and densities. As a first demonstration we have calculated spectra for the important L-shell ions Fe XVI-XIX in a complete collisional-radiative model under conditions appropriate to solar coronal plasmas. Using methods presented here, we have synthesized the X-ray spectra of solar flares and active regions over the wavelength interval 13 - 18 Å. The atomic model, which includes 705 atomic energy levels, is the largest and most detailed of its kind. In this introductory paper, we discuss the effects of dielectronic recombination on the spectrum of Fe XVII and present a new technique whereby the $3s$ lines can be used as a diagnostic of the electron temperature. Also included are new values for the rate of resonance excitation of the $n=3$ Fe XVII excited states. These rates are lower than those previously obtained and suggest that resonance excitation does not contribute significantly to the population kinetics. Finally, we present a direct comparison of our a calculated model spectrum with data from a solar flare.

1. INTRODUCTION

The new generation of high-spectral-resolution X-ray telescopes (AXAF, XMM) will provide data much too detailed to be properly analyzed in the context of existing plasma emission models. The current models suffer from either an insufficient treatment of the atomic physics or from their reliance on the assumption of collisional equilibrium. These deficiencies are especially apparent in the cases of X-ray binaries and active galactic nuclei where the sites of line emission are likely to be photoionized and far from collisional equilibrium so that line emissivities cannot be characterized by the local electron temperature.

We have embarked upon an atomic modeling program to develop plasma diagnostics for application to the phenomenology of cosmic X-ray sources. Beyond the augmentation of plasma emission models commonly in use, our program will extend diagnostic techniques to a much wider variety of astronomical environments.

Motivated by the high quality of extant solar X-ray data and the need to establish a point of reference for our model spectra, the subject of our first study is the L-shell spectrum of iron under solar flare conditions. Line emission from the ions of interest, Fe XV-XXIV, are known to comprise major portions of the X-ray spectra of solar flares and solar active regions (e.g., Phillips *et al.* 1982). Indeed, nearly the entire solar flare spectrum from 14 - 18 Å can be synthesized from Fe XVI - XIX (see Figure 1). Since the Fe XVII L-shell lines are especially prominent features in the soft X-ray band of many cosmic sources, a considerable amount of effort has gone into a careful analysis of the relevant atomic physics. Comparably detailed analyses of the remaining Fe L-shell ions are underway. We are extending our calculations to the low-density, multi-keV temperature regime appropriate to young supernova remnants. Also, processes involving an external source of ionizing radiation and the creation of highly excited $1s$-hole states are now being coupled into the code in order to address the problem of line formation in accretion-powered X-ray sources.

2. THE MODEL

Only a brief description of the model will be given here. The atomic structure, radiative transition rates, and collisional rates are computed using a sophisticated atomic physics package originated by M. Klapisch and A. Bar-Shalom of Hebrew University (Klapisch 1971; Klapisch *et al.* 1977; Bar-Shalom *et al.* 1988). The atomic structure is computed *ab initio* using the relativistic, multi-configuration, parametric potential method. The collisional rates are computed in the relativistic distorted wave approximation. The model is collisional-radiative, as opposed to coronal, i.e., all possible level-to-level transition rates, collisional and radiative (e1, e2, m1, m2), are calculated and included in the rate equations. The model consists of all singly excited levels of Fe XVIII and Fe XIX for n=3, all singly excited levels of Fe XVII for n=3 and 4, all doubly excited levels of Fe XVI of the form $2p^53l3l'$ and $2s2p^63l3l'$, and the set of configurations of the form $2p^63l3l'$ of Fe XV for a total of 705 levels.

Excited states can also be populated by processes involving adjacent ion stages. Included are the results of detailed calculations of dielectronic recombination into excited levels of Fe XVII originating from the three levels of the Fe XVIII ground configuration and summed over the Fe XVII $2p^43l3l'$, $2p^43l4l'$ and $2s2p^53l3l'$ doubly excited states (Chen 1988). The Fe XVI model includes the 237 doubly excited levels of the form $2p^53l3l'$. We have included the satellites to Fe XVII resonance lines which are formed by radiative relaxation of these doubly excited states (Oreg 1988). Resonance excitation of the n=3 singly excited states of Fe XVII begin with a radiationless electron capture onto the Fe XVII ground state into a doubly excited state of Fe XVI ($3lnl'$ for n=6-14 and $2p^54l4l'$) which then autoionizes to an Fe XVII singly excited state. New rates for this process have been calculated (Goldstein 1988) and included in the rate equations.

3. RESULTS

Earlier theoretical treatments of the Fe XVII spectrum (e.g., Smith *et al.* 1985) have not accounted for the effects of dielectronic recombination from Fe XVIII. We have found that, in addition to regulating the ionization balance at coronal temperatures, dielectronic recombination can have a pronounced effect on the line emission. For example, at $T_e = 250$ eV (log $T_e = 6.45$), approximately one-third of the flux into Fe XVII excited states originates from the ground state of Fe XVIII by way of dielectronic recombination. Thus, Fe XVII levels which are populated substantially by cascade, such as $2s^2 2p^5 3s$, can be strongly coupled to the ground configuration of Fe XVIII. However, the collisionally populated $2s^2 2p^5 3d$ levels are largely unaffected by processes involving the Fe XVIII population. Consequently, the nine Fe XVII line ratios of the form $I(3s)/I(3d)$ must be treated as functions of both T_e and n_1^F/n_1^{Ne}. Let us refer to the latter quantity as R and define a function, Φ, such that $I(3s)/I(3d) = \Phi(T_e, R)$ for a given pair of lines.

By taking the ratio of two collisionally excited lines from neighboring ion stages, one can obtain a simple temperature-dependent expression for the relative ion fractions. The most useful lines of this kind are the Fe XVII 3d line at 15.01 Å and the Fe XVIII two-line blend at ~14.20 Å. One obtains an expression of the form $R(T_e) = \Gamma(T_e) \times I(14.20)/I(15.01)$. The function, $\Gamma(T_e)$, involves collisional rate coefficients and a rate coefficient which describes the rate of production of unresolved dielectronic recombination satellites to the 15 Å resonance line. The importance of this latter effect has been pointed out by Raymond and Smith (1986). We are presently performing more detailed calculations of this process. Simultaneous solution of the expressions for $\Phi(T_e, R)$ and $R(T_e)$ will yield both the electron temperature and the relative ionization fractions of Fe XVII and Fe XVIII. Of course, this assumes that the emitting regions of the two ions are spatially coincident. In principle, a similar method can be used for the remaining ion stages.

New calculations of resonance excitation (Goldstein 1988) of the $2p^5 3s$ states of Fe XVII suggest that this process is not as effective at populating these levels as had been predicted by previous calculations (e.g, Smith *et al.* 1985). Using the old rates, temperatures inferred from line ratios of the form $I(3s)/I(3d)$ are too high since the rate for resonance excitation is a decreasing function of temperature. However, if dielectronic recombination from Fe XVIII is also accounted for, the effect on the temperature estimate can be partially offset. This effect manifests the temperature dependence of the collisional excitation rate coefficients, rather than the dielectronic rate coefficients which do not change appreciably over the temperature range of interest. In other words, for a given R (see above), one generally expects ratios of the form $I(3s)/I(3d)$ to decrease with temperature because the excitation rates for the collisionally populated $3d$ levels are increasing over the solar coronal temperature range.

Finally, we present a model spectrum for comparison with a spectrum from a solar active region (Figure 1, Rugge and McKenzie 1985). We defer quantitative conclusions to a later paper (Liedahl, *et al.* 1988) where the unresolved satellites to the

15 Å will be accounted for. For the sake of discussion, the model spectrum was calculated with $T_e=250\ eV$ and $n_e=10^{11}$. Note that this part of the spectrum contains an important contribution from O VIII Lyman β at 16 Å which has not been included in our model.

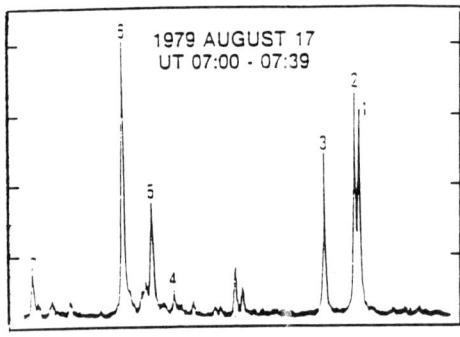

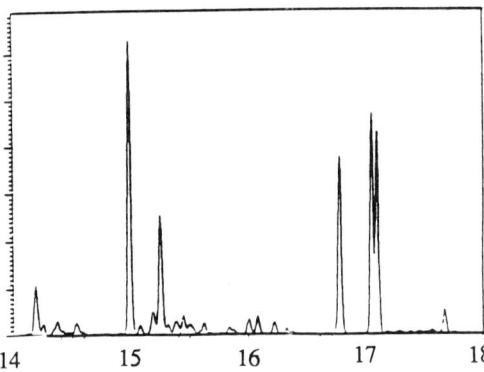

Figure 1. *Top:* Solar flare spectrum (Rugge and McKenzie 1985). The lines numbered 1-6 are Fe XVII lines and line 7 is the Fe XVIII two-line blend referred to in the text. *Bottom:* Model spectrum with $T_e=250\ eV$ and $n_e=10^{11}cm^{-3}$.

4. REFERENCES

Bar-Shalom, A., Klapisch, M., Oreg, J. 1988, *Phys. Rev. A* **38** 1773.
Chen, M. 1988, private communication.
Goldstein, W.H., This volume and Talk presented at the IAU Colloq. No. 115 on "Recent Advances in Atomic Modeling", Cambridge, Massachusetts, USA, Aug 1988.
Klapisch, M. 1971, *Computer Phys. Comm.* **2**, 239.
Klapisch, M., Schwab, J.L., Fraenkel, B.S., and Oreg, J. 1977, *J. Opt. Soc. Am.* **61** 148.
Liedahl, D.A., Kahn, S.M., Goldstein, W.H., and Osterheld, A.L. 1988, in preparation.
Oreg, J. 1988, private communication.
Phillips, K.J.H., *et al.* 1982, *Ap. J.*, **256**, 774.
Raymond, J.C., and Smith, B.W. 1986, *Ap. J.*, **306**, 762.
Rugge, H.R., and McKenzie, D.L. 1985, *Ap. J.*, **297**, 338.
Smith, B.W., Raymond, J.C., Mann, J.B., and Cowan, R.D., 1985, *Ap. J.*, **298**, 898.

MEASUREMENT OF CARBON ION PHOTOABSORPTION CROSS SECTIONS USING LASER PLASMAS

B. Wargelin[1], S.M. Kahn[1], W. Craig[2], and R. London[2]

[1] Department of Physics and Space Sciences Laboratory,
University of California, Berkeley, CA 94720, USA
[2] Lawrence Livermore National Laboratory, PO Box 808,
Livermore, CA 94550, USA

ABSTRACT. Laser plasmas are well-suited to studies of ionic photoabsorption because they can provide highly ionized, low temperature plasmas of high column density, as well as bright, compact continuum X-ray sources which can illuminate the plasma under study. In our experiment, continuum X-rays from a gold laser plasma are partially absorbed as they traverse a carbon plasma and are then dispersed by a grazing incidence reflection grating. An X-ray imaging camera records both the absorbed and unabsorbed spectra simultaneously for later computer analysis to determine the photoabsorption cross sections for each carbon ion species.

1. INTRODUCTION

Absorption spectra in the soft X-ray band are extremely important for studying cosmic accretion-powered sources as probes of the cool, circumsource, accreting material. In such photoionized nebulae (e.g. active galactic nuclei, cataclysmic variables, and X-ray binaries), a powerful, compact, X-ray emitting core photoionizes cooler surrounding gas, producing a series of ionization fronts. Electronic level populations in the ions are largely determined by the complex mechanisms of recombination cascades and photoexcitation, instead of collisional effects and spontaneous radiative decay as found in hot, optically thin, thermally emitting sources such as stellar coronae and supernovae remnants.

Future soft X-ray missions such as AXAF and XMM will provide abundant high resolution spectral data, but our ability to interpret these data may be severely limited by uncertainties in atomic physics and radiation transfer. This is especially true for photoionized nebulae where the complex processes described above are at work. Accurate photoabsorption cross sections would make calculation of ion column densities straightforward, thus constraining densities, temperatures, and source geometry, but experimental determination of these cross sections is very difficult because of the problems in producing and maintaining large column densities of highly ionized material. There are a few measurements for multiply-charged low Z ions, notably Jannitti *et al.* (1986), but available values come almost exclusively from theoretical efforts (using Hartree-Slater central field approximations, etc.) such as Reilman and Manson (1979). These calculations are generally accurate to within 10% or 20%, but can be off by factors of more than two near delayed maxima and Cooper minima. Furthermore, such calculations do not include autoionization and other subtle but important effects.

2. LASER PLASMAS

Laser plasmas are a very efficient means of producing large column densities of highly ionized atoms; of order 10% of the input laser energy is used in ionizing and exciting the plasma ions (Eidmann and Kishimoto 1986). In addition, one ionization state can often be selected to dominate the others (particularly for low Z materials) by varying the parameters of the laser pulse.

Laser plasmas also have relatively low kinetic temperatures, much like the photoionized gas surrounding an accretion source. This is because the recombination time scale for most ion species is usually much longer than the laser plasma cooling time (typically a few nanoseconds for a 1 ns pulse). During this cooling phase, the ions cascade down to their ground states. After the plasma has cooled, but before the ions have begun to recombine (usually a few tens of nanoseconds), the plasma is highly overionized relative to its kinetic temperature, and there is no recombination line emission.

As an example, one-dimensional computer simulations predict that a 10 Joule, 1 ns, 1064 nm (infrared) laser pulse focussed on a 10 mm × 100 μm × 1000 Å carbon foil will produce: a kinetic temperature peak of 70 eV, cooling to 10 eV after 10 ns; a 500 μm diameter cylinder of carbon plasma with 82% He-like and 17% Li-like carbon ions; and a He-like column density of 10^{19} ions/cm^2, or an optical depth of 2 at 20 Å.

Because of their excellent conversion efficiency and nearly continuum emission, high Z (e.g. Au and Ta) laser plasmas can be used as intense, compact X-ray sources. Nearly 50% of the laser pulse energy may be converted into X-ray emission during the plasma cooling phase (Eidmann and Kishimoto 1986). Such a bright source can shine through a cool, non-recombining, highly ionized laser plasma (just as in an accretion-powered photoionized nebula) and provide absorption spectra.

3. EXPERIMENTAL PROCEDURE

The basic approach of this experiment is much like that of Janniti *et al.* (1988), but with certain improvements which eliminate the largest sources of uncertainty in their data, namely plasma nonuniformity and shot-to-shot repeatability. Our first experiment is with carbon because it is easy to work with, being a solid.

A 1 nanosecond laser pulse of several Joules (the exact energy depending on the ionization state desired) is focussed on a 100 μm × 10 mm line on the carbon foil producing a cylinder of plasma with minimal density and temperature gradients in the center (Figure 1). After a few ns, when the plasma has cooled and thermal emission has died off (the exact timing to be determined with use of the X-ray streak camera described later), another 100 ps pulse of a few Joules is focussed on the end of a 50 μm diameter gold fiber, producing an intense continuum X-ray backlighter. These X-rays pass through the carbon plasma where they are partially absorbed, and are then dispersed by a concave grazing incidence reflection grating (Figure 2). Depending on the angle towards the backlighter, the X-rays striking the grating may have passed through the carbon plasma (absorption) or only alongside the plasma cylinder (no absorption). In this way, an X-ray camera can record both the absorbed and unabsorbed spectra at the same time, avoiding the problem of shot-to-shot variation in the backlighter spectrum.

4. EXPERIMENTAL EQUIPMENT

This experiment is conducted at the JANUS laser facility of the Lawrence Livermore National Laboratory. One beam of the Nd-YAG (1064 nm) laser generates the gold backlighter plasma, while the other creates the cylinder of carbon plasma. Apart from the lasers themselves, all experimental equipment is operated within a large vacuum chamber with window ports to admit the laser beams. The laser beam focussing lenses, target rod, and diffraction grating are all mounted on remotely controllable translation and rotation stages.

The carbon and gold targets are mounted on a stainless steel target rod with a 1.0 mm wide slot along the cylinder at the end of the rod (Figure 1). Carbon foils as thin as 400 Å are easily suspended across the slot by a flotation technique. Behind the foil is a brass mount for the gold fiber. The mount permits positioning of the fiber tip to an accuracy of better than 100 μm. One laser beam is focussed in a line on the carbon foil while the other beam (coming from the opposite direction) focusses on the end of the gold fiber through a hole in the target rod.

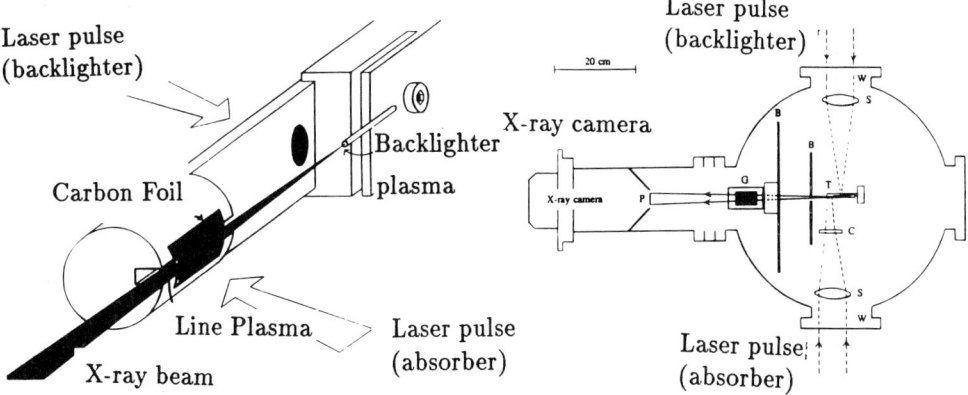

Figure 1. Target rod, 1.4× actual size. The laser pulse incident on the carbon foil creates a thin cylinder of absorbing plasma. A few nano-seconds later, the other beam shoots at the gold backlighter. X-rays from the backlighter pass through and to either side of the absorbing carbon plasma, and are dispersed by the reflection grating. The absorbed and unabsorbed spectra are then recorded simultaneously by the X-ray camera.

Figure 2. Top view of experimental setup. Solid arrowed lines are X-rays from backlighter. W window ports for incident laser beams; S spherical lenses for focussing beams; C cylindrical lens for line focussing; T target rod holding gold fiber and carbon foil; B baffles to stop charged particles and stray light; G grazing incidence reflection diffraction grating; P X-ray camera photocathode.

X-rays from the gold backlighter are dispersed by a variable line spacing, flat focal plane, grazing incidence diffraction grating (Kita *et al.* 1983). A 2400 line/mm $1°$ incidence grating provides coverage from 15 to 100 Å, and a 1200 line/mm $3°$ grating covers 50 to 300 Å. The grating is mounted on vertical and horizontal stages for focussing the spectrum on the X-ray camera photocathode.

Two X-ray cameras are used, a streak camera for time-resolved studies of the laser plasmas, and a gated X-ray imager (GXI) for two-dimensional data recording. In both cameras, X-rays hit a thin photocathode (potassium bromide on 1000 Å of Lexan) at the front of the camera, releasing electrons. A positively biased, fine copper mesh immediately behind the photocathode attracts the electrons, preserving the "image" of the X-rays on the photocathode. An electron optics system focusses the electrons onto a phosphor where they are "converted" back into a visible image. The image is intensified by a microchannel plate and recorded on calibrated film.

In the streak camera, a narrow slit is placed in front of the photocathode. As part of the electron optics system, a ramped bias voltage can be triggered by the laser pulse (using a photodiode) to sweep the electron "image" of the slit across the phosphor. Time-resolved pictures of the spectrum are used to determine the relative timing between the pulses that create the carbon plasma and the X-ray backlighter--a delay long enough to let the backlighter dominate any emission from the carbon plasma, but not so long that the carbon ions begin to recombine.

A gated X-ray imaging camera (GXI) records the final data by taking a two-dimensional snapshot of the spectrum (wavelength versus angle toward the backlighter). A 100 picosecond exposure is triggered by the second laser pulse so that the backlighter is at its brightest. Such short exposures are made by briefly applying a large bias across a thin metal film on the front of the phosphor, allowing electrons to penetrate the thin coating, reach the phosphor, and be "converted" back to visible photons. The spatial resolution of the GXI approaches 100 μm and limits our spectral resolution to about $\lambda/\Delta\lambda = 200$.

5. DISCUSSION

We have devised a method for measuring photoabsorption cross sections of multiply-ionized atoms that is simple and reliable. The carbon experiment was being run at the time of submission of this paper, and will provide much needed data for the interpretation of high resolution cosmic X-ray spectra. Experiments for other astrophysically abundant elements will be conducted in the near future.

REFERENCES

Eidmann, K., and Kishimoto, T. 1986, *Appl. Phys. Lett.*, **49**, 377.
Jannitti, E., Nicolosi, P., and Tondello, G. 1986, *Physica Scripta*, **36**, 93.
———. 1988, *Proc. of IAU Colloq. No. 102 on UV and X-Ray Spectroscopy of Astrophysical and Laboratory Plasmas*, ed. F. Bely-Dubau and P. Faucher, *Journal de Physique*, **49** Coll. C1 Suppl. 3, C1-71.
Kita, T., Harada, T., Nakano, N., and Kuroda, H. 1983, *Appl. Opt.*, **22**, 512.
Reilman, R. F., and Manson, S. T. 1979, *Ap. J. Suppl.*, **40**, 815.

TOKAMAK PLASMAS: A PARADIGM FOR CORONAL EQUILIBRIUM AND DISEQUILIBRIUM

Richard D. Petrasso
Plasma Fusion Center, MIT, Cambridge, MA 02139

ABSTRACT. Tokamaks operate over a wide parameter space, allowing access to plasma conditions relevant to astrophysical plasmas. For high electron density discharges, for example, the central electron density and temperature are $\sim 3 \times 10^{14} \text{cm}^{-3}$ and ~ 1.5 keV, and the central plasma region is in coronal equilibrium. Towards the edge of the plasma, however, many ion species will be far out of coronal equilibrium. A novel feature of the edge region is the seemingly contradictory property that it is, simultaneously, both a strongly recombining and a strongly ionizing plasma. Recent tokamak observations of strongly recombining plasmas also show that the G parameter (the ratio of forbidden plus intercombination to resonance lines) is larger by a factor of 3 than the ratio of statistical weights of the triplet-singlet series. Such observations can be of direct consequence to the interpretation of non-equilibrium astrophysical plasmas.

There are several features of tokamak plasmas relevant to astrophysical ones. Primarily this is due to the wide variety of plasma conditions that can be readily accessed (Artsimovich 1972; Nuclear Fusion 25 1985). For example, high temperature ($T_e \sim 3$ keV $\simeq 35 \times 10^6$ °K) plasmas are routinely obtained by running high field (8-10T), high current (~ 400 kA), and low electron density ($\leq 5 \times 10^{13}$ cm^{-3}) discharges (Parker et al. 1985). At these high temperatures, impurities such as titanium injected into a background plasma of hydrogen, will, upon reaching the plasma center, be ionized down to helium-like ions (Ti^{+20}). Of course, near the plasma edge where $T_e \sim 30$ eV, only low-charge states of Ti will dominate. Even here, however, there will reside a small fraction of highly ionized titanium ions that have rapidly transported themselves from the center. Such highly stripped ions are — by orders of magnitude — out of coronal equilibrium, i.e. the edge plasma is strongly recombining. What is novel, however, is that the same physical region of the plasma is simultaneously a strongly ionizing one too! The reason for this is that a small fraction of slightly stripped Ti ions pierce through the edge region in the direction of increasing temperature and, on the basis of the local electron temperature, possess too many bound electrons, i.e. they are underionized. It is through such rapid transport processes that the same physical location will have both under- and overionized ions, in steady state, but far out of coronal equilibrium. Superficially this situation appears unique in comparison to non-equilibrium astrophysical plasmas, which are often characterized as being either ionizing (Canizares et al. 1983) — such as shock or flare-heated plasmas — or recombining, but not both. However, there may well be analogous ionizing-recombining (I-R) regions in astrophysical settings. One that suggests itself is the transition zone where the temperature gradient is extremely large: If local ion transit times through this zone are sufficiently small compared to characteristic ionization and recombination times, then it will be an I-R plasma. Another possible candidate is the boundary, or high-temperature gradient region, of a flare. Even if such I-R regions do exist astrophysically, one is still faced with the rather formidable task of confirming this observationally.

In order to make these comments about (non)equilibrium tokamak conditions more quantitative, it is useful to examine germane characteristics of a tokamak, such as MIT's Alcator-C (Parker et al. 1985). A tokamak is a torus-shaped structure with toroidal magnetic flux surfaces containing the plasma. Fig. 1a shows a cross-sectional cut through the small-radius portion of the torus; as depicted, the coronal plasma (CP) occupies the central region and the ionizing-recombining plasma (I-RP) the outer annulus. (Of course there must be a third transition region between the two, but we will ignore such details here.) Directly below Fig. 1a is the electron temperature and density profile that characterizes local values of these parameters along the horizontal "cut" in Fig. 1a. Both the temperature and density are centrally peaked (1500 eV and 4×10^{14} cm^{-3}) and monotonically decrease towards the plasma edge (about 16 cm from the plasma center). As shown, their exact functional dependence is very different, the electron temperature being more peaked. These profiles are important since they are needed to calculate the characteristic atomic times for ionization and recombination, i.e. $\tau_A \simeq 1/[n_e(\sigma v)_A]$, where $(\sigma v)_A$ is the temperature dependent ionization or recombination rate coefficient appropriate for a given ion species. Fig. 1c shows the resulting spatial profiles of the atomic (τ_A) and ion transit or confinement time, τ_t. (τ_t is a semi-local function that is again evaluated along the horizontal cut of Fig. 1a.) The essential feature is that in the CP region, the plasma confinement time is longer than the atomic relaxation time, i.e. $\tau_A \lesssim \tau_t$. In contrast, for the I-R region the opposite is true, i.e., $\tau_t \lesssim \tau_A$.

Fig. 2 is a specific illustration of measured silicon ion profiles (solid curves) in the CP region, along with the corresponding coronal predictions (Petrasso et al. 1982). For this plasma discharge in the Alcator-C tokamak, a small amount of Si was injected into a background plasma of deuterium. Of importance to our considerations here is the fact that $\tau_A \sim 2$ ms for highly stripped Si, while, for the same ions, $\tau_t \sim 20$ ms. Within experimental error, deviations of the measured profiles from coronal equilibrium predictions were undetectable.

In contradistinction to the central coronal plasma, the edge plasma is strongly

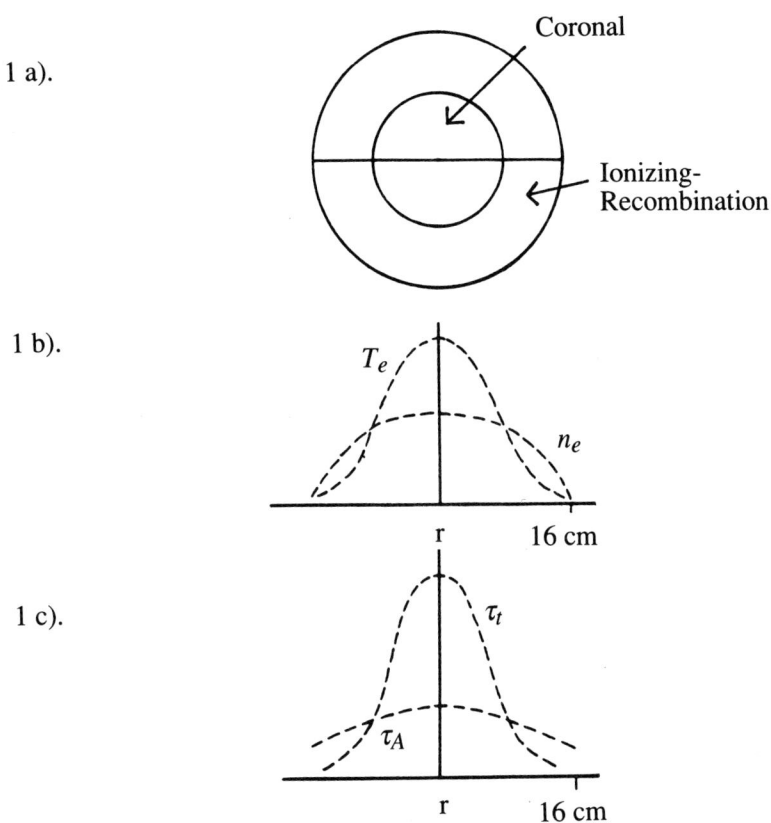

Fig. 1a.—A cross-sectional cut through the small radius portion of a tokamak plasma. From plasma center to very edge is about 16 cm. For high-electron density discharges, the central plasma is in coronal equilibrium. The outer annulus, or "edge" region, is simultaneously both an ionizing and recombining plasma.

Fig. 1b.—The electron temperature (T_e) and density (n_e) profiles along the horizontal cut of Fig. 1a. In this instance, the peak values of T_e and n_e are about 1.5 keV and $4 \times 10^{14} cm^{-3}$. The electron density profile is usually much broader than the temperature profile. The distance between the plasma center and outer edge is about 16 cm (based on Greenwald et al. 1984).

Fig. 1c.—A qualitative picture of the characteristic atomic times for ionization or recombination (τ_A), and the ion transit or confinement time (τ_t). The values are estimated along the horizontal cut of Fig. 1a. In the coronal region, $\tau_A \leq \tau_t$. In the ionizing-recombining region, $\tau_A \geq \tau_t$. At the plasma center, for example, $\tau_A \sim 2$ ms for highly stripped Si while, in contrast, $\tau_t \sim 20$ ms for the same ions.

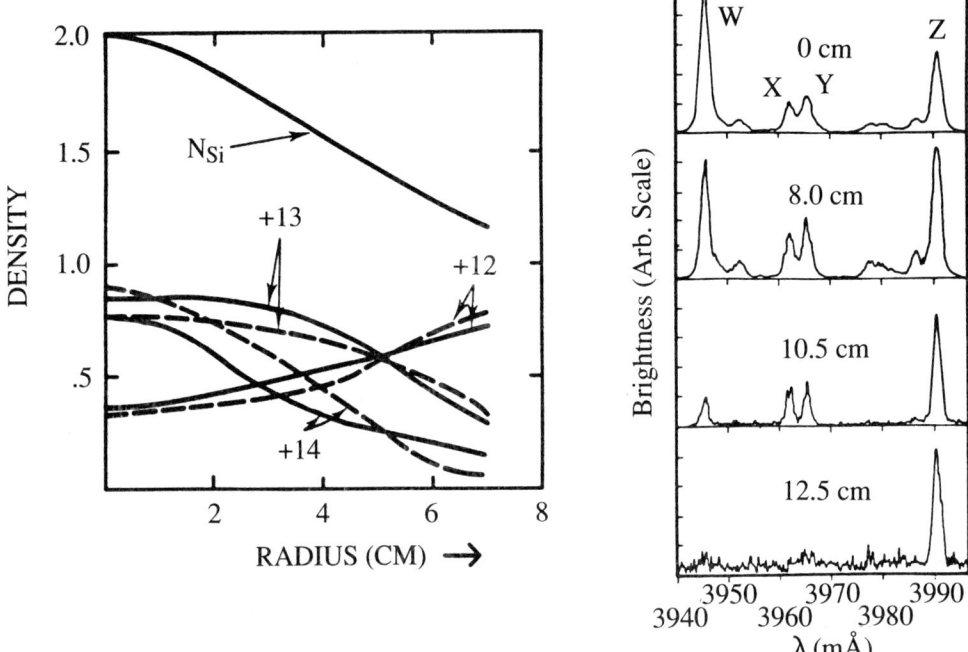

Fig. 2. Comparison between measured charged state Si profiles and the predictions of a coronal equilibrium model. Solid curves: Observed fully stripped (+14), H-like (+13), and He-like (+12) Si ion density profiles and their sum ($N_s i$). The central electron density and temperature are 3.4×10^{14} cm^{-3} and 1.3 keV; the corresponding atomic relaxation times (τ_A) for highly stripped Si are about 2 ms. In contrast, the transit or confinement time for these ions, τ_t, is about 20 ms (from Petrasso et al. 1982).

Fig. 3. Spectra of helium-like argon at four different radial locations (0 through 12.5 cm), showing the resonance (W), intercombination (X and Y), and forbidden (Z) lines. (The vertical scales are arbitrary.) Local electron temperatures at the four locations are 1650, 550, 250 and 100 eV. (For these Alcator-C plasma discharges, the plasma minor radius was 12.5 cm.) Particularly intriguing – and still quantitatively unexplained – is the data at 12.5 cm which show the forbidden (Z) line to be completely dominant. The corresponding G value [G=(X+Y+Z)/W] is $\geq$ 10, which is much larger than that expected simply on the basis of the statistical weights of the triplet-singlet series (i.e. G=3) (from Rice et al. 1987).

out of coronal equilibrium. John Rice and coworkers observed helium-like argon (Ar^{+16}) in the I-R region of the Alcator-C tokamak, and Fig. 3 shows a radial scan of the plasma with a crystal spectrometer which resolves the W (resonance), X and Y (intercombination), and Z (forbidden) lines of this ion (Rice et al. 1987). The G parameter [G=(Z+X+Y)/W], widely used in astrophysics as a plasma diagnostic (Jordan and Veck, 1982; Canizares et al. 1983; Canizares 1988; Mason 1988), is determined from these measurements. For the observations very near the plasma edge (the 12.5 cm case of Fig. 3), the spectrum is totally dominated by the forbidden line. In fact, the corresponding G value is about 10 or greater. In analogous tokamak experiments at the Princeton Plasma Laboratory, but using helium-like titanium (Ti^{+20}), a G value of about 10 was also obtained (Bitter et al. 1985). In contrast, for strongly recombining plasmas such as these, it is commonly believed that the G value will asymptote to a value of order 3, simply the ratio of statistical weights of the triplet-singlet series. Thus the mechanism by which the $2\,^3S$ level, which corresponds to the upper state of the forbidden transition, becomes so dominantly populated is not, at present, quantitatively understood. It has been suggested, however, that the combination of charge-exchange recombination plus preferential cascading into $2\,^3S$ state might resolve this enigma (Kallne et al. 1984). Therefore, such laboratory measurements as these could bear directly on the interpretation of plasma parameters in non-equilibrium astrophysical plasmas.

In summary, high density tokamak plasmas can be qualitatively divided into two regions, a central coronal region and an edge, ionizing-recombining one. In the coronal region, the atomic times for ionization and recombination are smaller than the characteristic ion confinement or transit times. In the ionizing-recombining region, the opposite prevails. In addition, because of the control that can be exercised over plasma conditions and impurities, a number of problems of astrophysical interest can be conveniently addressed in tokamak plasmas, and commonly used astrophysical diagnostics (e.g. the G parameter) can be subject to laboratory verification.

ACKNOWLEDGMENTS

For materials used in this presentation, I thank Chi-kang Li, Robert Granetz, John Rice and Kevin Wenzel. I gratefully acknowledge many helpful comments from Xing Chen and Dieter Sigmar. This work was supported by DOE contract, DE-AC02-78ET51013

REFERENCES

Artsimovich, L.A. 1972, Nuclear Fusion **6**, 215.
Bitter, M., Hill, K.W., Zarnstorff, S., von Goeler, S., Hulse, R., Johnson, L.C., Sauthoff, N.R., Sesnic, S., and Young, K.M. 1985, Phys.Rev.A **32**(5), 3011.

Canizares, C.R., Winkler, P.F., Markert, T.H., and Berg, C. 1983, in Supernova Remnants and their X-ray Emission, J. Danziger and P. Gorenstein (eds.), IAU., 205.

Canizares, C.R. 1988, IAU Colloquium 115 (this conference), P. Gorenstein and M. Zombeck (eds).

Greenwald, M., Gwinn, D., Milora, S., Parker, J., Parker, R., Wolfe, S., Besen, B., Camacho, F., Fairfax, S., Fiore, C., Foord, M., Gandy, R., Gomez, C., Granetz, R., LaBombard, B., Lipschultz, B., Lloyd, B., Marmar, E., McCool, S., Pappas, D., Petrasso, R., Pribyl, P., Rice, J., Schuresko, D., Takase, Y., Terry, J., and Watterson, R. 1984, Phys.Rev.Lett. **53**(4) 352.

Kallne, E., Kallne, J., Dalgarno, A., Marmar, E.S., Rice, J.E., and Pradhan, A.K. 1984, Phys.Rev.Lett. **52**, 2245.

Jordan, C. and Veck, N.J. 1982, Solar Phys. **78**, 125.

Mason, H. 1988, IAU Colloquium 115 (this conference), P. Gorenstein and M. Zombeck (eds.).

Nuclear Fusion **25**(9), 1985; this volume contains several articles on tokamak programs throughout the world.

Parker, R., Greenwald, M., Luckhardt, M., Marmar, E.S., Porkolab, M., and Wolfe, S.M. 1985, Nuclear Fusion **25**(9), 1127; references therein.

Petrasso, R.D., Seguin, F.H., Loter, N.G., Marmar, E., and Rice, J. 1982, Phys.Rev.Lett. **49**, 1826.

Rice, J.E., Marmar, E.S., Kallne, E., and Kallne, J. 1987, Phys.Rev. A **35**(7), 3033; references therein.

DISCUSSION-R. Petrasso

D. Sigmar: Tokomak experiments show 3X larger than predictions of coronal equilibrium theory for the Li-like to He-like ion ratio [M. Bitter et al. 1985 (Figure 10)]. Can impurity transport or other possibilities account for it?

R. Petrasso: An error or inaccuracy in the atomic physics calculations could account for the difference. Several workers have suggested that the ionization rate of Li-like ions should be reduced by about 50% with respect to the Lotz value. It is important to resolve this issue since astrophysicists commonly use the Li- to He-like ratio as a diagnostic to discern the non-equilibrium ionization state of astrophysical plasmas (Jordan and Veck 1982; Canizares et al. 1983).

2. Magnetic Effects

LOW ENERGY LINES IN SPECTRA OF GAMMA BURSTS

G.S. Bisnovatyi-Kogan, A.F. Illarionov

Space Research Institute, Moscow, USSR

ABSTRACT. We connect the phenomenon of gamma ray bursts with nuclear explosions on the old neutron stars. The matter of the neutron star in the non-equilibrium layer at depths of 30 m $\leq h \leq$ 100 m consists of superheavy ($A \geq 300$) nuclei with a surplus of neutrons ($A/Z = 3 \div 4$). These nuclei are metastable and exist only at high pressure. After the starquake some of the matter from non-equilibrium layer may move upwards and its nuclei become unstable. The β-decay is followed by a chain reaction of fission and nuclear explosion. The gamma ray burst is observed as radiation of the star surface heated to high temperature. Some mass may be ejected, forming expanding cloud. It consists mainly of the iron Fe^{56} with small ($\leq 1\%$) additions of heavy elements (Ba, 1, ...) arising from the fission. The passage of stellar gamma radiation through the expanding plasma clouds leads to the formation of short-lived spectral features. Strong absorption of the soft gamma rays on K-electrons of Fe^{56} must be observed in the early stages. The gamma quanta with energies ε = 40–70 keV beyond the K-edge of the heavy elements (Ba, I, ...) are absorbed in the later stages. A wide κ_α line (ε_α = 30 keV) appears simultaneously. The free–free emission of expanding hot plasma cloud may be observed as a short flash in optical band.

INTRODUCTION

The short rising time of the cosmic gamma ray bursts ($\tau \leq 2 \times 10^{-4}$ s) and hard spectrum (ε_0 = 0.1 $\div$ 1 MeV) could not be explained by the model of thermonuclear explosion in the matter of neutron star envelope gained by accretion in a binary system. Really this model gives larger rising times ($\tau \geq 0.1$ s) softer ($\varepsilon_0 = 5 \div 10$ keV) spectra with a maximum in X-ray region. The optical companions needed for accretion are not found in gamma ray burst sources.

We follow an alternative model based on nuclear explosions (fission), which happen in the undersurface layers of old ($\geq 10^7$ yr) single neutron stars. A very short rising time is possible to explain with this model. The analysis of spectral features of gamma ray bursts permits us to follow the process of the explosion and of the interaction of gamma radiation with the matter ejected from the neutron star by the shock wave.

Observations of gamma ray bursts in the "Konus" experiment have shown in some events the presence of spectral features in the energy range (see Fig. 1) ε = 30–70 keV (Mazets *et al.*, 1980, 1981; Golenetskii *et al.*, 1983).

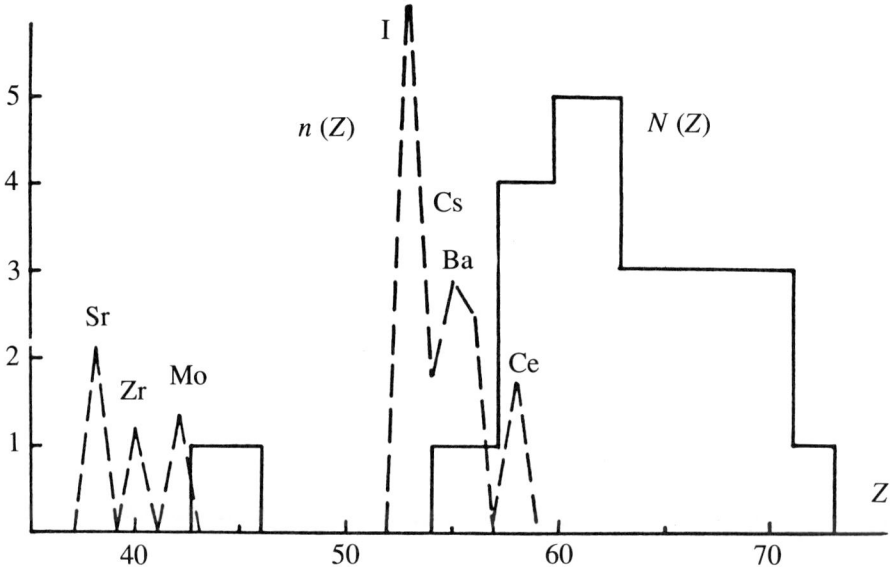

Fig. 1. Histogram of the distribution of the number of gamma ray bursts N on the energy ε, reduced to the axis Z according to relation $\varepsilon = 13.6Z^2$ eV (full line). The histogram $n(Z)$ of the element production in the fission of U^{235} as a function of the atomic number Z (dashed line).

The attempt to interpret these lines by cyclotron absorption (Mazets and Golenetskii, 1987) leads to unlikely large values of magnetic field

$$H \approx \varepsilon \frac{mc}{eh} \approx (3\text{–}7)\,10^{12}\,\text{Gs}.$$

The majority of the lines (70%), see Fig 1, have energies ε > 50 keV, corresponding to $H > 5 \times 10^{15}$ Gs. (An account of red shift on the surface of neutron star increases the estimated values of H.) Such values are too large in comparison with the fields (Rahakrishnan, 1982) of young neutron stars–radiopulsars ($H \approx 10^{12}$ Gs). Isotropic distribution of the gamma ray burst sources permits us to connect them with close (≤ 100 pc), old neutron stars with moderate magnetic fields ($H \approx 10^8 \div 10^{10}$ Gs). The absence of the motion of absorbance features across the spectrum and their rapid disappearance against the continuum background of gamma ray burst spectrum cannot be explained satisfactorily by the cyclotronic hypothesis of origin of these lines.

PHOTOABSORPTION OF GAMMA RAYS BY HEAVY ELEMENTS WITH Z = 40 ÷60

Rapid cooling of matter in conditions of high density $p \approx 10^{11}$ g/cm^3 leads to the formation of superheavy nuclei near the neutron drip line with atomic weight $A \approx 4Z$ and the charge

$$Z = 7 [33-0.511 (\rho/10^6 \mu_e)^{1/3}], \mu_e \approx 4,$$

(Bisnovatyi-Kogan and Chechetkin, 1974). These nuclei are stable in the non-equilibrium layer at densities

$$6 \times 10^{10} \text{ g/cm}^3 \approx \rho_1 < \rho < \rho_2 \approx 10^{12} \text{ g/cm}^3$$

At low densities $\rho < \rho_1$ the superheavy nuclei emit electrons and antineutrinos (β-decay), become unstable, divide into two non-equal parts (like the fission of U^{235}) and a chain reaction occurs (Bisnovatyi-Kopgan and Chechetkin, 1983). We believe that starquakes induce the movement of superheavy nuclei to lower density regions and thus fission chain reactions (see Fig. 2).

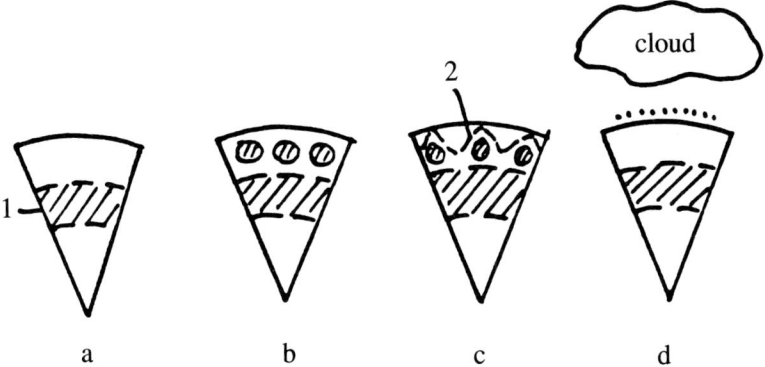

Fig. 2. Schematic picture of gamma ray burst formation
(a) non-equilibrium layer (1) in the static envelope
(b) after starquake
(c) after nuclear explosion (2) – the shock waves
(d) hot surface and expanding cloud

Table 1. The charges and ionization potentials of the most abundant nuclei formed in the fission

	Fe	Sr	Zr	Mo	I	Cs	Ba	Ce
Z	26	38	40	42	53	55	56	58
I keV	9.3	20	22	24	38	41	43	46

The shockwave from the nuclear explosion comes to the surface and ejects some matter in a form of an expanding cloud. Its radius $R \approx vt$ increases with a velocity $v \approx \sqrt{(2GM/R_0)} \approx (0.1–0.5)c$. The composition of the cloud is mainly iron Fe^{56} with small ($\leq$ 1%) additions of the products of nuclear fission (see Table 1). The initial spectrum of gamma radiation is distorted due to the passage of gamma rays through the cloud to the observer. Initially the optical depth of the iron cloud is high beyond the K-edge $\varepsilon \geq I_{Fe} = 13.6Z^2$ eV ≈ 9.3 keV because of photoabsorption. The cloud is transparent only for the hard component of gamma radiation (see Fig. 3), because the cross-section of absorption is high only near the edge and decreases with energy as

$$a(\varepsilon) \approx \frac{a_H}{Z^2} \left(\frac{I_Z}{\varepsilon}\right), \quad \varepsilon \geq I_Z,$$

where $a_H = 6.3 \times 10^{-18}$ cm^2 is the cross-section of the hydrogen photoionization.

The degree of matter (Fe) ionization rises during the expansion and the cloud becomes more transparent. The K-electrons of heavy elements like Ba still remain connected with nuclei. The optical depth of the cloud due to Ba K-electrons is $\tau = 2\, h_{Ba}\, a_{Ba}\, R$. The value $\tau > 1$ for quanta with energies $\varepsilon > I_{Ba} = 43$ keV, when the cloud is transparent for photoabsorption on Fe, if $n_{Ba}/n_{Fe} \geq 10^{-3} (M_{20}/L_{38}^2)^{2/9}$. Here $M = M_{20} \times 10^{20}$ g is the mass of

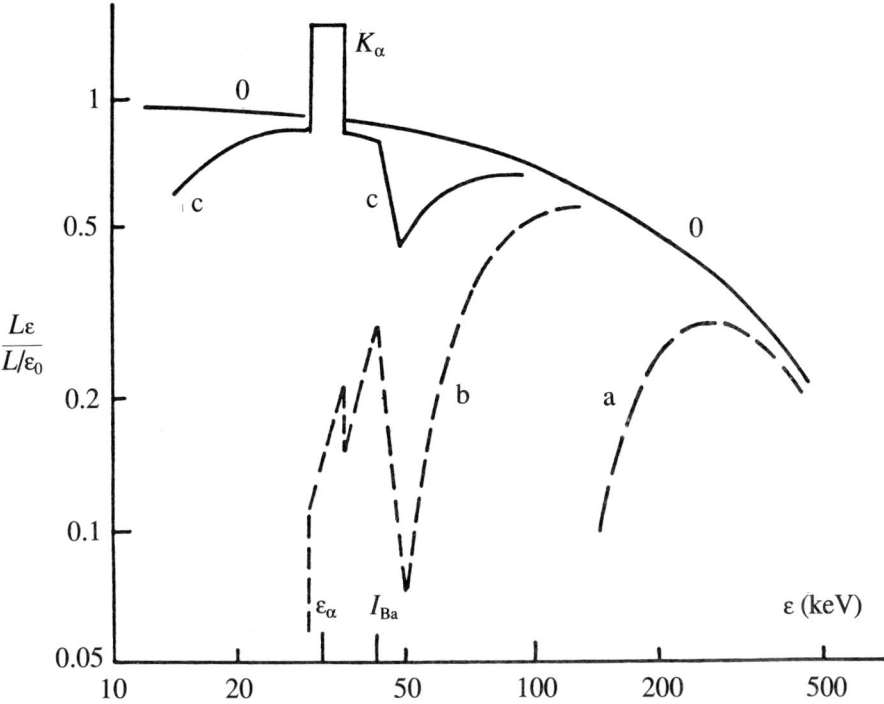

Fig. 3. The time evolution of the absorption line formed in the spectrum of gamma ray burst due to presence of Ba ions in the Fe plasma of the expanding cloud. The following cloud parameters are chosen: $M_{20} = L_{38} = 1$, $n_{Ba}/n_{Fe} \approx 1/300$. The time moments are $t_a = t_{1/2}$, $t_b = t_1 = 0.45$ s, $t_c = 2t_1$, 0 – the initial gamma ray spectrum. At the moment t_a the optical depth to photoabsorption of quanta with energy $\varepsilon = I_{Ba} = 43$ keV is equal to $\tau_{01} = 60$ on the Fe and $\tau_1 = 9$ on the Ba. At the moment t_b we have $\tau_{01} = 1$, $\tau_1 = 2,3$ and at the moment t_c we have $\tau_{01} = 0.015$, $\tau_1 = 0.6$.

the cloud and $L = L_{38} \times 10^{38}$ erg/s is the luminosity. The narrow absorption detail is formed beyond the K-edge of Ba which is shifted when expansion is included

$$\varepsilon_{abs} \approx I_{Ba}(1 + v/c) \approx 50 \text{ keV}.$$

A considerable number (~ 70%) of the gamma quanta absorbed by Ba ions are reradiated in the form of the K_α line, forming an additional emission feature near $\varepsilon_\alpha \approx 30(1 + v/c) \approx 35$ keV in the spectrum of gamma ray burst (see Fig. 3). All spectral features gradually disappear during expansion of the cloud. The soft free–free emission of expanding iron plasma is important in the early stages of the burst and may be observed in optics as a short flash with a duration of ~ 0.5 s and maximum luminosity $L_{opt} \approx 5 \times 10^{-4} L_\gamma$ (Bisnovatyi-Kogan, Illarionov, 1986). The optical flash must appear in the stage proceeding the formation of spectral features in the gamma band.

REFERENCES

Bisnovatyi-Kogan G.S., Chechetkin V.M., 1974, *Ap. Sp. Sci.*, **26**, 25.
Bisnovatyi-Kogan G.S., Chechetkin V.M., 1983, *Ap. Sp. Sci.*, **89**, 447.
Bisnovatyi-Kogan G.S., Illarionov A.F., 1986, *Astron. Zh.*, **63**, 984.
Golenetskii S.V. *et al.*, 1983, preprint F.T.I. N 819.
Mazets E.P. *et al.*, 1980, preprint F.T.I. N 657.
Mazets E.P. *et al.*, 1981, preprint F.T.I. N 719.
Mazets E.P., Golenetskii S.V., 1987, *Itogi nauki i techniki, Astronomy,* **32**, 16.
Radhakrishnan V., 1982, *Contemp. Phys.* **23**, 207.

DISCUSSION-G. Bisnovatyi-Kogan

N. Itoh: According to your model, γ-ray bursts are caused by neutron star quakes. Did you investigate the correlation between γ-ray bursts and pulsar glitches?

G. Bisnovatyi-Kogan: You are right, that pulsar glitches in our model must be correlated with gamma ray bursts. But these bursts must be very faint, because of long distance to the known pulsars. We suppose, that the average distance to the observed gamma ray burst sources is about 100 pc. There have been indications of sporadic emission from the Crab pulsar in the region of 30–300 keV.

G. Ricker: How would your model account for the two absorption features, one at ~ 29 keV and the other precisely 2× higher in energy, recently reported in the same GRB by the Ginga group (Murkami et al 1988)?

G. Bisnovatyi-Kogan: The precision of measurements is not enough for statement that the ratio of energies is precisely two. The observed two lines may be connected with absorption on the two most abundant elements produced in the fission (see fig. 1).

H. van der Woerd: From the absorption feature you can make an estimate of the mass of cloud. The cloud has a velocity of ~ 0.15 C, and so you can calculate the energy of the expansion. How does this energy compare to the total energy in the burst?

G. Bisnovatyi-Kogan: In order to overcome the gravitation of the neutron star, the cloud of the mass M_c must obtain kinetic energy of the order of $0.2\,M_c^2$. If the heat capacity of fission is ~ $10^{-3}c^2$, then you need $M_f > 100\,M_c$ to take part in chain reaction. For efficiency of transformation of the thermal energy into kinetic energy, 'y', the total mass M_f $\frac{100\,M_c}{y}$ and total energy of explosion Eb ~ $\frac{0.2\,M_c c^2}{y}$. Almost all this energy, E_b is emitted during gamma-ray burst. For $M = 10^{19}$ g, y = 0.3 we have $E_b = 6 \times 10^{39}$ erg.

THE EFFECT OF MAGNETIC FIELDS UPON

COSMIC X-RAY SPECTRA

P. Mészáros

Pennsylvania State University, University Park, PA, USA

ABSTRACT. The effect of strong magnetic fields (B $\gtrsim 10^{11}$ *Gauss*) upon various atomic line emission mechanisms in the X-ray range is considered, in particular for H and H-like or He-like ions, and a discussion of the detectability and significance of possible measurements is given. The cyclotron mechanism, the one- and two-photon scattering and the bremsstrahlung effects in a strong B are reviewed, as well as the role they play in determining X-ray spectra. These considerations are applied to typical models of X-ray pulsars and Gamma-ray bursters, contrasting observations of magnetic related features to the present theoretical understanding of these objects.

1. INTRODUCTION

Magnetic effects in the X-ray spectrum of cosmic sources are of interest in some white dwarf sources ($B_* \lesssim 10^{7-8} Gauss$), and are of major importance in neutron star sources such as accreting or rotation powered pulsars ($B_* \gtrsim 10^{11-12} Gauss$) and possibly Gamma-ray bursters. In this paper we shall concentrate on the latter type of high field objects. In these sources the magnetic field plays the leading role in determining the pulsations by which these objects are identified, and in those which are in binaries this allows one to determine such fundamental parameters as the neutron star mass, c.f. Joss and Rappaport, S., 1984, Mészáros, 1984. In most of these sources the cyclotron ground harmonic is in the X-ray range and the magnetic effects play a major role in the production of photons and in shaping both the line and the continuum spectrum. In what follows we shall describe the current understanding of the physics of strong magnetic fields upon the plasma emission mechanisms, as well as upon the radiative transfer, and how this affects the modeling of accreting pulsars and Gamma-ray bursters.

2. ATOMIC LINES IN A STRONG B-FIELD

In a strong magnetic field, the shape of the probability distribution of the electrons around a nucleus gets elongated along the field lines. A rough criterion for the onset of this is obtained by comparing the energy of the free electron ground cyclotron harmonic, $E_c \simeq 11.6 B_{12} keV$ with the approximate field free line energy $E_l(Z) \simeq 0.0136 \epsilon Z^2 keV$, where ϵ is the line energy in Rydberg units. Magnetic effects are important for atoms whose nuclear charge Z is

$$Z \lesssim 29 (B_{12}/\epsilon)^{1/2} . \qquad (1)$$

2.1 The Hydrogen Atom

In the simplest case of the Hydrogen atom, a non-perturbative approach to the magnetic field effects leads to replacing the usual n_p, l, m quantum numbers with a new set n, m, ν, where n is the Landau quantum number, m is the z-projection (along B) of the angular momentum, and

ν is the number of nodes of the longitudinal wave function, c.f. Canuto and Ventura (1973). For $B \gtrsim 10^{11-12}G$, the atom is usually in the state $n = 0$, and the line structure is characterized by the m-bands, the lowest of which are $m = 0, -1, -2, ...$ Within each m-band there is a tightly bound $\nu = 0$ state, whose binding energy is increasingly stronger with increasing field (compared to the field-free H), and a series of $\nu > 0$ states which are called "hydrogen-like", whose energies look like blue-shifted and red-shifted Lyman, Balmer, etc. states, with however different energy and intensity ratios, c.f. Rösner, et. al., 1984. As the magnetic field is increased gradually from zero, the energy levels split first into the usual linear, then quadratic Zeeman sublevels, and as the field is further increased they gradually rearrange themselves into a completely different pattern characterized by the magnetic quantum numbers n, m, ν. For the Hydrogen atom this occurs above $\beta = (B/4.7\ 10^9 G) \gtrsim 1$. The dipole transitions occur for $\Delta m = 0, \pm 1$, but the $\Delta m = 0$ are the strongest, since the $\Delta m = \pm 1$ involve the transverse dipole moment, which is supressed by the magnetic field, e.g. Wunner and Ruder, 1980. The strengths are somewhat lower than in the field-free case, but not much. The $\Delta m = 0$ transitions are linearly polarized and have a directionality $\propto sin^2\theta$, while the $\Delta m = \pm 1$ are circularly polarized and $\propto (1 + cos^2\theta)$, as in the Zeeman effect. The tightly bound transitions (involving $\nu = 0$) can be in the range 0.2-0.5 keV for $B_{12} \gtrsim 1$, while the hydrogen-like ($\nu > 0$) transitions are in the eV range. The former would be accessible to AXAF or ROSAT, while the latter would be in the province of HST or HUT.

2.2 Hydrogen-like and He-like Heavy Ions

The pioneering work in calculating energy levels of various atomic species in pulsar magnetic fields dates back to Ruderman and Sutherland (1975), and a fair amount of detailed work has been done since then. For hydrogen-like ions, the energy levels and transition strengths can be scaled from those of the hydrogen atom according to

$$E_m(Z, B) = Z^2 E_m(Z = 1, B/Z^2) \quad , \quad f_{\tau,\tau'}(Z, B) = f_{\tau,\tau'}(Z = 1, B/Z^2) \ ,$$

c.f. Ruder et. al.(1981). The level structure is similar to that of the previously described H-atom. For Fe^{25+}, most of the low lying levels are in the keV range, with significant departures from the field-free values occuring for $\beta \gtrsim 20$, or $\gtrsim 10^{11}G$. The strongest transition is the (magnetic) 001 $\to$ 000, which for fields of order $10^{12}G$ would produce about $10^{15}s^{-1}$ photons per atom, assuming the level population on average to be one electron. These are fairly substantial deexcitation rates. As in the H-atom itself, there are also continuum limits, e.g. a Lyman-like, Balmer-like, etc limits.

A procedure for calculating the energy levels of He-like atoms in a strong field based on the Hartree-Fock approach has been discussed by Pröschel et. al.(1982). They give the ground state energies of all He-like ions up to and including Fe^{24+}, as a function of the magnetic field strength. For fields in excess of about $5\ 10^{11}G$ ions heavier than C^{4+} have ground state energies above keV values, increasing with field strength. Thus, at $B \simeq 10^{12}G$, Fe^{24+} has its ground state energy at 32.1 keV. A different method of calculation, based on a density functional, has also been discussed by Kössl et. al.(1986).

The most basic observational limitation to the detection of atomic X-ray lines is that one needs to avoid smearing by Compton scattering, and this favors the observational candidates with lower luminosities, $L_x << L_{Edd}$. As an example (Ruder et. al., 1981) consider the low luminosity X-ray pulsar 4U0900-40, where if one takes an emitting area about $10^{10}cm^2$, a distance about 1 Kpc, a solar abundance of Fe and estimating from Saha's law the ratio of excited to ground level population to be about 10^{-4}, one gets an expected line photon flux of about $10^{-3}cm^{-2}s^{-1}$, with $\Delta E/E < 0.1$. This effect should be detectable with present instrumentation.

3. CYCLOTRON AND CONTINUUM RADIATION MECHANISMS

Free electrons have unquantized momenta along the magnetic field (z-axis) but are restricted in the transverse plane to discrete energy states, the Landau levels. In units of the critical magnetic field $B_q = (m^2c^3/e\hbar) = 4.413\ 10^{13}G$ (at which the ground cyclotron frequency equals the electron rest energy), the energy levels are given by

$$E = \left[p^2c^2 + m^2c^4(1 + 2j\frac{B}{B_q})\right]^{1/2} \quad (2)$$

where $j = n+s+\frac{1}{2} = 0,1,2,..$, $s = \pm\frac{1}{2}$ and $n = 0,1,2,...$ In the non-relativistic limit, the levels are evenly spaced, with ground level energy energy equal to the separation of subsequent levels given by

$$\hbar\omega_c = \frac{\hbar eB}{mc} = 11.6 B_{12}\ keV\ . \quad (3)$$

This quantizing aspect of the magnetic field has a profound influence upon all elementary processes involving typical energies not much larger than given by equation (3). Another significant departure from non-magnetic conditions is that the plasma becomes anisotropic, all rates and cross sections depending not only on energy but also on direction of the photon or particle respect to the magnetic field direction. In addition, for all processes involving photons, a significant polarization dependence is introduced for photon frequencies $\omega \lesssim \omega_c$ given by equation (3). It is easy to visualize the physical basis of this dependence, if one considers that a photon polarized with electric vector parallel to the field (ordinary photon) can interact with the electron as essentially as in the absence of a field (since it moves the electron along B) whereas a transverse electric field (extraordinary photon) tries to move the electron in the plane transverse to B, where the phase space of the electron is limited by the discrete Landau orbitals unless if $\omega \gg \omega_c$.

3.1 Cyclotron and Bremsstrahlung Processes

The frequency of collision of electrons with other charged particles, for instance, becomes a jagged rather than a smooth function of the particle energy of motion, exhibiting cyclotron resonances affecting both the Coulomb thermalization rates as well as the collisional excitation rate responsible for the bremsstrahlung process (Ventura, 1973, Bussard, 1980). The radiative deexcitation rates are extremely fast, $t_{rad} \simeq 10^{-18}(B/B_q)^{-2}s$, shorter than collisional excitation rates by many orders of magnitudes under typical X-ray pulsar densities and temperatures. The usual cyclotron or synchrotron radiation rates, calculated using an equilibrium excited level population, do not usually apply since at keV temperatures the electrons are predominantly in the ground level $n = 0$, and return there radiatively as soon as excited. The equivalent of the "cyclotron" emission mechanism, related by Kirchoff's law to a corresponding absorption coefficient, is therefore the Bremsstrahlung process (Nagel and Ventura, 1983). This consists of both a continuum part, related to knock-on excitations in the longitudinal component of the electron momentum, and to transverse excitations which give rise to cyclotron resonances, e.g. Pavlov and Panov (1976). This is the major source of seed photons both at the resonances and in the continuum.

3.2 Compton Scattering

Comptonization, or the reshuffling of photons in energy space by Compton interactions is a major effect in X-ray sources, whether magnetized or not, since the scattering opacity usually dominates over the absorption and emission opacities. A recent review of this process has been given by Pavlov et. al.(1988). In a magnetic field, the scattering opacity is both angle and frequency dependent (Canuto et. al., 1973), and has a rather interesting dependence on the polarizability of the plasma and of the vacuum provided by virtual $e^+ - e^-$ whose density depends on the magnetic field strength (Gnedin et. al., 1978, Mészáros and Ventura, 1978, 1979). The energy of the resonances depends also on the angle θ between photon and field, which contributes to spread the resonance in the diffusion regime, in addition to the thermal Doppler broadening (which in this case is θ-dependent, thermal motions being along the z-axis). Ordinary and extraordinary polarization photons have rather different scattering cross sections at $\omega \ll \omega_c$, the extraordinary being supressed by a factor typically of the order $(\omega/\omega_c)^2$ respect to the Thomson value. Relativistic effects can become important at temperatures above about 10 keV, c.f. Herold, 1979, Bussard et. al., 1986, Daugherty and Harding, 1987.

3.3 Double Compton and Two-photon Processes

In general two-photon processes are down by one factor of α_f respect to the corresponding single photon processes. However, in a magnetic field the cross section is resonant at particular photon frequencies, which can compensate partly for the decrease in cross section. In fact, the one photon scattering starting with an electron in the ground Landau level and a photon not far from resonance, which leaves the electron in the first excited state with emission of a soft photon $(\omega_c, 0 \to \omega_s, 1,$ where $\omega_s \ll \omega_c)$, has a cross section roughly comparable with that of the normal $(\omega, 0 \to \omega', 0)$ scattering. The excited final electron deexcites very fast by emission of a second, resonant photon (Bussard, Mészáros and Alexander, 1985), and this provides a substantial source of soft photons, in addition to Bremsstrahlung. This process is related to the two-photon decay from an excited state (Kirk and Melrose, 1986a,b). At low frequencies, the effectiveness of this mechanism as a photon source is limited by the inverse process, saturating at the blackbody level, but it can still exceed the bremsstrahlung effect if the radiation density is high enough. Detailed calculations of the cross sections as a function of energy, angle and field strength in a thermal plasma have been given by Bussard et. al., (1986), including the effect of higher resonances in the relativistic case. The cross section behaves $\propto \omega^{-1}$ at frequencies below the resonance, starting at a value below the thermally broadened resonance peak.

4. SPECTRAL EFFECTS IN ACCRETING X-RAY PULSARS

Accreting pulsars make up a large fraction of all known galactic X-ray sources, several dozen having been measured. The spectra are typically hard, often a power-law extending up to a maximum energy $E_m \simeq 20 - 50$ keV followed by a dropoff at higher energies. The pulse shapes, believed to be caused by magnetic beaming, are energy dependent. Lines at around 6.5-7 keV have been measured in a number of them (White, Swank and Holt, 1982), but these are believed to be K_α fluorescence lines of Fe in lower ionization stages ($< Fe^{18+}$), which would arise far from the surface of the star (since no gravitational redshift is seen), possibly in the Alfvén surface or the wind from the companion. No magnetically shifted atomic or ionic lines have been positively identified thus far. On the other hand, cyclotron lines have been reported in two of them, Her X-1 (Trümper et. al., 1978) and 4U0115+63 (Wheaton et. al., 1979, White et. al., 1982), indicating field strengths B_{12} of about 3.3 and 1.1 respectively. Radiative transfer calculations in the coherent scattering limit (neglecting comptonization) allow one to calculate energy dependent pulse shapes at energies below the resonance from simple emission regions ,e.g. caps or columns, c.f. Kanno, 1980, Nagel, 1981a, Kaminker et. al., 1982, 1983. Since the pulse (or beam) shapes depend on the ratio of the frequency to the cyclotron frequency, it is possible to attempt to infer the field strength in other sources where cyclotron lines are not seen, e.g. Kanno, 1980, Mészáros, 1982, Kii et. al., 1986. The energy and strength of the cyclotron features are pulse phase dependent (Pravdo et. al., 1979, Voges et. al., 1983), an effect which is understandable in terms of the angle dependence of the Doppler shifting of the resonance and the angle dependence of the continuum and line flux caused by the magnetic angle and frequency dependence of the opacities. To properly model this one needs to include Comptonization as well as angular effects in the radiative transfer, c.f. Mészáros and Nagel (1985a,b). Using flat space propagation (i.e. neglecting general relativistic effects such as light bending), this model comparison indicates that the Her X-1 radiation is of a pencil beam type, in order to reproduce the observed decrease of line energy with increasing phase. Other indications favoring a pencil beaming pattern at some distance from the neutron star have been put forward by Trümper et. al., 1986, based on the multiple peak structure of Her X-1. Interestingly, from simplified radiation hydrodynamic calculations (Basko and Sunyaev, 1976, Wang and Frank, 1981) one might conclude that higher luminosity pulsars such as Her X-1 ($L_x \gtrsim 0.1 L_{Ed}$) would have a radiation deceleration shock, and therefore an accretion column, rather than a low profile cap, a configuration which would naively lead to a fan-beaming pattern. More detailed radiation hydrodynamic calculations including magnetic anisotropies and resonance effects are needed to confirm this, c.f. Klein et. al., 1985. However, if it proves that close to the star the radiation is emitted sideways (fan), gravitational light bending can distort this into a pencil beam at $R \gtrsim few R_*$, if the neutron star is more compact than about 2-2.2 Schwarzschild radii, c.f. Mészáros and Riffert, 1988. This would also reproduce the observed pulse phase dependence of the cyclotron line energy observed, and reconcile the expected presence of a stand-out column with an effectively observed pencil beam at large distances.

One of the best tools for investigating the geometry of the emission regions of compact objects is X-ray polarimetry, which can probe the directionality and the type of physical radiation mechanism of a variety of sources where imaging is impossible due to the small scales involved (Rees, 1975). This is particularly true for strongly magnetized X-ray sources, where the emission and transfer processes are strongly polarization dependent (Gnedin and Sunyaev, 1974) and a number of relevant calculations and estimates have been presented by Mészáros, et. al.(1988). In accreting pulsars, the degree of linear polarization obtained from simple accretion columns and polar caps can get up to 70-80 % near the cyclotron energy, and is very large also in the continuum, being a strong function of the pulse phase. It is possible to distinguish between pencil and fan beam radiation patterns because of the different correlation between flux and polarization degree maxima as a function of phase. Rocket borne detectors with current design scattering polarimeters (Novick et. al., 1985) could measure polarizations down to 2 % in Her X-1 in about $2\ 10^5\ s$, while instruments on board a space platform with a photon collector could achieve similar sensitivities in extragalactic sources. This would be decisive for determining whether the emission mechanism of AGNs is indeed non-thermal (e.g. inverse synchro-Compton) or quasi-thermal (e.g. a thick accretion disk), since these alternatives have drastically different polarization predictions. It would also provide invaluable information about the magnetic field direction and/or the emission region geometry.

5. MAGNETIC SPECTRAL EFFECTS IN GAMMA-RAY BURSTERS

The evidence for Gamma-ray bursters (GRBs) being magnetized neutron stars is strongly suggestive, although not universally accepted. The partial similarity in temporal behavior to X-ray bursters, the presence of pulsations in at least one object (GRB 030579) and the observation of low energy (30-60 keV) cyclotron-like features in about a third of the GRBs detected with KONUS (Mazets et. al., 1981) argue for this, as well as the fact that a magnetic field might be better in explaining why γ-rays and not X-rays are observed (smaller effective area, lack of thermal equilibrium over the short time of the burst, etc.), c.f. Liang, 1982, Woosley, 1984. Cyclotron lines have been also reported from SMM in one object (Dennis et. al., 1982), and most recently from TENMA in two cases (Murakami et. al., 1988). However the interpretation is complicated due to the lack of significant X-ray emission, and the apparent small upper limits ($B_{12} \lesssim 0.2 - 0.4$) on field strength inferred (Matz et. al., 1985) from SMM burst statistics at 1-10 MeV, based on the expected effectiveness of the magnetic one-photon pair creation in a field of the magnitude indicated by the low energy line features. This prompted a reevaluation of the magnitude of the field and the location of the emission in simple models (Zdzierski, 1987, Lamb, 1988), and led to the investigation of more involved or alternative models. Spectral models based on synchrotron emission were most recently calculated by Brainerd and Lamb, 1987, and Brainerd, 1988.

The problem of γ-ray propagation in a relativistic magnetosphere with an arbitrary dipole field has been recently considered by Riffert, Mészáros and Bagoly (1988). Interesting features arise from the angle and frequency dependence of the absorption threshold ($\omega_t > 2mc^2/sin\theta$) and the fact that the photon paths bend in the gravitational field, as well as being redshifted. The magnitude of the bending of course depends on the compactness of the neutron star. The escaping beams of high energy radiation are broader for compact neutron stars, and for smaller fields. However, at a given field, even a moderately compact star gives a significantly broader escape beam than inferred from simple flat space calculations. Furthermore, as one increases the field beyond about $B_{12} \simeq 0.25$, the beam shapes saturate, no longer narrowing as one increases further the field strength, this saturation effect arising simply from the double exponential nature of the angle dependent transport problem (the escape is proportional to $e^{-\tau}$ and the cross section is also exponential in θ, ω). A theoretical simulation of the expected SMM statistics (Mészáros, Bagoly and Riffert, 1988) indicates that if one assumes that all stars sampled have the same magnetic field, then this field indeed should be less than about $B_{12} \lesssim 0.4$, in agreement with Matz et. al., 1985. However, it is more likely that one is sampling a distribution of field strengths, if for no other reason that the latter may decay in time. If one has a combination of low field and high field objects, then due to the saturation effect mentioned above, the sample could be hiding a considerable fraction of rather large field objects, since their beam sizes are not correspondingly narrower. A simulation of expected SMM statistics in this case indicates that the SMM and KONUS statistics can be compatible with each other, with a considerable ($\gtrsim 1/3$) fraction of neutron stars having fields $B_{12} \gtrsim 5$. In this case, an understanding of the low energy spectrum involving cyclotron lines will require further work.

Acknowledgements: This research has been supported in part through NSF grant AST 85-14735 at Penn State.

REFERENCES

Basko, M.M. and Sunyaev, R.A., 1976, MNRAS, 175, 395.

Bussard, R.W., 1980, Ap.J., 237, 870

Bussard, R.W., Mészáros, P. and Alexander, S., 1985, Ap.J.(Lett), 297, L21

Bussard, R.W., Alexander, S. and Mészáros, P., 1986, Phys.Rev., D34, 440

Brainerd, J.J. and Lamb, D.Q., 1987, Ap.J., 313, 231

Brainerd, J.J., 1988, *Ap.J.*, in press.

Canuto, V., Lodenquai, S. and Ruderman, M., 1971, Phys.Rev., D3, 2303

Canuto, V. and Ventura, J., 1977, Fund.Cosm.Phys., 2, 203

Daugherty, J. and Harding, A.K., 1986, Ap.J., 309, 362

Dennis, B. et. al., 1982, AIP Conf.Proc. No. 77, p153.

Gnedin, Yu.N. and Sunyaev, R.A., 1974, Astron.Ap., 36, 379.

Gnedin, Yu.N., et. al., 1978, JETP (Lett), 27, 305

Herold, H., 1979, Phys.Rev., D19, 2868

Joss, P. and Rappaport, S., 1984, A.R.A.A., 22, 537

Kaminker, A.D., Pavlov, G.G. and Shibanov, Yu.A., 1982, Ap.Sp.Sci., 86, 249

Kanno, S., 1980, PASJ, 32, 105.

Kii, T. et. al., 1986, PASJ, 38, 751

Kirk, J. and Melrose, D., 1986a,b, Astron.Ap.

Klein, R., Arons, J. and Lea, S., 1985, report at *Los Alamos Workshop on Time Variability in X- and Gamma Ray Sources*, Taos, NM.

Kössl, D. et. al., 1986, Astron.Ap.

Lamb, D.Q., 1988, in *Nuclear Spectroscopy of Astrophysical Sources*, AIP Conf.Proc., N. Gehrels and G. Share, eds. (AIP, New York) in press

Liang, E.P., 1982, *Nature*, 299, 321.

Matz, S.M., et. al., 1985, *Ap.J. (Letters)*, 288, L37

Mazets, E.P. et. al., 1981, *Nature*, 290, 378.

Mészáros, P. and Ventura, J., 1978, Phys.Rev.Lett., 41, 1544; 1979, Phys. Rev., D19, 3565

Mészáros, P., 1984, Space Sci.Rev., 38, 325

Mészáros, P. and Nagel, W., 1985a,b, Ap.J., 298,147; 299, 138.

Mészáros, P., et. al., 1988, Ap.J., 324, 1056.

Mészáros, P. and Riffert, H., 1988, Ap.J., 327, 712.

Mészáros, P., Bagoly, Z. and Riffert, H., 1988, preprint.

Nagel, W. and Ventura, J., 1983, Astron.Ap., 118, 66

Nagel, W., 1981, Ap.J., 251, 188

Novick, R., Chanan, G. and Helfand, D., 1985, in *Cosmic X-ray Spectroscopy Mission* (Paris: European Space Agency), ESA SP-239, p.265

Murakami, T. et. al., 1988, *Nature* (submitted)

Pavlov, G.G. and Panov, A.N., 1976, JETP, 44,300

Pavlov, G.G., Shibanov, Yu.A. and Mészáros, P., 1988, preprint

Pröschel, P. et. al., 1982, J.Phys., B15, 1959

Rees, M., 1975, M.N.R.A.S., 171, 457.

Riffert, H., Mészáros, P. and Bagoly, Z., 1988, *Ap.J.*, in press

Rösner, W. et. al., 1984, J.Phys., B17, 29

Ruderman, M. and Sutherland, P., 1975, Ap.J., 196, 51

Ruder, H., et. al., 1981, Phys.Rev.Lett., 46, 1700

Trümper, J. et. al., 1978, Ap.J.(Lett.), 219, L105.

Trümper, J., 1987, in *Very High Energy Gamma Rays in Astronomy*, NATO ASI Series vol.199, ed. K. Turver (Reidel, Dordrecht, Holland), p.7.

Ventura, J., 1973, Phys.Rev., A8, 3021.

Voges, W. et. al., 1982, Ap.J., 263, 803.

Wang, Y.M. and Frank, J., 1981, Astron.Ap., 93, 255.

White, N., Swank, J. and Holt, S., 1983, Ap.J., 270, 711.

Wunner, G. and Ruder, H., 1980, Ap.J., 242, 828

Woosley, S., 1984, in *High Energy Transients in Astrophysics*, AIP Conf.Proc. 115, ed. S. Woosley (AIP, New York)

Zdziarski, A., 1987, in *Proc. 13th Texas Symp. Relat. Astrophysics*, ed. M. Ulmer (World Scientific, Singapore), p. 563.

DISCUSSION-P. Meszaros

G. Bisnovatyi-Kogan: Neutron star radius depends on the equation of state. Will your conclusion about the possibility of large magnetic field in gamma-ray bursts due to gamma-ray effects survive for large neutron star radius?

P. Meszaros: For $R \sim < 5\ R_{schwarschild}$ we get a poorer fit to the SMM detection rates. Smaller radii, between 2 and 2.5, provide better fits.

J. Krolik: Did you consider the effects of two possibly important complications, the effects of the radiation field in populating the higher harmonics, and the magnetic effects in changing the radiation pressure entering estimates of when a radiation shock appears?

P. Meszaros;: We are currently in the process of addressing the first question with Steve Alexander. The second problem requires a careful numerical treatment of the radiation hydrodynamics with magnetic opacities, on which Klein, Arons and Lea are currently working.

N. Itoh: You assumed an exponential decay of neutron star magnetic fields in your model of γ-ray bursts. What kind of timescale did you take for the exponential decay?

P. Meszaros: We did not use a specific decay law. That one as an example. The important thing is that we are likely to be sampling neutron stars with a variety of field strengths.

EFFECT OF VACUUM POLARIZATION IN A STRONG MAGNETIC FIELD AND SPECTRAL FEATURES OF X-RAY SOURCE EMISSION

Yu. N. Gnedin

USSR Academy of Sciences Central Astronomical Observatory
196140 Leningrad, Pulkovo, USSR

ABSTRACT. In strong magnetic fields of neutron stars electron–positron vacuum behaves as an anisotropic medium. Vacuum influences the generation and propagation of electromagnetic radiation in plasma and changes the spectrum of radiation. As a result the change of cyclotron lines shape and appearance of specific "vacuum lines" should be observed in X-ray spectra of accreting neutron stars.

KEYWORDS. X-ray astronomy; X-ray spectroscopy; pulsars; black holes.

One of the recent remarkable achievements in Astrophysics is the discovery of neutron stars and magnetic white dwarfs which have rather strong magnetic fields $B \sim 10^{10} \div 10^9$ G and $10^6 \div 10^9$ G, respectively. These magnetic fields are much higher than those achieved in laboratory conditions ($10^6 \div 10^7$ G) or those "ordinary" stars ($\leq 10^4$ G). Such a large difference gives rise to new qualitative effects in the processes involving interaction of radiation with matter (see Pavlov and Gnedin, 1984).

First, they are related to quantization of electron motion transverse to the magnetic field lines. Quantization becomes important if the characteristic energy of electrons appears to be smaller than the distance between Landau levels $\hbar\omega_B = 11.6\, B_{12}$ keV. This condition holds for thermal electrons if

$$C \equiv \frac{\hbar\omega_B}{kT} = 1.3\, B_{12}/T_8 \geq 1 \qquad (1)$$

Second, the enormous magnetic fields lead also to the appearance of the relativistic quantum effects which are characterized by a rather large ratio:

$$\hbar\omega_B / m_e c^2 \equiv B/B_c \qquad \text{where } B_c = 4.41 \times 10^{13}\, G \qquad (2)$$

As an example we might present (Fig. 1) the creation of an electron–positron pair (a), the splitting of a photon into two photons (c) and the so-called polarization of electron–positron vacuum (b).

For the case (b) if $\hbar\omega < 2mc^2$ the one-photon creation of real pairs is forbidden by the energy conservation law even in a magnetic field. However the magnetic field affects the virtual electron–positron pairs which are produced due to the photon propagation (Fig. 1b). As a result, the magnetized vacuum behaves with respect to the propagation of photons like an anisotropic medium.

The propagation of radiation in a magnetized vacuum may be described in terms of two normal waves with different polarization and different refractive indices.

$$n_{1,2} = 1 + \frac{1}{90\pi} \frac{e^2}{\hbar c} \left(\frac{B}{B_c} \sin\theta\right)^2 [(7)_1, (4)_2] \qquad (3)$$

at $B \ll B_c$ and $\hbar\omega \ll m_e c^2$.

The normal waves are linearly polarized. Under the real physical conditions of the X-ray sources one needs to take into account "vacuum" contribution to the usual plasma effects. It means that the vacuum effects are determined by the ratio of the vacuum polarizability to the plasma polarizability. In the region of the cyclotron resonance $\omega \approx \omega_B$ this ratio is

$$\frac{\alpha_V}{\alpha_{Pl}} \equiv V = \frac{1}{60\pi^2 N} \left(\frac{m_e c}{\hbar}\right)^3 \left(\frac{\hbar \omega_B}{m_e c^2}\right)^4 \cong (3 \times 10^{28}/N)(B/B_c)^4 \quad (4)$$

where N is the electron density. For $\omega \ll \omega_B$ the ratio is

$$\alpha_V/\alpha_{Pl} \sim V(\omega/\omega_B)^{B_2}$$

and for $\omega \gg \omega_B$

$$\alpha_V/\alpha_{Pl} \sim V(\omega/\omega_B)^4$$

The value $V \sim 1 \div 10^5$ follows from estimates of electron densities and magnetic fields in the emission regions of X-ray pulsars: $B \sim 0.1 B_c$, $N \sim 10^{19} \div 10^{24}\,\text{cm}^{-3}$. Thus, the vacuum polarization in X-ray sources has a strong effect on plasma radiation.

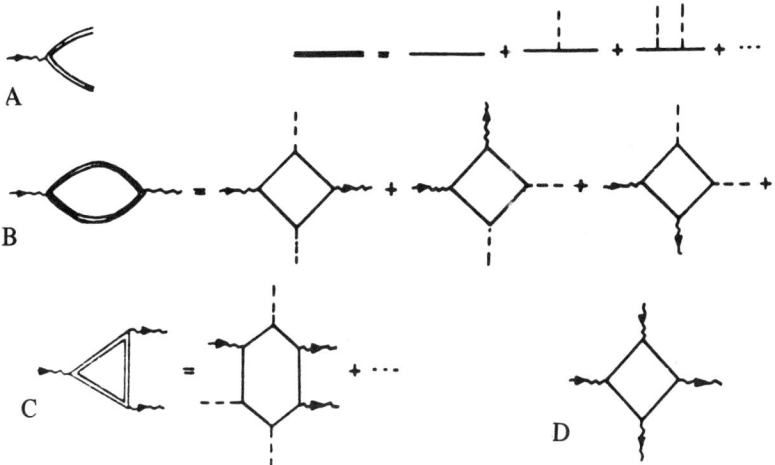

Figure 1. An example of the interaction of photons with the electromagnetic field. (a) Pair production by a photon in a magnetic field; (b) vacuum polarization by a magnetic field (three lower-order diagrams illustrating the interaction of an electron or positron with the magnetic field are shown on the right side); (c) splitting of a photon into two photons (one of the lower-order diagrams illustrating the interaction with the magnetic field is shown on the right side); (d) one of the lower-order diagrams for photon–photon scattering. The wavy line represents a photon, the single solid line represents a free electron or a free positron, the double line denotes an electron or a positron in a magnetic field, and the dashed line represents a magnetic field "quantum".

Though the vacuum phenomena affect mainly the polarization of radiation nonetheless the X-ray line positions and shapes are also determined by vacuum effects, especially for the optically thick plasma. The most pronounced effects arise in the range of wavelengths where the contribution from the vacuum is compensated by that from the plasma. There are frequencies $\omega_{c1,2}$ at which the effects of the vacuum and plasma on the linear polarizaion of normal waves cancel out each other: for $V \gg 1$

$$\omega_{c1} \approx \omega_B \left(1 - \frac{1}{2V}\right); \quad \omega_{c2} \approx \omega_B / \sqrt{V} \tag{5}$$

These frequencies are close to the points of intersection of dispersion curves for the absorption coefficients and refraction indices (see Fig. 2).

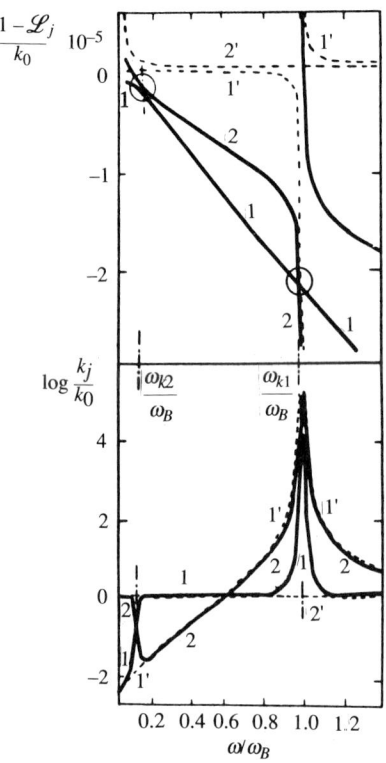

Figure 2. Frequency dependence of the refractive indices and absorption coefficients of normal waves for $V = 100$, $\theta = 85°$. The numbers 1 and 2 near the curves denote the number of the wave j (1' and 2' for the "vacuumless" plasma). $K_0 = (c/2\omega) N\sigma_T = K_d'$ ($B = 0$). The regions of enhanced linear transformation of normal waves in an inhomogeneous plasma are circled.

Although $d\hbar\omega \ll m_e c^2$ the vacuum cannot absorb the photons directly it has nonetheless a strong effect on the absorption of radiation by plasma. This results from the fact that the absorption of radiation is an anisotropic medium depends essentially on its polarization which is changed by the vacuum (Fig. 3).

The main result due to the vacuum effect is the increasing of the contribution from second normal wave near cyclotron resonance. In the usual thermal plasma cyclotron resonance absorption is practically determined by the first normal wave: $K_2 \sim \beta K_1$, where $\beta = (2kT/mc^2)^{1/2}$. In the vacuum case (V » 1) the cyclotron absorption coefficients are to be of the same order where $K_2 \sim K_1$ (Fig. 4). This striking difference from the vacuumless plasma is due to the change in the selection rules for the absorption of normal waves.

As a result the behaviour of absorption coefficients K_j near the first cyclotron resonance ω_B turns out to be rather complex (Fig. 5). The profiles of $K_j(\omega)$ acquire a complex shape, additional points ω_K at which K_j or refraction indices n_j crossing appear, additional shifts and dips occur as a result of vacuum polarization.

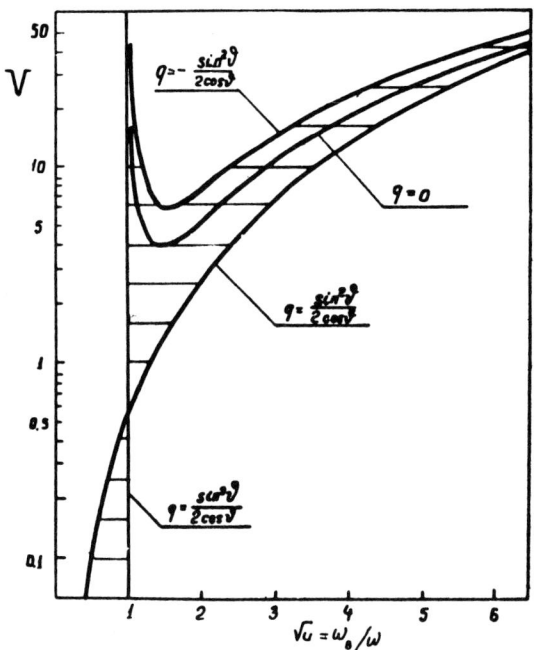

Figure 3. The regions of quasi-transverse and quasi-longitudinal (the hatched area) propagation in the "plasma + vacuum" system. The vacuum polarization changes considerably the polarization of normal waves above the curve $q = \sin^2\theta/2\cos\theta$.

The quantizing field ($b \geq 1$) causes the cyclotron resonance to shift by $S^2\omega_B$ ($B/2B_c$) $\sin^2\theta$ and the quantity

$$\xi = S \frac{B}{B_c} \tanh \frac{C}{2} \sin^2 \theta \tag{6}$$

which takes into account the contribution of M1 transitions with spin flip to increase. Furthermore, as the magnetic field is increased at a fixed temperature, the absorption at a higher-order harmonics increases $\sim B^{S-1}/S$ at $b \gg 1$ due to the change of the thermal energy kT of the transverse motion of electrons by the quantum energy of Landau levels $\hbar\omega_B$ and due to the decrease in the role of the induced transitions. This situation is valid only for not too high harmonics $S \ll B_c/B$ when the profiles do not overlap each other. As

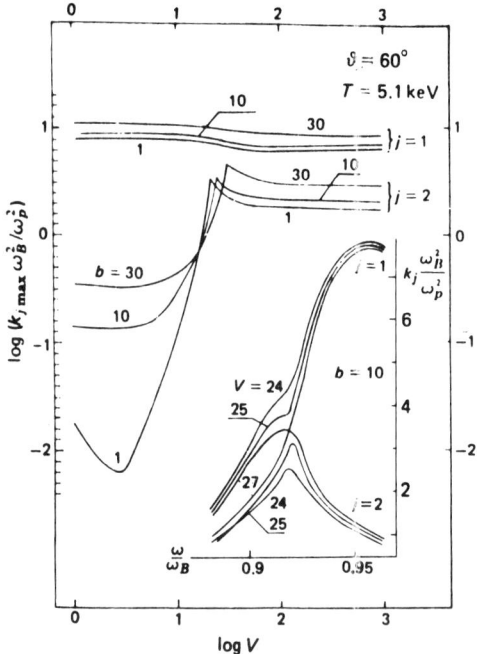

Figure 4. Top: Dependence of the maxima of the absorption coefficients of the normal waves neart the main resonance on the paramater V for $\theta = 60°$, $\beta = 0.141$ and different values of b indicated on the curves. The maximum values of K_2 are attained at $V = V_0$. The curves K_j max (V) intersect at lower values of θ. Bottom-right: A change in the profiles of $k_j(\omega)$ occurring when V passes through $V_0 = 25.2$ at $b = 10$.

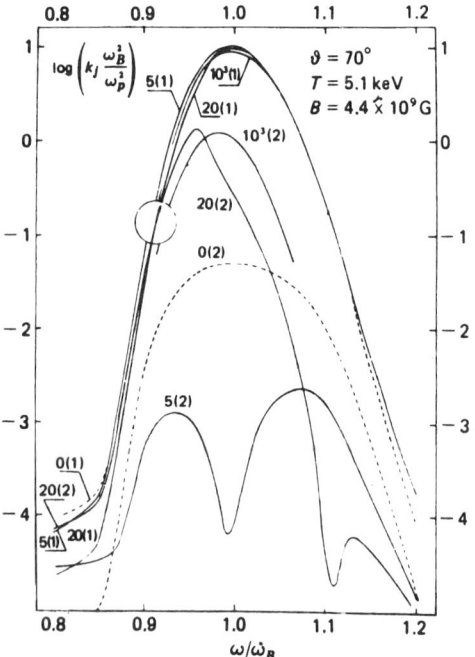

Figure 5. The effect of vacuum polarization on the absorption of normal waves near the first resonance at $\beta = 0.141$ and $\theta = 70°$ for $b \ll 1$. The circle indicates the point of intersection of the k_j curves for $V = 20$.

b is increased the resonances overlap each other at lower S due to quantum shift.

The vacuum polarization has no effect on the intensity of radiation of optically thin plasma, since the sum $k_1 + k_2$ is independent of V. Therefore the vacuum changes the frequency and angular dependences of the polarization characteristic of radiation. In contrast with this case the vacuum changes both the polarization properties and the intensity of optically thick plasma radiation because the intensity $I = I_1 + I_2$ depends nonlinearly on absorption coefficients K_j.

For example, the peaks caused by the vacuum polarization at $\omega \approx \omega_B (1-2/V)$ and at $\omega \approx \omega_B V^{-1/2}$ in the frequency dependence of the smaller absorption coefficient in a cold plasma lead to additional absorption lines in the spectrum of the emitted radiation (Fig. 6). One of these lines appears near the cyclotron frequency ω_B. In a cold plasma without scattering it arises only due to vacuum polarization, although both thermal motion and the conversion of normal waves during scattering give rise to the cyclotron line without allowance for the vacuum effect. The other line at $\omega \approx \omega_B V^{-1/2}$ whose position depends on the plasma density appears exclusively because of vacuum polarization.

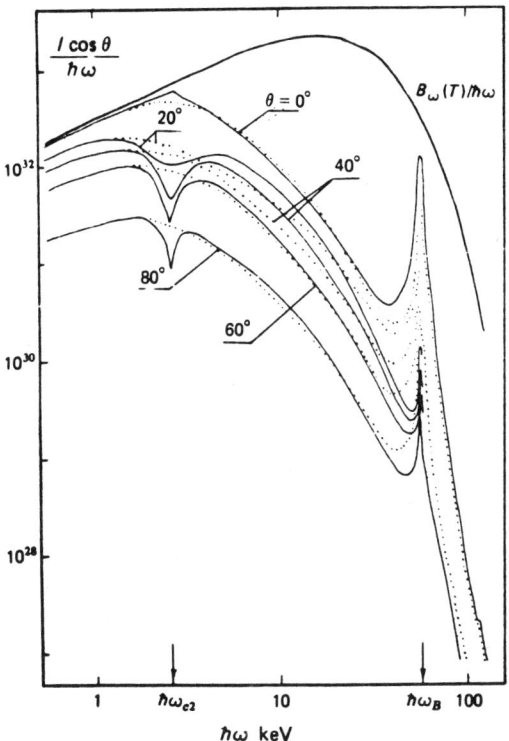

Figure 6. Spectra of partial fluxes from a semi-infinite plasma calculated with and without allowance for vacuum polarization (V = 480 and V = 0 solid and dashed curves, respectively) for kT = 10 keV, B = 5 × 10^{12} G, N = 10^{22} cm^{-3} and different angles θ between the normal and wave vector. Magnetic field is perpendicular to the surface.

For a simple commonly accepted model of X-ray pulsar (Lamb et al., 1973) one can obtain

$$V(r) = 150 \frac{B_{12}^{24/7} R_6^{67/7}}{L_{37}^{5/7} r_6^{19/2}} (M/M_\odot)^{19/4}$$

$$\hbar\omega_{c2} \cong 0.88 \, L_{37}^{5/14} \, r_6^{19/4} \, M_\odot^{19/28} / B_{12}^{5/7} \, R_6^{67/14} \, M^{19/28} \text{ keV} \qquad (7)$$

For Her X-1 the value $\hbar\omega_{c2} \sim 0.1$ keV. For 4U 0115 + 634, if the feature near 20 keV is assumed to be a cyclotron absorption line, the value $\hbar\omega_{c2} \approx 0.55$ keV. In other models these values may be larger. For example, it may be suggested that the absorption features in spectrae of γ-ray bursts discovered by Mazetz et al. (1981) can be possibly explained as the vacuum features. One can suppose that the peculiar emission near 12 keV observed in the spectrum Her X-1 by McCray et al. (1982) is a vacuum feature near ω_{c2}. Thus we would have $V \approx 3400$ in the emission region. It gives the electron density value $N = 3 \times 10^{21}$ cm^{-3} for $B = 6 \times 10^{12}$ G. It should be noted that a simultaneous detection of vacuum and cyclotron features make it possible to uniquely determine the magnetic field and the electron density in the emission region.

REFERENCES

Lamb, F.K., Pethick, C.J. and Pines, D. (1973), Astrophys. J., **184**, 271.
Mazets, E.P., Golenetskii, S.V., Aptekar', R.L., Gur'yan, Yu.A. and Il'inskii, V.I. (1981), Nature, **290**, 378.
McCray, R. A., Shull, J.M., Boyton, P.E., Deeter, J.E., Holt, S.S. and White, N.E. (1982), Astrophys. J., **262**, 301.
Pavlov, G.G. and Gnedin, Yu.N. (1984), Astrophys. Space Physics, v. 3, 197–253.

MERGING OF IRON LINES IN SPECTRA OF X-RAY BURST SOURCES

J. Madej

Astronomical Observatory of the Warsaw University,
Warszawa, Poland,
   and
Department of Physics, Queen's University, Kingston,
Ontario, Canada.

ABSTRACT. Model atmospheres and synthetic spectra of neutron star of the effective temperature 10^7 K are presented. All the iron spectral features in the energy range 6 - 10 keV (uncorrected for redshift) are washed out by instrumental or intrinsic broadening, which leaves only a single line of the equivalent width comparable with the observations.

1. INTRODUCTION

Waki et al. (1984) detected prominent 4.1 keV absorption line in spectra of MXB 1636-536 (cf. also Turner and Breedon 1984), and suggested that the line represents resonance $K\alpha$ transition in highly ionized iron ions in stellar atmosphere with energy around 6.7 keV, if measured on the neutron star surface. The equivalent width (W_E) of this line was very large, and exceeded 200 eV in some bursts. Existence of the 4.1 keV line was also confirmed in few other X-ray bursters.

No other spectral features were credibly detected in X-ray burst spectra. Consequently, identification of the line with $K\alpha$ transition of iron is strictly arbitrary, but - if correct - would be a direct measure of the mass to radius ratio, M/R, of a neutron star. In this paper we examine the appearance of all iron spectral features in model atmosphere of a neutron star, when observed with X-ray detectors of limited spectral resolution.

2. SPECTRAL FEATURES

This paper is based on set of nongrey model atmospheres in radiative and hydrostatic equilibrium, computed for $T_e = 10^7$ K, log g between 14 and 15, and with various abundance of iron. The most important opacities included in the models are: electron scattering (taken in Thomson approximation), and free-free absorption of fully ionized hydrogen and helium. Contribution of iron opacity to the models (in LTE) included b-f and f-f opacity of iron in the highest four stages of ionization, and blanketing by resonance lines of He-like and H-like ions. More extensive description of the models is given elsewhere (Madej 1988a; 1988b).

Careful line profile calculations show, that equivalent widths (W_E) of the resonance lines of iron can approach 400 eV in such conditions, with iron abundance increased only 10 times above the solar value (Madej 1988a; 1988b). However, this result should be revised when the assumption of local thermodynamic equilibrium (LTE) in the equation of state of iron is removed. After correction for gravitational redshift with $q \approx 0.6$, the observed equivalent widths W_E still exceed 200 eV, which is fully compatible with the observations.

One can note, that such values of W_E correspond to two resonance lines of heliumlike and hydrogenic iron, which are formed in neighboring atmospheric layers with different temperatures, and partly overlap each other on the continuum dominated by electron scattering. Additional (instrumental or intrisic) broadening assumed in the above papers causes merging of both lines into a single spectral feature.

If we accept presence of resonance lines in the X-ray spectrum, then one has to explain simultaneously the following questions:
A. Why we do not see lines corresponding to transitions between higher excited levels in both iron ions?
B. Why we do not see two ionization (bound-free) jumps at 8.8 keV and 9.28 keV (times redshift factor), which could eventually merge into a single absorption edge?

3. CALCULATIONS AND RESULTS

Ready for use model atmospheres were subsequently used for determination of the emergent monochromatic flux in many (almost 200) energies in range 5.4 - 9.4 keV. (Gravitational redshift is not included here). Spectrum synthesis included opacity of the 4 lowest lines of the singlet series ($1s$ $^1S_0$ - $1s\,np$ $^1P_1^0$) of He-like iron, and 5 doublet lines of H-like iron ($1s$ $^2S_{1/2}$ - np $^2P_{1/2,\,3/2}^0$). These are the only series of allowed lines arising from the ground levels in both ions. Corresponding wavelengths and oscillator strengths are taken from Mewe et al. (1985).

Line opacity profiles were obtained as careful numerical convolution of natural, thermal and Stark profiles. Thermal and natural broadening profiles were computed in a standard way (Mihalas 1978, cf. also Madej 1987b), and the pressure (Stark) profiles were estimated according to Griem (1974). (cf. Chap. IV.6 of his book.) Final computations of the flux were done with the Feautrier method (Mihalas 1978).

Fig. 1 shows hard energy part of the synthetic spectrum of an X-ray burster with $T_e = 10^7$ K, assuming low gravity log g = 14.0, and iron number abundance 10 times the solar value (DOTTED LINE). Horizontal axis gives decimal logarithm of energy (in keV), and vertical axis gives logarithm of the monochromatic flux (in cgs units). Assumed abundance of iron produces well developed pair of the resonance lines (the leftmost lines in the Figure). The resonance line of He-like iron (6.70 keV) is more prominent than neighboring resonance line of hydrogenic iron (6.97 keV). Huge b-f edges of both ions practically coincide, and decrease flux by more than two orders of magnitude.

The above example demonstrates spectrum of X-ray burster atmosphere in perfect hydrostatic equilibrium, seen on the neutron star surface with perfect spectral resolution. Spectrum at infinity can be immediately obtained rescaling horizontal scale by gravitational redshift factor. However, comparison with the existing observational data should be done taking into account limited spectral resolution of X-ray detectors, which is still rather poor.

Theoretical spectrum from Fig. 1 (dotted line) has been numerically folded with the normalized gaussian profile of the full width at half maximum (FWHM) equal 0.7 keV (about 10% of the energies in Fig. 1). Such a FWHM very roughly corresponds to spectral resolution of TENMA proportional counters (Koyama et al 1984). Smoothed spectrum (SOLID LINE) drastically changes: all the higher lines merge practically perfectly into continuum and only pair of resonance lines remains visible as a single, broad spectral feature. At the same time huge bound-free opacity jump gets invisible. Smoothing procedure changes it into steep continuum, which is very similar to high energy branch of a blackbody.

5. CONCLUSIONS

Detailed model atmosphere and line profile computations in LTE show, that broad, single absorption features observed in some X-ray burst spectra can be interpreted as merging resonance lines of both heliumlike and hydrogenic iron ions, at least in favourable effective temperatures. Iron lines arising from transitions to higher excited levels, as well as b-f ionization edges, are efficiently washed out due to limited spectral resolution of satellite X-ray detectors and are not visible. (Existence of strong turbulences in surface layers of a burster can cause the same smoothing effect).

This conclusion supports identification of the strongest 4.1 keV line with redshifted lines of iron, i.e. this means that the assumption of atomic origin of the single X-ray line is not in contradiction with observations. Results of this paper do not imply, however, that iron is the necessary source of the 4.1 keV line.

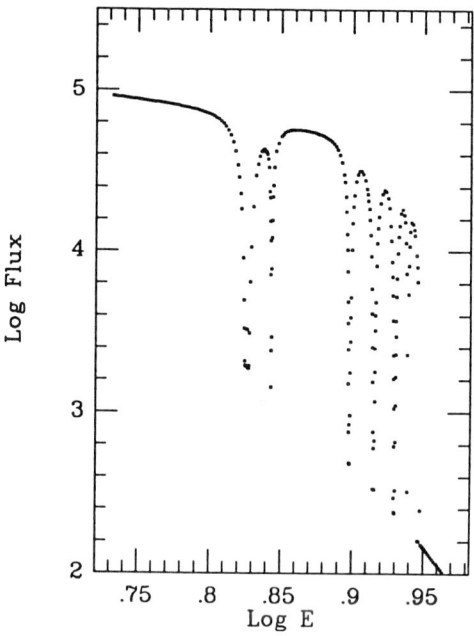

Fig. 1. Synthetic iron features in spectrum of X-ray burster with $T_e = 10^7$ K, log g = 14.0, and iron abundance = 10 x solar value (dotted line). Solid line represents spectrum seen by X-ray detectors.

ACKNOWLEDGEMENTS

This research has been partly supported by grant CPBP-01.20 from the Polish Academy of Sciences.

REFERENCES

Griem, H.R. 1974, Spectral Line Broadening by Plasmas, Academic Press, New York and London.
Koyama, K. et al. 1984, Publ. Astr. Soc. Japan, 36, 659.
Madej, J. 1988a, Adv. Space Res., 8, (2)481.
Madej, J. 1988b, Astron. Astrophys., in press.
Mewe, R., Gronenschild, E.H.B.M., and van den Oord, G.H.J. 1985, Astron. Astrophys. Suppl., 62, 197.
Mihalas, D.M. 1978, Stellar Atmospheres, W.H. Freeman and Co., San Francisco.
Turner, M.J.L., and Breedon, L.M. 1984, Monthly Notices Roy. Astron. Soc., 208, 29P.
Waki, I. et al. 1984, Publ. Astr. Soc. Japan, 36, 819.

DISCUSSION-J. Madej

T. Kallman: The degree of gravitation redshift needed to explain the observed line energy must constrain the neutron star equation of state. Have you thought about this?

J. Madej: No, I did not go so far, because results of my research do not prove strictly that 6.7 - 7.0 keV lines of iron are seen as the redshifted 4.1 keV line in bursters. There exists only indication for that. Moreover, additional effects can perhaps contribute to the atomic line shift (e.g. magnetic fields, even less than 10^{12} g), making determination of the redshift inaccurate.

A. Fabian: The enormous photoelectric edges you predicted would mean few photons observed $\sim > 7$ keV. This is not the case, (see also Foster et al, *M.N.R.A.S.*, 1987, **228**, 259).

J. Madej: Bound-free iron edges predicted in synthetic spectrum of a burster (cf. dotted line in Fig. 1) cannot be detected observationaly. After folding with some schematic detector response function such edges vanish and an observer can see very smooth continuum spectrum (solid line in the Figure). As a result of such broadening, number of photons predicted above 7 keV gets significant.

ROLE OF MAGNETOSPHERIC PLASMA PHYSICS FOR UNDERSTANDING COSMIC PHENOMENA

Indra Mohan Lal Das
Department of Physics, University of Allahabad,
Allahabad-211 002, India

Abstract

Cosmic phenomena occur in the remote regions of space where in-situ observations are not possible. For proper understanding of these phenomena laboratory experiments are essential, but the in-situ observations of magnetospheric plasma provides even a better background to test various hypothesis of cosmic interest. This is because the ionospheric-magnetospheric plasma and the solar wind are the only cosmic plasmas accessible to extensive in-situ observations and experiments.

1. Introduction :

Astrophysical or astronomical phenomena occur in the remote regions of the universe where in-situ observations are not possible. They involve matter in extreme states of temperature and/or density usually in a magnetised plasma state. Since in-situ diagnosis of cosmical objects is not possible, our understanding of astrophysical phenomena will depend critically on the reliability of interpretational techniqes of obervational data.

Our nearby environment is dominated by three ordinary states (viz.solid, liquid and gas). However, on the cosmic scale the situation is just opposite. Almost all the matter in the universe is in plasma state. The nearest shore of this cosmical plasma ocean is located in our own near space environment. The near space environment comprising ionosphere-magnetosphere of Earth and nearby interplanetary space provides a rich variety of naturally occuring plasma : cool, weakly ionised, collision dominated, gravity confined plasma as well as hot, fully ionised and virtually collisionless, magnetically confined plasma. Although there is enormous difference in linear scales as well as time scales involved in the working of astrophysical/cosmical systems and near space plasma, a careful extrapolation of results to cosmic scales have been successfully used to explain many distant cosmic phenomena. The theoretical studies in conjunction with the in-situ investigations of easily accessible magnetospheric plasma provides a more reliable background to test various hypothesis of cosmic physics. This is why astrophysics has made many remarkable discoveries during the last few decades with the development of highly shophisticated space science and technology.

2. Cosmic physics via magnetospheric research :

For proper understanding of astrophysical phenomena theoretical interpretations of observational data relying on certain simplifying assumptions must be guided by some emperical knowledge of how the real plasma behave in nature. It is in this context that the significance of magnetospheric plasma research can not be ignored. The terrestrial ionosphere-magnetosphere can be regarded as a sample of the more distant astrophysical/cosmical plasma. History of magnetospheric research amply illustrates how the in-situ observations of magnetospheric plasma have dramatically changed our perceptions of cosmic phenomena. We will discus here only few examples.

The early magnetospheric physicists regarded magnetosphere as essentially collisionless with infinitely large conductivity and hence unable to support magnetic field-aligned potential drops/electric fields. However, subsequent spacecraft measurements have proved beyond doubt that this is not true. The significance of magnetic field-aligned electric currents and fields is the realisation that the classical concept of frozen-in-magnetic filed has only limited validity. The frozen-in-magnetic field concept previously used to be a fundamental postulate of cosmic electrodynamics. As a consequence of violation of frozen-inmagnetic field concept existence of magnetic field-aligned potential drops/electric fields becomes a reality. There are several mechanism which can support such field-aligned electric fields viz. anomalous resistivity, collisionless thermo-electric effect, magnetic-mirror effect and potential double layers (Falthammar, 1978).

All known astrophysical plasma is magnetised. This plasma is typically of low density i.e. the mean free path being much larger than the scale size of the phenomena considered and hence electric currents flow essentially along magnetic field lines which is usually inhomogeneous. Thus, conditions favourable for formation of double layers and magnetic-mirror supported electric fields may not be very uncommon in cosmical plasmas. This will have a very far reaching consequence. Many explosive events observed in cosmic physics such as magnetic substorm, solar flare (releasing energy of the order of

10^{22}W), double radio sources (emitting power of the order of 10^{35}W) can be attrributed to double layer formations. Some flare star radio outbursts contain orders of magnitude more power than a solar flare but the time scale and high degree of polarisation suggest that the basic mechanism may be similar (Spangler and Moffet, 1976). Similarly a model for pulsar emission due to Sturrock (1971) based on two-stream instability observed in geomagnetosphere is capable of explaining the γ-and X-ray bursts observed from Crab pulsar (Kennel, 1975). Of interest to this pulsar emission model are the mechanism of sheath particle acceleration and current driven plasma wave instability which is very common in magnetosphere. Filamentary structures so abundantly observed in cosmic plasmas (interstellar clouds, interstellar medium, cometary tails, magnetic flux ropes, solar prominences, spicules, coronal streamers etc.)can also be attributed to field-alig-currents. Alfven (1981) provides an excellent treatise on these problems.

Space explorations have found number of boundary layers separating regions of different magnetisation, density, temperature, electron density and even chemical composition such as magnetopause, magnetotail sheet, heliospheric equatorial sheet etc. Since such properties can not be attributed to only those regions of space which are accessible to man made spacecrafts, one can reasonably assume that the interstellar and intergalactic space, in general have a cellular structure. As a consequence of this cellular structure of space one can conclude that the cosmic space is divided into large number of cells. Then the demand for symmetry is satisfied only if half of these cells are filled with ordinary matter while other half contains antimatter separated by the socalled Leidenfrost layers (Lehnert, 1977). However, size of such structures is difficult to derive theoretically and impossible to observe directly with the present day science and technology.

An important but puzzling problem in astrophysics is how the energetic (sometimes extremely energetic) radiations are produced and matter is accelerated to very high velocities (sometimes approaching the velocity of light). Cosmic ray particles with energies upto 10^{19} eV or more have been observed. Advances in γ- and X-ray astronomy have demonstrated existence of large number of astrophysical objects emitting high energy photons. Some of these acceleration/energisation mechanism responsible for cosmic radiation/galactic cosmic rays can be studied in the magnetosphere. Figure 1 taken from Alfven (1981) gives a survey of some important plasma phenomena in laboratory and cosmos.

3. Conclusion:

Results of magnetospheric plasma research have led to drastic revision of views on cosmic plasma. As nature believes in simplicity, one can assume that the basic properties of plasma to be the same everywhere from laboratory to cosmos. Radio emissions from solar chromosphere and corona and from cosmic plasma beyond our solar system are the result of the same plasma wave processes. Hence, efforts should be made to put all the available knowledge in a unified framework.

References:

Alfven, H. (1981); Cosmic Plasma, D. Reidel Pub. Co., Dordrecht, Holland.
Falthammer, C-G. (1978); J. Geomag. Geoelectr., 30, 419.
Kennel, C.F. (1975); Comments on Astrophys. Space Sci. 6, 71.
Lehnert, B. (1977); Astrophys. Space Sci., 53, 459.
Spangler, S.R. and Moffett, T.J. (1976); Ap.J., 205, 479.
Sturrock, P.A. (1971); Ap.J., 164, 529.

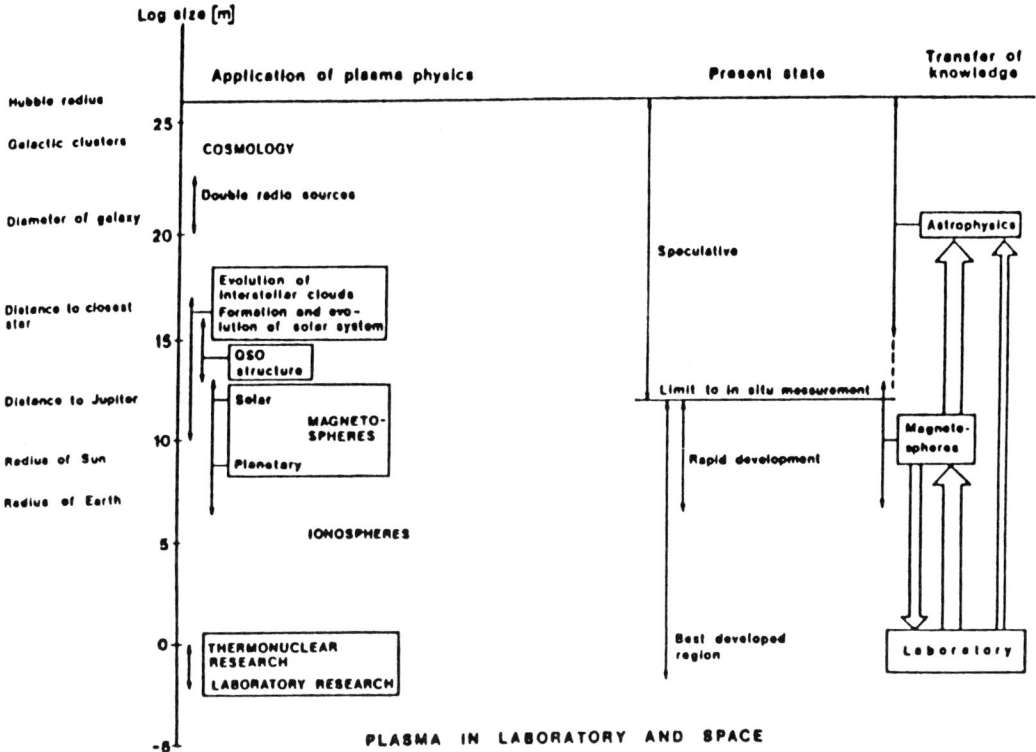

Fig. 1 The diagram indicates regions which can be explored by high quality diagnostics and those which cannot. The transfer of knowledge gained from laboratory experiments to magnetosphere research is now supplemented by a transfer in the opposite direction. Knowledge of more distant regions will also be obtained in this way.

3. Stellar Coronae

GOALS FOR THE APPLICATION OF HIGH-RESOLUTION X-RAY
SPECTROSCOPY TO THE DIAGNOSIS OF STELLAR CORONAL PLASMAS

Jeffrey L. Linsky[*]

Joint Institute for Laboratory Astrophysics
University of Colorado and National Institute of Standards
and Technology, Boulder, CO, USA

Don Quixote - The mission of each true knight...
            his duty - nay, his privilege!
            To dream the impossible dream,
                ...
            To reach the unreachable star,
            Though you know it's impossibly high,
            To live with your heart striving upward,
            To a far, unattainable sky!
                    (Man of La Mancha - Wasserman et al. (1966))

ABSTRACT

I provide examples of how high-resolution x-ray spectra may be used to determine the temperature and emission measure distributions, electron densities, steady and transient flow velocities, and location of active regions in stellar coronae. For each type of measurement I estimate the minimum spectral resolution required to resolve the most useful spectral features. In general, high sensitivity is required to obtain sufficient signal-to-noise to exploit the high spectral resolution. Although difficult, each measurement should be achievable with the instrumentation proposed for AXAF.

1. PERSPECTIVE

May I share with you my nearly impossible dream in the hope that its description will speed its fulfillment within our "scientific" lifetimes. I dream that we will extend the powerful diagnostic techniques for ultraviolet and optical spectra, which are now used to infer the properties of astrophysical plasmas at $\log T = 3.5$-5.2, to the x-ray domain where hot coronal plasmas ($\log T = 6$-8) emit the bulk of their radiation. The fulfillment of this dream requires the combination of high sensitivity and spectral resolution that will become feasible with AXAF and with later missions like XMM. As I shall describe in some detail, the scientific results can be enormous, although some experiments will push technology to its limits.

---

[*]Staff Member, Quantum Physics Division, National Institute of Standards and Technology

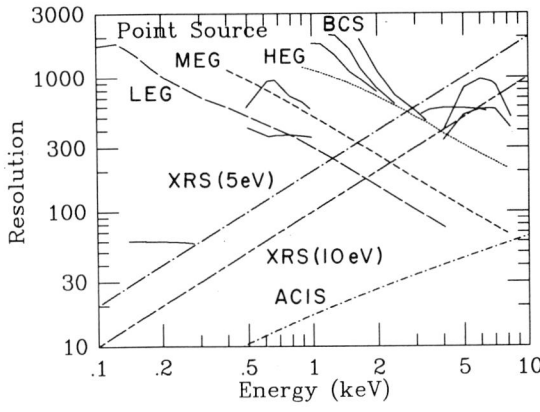

Figure 1. Anticipated resolving power $R = E/\Delta E$ of the AXAF spectrometers for point sources. The acronyms for the individual instruments are defined in the text. The solid lines refer to different crystals in the BCS. The resolving power of the XRS is given for assumed energy resolutions of 5 eV and 10 eV.

AXAF's spectrometers and detectors have not yet been fully developed, but we can anticipate what their final capabilities should be if no insurmountable difficulties are encountered. The six AXAF spectroscopic instruments (the Bragg Crystal Spectrometer = BCS, Low Energy Transmission Grating Spectrometer = LEG, Medium Energy Grating = MEG, High Energy Grating = HEG, X-ray Spectrometer (Microcalorimeter) = XRS, and Advanced CCD Imaging Spectrometer = ACIS) are described in papers presented at a special session of the AAS in January 1986 (Canizares et al. 1987; Brinkman et al. 1987; Holt 1987; and Nousek et al. 1987). The anticipated capabilities of these instruments are still evolving, and I am indebted to Claude Canizares and Tom Markert for providing the current compilation of the anticipated resolving power (Figure 1) and effective area (Figure 2) as a function of energy for these instruments.

In this talk I shall concentrate on the advantages of the high spectral resolution of AXAF, although the satellite will also have excellent high-throughput imaging capabilities. The high spectral resolution capabilities of XMM and other future missions are not yet determined and will not be discussed further. The scientific objectives that I shall shortly describe require high spectral resolution with a

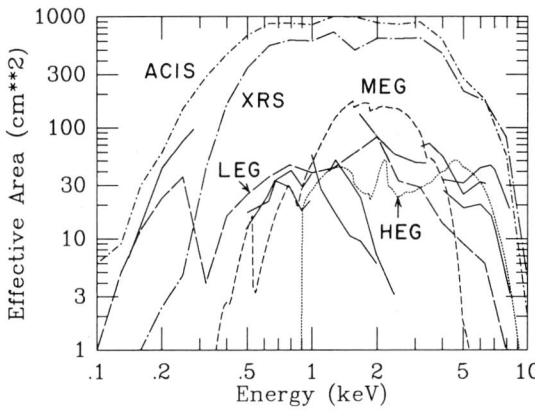

Figure 2. Anticipated effective areas for the AXAF spectrometers. The acronyms for the individual instruments are defined in the text. The solid lines refer to different crystals in the BCS.

resolving power $R = E/\Delta E > 500$. By this criterion the high-resolution instruments are the LEG($E < 0.5$ keV), MEG($E = 0.4$-0.8 keV), HEG($E = 0.9$-2.2 keV), BCS($E = 0.95$-1.9 and 3-8 keV), and the XRS($E > 2.5$ keV, assuming that the XRS will achieve an energy resolution of 5 eV). Thus high resolution spectroscopy is feasible over the full spectral range of AXAF if all of the above instruments are confirmed and perform as anticipated. Better still, $R > 1000$ should be achieved in limited spectral intervals ($E < 0.2$ keV, 1-2 keV and > 5 keV), and $R \sim 2000$ may be achieved near 0.1 keV and near 1 keV.

High resolution without high sensitivity has limited utility. The effective area plots in Figure 2 show that $A_{eff} > 10$ cm^2 for at least one of the high-resolution ($R > 500$) spectrometers over nearly the full energy range. In addition, A_{eff}(MEG) ~ 100 cm^2 near 1 keV and A_{eff}(ACIS) ~ 1000 cm^2 between 2.5 and 4 keV where $R > 500$. Thus high-throughput, high-sensitivity spectroscopy appears to be feasible over limited spectral intervals. By comparison, the Einstein SSS had $R \sim 7$ and $A_{eff} \sim 100$ cm^2 at 1 keV and the 500 line/mm transmission grating had $R \sim 50$ and $A_{eff} \sim 1$ cm^2 over its useful energy range (Giacconi et al. 1979). While the crystals in the BCS have $A_{eff} \sim 30$-100 cm^2 over much of their energy range, the spectral interval observed at any one time is small, so that the values of A_{eff} must be divided by the number of energy bins to be scanned in order to compare the BCS with other instruments that are spectrally multiplexed.

I shall now quantify the required high sensitivity needed to take advantage of high spectral resolution. A useful study of this question is provided in a paper by Landman, Roussel-Dupre and Tanigawa (1982). They computed the statistical uncertainties in the amplitude (A), line-center wavelength (B), and Doppler width (C) in a least-squares fit of an observed spectral line by a Gaussian

$$y(x) = A \exp\{-[(x-B)/C]^2\}, \qquad (1)$$

where $y(x)$ is the flux at a given wavelength x. They showed that for normally distributed errors in which the noise is independent of x, the uncertainties in the measured values of A, B and C are given by

$$S_A/A, \; S_B/C, \; S_C/C \propto (S/N)^{-1} (dx/C)^{1/2}, \qquad (2)$$

where dx is the spectral resolution and the S/N ratio is measured at line center using the signal per dx interval. Thus to measure the line shift and width to, say, 10% requires at least 100 counts per resolution element, provided the spectral line is resolved (dx < C). Conversely, low S/N data provide very uncertain values for all three Gaussian fit parameters even when the line is resolved.

To estimate whether the AXAF high-resolution spectrometers can obtain a sufficient number of counts in important spectral lines of bright coronal sources, the effective area of the AXAF instruments must be multiplied by the computed flux for the assumed stellar emission measure distribution. Several plasma emission codes exist, such as those of Mewe and Gronenschild (1981), Doschek and Cowan (1984), Shull (1986), and Raymond and Smith (1977).

Doschek and Cowan (1984) have computed the photon rates at Earth between 10Å (E = 1.24 keV) and 200Å (E = 0.062 keV) for two bright RS CVn systems (Capella and UX Ari) using approximations to their emission measure distributions derived by Swank et al. (1981) from Einstein SSS data. Two of their simulated spectra for Capella with an instrumental resolution of 0.25Å (R increases from 40 to 300 in the wavelength interval) are shown in Figure 3. The counts per 0.25Å bin are for a 10 second integration when A_{eff} = 40 cm^2. Thus an integration time of only one minute is sufficient to provide >100 counts for the brighter lines in these spectral regions, indicating that the sensitivity is sufficient to take advantage of high resolution spectroscopy. The spectral resolution of R = 40-300 in this simulation is inadequate, however, because a great many of the apparently single emission lines are blends as indicated by comparison with the higher resolution emission rate spectra (see Figure 3). Their simulated spectra of the important Fe XXIV 192.0Å line at different resolutions shows that R = 800 is needed to resolve this line. The value of R = 800 may be typical of what is required, although the resolution required to resolve specific lines must be determined on a case by case basis.

Another example of the sensitivity of the AXAF spectroscopic instruments is Linsky's (1987) computation of the total line counts in 1000 seconds for the 17 brightest lines (or close blends) between 10Å and 14Å using the Doschek and Cowan (1984) plasma emissivities and the Swank et al. (1981) emission measure distribution for another RS CVn system -- V711 Tau (=HR 1099). The numbers of counts in 1000 seconds lie in the range 20-620 for observations with the MEG and 60-1700 for observations with the XRS. The number of counts for the BCS should be roughly 1/4 that of the MEG if the lines have FWHM=300 km/s and the BCS steps through the line profiles. Several hundred counts per line are feasible in only 100 seconds for the brightest line (FeXX 12.82Å) observed with the XRS, but 40,000 seconds are required for the faintest of the 17 lines (NeIX 11.56Å) observed with the BCS. Thus many lines are bright enough to effectively utilize high-resolution spectroscopy.

The sensitivity and spectral resolution of the AXAF instruments, as well as the spectrometers that may fly on XMM and Spektrosat, far exceed the capabilities of the Einstein and Exosat spectrometers, previously described by Schmitt and Pallavicini.

2. APPLICATION OF SPECTROSCOPIC DIAGNOSTICS TO THE MEASUREMENT OF THE THERMODYNAMIC PROPERTIES OF STELLAR CORONAL PLASMAS

The availability of high-resolution spectra of the quiet and flaring Sun obtained with the P78-1, SMM and Hinotori spacecraft and with laboratory sources has stimulated the development of diagnostics to measure the temperature, emission measure distribution, electron

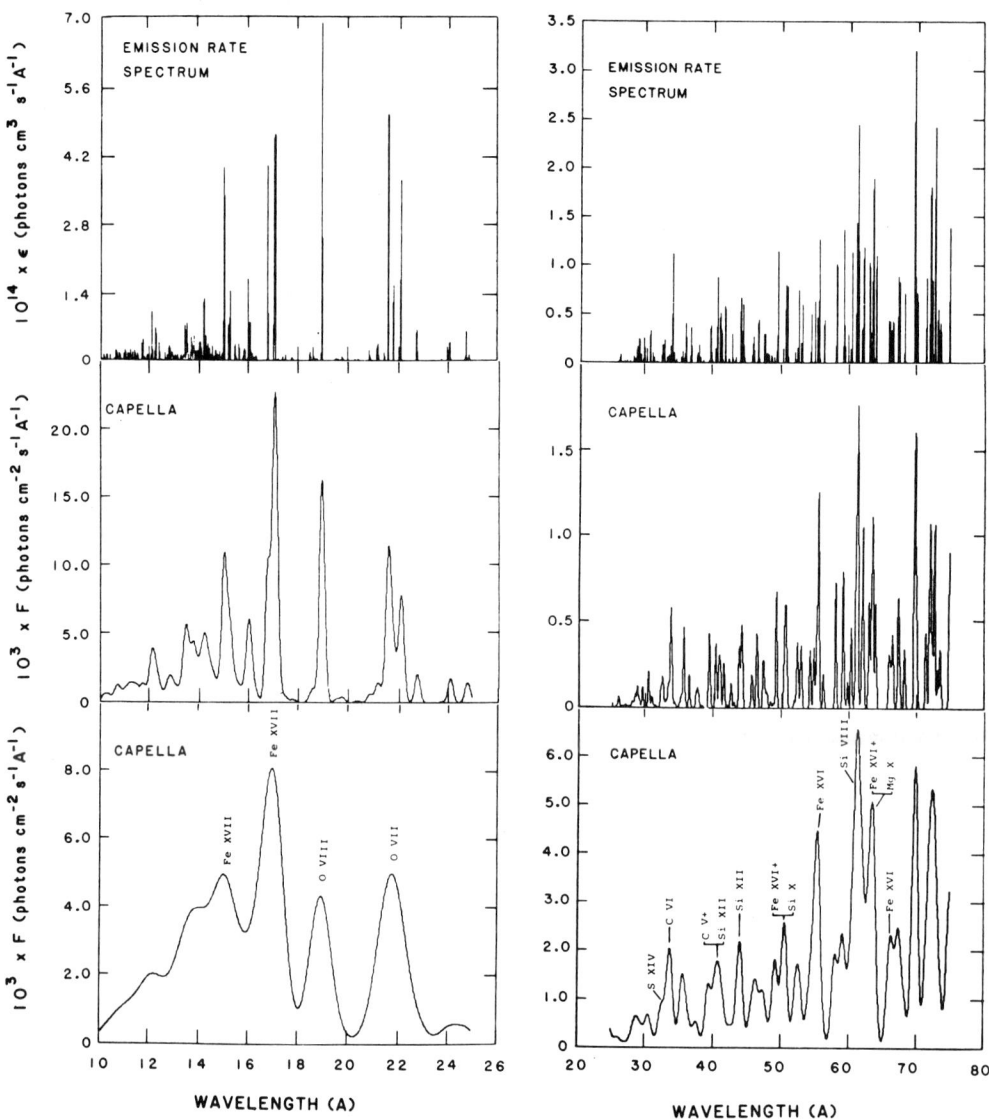

Figure 3. Emission rate spectra (fluxes per unit emission measure, instrumental resolution 0.02Å) for two wavelength regions are compared with simulated spectra for Capella with instrumental resolutions of 0.25Å (middle plots) and 1Å (lower plots). (From Doschek and Cowan 1984.)

density, and departures from coronal ionization equilibrium. Space does not permit a comprehensive review of this large topic, and I shall emphasize a few important diagnostics which provide critical tests of the spectral resolution and sensitivity required for the proper diagnosis of stellar coronal plasmas. I direct the reader to Mason's review in these Proceedings as well as to the reviews by Feldman (1981), Dere and Mason (1981), and Dupree (1978).

2.1 Temperature and Emission Measure Diagnostics

The first evidence that stellar coronal plasmas are not isothermal was provided by the SSS low resolution ($\Delta E = 140$ eV) spectra of seven RS CVn systems and Algol that Swank et al. (1981) showed should be fit by at least two-temperature plasmas, with the emission measure, EM(T), peaked near logT = 6.8 and 7.5-8.0. The SSS spectrum of the M dwarf system Wolf 630 AB (dM3.5e+dM3.5e) has also been fit with a two-temperature plasma with peaks in the EM(T) at logT = 6.8 and >7 (Swank and Johnson 1982). In their study of EXOSAT transmission grating spectra of two RS CVn systems (Capella and σ^2 CrB), Lemen et al. (1988) concluded that both spectra are better fit by an EM(T) distribution peaked near logT = 6.7 and 7.5 than by a continuous distribution. The Mewe et al. (1982) analysis of the Einstein Objective Grating Spectrometer (R= 15-75) spectrum of Capella also results in a two-temperature fit with logT = 6.7 for the cooler component but logT = 7.0 for the hotter component. The analyses of moderate resolution spectra of the RS CVn systems are thus in close agreement except for the less well determined hotter component.

Majer et al. (1986), however, fit the much lower resolution IPC spectra of seven RS CVn systems in common with the SSS data set also with two-temperature plasmas, but in all cases both the cool and hot components have derived temperatures a factor of 3 lower. While the analysis of higher resolution data must be given greater weight, discrepancies among the conclusions of different authors, the absence of a compelling argument for thermal bifurcation, and the absence of steady-state plasma in either temperature range in the solar corona suggest that the two-temperature distributions may provide an incomplete picture of the plasma properties in stellar coronae. The only way to settle this question is to obtain and analyze higher-resolution spectra with good signal-to-noise. This last point is important, because the Einstein Bragg Crystal spectra of Capella have a resolution R ~ 150 but sufficient sensitivity to detect only three lines (O VIII 18.97Å, Fe XVII 15.01Å, and FeXX 12.83Å) in 60,000 seconds observing. Vedder and Canizares (1983) were able to derive from these data a mean temperature (logT = 6.8) for the plasma but no information on the temperature distribution.

A simple method for deriving the temperature of an assumed isothermal coronal plasma is to compare the observed fluxes in collisionally-excited lines of different ions of the same element with flux ratios computed assuming coronal steady-state ionization and excitation equilibrium. A virtue of this method is that spectral lines may be selected on the basis of their strength and absence from obvious blends. Since the ionization ratio technique can use the brightest and least blended among a large number of possible lines, the simulated spectra in Figure 3 suggest that R = 200 may be sufficient to find enough suitable lines, although higher resolution would be desirable. This method is limited, however, by inaccurate atomic data and uncertain assumptions underlying the ionization equilibrium calculations. Even worse, such calculations assume the unlikely physical situation that the plasma is homogeneous and isothermal, that the electron velocity distribution is maxwellian without a high energy tail, that the ionization is steady state and in balance with the local electron temperature and density, and that photoionization is unimportant. The ionization equilibrium in the outer solar corona is frozen-in, however, since the time scale for the wind to transport plasma to regions of significantly different temperature and density exceeds the time scale for ionization and recombination (Owocki, Holzer, and Hundhausen 1983). Also, transient ionization occurs in flares (Doschek and Tanaka 1987), and flaring may be responsible for the bulk of coronal heating. Even when the standard coronal ionization equilibrium assumptions are valid, the temperatures at which specific ions have peak abundance often differs by $\Delta \log T = 0.1$ among different calculations as a result of different assumed ionization and recombination rates.

Uncertainties in the ionization equilibrium may be avoided by measuring the flux of two spectral lines of the same ion for which their ratio is sensitive to temperature rather than density. This occurs, for example, when both lines are collisionally excited from the ground state (or a common lower level) and de-excitation is radiative rather than collisional. Their flux ratio for an isothermal plasma will only depend upon atomic cross-sections and the Boltzmann factor, $\exp[-(E_2-E_1)/kT_e)$, where E_1 and E_2 are the excitation energies of the two upper levels relative to the lower state. One particularly useful ratio is F(3p-3s)/F(3d-3p) of the NaI isoelectronic sequence (including FeXVI 251Å/263Å), because the two lines are close in wavelength (cf. Flower and Nussbaumer 1975) and thus instrumental flux calibration uncertainties should be minimal. Additional line ratios are summarized in the previously listed reviews. The spectral resolution required is set by the need to avoid significant blends and is, therefore, different for each line pair. A typical resolution needed for medium strength lines is about R = 800 as shown in the Doschek and Cowan (1984) simulated spectra of the region near the important FeXXIV 192.0Å line.

A third method for inferring plasma temperatures uses the ratio of line fluxes within the K shell transition blend of He-like ions such as CaXIX (3.18-3.20Å) and FeXXV (1.85-1.87Å). The F(j)/F(w) ratio is particularly important, because the fluxes of both the dielectronic satellite line (j) and the collisionally-excited resonance line (w) depend upon the population of the He-like ion. Thus the ratio depends upon the electron temperature (T_e) rather than the electron density or ionization equilibrium. On the other hand, the F(q)/F(w) ratio can be used to measure departures from coronal ionization equilibrium, because line q is an inner-shell collisional excitation line of the Li-like ion, whereas line w is the collisionally-excited resonance line of the He-like ion. For a more complete discussion of the atomic physics of these transitions and applications to the diagnosis of solar coronal plasmas see, for example, Feldman (1981). The important point of this discussion is that the typical splitting of lines in the CaXIX 3.19Å region is 0.0040Å and in the FeXXV 1.86Å region is 0.0024Å. Thus R = 800 is required to diagnose both regions. Both the BCS and the XRS (for an energy resolution ΔE = 8 eV) have sufficient resolution at 1.86Å (6.6 keV), but only the XRS (for ΔE = 5 eV) has sufficient resolution at 3.19Å (4.0 keV). The anticipated effective area of the XRS at 4.0 keV is 700 cm^2 compared with 200 cm^2 at 6.6 keV and 25 cm^2 per resolution element for the BCS at this energy. Thus the XRS would be the instrument of choice and the Ca XIX feature the appropriate temperature diagnostic if the XRS can achieve the required resolution ΔE = 5 eV.

2.2 Electron Density Diagnostics

The determination of EM(T) alone is inadequate to model a stellar corona because the unknown density or pressure means that the emitting volume and its geometry are unknown, and the plasma may not be homogeneous or in hydrostatic equilibrium. These two approximations are very unlikely if magnetic loops control the geometry and confine high density structures as occurs in the solar corona. Thus one needs density-sensitive diagnostics that cover the important temperature range to constrain the models. Ratios of lines for a given ion are density sensitive when the radiative de-excitation rates for two transitions to the same lower level are very different. Thus, over a range of density de-excitation transitions are predominately radiative in one transition but predominately collisional in the other.

One can understand the essential physics of the line ratio method by considering an ion with a ground state (1) and two excited states (2 and 3) which are, for illustrative purposes, only connected to the ground state. If we assume that the lines are optically thin and radiative excitation is unimportant, then the statistical equilibrium equation for each excited state is

$$n_u(A_{u1} + n_e C_{u1}) = n_1 n_e C_{1u} \qquad (3)$$

and the emergent flux in each transition is

$$F_{u1} = A_{u1} h\nu n_u. \qquad (4)$$

The line flux ratio, $r = F_{21}/F_{31}$, is then

$$r = (A_{21}/A_{31}) * [n_e C_{12}/(A_{21} + n_e C_{21})] / [n_e C_{13}/(A_{31} + n_e C_{31})]. \qquad (5)$$

This ratio will be density sensitive within some density range when the radiative de-excitation rate for one transition is much larger than for the other. A good example is the BeI isoelectronic sequence ion CIII (cf. Keenan et al. 1984) for which the optically allowed (2s2p ^{1}P- 2s^2 ^{1}S) 977Å resonance line has a spontaneous de-excitation rate A_{31}=1.7

10^9 s^{-1}, whereas the 1908Å intersystem transition (2s2p $^3P_0$ - 2s^2 ^1S) has a rate A_{21}=190 s^{-1}. In the low density ("coronal") limit ($A_{21} > n_e C_{21}$ and $A_{31} > n_e C_{31}$), the line flux ratio is simply r=C_{12}/C_{13}, which is independent of density. In this limit the fluxes of both lines are proportional to the emission measure and thus behave as "allowed" lines. On the other hand, in the high density limit ($A_{21} < n_e C_{21}$ and $A_{31} < n_e C_{31}$), the line flux ratio is also a constant, because both transitions are collisionally de-excited and behave as "forbidden" transitions. The interesting regime occurs at intermediate densities when $A_{21} < n_e C_{21}$ but $A_{31} > n_e C_{31}$. Now the line flux ratio depends explicitly on n_e,

$$r = A_{21}C_{12}/n_e C_{21}C_{13} \sim 1/n_e. \quad (6)$$

One can define a critical density, $n_e(cr) = A_{21}/C_{21}$ which is a practical upper limit to the density for which r is a useful density diagnostic.

A number of useful line ratios for FeXVIII-XXVI and other highly ionized species are available in the x-ray region as described in Feldman's (1981) review, although collisional and radiative transitions among excited levels often result in a more complex situation than the simple case just described. Of particular importance are the HeI sequence ions for which Gabriel and Jordan (1969) showed that ratios of the forbidden (1s^2 $^1S_0$ - 1s2s $^3S_1$) to the intercombination (1s^2 $^1S_0$ - 1s2p $^3P_1$) lines are density sensitive. The important ratios for the HeI sequence are listed in Table 1 with data from Pradhan and Shull (1981). Included in the table are the wavelengths, temperatures (T_m) at which each ion has maximum abundance, and range of electron densities over which the line ratio is density sensitive. The OVII lines are routinely used as density diagnostics for solar flares, and the lines of CV through NeIX should be useful for studying stellar coronae and flares.

Table 1. Density-Sensitive Line Ratios

Ion	Isoelectronic Sequence	Line Wavelengths (Å)	Transition	log(T_m)	Useful Range in log (n_e)	Ref.
C V	He I	40.73/41.47	$^1S_0$-$^3S_1$/$^1S_0$-$^3P_1$	5.5	> 8.8	1
O VII	He I	21.80/22.02	"	5.9	>10.5	1
Ne IX	He I	13.55/13.70	"	6.2	>11.8	1
Mg XI	He I	9.23/9.32	"	6.4	>12.7	1
Si XIII	He I	6.69/6.74	"	6.6	>13.6	1
Ca XIX	He I	3.19/3.21	"	7.4	>15.4	1
Fe XXV	He I	1.86/1.87	"	7.7	>16.8	1
Fe XXII	B I	114.41/117.18	$^2P^o_{3/2}$-$^2P_{3/2}$/$^2P^o_{1/2}$-$^2S_{1/2}$	7.0	12-14	2
Fe XXII	B I	11.93/11.74	$^2P^o_{3/2}$-$^2D_{3/2,5/2}$/$^2P^o_{1/2}$-$^2D_{3/2}$	7.0	12-14	2
Fe XXI	C I	142.14/128.74	$^3P_1$-$^3D_2$/$^3P_0$-$^3D_1$	7.0	11-13	3
Fe XXI	C I	102.35/128.74	$^2P_2$-$^3S_1$/$^3P_0$-$^3D_1$	7.0	11-13	3
Fe XXI	C I	12.38/12.313	$^3P_2$-$^3D_3$/$^3P_0$-$^3D_1$	7.0	11-13	3

References: (1) Pradhan and Shull (1981); (2) Mason and Storey (1980); (3) Mason et al. (1979).

Since the separation of these lines increases from $E/\Delta E = 60$ for CV to 90 for NeIX, a modest resolution of $R = 150$ would be adequate. The higher ions in the sequence are unlikely to be useful as their critical densities are very large. Some line ratios of FeXXI and FeXXII used for analyzing solar flares are listed in Table 1.

3. APPLICATION OF SPECTROSCOPIC DIAGNOSTICS TO THE MEASUREMENT OF THE FLOWS AND GEOMETRY OF STELLAR CORONAL PLASMAS

3.1 Winds and Downflows

Is it feasible to measure the expansion velocities of stellar coronae with high-resolution x-ray spectroscopic techniques? To answer this question we must understand the dynamics and geometry of stellar coronae. Since we know very little about these properties in stellar coronae, it is useful to start with the Sun. The solar wind passing the Earth ranges in velocity from roughly 400 km/s up to 800 km/s in high speed streams. The steady-state mass loss rate at the equator is roughly 2×10^{-14} $M_\odot$/yr with some 5-10% of the total occurring in sudden ejections of material in the form of coronal mass ejections (CMEs) which are in part flare related. At present no spectroscopic indicator of mass loss is apparent in the spectrum of the Sun viewed as an unresolved star. In fact, the only spectroscopic evidence of expanding hot plasma comes from the measurements of the OV 629Å and MgX 625Å lines by Orrall, Rottman and Klimchuk (1983) and previous experimenters. They measured a mean blueshift of about 8 km/s in the MgX line, which is formed at $\log(T) = 6.15$, above a coronal hole, but redshifts of this line in the surrounding active region. In general, these and previous studies found an anticorrelation between the flow direction and line intensity consistent with the picture of upflows in low-density, magnetically-open structures (coronal holes) and downflows in high-density, magnetically-closed regions (active regions). The upflow velocities are far smaller than the terminal velocity of the solar wind, because the MgX line is formed low in the corona where the density is much larger than at the critical point. If the Sun were observed as a point source, even this small velocity would be difficult to detect given the redshifts of the much brighter active regions.

The solar example does not provide optimism for the detection of coronal winds, but the solar mass loss rate is as much as eight orders of magnitude smaller than that measured in some luminous cool stars. In her recent review of mass loss from cool stars, Dupree (1986) summarized the spectroscopic evidence for outflowing material from cool giants and supergiants, pre-main sequence stars, and hybrid chromosphere stars. These data consist of blue-shifted absorption features or P Cygni-shaped profiles of lines formed only at relatively low temperatures, including MgII, CaII, HI, HeI, and neutral atoms formed in circumstellar envelopes. Outflow velocities approaching 200 km/s have been detected in the hybrid star α TrA (Hartmann et al. 1985) and in the GIb-II star 22 Vul (Reimers and Che-Bohnenstengel 1986), but the higher temperature lines of CIII and SiIII in an IUE spectrum of α TrA (Brown et al. 1986) show no significant blueshifts, indicating that the hotter material does not participate in the outflow. For the more active stars including the RS CVn binaries, main sequence stars, and the G2 Ib-IIa star β Dra, high-resolution IUE spectra of lines formed at $\log(T) = 4.0 - 5.0$ exhibit a trend of increasing red shift (downflow) with temperature of formation. Mariska (1988) explains these data by asymmetric heating in magnetic loops that produces steeper temperature gradients and thus less emission measure for the upflowing plasma than for the downflowing plasma. Thus, the net spectrum of a star covered with such loops would show redshifted lines, even though there is no net flow of material. These calculations predict that lines formed at $\log(T) > 5.3$ will be blueshifted rather than redshifted, and x-ray spectra are needed to test this prediction. The maximum redshifts for ultraviolet emission lines measured by Ayres et al. (1988) in active stars like Capella are about 20 km/s.

Could either winds or downflows in coronae be detected with x-ray spectroscopy? Using Eq. (2) with an assumed line width $C = 150$ km/s and $S/N = 10$, I find that a resolution $R = 1000$ is sufficient to measure line shifts as small as 20 km/s. This quantity scales as $R^{1/2}/(S/N)$. Thus only modest spectral resolution and S/N are required to measure x-ray lineshifts comparable with the ultraviolet results. However, the determination of flow properties in stellar coronae and how they relate to the flow pattern lower in a stellar atmosphere requires considerable theoretical analysis.

3.2 Transient Flows

Blueshifted components in coronal emission lines are commonly seen during the early stages of solar flares. These components detected in the CaXIX 3.176Å line have been interpreted by Antonucci et al. (1984, 1987) and others as indicating 10^7 K plasma upflows with velocities as large as 400 km/s at the impulsive stage of a flare decreasing to essentially zero in 3-10 minutes near the peak of the thermal x-ray emission. A common interpretation of this phenomenon is that nonthermal electrons accelerated during the impulsive event heat and evaporate the chromosphere at the footpoints of magnetic loops, producing the rapid expansion of 10^7 K plasma up the loops. These upflows have never been detected in stellar flares. Flows during flares on RS CVn stars are detected in chromospheric lines like Hα and MgII k, but the line shifts and additional components are generally to the red indicating downflows (e.g. Simon et al. 1980; Linsky et al. 1988).

How are these flows related to the solar flare evaporative flows detected in x-rays, and do evaporative flows also occur during stellar flares? Since the spectroscopic signature of the upflows is the presence of blueshifted components adjacent to the main emission line, a resolution of 200 km/s (R = 1500) is required and R = 3000 is desirable. In addition, a time resolution of 5 minutes or better is needed if the solar example is pertinent. These requirements are stringent but can be met with the AXAF LETGS. Linsky (1987) estimates that the LETGS will obtain 84 counts in 5 minutes in the FeXVI 63.71Å line for a flare similar to that observed from AT Mic (dM4.5e+dM4.5e) by Kahn et al. (1979).

3.3 Doppler Imaging

The presence of bright or dark structures on the surfaces and above the limbs of stars can be detected by timing and spectroscopic techniques. The presence of dark spots on the photospheres of active stars, for example, is indicated by photometric variations with the stellar rotational period. Similarly, the presence of discrete, bright active regions in stellar chromospheres is indicated by the increase and decrease in emission line flux as an active region appears and then disappears over the limb. The correlation of photometric decline with MgII and CIV flux increase and the reverse behavior half a rotational period later led Rodono et al. (1987) to conclude that during October 1981 a bright active region was located above a large spot complex on the RS CVn star II Peg.

Photometric and spectrophotometric techniques provide only crude data on the location of spots and active regions on a star. A more precise technique is Doppler imaging which analyzes the changing Doppler shift of a discrete region on a rapidly-rotating star as the region rotates across the stellar disk to determine both its longitude and latitude (from the amplitude of the radial velocity changes). Vogt and Penrod (1983) and Vogt et al. (1987) have developed this technique to obtain images of RS CVn systems with several dark spot groups. Walter et al. (1987) and Neff et al. (1988) have used these techniques to locate two or three bright active regions in the chromosphere of the K0 IV star of their favorite RS CVn-type system AR Lac.

Can the Doppler imaging technique be applied to x-ray spectra to derive images of stellar coronae? This is a very demanding task as the equatorial rotational velocities of these systems are generally less than 70 km/s and most active dwarf stars rotate more slowly. To test the feasibility of this technique, we plot in Figure 4 profiles of the NeX 12.20Å line from the K0 IV star in AR Lac (vsini - 72 km/s) assuming that half the flux is from a structure with a Doppler shift of 100, 200, and 300 km/s relative to the star for assumed resolving powers R - 1000, 2000 and 3000. The Doppler shifts are for a corotating structure located in the equatorial corona 0.4, 1.8, and 3.2 stellar radii above the limb. These calculations indicate that with good S/N and a stable wavelength scale a resolution of only 1000 is required to identify the closest structure, but a resolution of 3000 would locate the active region reliably and permit its study as a function of phase. Thus the goal of Doppler imaging of stellar coronae is probably not an impossible dream.

4. CONCLUSIONS

Each type of measurement described in this paper places different requirements on the instrumental spectral resolution. In some cases

106 Linsky

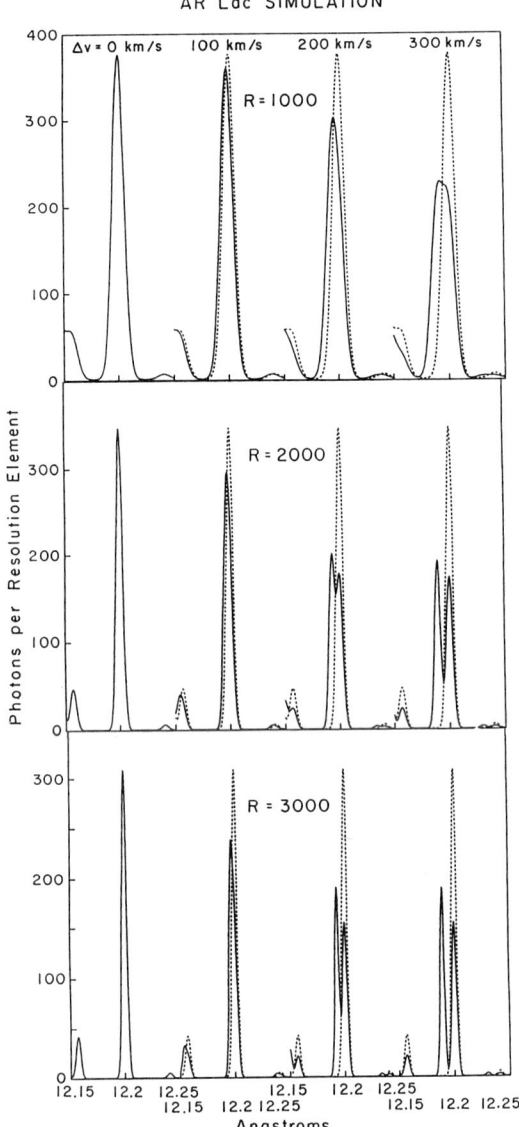

Figure 4. Simulated profiles of the NeX 12.20Å line from the K0 IV star in AR Lac. Each line profile is a convolution of 20 km/s turbulence, 72 km/s rotation, and an instrumental resolution R = 1000 (top panel), 2000 (middle panel), or 3000 (bottom panel). Half of the emission is from the stellar disk and half from an active region with a Doppler shift of Δv = 0, 100, 200, or 300 km/s towards the observer. The dashed profiles are for Δv = 0 km/s to compare with the profiles including a Doppler-shifted active region. The calculation assumes emission measures from Swank et al. (1981) and $A_{eff}\Delta t = 1 \times 10^6$ cm^2 s, where A_{eff} is the instrumental effective area and Δt the integration time. Other lines are present at 12.16 and 12.24Å.

Table 2. Summary of Spectral Resolution Requirements

Measurement	Minimum Resolution Required (R=E/ΔE)	Desirable Resolution
Measure coronal temperature by the ionization ratio technique	200	400
Measure coronal temperature from line ratios for a given ion	800	800
Measure coronal temperature and departures from coronal ionization equilibrium using the Ca XIX 3.18-3.02Å or Fe XXV 1.85-1.87Å features	800	1200
Measure electron densities from He sequence ions	150	400
Measure flow velocity of 20 km/s when S/N=10	1000	1000
Measure transient upflows during flares	1500	3000
Doppler image active regions in the coronae of RS CVn systems	1000	3000

there are also preferred spectral regions. Table 2 summarizes the minimum resolutions required for each of the measurements. For most measurements, R = 1000 should provide usable information for at least some stellar coronae, although higher resolution is desirable and high signal-to-noise is mandatory. Although difficult, each measurement should be achievable for at least a few stars with the instrumentation proposed for AXAF.

I wish to thank Drs. C. Canizares and T. Markert for providing Figures 1 and 2, and Drs. G. A. Doschek and R. D. Cowan for allowing my use of Figure 3. I also thank Mr. A. Veale for computing the Doppler-imaged profiles in Figure 4. This work is supported by NASA grant H-80531 to NIST.

5. REFERENCES

Antonucci, E., Gabriel, A. H., and Dennis, B. R. 1984, Ap. J., 287, 917.
Antonucci, E., Dodero, M. A., Peres, G., Serio, S., and Rosner, R. 1987, Ap. J., 322, 522.
Ayres, T. R., Jensen, E., and Engvold, O. 1988, Ap. J. Supp., 66, 51.
Brinkman, A. C., van Rooijen, J. J., Bleeker, J. A. M., Dijkstra, J. H., Heise, J., De Korte, P. A. J., Mewe, R., and Paerels, F. 1987, Astro. Lett. Comm., 26, 73.
Brown, A., Reimers, D., and Linsky, J. L. 1986, New Insights in Astrophysics, ESA SP-263, 169.
Canizares, C. R., et al. 1987, Astro. Lett. Comm., 26, 87.
Dere, K. P. and Mason, H. 1981, in "Solar Active Regions", ed. F. Q. Orrall (Boulder: Colorado Assoc. Univ. Press), p. 129.
Doschek, G. A. and Cowan, R. D. 1984, Ap. J. Suppl., 56, 67.
Doschek, G. A. and Tanaka, K. 1987, Ap. J., 323, 799.
Dupree, A. K. 1978, Adv. Atomic Mol. Phys., 14, 393.

Dupree, A. K. 1986, Ann. Rev. Astron. Ap., 24, 377.
Feldman, U. 1981, Physica Scripta, 24, 681.
Flower, D. R. and Nussbaumer, H. 1975, Astron. Ap., 42, 265.
Gabriel, A. H. and Jordan, C. 1969, M.N.R.A.S., 145, 241.
Giacconi, R. et al. 1979, Ap. J., 230, 540.
Hartmann, L., Jordan, C., Brown, A., and Dupree, A. K. 1985, Ap. J., 296, 576.
Holt, S. S. 1987, Astro. Lett. Comm., 26, 61.
Kahn, S. M., Linsky, J. L., Mason, K. O., Haisch, B. M., Bowyer, C. S., White, N. E., and Pravdo, S. H. 1979, Ap. J., 234, L107.
Keenan, F. P., Berrington, K. A., Burke, P. G., Kingston, A. E., and Dufton, P. L. 1984, M.N.R.A.S. 207, 459.
Landman, D. A., Roussel-Dupre, R., and Tanigawa, G. 1982, Ap. J., 261, 732.
Lemen, J. R., Mewe, R., Schrijver, C. J., and Fludra, A. 1988, Ap. J., submitted.
Linsky, J. L. 1987, Astro. Lett. Comm., 26, 21.
Linsky, J. L., Neff, J. E., Brown, A., Gross, B. D., Simon, T., Andrews, A. D., Rodono, M., and Feldman, P. A. 1988, Astron. Ap., in press.
Majer, P., Schmitt, J. H. N. N., Golub, L., Harnden, F. R. Jr., and Rosner, R., 1986, Ap. J., 300, 360.
Mariska, J. T. 1988, Ap. J., 334, 489.
Mason, H. E., Doschek, G. A., Feldman, U., and Bhatia, A. K. 1979, Astron. Ap., 73, 74.
Mason, H. E. and Storey, P. J. 1980, M.N.R.A.S. 191, 631.
Mewe, R. and Gronenschild, E. H. B. M. 1981, Astron. Ap. Suppl., 45, 11.
Mewe, R., et al. 1982, Ap. J. 260, 233.
Neff, J. E., Walter, F. M., Rodono, M., and Linsky, J. L. 1988, Astron. Ap., in press.
Nousek, J. A., Garmire, G. P., Ricker, G. R., Collins, S. A., and Reigler, G. R. 1987, Astro. Lett. Comm., 26, 35.
Orrall, F. Q., Rottman, G. J., and Klimchuk, J. A. 1983, Ap. J., 266, L65.
Owocki, S. P., Holzer, T. E., and Hundhausen, A. J. 1983, Ap. J., 275, 354.
Pradhan, A. K., and Shull, J. M. 1981, Ap. J., 249, 821.
Raymond, J. C., and Smith, B. W. 1977, Ap. J. Suppl., 35, 419.
Reimers, D., and Che-Bohnenstengel, A. 1986, Astron. Ap., 166, 252.
Rodono, M., et al. 1987, Astron. Ap., 176, 267.
Shull, J. M. 1986, private communication.
Simon, T., Linsky, J. L., and Schiffer, F. H. III 1980, Ap. J., 239, 911.
Swank, J. H., White, N. E., Holt, S. S., and Becker, R. H. 1981, Ap. J., 246, 208.
Swank, J. H., and Johnson, H. M. 1982, Ap. J., 259, L67.
Vedder, P. W., and Canizares, C. R. 1983, Ap. J., 270, 666.
Vogt, S. S. and Penrod, G. D. 1983, Publ. Astron. Soc. Pacific, 95, 565.
Vogt, S. S., Penrod, G. D., and Hatzes, A. P. 1987, Ap. J., 321, 496.
Walter, F. M., Neff, J. E., Gibson, D. M., Linsky, J. L., Rodono, M., Gary, D. E., and Butler, C. J. 1987, Astron. Ap., 186, 241.
Wasserman, D., Darion, M., and Leigh, M. 1966, "Man of La Mancha" (New York: Random House).

DISCUSSION-J. Linsky

S. Serio: You did mention non-ionization equilibrium. Where can it be observed in stellar coronae? In fact, SMM does not see NIE in solar flares because higher sensitivity would be needed.

J. Linsky: We should look for departure from coronal ionization equilibrium during stellar flares by whatever diagnostics are available. One such diagnostic is the $F(q)/F(w)$ ratio of the inner-shell collisional excitation line of the Li-like ion to the collisionally-excited resonance line of the He-like ion. These lines are located near 3.9 Å for C_a and 1.86 Å for Fe.

R. Pallavicini: Just to give the proper credit to the authors, I have to say that the comparison of x-ray temperatures derived from IPC and SSS observations is due to the work of P. Majer, J. Schmitt and Collaborators (Ap.J. 1986), and not to me.

The second comment I have is that high-resolution spectroscopy will also help understanding the physics of early-type stars, with regard to both line broadenings (as expected from turbulent winds) and Doppler shifts due to wind expansion.

J. Linsky: I agree with both of your comments.

S. Kahn: 1) Do the Spectral predictions for UX Ari and Capella include interstellar absorption out at 19Å ? and 2) Many of your suggested "experiments" push the instrumental limits. Do you think there are an equal number of exciting projects for coronae which are a little less demanding?

J. Linsky: 1) The calculated spectra I showed for Capella and UX Ari do not include interstellar absorption. For Capella the hydrogen column density is about 2×10^{18} cm^{-2}, which corresponds to interstellar optical depth unity at about 280 Å and interstellar optical depth about 0.3 at 190 Å. I do not know what the interstellar hydrogen column density is towards UX Ari. 2) While I have concentrated on the more demanding spectroscopic studies, there are important studies that can be made with lower spectral resolution data. For example, with a spectral resolution $E/\Delta E = 10\text{-}100$ and large effective area, comparison of observed with theoretical spectra should lead to far more accurate distributions of emission measure with temperature than could be obtained with the Einstein SSS or the EXOSAT TGS. Such data could be obtained for a wide variety of stars, and binary systems, for example, with high time resolution or the brighter transient sources.

X-ray Spectroscopy across the HR-Diagram

J.H.M.M. Schmitt
Max-Planck-Institut für Extraterrestrische Physik
8046 Garching bei München
Federal Republic of Germany

Abstract. X-ray spectra of stellar X-ray sources taken with high, moderate and low spectral resolution are discussed. Only low resolution spectra are available for a sufficiently large number of late-type stars. It is shown that high temperature plasmas ($log\ T > 7$) produce the dominant emission component in the coronae of red dwarfs as well as yellow giants, while the coronae of F stars - usually - have X-ray temperatures similar to those found in the quiet Sun. The need for density diagnostics of stellar coronae is stressed and it is argued that the waveband region between 90 - 140 Å is particularly well suited for this purpose.

1. Introduction

The discovery that X-ray emission from normal stars is an ubiquitous phenomenon found thoughout the HR-diagram is certainly one of the outstanding results of last decade's research in X-ray astronomy (cf., Vaiana *et al.* 1981). X-ray emission, detectable with modest-sized imaging X-ray telescopes, is not restricted to peculiar and more or less exotic types of stars; on the contrary, except for main-sequence A dwarfs and late-type giants and supergiants, it is found in stars of all spectral types and luminosity classes (for recent reviews of the subject see Rosner, Golub and Vaiana (1985), Linsky (1985), Schmitt (1988) and references therein).

While the existence of coronal X-ray sources throughout the HR-diagram has been well established, the spectral properties of the emitted X-ray radiation are only poorly known. The vast majority of the stellar observations was taken with little or no spectral resolution and rather modest signal-to-noise ratio (SNR). Only a relatively small number of coronal sources was observed with modest spectral resolution with the *Einstein Observatory* Solid State Spectrometer (SSS) and the EXOSAT and GINGA proportional counters; spectra with resolving power $\lambda/\Delta\lambda \geq 30$ were obtained for an exceedingly small number of sources with the *Einstein Observatory* Objective Grating Spectrometer (OGS) and Focal Plane Crystal Spectrometer (FPCS) and the EXOSAT transmission grating (TG). While solar soft X-ray spectra are available with resolving powers of $\lambda/\Delta\lambda \sim 10000$ and good SNR, the resolving power of the best FPCS spectra of coronal X-ray sources is of the order ~ 100 and very modest SNR.

FPCS spectra have been published for the RS CVn systems Capella (Vedder and Canizares 1983) and $\sigma\ CrB$ (Agrawal, Markert and Riegler 1985). Because of the relatively small band passes of individual crystals many band scans must be performed in sequence, and taking a spectrum over a larger energy range is rather time-consuming. Consequently, the available spectra cover only a relatively small energy range, typically centered on some interesting lines, in the case of $\sigma\,CrB$, the energy range between 800 - 840 eV and 986 - 1016 eV which contain lines of iron in ionisation stages Fe XVII to Fe XXII. The FPCS spectra of both Capella and $\sigma\,CrB$ are consistent with an isothermal plasma emitting at a temperature $log\,T \sim 6.85$.

OGS transmission grating spectra are available for Capella (Mewe et al. 1982), EXOSAT transmission grating spectra have been published for Capella and $\sigma\,CrB$ as well as the nearby F star Procyon (Schrijver 1985; Mewe et al. 1986); in addition grating spectra are available for the hot white dwarf stars HZ 43, Sirius B and Feige 24 (Paerels et al. 1986 a,b,c). Grating spectra cover a broader spectral range with a spectral resolution below that of crystal spectrographs; the useful bandpass is determined by the telescope/grating/detector geometry as well as the detector efficiency. While the OGS spectrum of Capella extends out to only $\sim 40\,$ ÅÅ , the EXOSAT TG spectra extend out to 200 Å (or more), covering in particular the range between 90 - 140 Å which shows extremely strong Fe lines produced in $logT \sim 7$ plasmas as well as the region around 170 Å which contains the strongest spectral line emitted by the quiescent solar corona. The transmission grating spectra typically require two spectral components for a satisfactory fit, or alternatively, a continuous emission measure distribution (Lemen et al. 1988).

The Solid State Spectrometer (SSS) had a resolution of about 160 eV in the energy range 0.5 - 4 keV; thus its resolving power increased with increasing energy with photon fluxes however decreasing because of the mirror cutoff and the shape of the incident spectra. Appoximately 20 coronal sources were observed, mostly RS CVn systems such as Capella and $\sigma\,CrB$, Algols and O stars; two M dwarfs (Wolf 630 and AD Leo) and an active G star ($\pi^1 Uma$) were also observed (Holt et al. 1979; Swank et al. 1981; Swank and Johnson 1983; Cassinelli and Swank 1983; Swank 1984). SSS spectra have lower spectral resolution than grating spectra, but the resolving power is still sufficient to detect the resonance lines of the helium-like ionisation stages of Mg, Si and S if present. The detection of these lines in the SSS spectrum of Capella by Holt et al. (1979) led in fact to the first reliable coronal temperature determination and clearly pointed out the need for - a presumably magnetic - confinement of stellar coronae.

Proportional counters provide even less energy resolution than solid state spectrometers ($\Delta E_{keV} \sim 0.4 E_{keV}$), and consequently only very strong, spectrally isolated line features can be detected. The 6.7 keV "iron line", in reality an extremely complex line blend (cf., section 3), has been detected in the RS CVn system UX Ari with both the EXOSAT ME (Pasquini, Schmitt and Pallivicini 1988) and GINGA LAC (Tsuru et al. 1988) proportional counters; however, at lower energies proportional counter spectra of coronal sources are essentially featureless and any interpretation relies on the use of theoretical spectral models. Most of the available data on coronal X-ray sources has been taken with proportional counters, i.e., the *Einstein Observatory* IPC which I will discuss in more detail in section 2.

A last method to obtain spectral resolution is the use of filter ratios, a method heavily employed with EXOSAT LE data (cf., Pallavicini et al. 1988). Here I shall not discuss filter ratios in any detail except for pointing out that the temperatures obtained from EXOSAT filter ratios are by and large consistent with the conclusions derived from *Einstein* data (see also Schmitt et al. 1988).

2. X-ray Spectroscopy throughout the HR-diagram

X-ray astronomers tend to think very positively about the notion of "spectral resolution" and do "spectroscopy" on data with almost any spectral resolution. From the point of view of an optical astronomer most of the X-ray data is color-photometric at best, very similar to UBV photometry at optical wavelengths. Since my task is to discuss the spectral properties of X-ray emission throughout the HR-diagram and since, as pointed out in section 1, the X-ray data with moderate or high spectral resolution are limited to less than a handful of coronal sources, such a spectral characterisation can only be attempted with the presently available low resolution data.

The physical interpretation of such low resolution X-ray data requires many assumptions on the form of the incident spectrum. Here I shall consider only spectra produced by plasma in collisional thermal equlibrium, although in many cases - especially in low SNR situations - other spectral forms can also produce acceptable fits. However, the higher resolution X-ray spectra obtained with the FPCS and the *Einstein* and EXOSAT transmission gratings clearly show emission lines and thus the thermal nature of the emitting plasma. At temperatures below $log\ T \sim 7$ most of the energy loss of such a plasma is carried by a few emission lines, which are however merged into a "pseudocontinuum" due to the low spectral resolution of the proportional counter.

Despite these limitations of proportional counter data Schmitt et al. (1989) carried out a survey of coronal temperatures of late-type stars observed with the *Einstein Observatory* IPC; they considered all X-ray sources detected with the IPC either as targets or serendipitously, identified with a bright ($m_V < 6.5$) and/or nearby star ($d < 25\ pc$) of spectral type F, G, K or M, containing more than 200 counts in its X-ray spectrum. Their sample was further subdivided into single stars and binaries as well into the various luminosity classes in order to study any dependence of X-ray temperature on these parameters. In the following I will describe some of the basic results obtained by Schmitt et al. (1989) in as much as they are relevant for future observations.

Results: One-component fits

In figure 1 I show - for acceptable one-component fits of single stars - the resulting X-ray temperature as a function of B-V color (open circles denote main sequence stars, filled squares subgiants, and filled circles giants). For M stars with B-V > 1 a large spread in coronal temperature is found, but as shown by Schmitt et al. (1989) these data points have low SNR. For stars with 0.2 < B-V < 1 there appears to be a discrepancy between main sequence stars and giants; main sequence stars are characterised by rather low temperatures between $2\ 10^6\ K$ and $5\ 10^6\ K$, quite reminiscent of coronal temperatures typically derived for the quiet Sun. On the other hand, the coronae of giants seem to be fairly hot, with temperatures in excess of 10^7 K, more reminescent of the flaring Sun. However, in most cases flaring is not the cause of the high temperature, rather the high temperature seems to be characteristic of the quiescent emission. Interestingly, this high temperature component does not appear to be accompanied by a corresponding low temperature component, as is the case for stars with B-V > 1. Therefore the coronae of giant stars appear to be quite isothermal (at least in our band pass) with extremely large temperatures, yet the transition from chromoshperic to coronal temperatures must be very abrupt with very little emission measure present at intermediate temperatures.

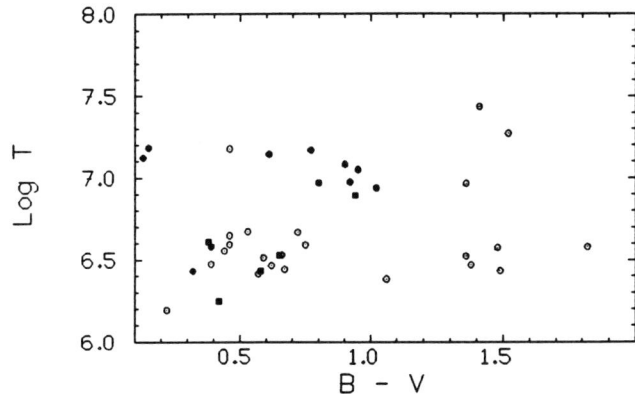

Fig.1: Coronal temperatures as a function of B-V color from single component fits for single stars; circles denote main sequence stars, filled squares subgiants and filled circles giants.

An obvious gap in the X-ray temperature distribution exists between $6.7 < log T < 6.9$, where absolutely no X-ray temperature measurements were found. From the spectral simulations by Schmitt et al. (1989) it is clear that if isothermal coronae with temperatures in the range $6.7 < log T < 6.9$ did exist, their temperatures could be determined with the IPC; therefore the gap must be due to both the intrinsic differential emission emission distribution of coronal X-ray sources and the spectral response of the IPC.

Results: Two-component fits

In figure 2 I plot - similarly to figure 1 for single stars only - the fitted two temperature components as a function of color for those cases where a one-component description turned out to be statistically unacceptable. From figure 2 it is immediately apparent that a two-component description is required only for main sequence stars redder than B-V $\sim$ 0.5; despite the availability of a few spectra of good SNR single giants never require two temperature components. The temperature components cluster around values of $log T \sim 7.2$ and $log T \sim 6.4$ with some dispersion; in most cases the errors in the temperature determination are so large that the deviations from these "standard" values are not significant. In any event, in many late type stars the appearance of a high temperature component is linked to a low temperature component, with the value of temperature showing little variation for IPC measurements. Schmitt et al. (1989) interpret this fact as evidence for the non-isothermality of stellar coronae; however, at the same time, the point out that derived spectral temperature components must not be taken as independent physical entities, but rather as "effective temperatures" in the sense discussed by

Underwood and McKenzie (1977).

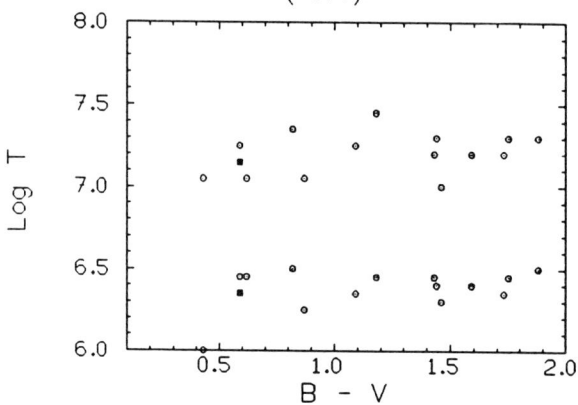

Fig.2: Coronal temperatures as a function of B-V color from two component fits for single stars; circles denote main sequence stars, filled squares subgiants and filled circles giants.

Results: Continuous emission measure distributions

If one accepts the idea that X-ray emission from late-type stars is produced by magnetically confined plasma, the differential emission measure distribution $Q(T)$ is approximately of the form $Q(T) \sim T^\alpha$, where the slope of the assumed emission measure distribution is related to the radiative cooling law. Spectral fits to such simply parameterised differential emission measure distribution functions involve three parameters: the power law slope α, the maximum temperature T_{max}, to which the emission measure distribution extends, and a normalisation constant. In figure 3 the results of fits assuming a continuous emission measure distribution are presented for single stars (symbols as in figure 1); the lower and upper panels show the fitted values of T_{max} and α as a function of color; note that the fit results are also displayed even if a one-component description provides an acceptable fit to the spectrum. In figure 3 much larger larger spread of temperatures than in the corresponding figure 1 is apparent. Thus figure 3 demonstrates the fact that the high temperature components found in a two-component fit do not need to reflect the largest possible temperature present in the stellar corona. Obviously, the temperatures resulting from these simple fits are model-dependent; however, they indicate, that the range of coronal temperatures encountered in stars may not be as narrow as a one- or two-component analysis seems to suggest.

Conclusions

Late-type main sequence stars (usually of spectral type K and M) always exhibit a high- and low-temperature component with the emission measure in the high temperature component being substantially larger than that in the low-temperature component. In essentially all cases there is evidence for plasma at temperatures in excess of 10^7 K. Most main sequence stars of earlier type (i.e., spectral type F or G) do not show this high-temperature component and are dominated by plasma at lower temperature; in fact, from the X-ray point of view they appear very much like the Sun in full disk observations outside of strong flares. The observed high-temperature plasmas do not appear to be directly associated with flares, i.e., they are found during periods of "quiescent" emission with no apparent variability present. Yellow giants on the other hand do show a high-temperature component which is not accompanied by low-temperature emission. The lack of this soft component is an intrinsic property of the spectrum; it is not caused by selection effects or absorption of the soft component by the interstellar medium. Two-temperature models do not provide the only possible way of obtaining acceptable fits (especially for M-type stars), rather the X-ray temperatures derived from low-resolution data such as the IPC data should be interpreted as effective temperatures but not necessarily as physical temperatures of the observed plasma.

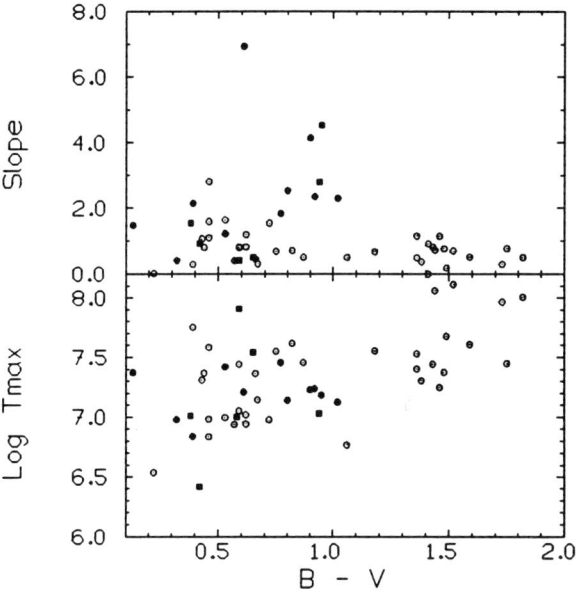

Fig.3: Maximum temperature T_{max} (lower panel) and power law slope α (upper panel) for continuous emission measure distribution fits.

3. Future observations: The case for XUV diagnostics

Why do we need plasma densities ?

Thermal X-ray spectra with their wealth of emission lines obviously call for high spectral resolution observations. For the study of coronal abundances, ionisation equlibria and temperature stratification one needs to identify and measure individual lines, for the study of coronal dynamics and Doppler imaging one even needs to resolve lines and measure their shapes. Instead of discussing the observational requirements for such measurements, I shall single out one specific item which - in my opinion - is of utmost importance for stellar X-ray astronomy, i.e., spectroscopic determination of plasma density.

The need for such density diagnostics is almost self-evident: Emission measure analysis - even when coupled with loop scaling laws - only constrains the product of density and filling factor. As a consequence, large tenuous coronal structures have the same spectral appearance as rather dense and compact structures. Using the solar analogy one would expect the latter to be the paradigm of stellar X-ray emission, while eclipse measurements of RS CVn binaries (Walter, Gibson and

Basri 1983; White *et al.* 1989) seem to indicate the presence of rather hot plasma on length scales comparable to the binary separation rather than a stellar radius, in apparent conflict with the solar analogy.

The same dichotomy between tenuous and compact structures applies to stellar flares. While modelling of many stellar flare events observed with the *Einstein* and EXOSAT satellites indicates relatively small structures with characteristic length scales of a stellar radius or less, some observations, most notably those of Stern, Underwood and Antiochos (1982) on the Hyades star HD27130 and by Tsuru *et al.* (1988) on the RS CVn system UX Ari, seem to indicate extremely large structures if one accepts that the observed decay times of these events have any relationship with the inferred radiative cooling time scales. Again, high resolution spectral diagnostics would provide a direct measurement of density and thus solve the problem.

Density diagnostics: Helium-like lines

The "classic" way to determine densities in a hot plasma is to measure the ratio of intercombination and forbidden lines in the triplets of helium-like ions (cf., Gabriel and Jordan 1969): in a high density plasma helium-like ions in the $1s2s\ ^3S_1$ state are de-excited collisionally and therefore the forbidden line disappears. By choosing a suitable ion one obtains density diagnostics at a variety of temperatures. However, with increasing atomic number the critical density N_{crit}, i.e., the density at which the $1s2s\ ^3S_1$ state starts to become significantly depopulated by collisions, becomes higher and higher; for example, for Si XIII with a peak formation temperature of ~ 6.6, one finds $log\ N_{crit} \sim 13.6$, a density probably not reached in solar flares and certainly not reached under quiescent conditions. I therefore conclude that density diagnostics with helium-like triplets in active stars where - as demonstrated by Schmitt *et al.* (1989) - most of the emission measure is located at temperatures of $log\ T \sim 7.0$ or higher, does not appear to be promising unless one postulates significantly higher coronal densities and pressures. While somewhat unlikely this last possibility can of course not be ruled out: active stars may have larger coronal magnetic fields, larger confining pressures and therefore higher densities than the solar corona.

Spectroscopy of Helium-like lines: What spectral resolution ?

In order to exclude any misunderstanding I wish to emphasize that I am **not** arguing that studies of the resonance lines of helium-like triplets are of no diagnostic use; on the contrary, these lines provide an extremely valuable plasma diagnostics for temperature, abundances and ionisation equilibrium in conjunction with the dielectronic recombination lines (satellite lines) present in such spectra. However, in order to extract the information contained in these lines extremely high resolving power is required as I will demonstrate momentarily. On SMM two crystals tuned to the resonance lines of Ca XIX at 3.18 Å and FeXXV at 1.85 Å were flown; in both cases the resolving power was of the order $\lambda/\Delta\lambda \sim 10000$; examples of these SMM crystal spectra are presented by Bely-Dubau *et al.* (1982) and Lemen *et al.* (1984).

In order to demonstrate the effects of spectral resolution on our capability to interpret spectra of helium-like resonance lines I have performed the following numerical experiment. I took a spectrum of "the iron line" as observed with SMM (figure 4, upper panel); the line in fact consists of a blend of many lines of iron ions in ionisation stages Fe XXV - Fe XXII which are discussed and identified by Lemen et al. (1984). In the next step the spectral resolution of the SMM spectrum was artificially decreased by replacing a point in the original spectrum by the mean intensity over a broader and broader energy range. The six spectra shown in figure 4 have resolutions of $\sim 0.7\ eV$, the instrumental resolution, in the top panel, $\sim 2.1\ eV$, $\sim 5\ eV$, $\sim 10\ eV$, $\sim 30\ eV$ and $\sim 60\ eV$ in the bottom panel. While in going from 0.7 eV resolution to 2.1 eV resolution little information is lost, the spectral features most important for temperature diagnostics become blended at resolutions of 5 eV, and are only partially resolved at 10 eV, a resolution thought to be feasible for calorimeters. At a resolution of 30 eV most of the satellite lines are gone, and at 60 eV, a resolution that may be within reach of CCD cameras, they are virtually indistinguishable. I therefore conclude that studies of helium-like resoncance lines for high Z atoms require dispersive spectroscopy (unless the spectral resolution of calorimeters can be pushed to the theoretical limits).

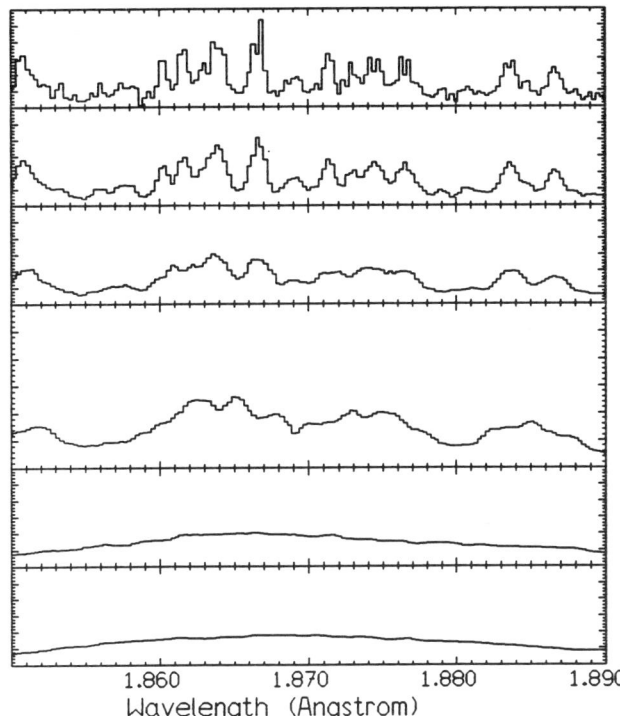

Fig.4: Effects of spectral resolution on 1.85 Å "iron line": top panel spectrum has (instrumental) resolution of $0.7\ eV$, the subsequent spectra have resolutions of $2.1\ eV$, $5\ eV$, $10\ eV$, $30\ eV$, and $60\ eV$ respectively.

Density diagnostics: Lithium-like to fluorine-like iron lines

For density diagnostics in active late-type stars one must observe lines formed at temperatures $logT \sim 7.0$; the only sufficiently abundant element not stripped off all but the K shell electrons is iron. Iron atoms in the ionsiation stages Fe XVIII (fluorine-like iron) to Fe XXIV (lithium-like iron) produce intense emission lines the waveband between 90 Å - 140 Å. The ground states of these ions belongs to the electron configuration $1s^2 2s^2 2p^k$ which splits up into a number of states; the energy separation between these states is typically of the order $\sim 10\ eV$. Various excitations from the ground state configuration are possible: for example, excitations into the $1s^2 2s^2 2p^{k-1} 3s$ configuration which require energies of $\sim 1\ keV$ and decay through electric dipole radiation; or excitations into the $1s^2 2s^1 2p^{k+1}$ configuration which require only $\sim 100\ eV$. Transitions from the latter configuration into the ground state configuration are not permitted in LS coupling, however, because of state mixing in high Z atoms, they do have significant transition probabilities.

Thus two types of transitions can occur: allowed transitions requiring high excitation energies and "forbidden" transitions, which can be excited by a much larger fraction of the electrons in the thermal pool. Which type of transition leads to higher photon fluxes cannot be determined a priori; however, an inspection of the line fluxes (in table 1) produced by a plasma in collisonal equilibrium as calculated by Mewe, Gronenschild and van den Oord (1985) shows that the photon fluxes in XUV lines exceed those in the corresponding X-ray lines by factors of 4 - 60; in table 1 I have compiled for the various ionisation stages of iron the peak formation temperature, the wavelengths of the strongest XUV and X-ray line, and the photon flux ratios of these lines in the low density limit.

Ionisation stage	T_{max}	λ_{XUV} Å	λ_{X-ray} Å	XUV/X-ray flux ratio
Fe XVIII	6.7	93.93	14.22	4
Fe XIX	6.8	108.37	13.52	20
Fe XX	6.9	132.85	10.09	19
Fe XXI	7.0	128.73	12.29	10
Fe XXII	7.0	135.78	11.77	20
Fe XXIII	7.1	132.84	11.74	65
Fe XXIV	7.2	192.03	10.64	58

Table 1: Comparison between XUV and X-ray line fluxes

The above transitions can be naturally utilised for density diagnostics. While in a low density plasma only the ground state is populated, the level populations of the ground state configuration must approach the Boltzmann limit as the density increases. Therefore, in a high density plasma lines from excited levels appear which are absent in a low density plasma; this is in contrast to the density diagnostics with helium like ions where one looks for the disappearance of the forbidden line. Since the ground state configuration usually splits up into a larger number of energy levels, one obtains density sensitive line ratios over a whole range of densities, typically starting at $\sim 10^{10}\ cm^{-3}$ up to $\sim 10^{15}\ cm^{-3}$.

Interstellar Absorption

A fundamental problem of XUV observations is of course interstellar absorption. While observations of even the nearest hot stars are difficult or impossible at XUV wavelengths I wish to point out that observations of the majority of cool stars of particular interest are not likely to be significantly affected by interstellar absorption.

For a quantitative discussion I consider an XUV line at wavelength λ and a fictitious X-ray line at 1 keV; unabsorbed, i.e., for $N_H = 0$ both lines are assumed to have a photon flux ratio r_0. Because of the exponential character of interstellar absorption the long wavelength parts of the spectrum are far more affected than the short wavelength parts, and therefore r will decrease with increasing N_H. In table 2 I present those distances out to which $r > 1$, i.e., where the photon flux in the XUV line is at least that of the X-ray line; for the conversion from N_H to distance a mean interstellar density of $n = 0.007\ cm^{-3}$ was assumed (Paresce 1984). While admittedly the assumption of a constant mean interstellar density is quite poor in view of the extreme degree of inhomogeneity of the interstellar medium, table 2 shows nevertheless that even for wavelengths $\lambda \sim 200 \text{Å}$ XUV lines produce larger apparent photon fluxes than the corresponding X-ray lines for the vast majority of the most easily observable cool stars. Therefore, for the purposes of observing cool stars, the finiteness of the XUV universe is of no great concern.

r_0	90 Å	110 Å	130 Å	170 Å	200 Å	240 Å
3	210 pc	120 pc	80 pc	40 pc	26 pc	16 pc
10	440 pc	260 pc	165 pc	80 pc	54 pc	34 pc
30	660 pc	380 pc	240 pc	120 pc	80 pc	50 pc
100	890 pc	520 pc	330 pc	160 pc	110 pc	68 pc

Table 2: Effects of absorption on XUV lines

Why not simply do it ?

For reasons unknown to me the wavelength range between 90 - 140 Å has been somewhat neglected observationally. On the Sun, this waveband range is only interesting during flares and the only available spectra (taken in 1969 with the OSO-5 satellite) seem to be those reported by Kastner *et al.* (1974). In the stellar case, the TG spectra of Capella and $\sigma\ CrB$ show the wealth of emission lines of highly excited iron in this wavelength band; by analogy we may expect to find similar spectra in all active late-type stars.

Laboratory measurements (Lawson and Peacock 1980) as well as theoretical studies in particular by H. Mason and collaborators have led to a complete listing and understanding of the term schemes and lines for the especially interesting iron lines in the fluorine to lithium isoelectronic sequences (Loulergue et al. 1984 for Fe XIX; Bhatia and Mason (1980) for Fe XX; Mason et al. 1979 for Fe XXI; Mason and Storey (1980) for Fe XXII; Bhatia and Mason (1981) for Fe XXIII; Hayes (1979) for Fe XXIV). The basic result coming out of this extensive effort is that the wavelength range between 90 - 140 Å is extremely well suited for temperature and density diagnostics of hot gas with temperatures of $\sim 10^7 K$; all the relevant lines and their atomic parameters are sufficiently accurately known. A concern is line blending; however, a spectral resolution of $\sim 0.1 \text{Å}$ or a resolving power of $\lambda/\Delta\lambda \sim 1000$ is generally thought to be sufficient to avoid ambiguities due to line blending.

It is clear that the spectroscopic goals, i.e., temperature and density diagnostics, can also be accomplished at higher energies at $\sim 1\ keV$; the observational requirements in terms of resolving power are the same. However, the XUV region appears advantageous for the following reasons: First, the same amount of emission measure will produce fewer X-ray photons than XUV photons; second, the XUV band contains also low temperature lines so that a wide temperature range - which we know to be present in cool stars - can be diagnosed at the same time; and third, a resolving power $\lambda/\Delta\lambda \sim 1000$ can - with currently available technology - only be obtained with crystals at energies of $1 keV$. While the proposed AXAF instrumentation (in particular the HETG) will be able to reach $\lambda/\Delta\lambda \sim 1000$, the same resolving power can be obtained at XUV wavelengths with present-day technology, i.e., the ROSAT mirror, a 1000 l/mm transmission grating and a curved micro-channel plate detector. It is therefore hoped that future missions with high resolution Wolter mirrors such as SPEKTROSAT (cf., Predehl 1989) and AXAF will be able to perform the required measurements.

References

Agrawal,P.C., Markert, T.H. and Riegler, G.R., 1985, *MNRAS*, **213**, 761.
Bely-Dubau, F. et al., 1982, *MNRAS*, **201**, 1155.
Bhatia, A.K. and Mason, H.E., 1980, *Astron. Ap.*, **83**, 380.
Bhatia, A.K. and Mason, H.E., 1981, *Astron. Ap.*, **103**, 324.
Cassinelli, J.P. and Swank, J.H., 1983, *Ap. J.*, , **271**, 681.
Gabriel, A.H. and Jordan, C., 1969, *MNRAS*, **145**, 241.
Hayes, M.A., 1979, *MNRAS*, **189** 55.
Holt, S.S. et al., 1979, *Ap. J. Lett.*, **234**, L65.
Kastner, S.O, Neupert, W.M. and Swartz, M., 1974, *Ap. J.*, **191**, 261.
Lawson, K.D. and Peacock, N.J., 1980, *J. Phys. B.: Atom. Molec. Phys.*, **13**, 3313.
Lemen, J.R., Phillips, K.J.H., Cowan, R.D., Hata, J. and Grant, I.P., 1984, *Astron. Ap.*, **135**, 313.
Lemen, J.R., Mewe, R., Schrijver, C.J. and Fludra, A. 1988 in preparation.
Linsky, J.L. 1985, *Solar Physics*, **100**, 333.
Loulergue, M., Mason, H.E., Nussbaumer, H. and Storey, P.J., 1985, *Astron. Ap.*, **150**, 246.

Mason, H.E. and Storey, P.J., *MNRAS*, **191**, 631.
Mason, H.E., Doschek, G.A., Feldman, U. and Bhatia, A.K., 1979, *Astron. Ap.*, **73**, 74.
Mewe, R. *et al.*, 1982, *Ap. J.*, **260**, 233.
Mewe, R., Gronenschild, E.H.B.M. and van den Oord, G,H.J., 1985, *Astron. Ap. Supp.*, **62**, 197.
Mewe, R., Schrijver, C.J., Lemen, J.R. and Bentley, R.D., 1986, *Adv. in Space Res.*, **6, No. 8**, 133.
Paerels, F.B.S., Bleeker, J.A.M., Brinkman, A.C., Gronenschild, E.H.B.M. and Heise, J., 1986, *Ap. J.*, **308**, 190.
Paerels, F.B.S., Bleeker, J.A.M., Brinkman, A.C., and Heise, J., 1986, *Ap. J.*, **308**, 190.
Paerels, F.B.S., Bleeker, J.A.M., Brinkman, A.C., and Heise, J., 1986, *Ap. J. Lett.*, **309**, L33.
Pallavicini, R., Monsignori-Fossi, B.C., Landini, M. and Schmitt, J.H.M.M., 1988, *Astron. Ap.*, **191**, 109.
Paresce, F., 1984, *Astron. J.*, **89**, 1022.
Pasquini, L., Schmitt, J.H.M.M. and Pallavicini, R., 1988, in **Activity in Cool Star Envelopes**, ed. O. Havnes, B.R. Pettersen, J.H.M.M. Schmitt, J.E. Solheim, 241.
Predehl, P., 1989, these proceedings
Raymond, J.C. and Smith, B.W. 1977, *Ap. J. Supp.*, **35**, 419.
Rosner, R., Golub, L. and Vaiana, G.S. 1985, *Ann. Rev. Astron. Astrophys.*, **23**, 413.
Schmitt, J.H.M.M., Pallavicini, R., Monsignori-Fossi, B.C. and Harnden, F.R., Jr., 1987, *Astron. Ap.*, **179**, 193.
Schmitt, J.H.M.M., 1988, in **Hot Thin Plasmas in Astrophysics**, ed. R. Pallavicini, NATO ASI Vol. 249, Kluwer Academic Publishers, 109.
Schmitt, J.H.M.M. *et al.*, 1989, in preparation.
Schrijver, C.J. 1985, *Space Sci. Rev.*, **40**, 3.
Stern, R.A., Underwood, J.H. and Antiochos, S.K., 1982, *Ap. J. Lett.*,
Swank, J.H., White, N.E., Holt, S.S. and Becker, R.H., 1981, *Ap. J.*, **246**, 208.
Swank, J.H. and Johnson, H.M., 1983, *Ap. J. Lett.*, **259**, L67.
Swank, J.H. 1984, in **The Origin of Nonradiative Heating/Momentum in Hot Stars**, ed. by A.B. Underhill and A.G. Michalitsianos, NASA Conference Publication 2358
Tsuru, T. *et al.*, 1988, in preparation.
Underwood, J.H. and McKenzie, D.L., 1977, *Sol. Phys.*, **53**, 417.
Vaiana, G.S. *et al.*, 1981, *Ap. J.*, **245**, 163.
Vedder, P.W. and Canizares, C.R., 1983, *Ap. J.*, **270**, 666.
Walter, F.M., Gibson, D.M. and Basri, G.S., 1983, *Ap. J.*, **267**, 665.
White, N.E. *et al.*, 1989, submitted.

SPECTROSCOPY OF STELLAR CORONAL SOURCES WITH THE MEDIUM ENERGY EXPERIMENT ON EXOSAT

R. Pallavicini[1], L. Pasquini[2], J.H.M.M. Schmitt[2] and G. Tagliaferri[3]

[1] Arcetri Astrophysical Observatory, Florence, Italy
[2] Max-Planck-Institut fur Extraterrestrische Physik, Garching, Germany
[3] ESTEC, EXOSAT Observatory, Noordwijk, The Netherlands

ABSTRACT. We summarize some of the results obtained by spectral analysis of EXOSAT ME observations of stellar coronal sources. We focus on time-resolved spectroscopy of stellar flares and on the determination of the temperature structure of quiescent RS CVn binaries

1. INTRODUCTION

The Medium Energy (ME) experiment on the EXOSAT Observatory provided pulse-height spectra of stellar X-ray sources over the range $\approx$ 1-10 KeV with a resolution $E/\Delta E \approx$ 3-5. This resolution does not allow individual lines to be resolved, but is already sufficient to determine the general shape of the spectrum and to search for the presence of the Iron line complex at 6.7 KeV. By fitting the observed spectrum with appropriate theoretical models, it is possible to derive information on physical parameters such as electron temperature and emission measure. Broad-band data over the spectral range 0.05-2 KeV were acquired simultaneously by the Low Energy (LE) experiment on EXOSAT in selected passbands. The LE data can be used in conjunction with the ME spectral data to constraint better the temperature stratification in the source. Details on the ME and LE experiments on EXOSAT can be found in White and Peacock (1988).

Stellar coronal data obtained with the EXOSAT ME include:

- quiescent emission from RS CVn and Algol-type binaries (about 15 sources);

- quiescent and/or flaring emission from a few other very active late-type stars (flare stars, PMS stars, etc; in total, about 5 sources);

- time-resolved spectra of flares in M dwarfs and RS CVn binaries (about 10 sources).

As can be seen from the list above, the limited sensitivity of the ME experiment allowed only a small number of stellar coronal sources to be studied spectroscopically at energies higher than $\approx$ 1 KeV. Nevertheless, these data are important because they give us a first glimpse of what could be obtained with higher sensitivity and higher spectral resolution; moreover, they have provided new interesting results which have increased appreciably our understanding of stellar coronal structure and variability. In this paper we will give some examples of the results that can be obtained from the analysis of ME data, focussing on time-resolved spectroscopy of stellar flares and on spectral analysis of quiescent emission from RS CVn binaries. Results of eclipse observations of binaries are reported by N. White elsewhere in this volume.

2. TIME-RESOLVED SPECTROSCOPY OF STELLAR FLARES.

The highly eccentric orbit of the EXOSAT Observatory allowed continuous monitoring of a source for periods as long as three days. This is particularly useful when studying time varying phenomena such as flares or when monitoring light modulations associated with orbital and/or rotational motions. Several flares were observed by EXOSAT that were sufficiently intense and long-lived to allow time-resolved spectroscopy during their evolution.

Probably the best example is the five hours-long flare observed from Algol on August 19, 1983, which was first studied by White et al. (1983). The flare started at $\approx$ 10:00 UT, peaked at $\approx$ 11:00 UT and than decayed steadily until $\approx$ 15:00 UT. We have reanalyzed this flare by subdividing it into five time intervals, one during the rise phase (from 10:00 to 11:00 UT), and the other four during the decay. ME spectra were accumulated over these five time intervals, after subtraction of a preflare quiescent spectrum. The emissivity model assumed was that of an optically-thin line+continuum thermal plasma as computed by Mewe, Gronenschild and van den Oord (1985). We find that the temperature was decreasing from a peak value of $(6.1 \pm 0.8) \times 10^7$ K to $(2.3 \pm 0.6) \times 10^7$ K in the late decay. During the same time interval the volume emission measure decreased from 6.7×10^{53} cm^{-3} to 1.1×10^{53} cm^{-3}. The average temperature during the rise phase (from 10:00 to 11:00 UT) was $(5.2 \pm 0.8) \times 10^7$ K and the emission measure was 3.0×10^{53} cm^{-3}. Although the count rate was not sufficiently high to resolve the temperature variations during the rise phase, these data already indicate that the plasma was first heating and then cooling during the flare evolution. The total energy released over the spectral band 0.1-10 KeV was 1.1×10^{35} ergs.

Another flare for which time-resolved spectra could be obtained throughout the flare was observed from the M dwarf flare star EQ Peg on August 6, 1985. Only the peak and the decay phase were observed in this case, from $\approx$ 06:30 UT, when the flare reached its maximum intensity, to $\approx$ 09:00 UT, when the emission returned to the quiescent level. Over this time interval, five different spectra could be accumulated that were fitted with a simple bremsstrahlung spectrum + a Fe line at 6.7 KeV. This model spectrum provided a good fit to the data, with a reduced χ^2 which was in all cases less than 1. Except for the last spectrum, obtained during the very late decay, the presence of an Fe line complex at 6.7 KeV appears necessary to fit the data satisfactorily. The same line complex was also prominent in the spectra of the Algol flare discussed above. During the decay of the EQ Peg flare, the temperature was found to decrease from $(4.4 \pm 0.4) \times 10^7$ K to $(1.7 \pm 0.4) \times 10^7$ K and the emission measure decreased from 2.1×10^{53} cm^{-3} to 3.0×10^{52} cm^{-3}. The total energy released by the flare during the observed period over the spectral band 0.1-10 KeV was $\approx 5 \times 10^{33}$ ergs.

There are only a few other flares observed by EXOSAT from M dwarf stars that could be analyzed spectrally at different times during their evolution. The best case is another long-lived event observed from the M dwarf eclipsing binary YY Gem on Nov 14, 1983. The temperature appeared to decrease slightly from 3.3×10^7 K at the flare peak to 2.4×10^7 K in the late decay, while the emission measure decreased over the same time interval from 1.8×10^{53} cm^{-3} to 4.5×10^{52} cm^{-3}. In all other cases only one or two spectra could be accumulated over the duration of the flare. The derived temperatures are in the range $\approx 2 \times 10^7$ K and $\approx 4 \times 10^7$ K, similar to typical temperatures observed in X-ray flares on the Sun. We stress, however, that the data sample is quite small: in addition to the two flares mentioned above (from EQ Peg and YY Gem), there were only three other flares from M dwarf stars for which the ME experiment could detect enough photons to allow reliable spectral fits: they were observed from the stars Wolf 630 (on Aug 25, 1985), AT Mic (on May 25, 1985) and again from EQ Peg (on Aug 29, 1984). Details of spectral fits of these flares will be given elsewhere (Pallavicini, Stella and Tagliaferri 1988).

3. QUIESCENT EMISSION FROM RS CVN STARS

A class of stellar sources which was more easy to observe with the EXOSAT ME is constituted by the RS CVn binaries. For these sources, it was possible to observe not only flares, but also quiescent emission. The count rates for the closest and brightest sources were high enough to allow meaningful spectral fits to be performed (this is usually not the case for quiescent emission of M dwarfs, which were either not detected by the ME during quiescent periods or were too faint to allow spectral analysis).

For each RS CVn star observed by EXOSAT we have analyzed at least one ME spectrum (Pasquini, Schmitt and Pallavicini 1988). The LE data obtained simultaneously in one or more soft X-ray filters were used in conjunction with the ME spectral data in order to determine the temperature stratification in the source. The data were fitted using a Raymond and Smith (1977) line + continuum thermal model and a fixed interstellar hydrogen column density (estimated for each source from its distance and the average value of the hydrogen density in the solar neighborhood given by Paresce 1984). We used one- and two-temperature thermal models and we evaluated the goodness of the fit by the usual χ^2 technique. The results (which refer to a sample of fifteen sources) can be summarized as follows. For spectra of sufficiently high S/N ratio, a two-temperature model is required to simultaneously fit the ME and LE data. The derived temperatures cluster around two well defined values, the lower one at $\approx 6\text{-}8 \times 10^6$ K, and the upper one at about $2\text{-}3 \times 10^7$ K. Only spectra with a low S/N ratio could be fitted satisfactorily with a one-temperature model.

The above results are similar to those obtained previously with the SSS and IPC experiments on board the EINSTEIN Observatory (Swank et al. 1981, Majer et al. 1986). However, as discussed by Majer et al. (1986) and Schmitt et al. (1988), an acceptable fit with a two-temperature model does not necessarily imply the existence in the source of two physically distinct regions in two different temperature regimes. MonteCarlo simulations of EINSTEIN IPC and EXOSAT ME spectral data (Pasquini, Schmitt and Pallavicini 1988) show in fact that acceptable two-temperature fits could also be obtained with a continuous emission measure distribution, i.e. with the presence of plasma at all temperatures from less than 10^6 K to several times 10^7 K. Such a continuous emission measure distribution could result from the presence of magnetically confined structures in the coronae of these stars.

In order to investigate whether the coronae of RS CVn binaries could be interpreted as constituted by a single family of coronal loops of the solar-type (i.e. in hydrostatic equilibrium and energy balance) we have fitted all spectra with a continuous emission measure distribution of the simple form EM $\sim (T/T_M)^\alpha$, where T_M is the maximum temperature in the loop and α is the slope of the distribution which in the solar case is expected to be ≈ 1. We found that this simple model is capable of fitting some of our RS CVn data, but not all of them. For instance, while we were able to fit with a continuous emission measure distribution the quiescent LE + ME spectrum of UX Ari, AR Lac, II Peg and Algol, no acceptable fit could be obtained for Capella, λ And, σ CrB and HR 1099. Moreover, even for those sources that were fitted with a continuous emission measure distribution, the derived α values were usually larger than expected from simple models of solar coronal loops.

The above findings indicate that substantial differences are likely to exist between the coronae of RS CVn stars and the coronae of the Sun and other single late-type stars. The results of our spectral fitting analysis suggest that either the coronae of most RS CVn stars involve more than one family of loops (as also suggested by eclipse observations, cf. White et al. 1988) or the coronal structures in these stars have a more complex emission measure distribution than the simple power-law assumed by us.

4. OTHER CORONAL SOURCES

The EXOSAT data base comprises a few other coronal sources for which spectral data could be obtained with the ME experiment. One of the most interesting observations is a flare from the A-type visual binary Castor (Pallavicini et al. 1988). This source has a nearby M dwarf flare star companion (YY Gem) and the two souces cannot be spatially resolved by the ME experiment. However, by comparing the ME light curve with the separate light curves of YY Gem and Castor obtained simultaneously by the LE experiment, it is possible to attribute unambiguously the observed flare events to either YY Gem or Castor. A flare from the latter source was detected on Nov 14, 1984 at 03:40 UT. Spectral analysis of the flare was done over three distinct time intervals throughout the flare evolution (peak, early decay, late decay). The results of the spectral fits, made using the optically-thin line+continuum thermal model of Mewe, Gronenschild and van den Oord (1985), shows that the temperature decreased from $\approx 50 \times 10^6$ K at the flare peak to $\approx 30 \times 10^6$ K in the late decay, while the emission measure decreased from 5.0×10^{53} cm^{-3} to 1.2×10^{53} cm^{-3}. The total energy released by the flare over the spectral band 0.1-10 KeV was 4.3×10^{33} ergs. The time scales and energy involved as well as the values of the derived parameters are similar to those typically observed from M dwarf flare stars. As discussed by Pallavicini et al. (1988) this suggests the possibility that the flare originated not from the A-type primary components of the Castor system, but rather from an unseen low-mass companion.

Another interesting case is a serendipitous stellar source that was observed to flare in the LE and ME detectors while EXOSAT was pointing to the Seyfert galaxy III ZW 2 (Tagliaferri et al. 1988). Since the source was quite faint, only one ME spectrum could be accumulated during the flare, yielding an average temperature of $\approx 4.5 \times 10^6$ K and an emission measure of 1.8×10^{54} cm^{-3}. The total energy released by the flare in X-rays was 1.9×10^{35} ergs, much larger than for typical flares observed in M dwarf stars and comparable to the total energy released by the flare on Algol. Tagliaferri et al. (1988) have argued that the flare likely originated from the secondary component of the visual binary HD 560 (a system formed by a B9V primary and a G0Ve secondary). From optical studies HD 560 B has been identified as a post-T Tauri star. This optical identification is consistent with the high quiescent X-ray luminosity ($\approx 10^{30}$ erg s^{-1}) and strong flaring activity observed from this source.

REFERENCES

Majer, P., Schmitt, J.H.M.M., Golub, L., Harnden, F.R.Jr. and Rosner, R.: 1986, Ap. J. 300, 360.
Mewe, R., Gronenschild, E.H.B.M. and van den Oord, G.H.: 1985, Astr. Ap. Suppl. 62, 197.
Pallavicini, R., Stella, L. and Tagliaferri, G.: 1988, to be submitted to Astr. Ap.
Pallavicini, R., Tagliaferri, G., Pollock, A., Schmitt, J.H.M.M. and Rosso, C.: 1988, submitted to Astr. Ap.
Paresce, F.: 1984, Astr. J. 89, 1022.
Pasquini, L., Schmitt, J.H.M.M. and Pallavicini, R.: 1988, submitted to Astr. Ap.
Raymond, J.C. and Smith, B.W.: 1977, Ap. J. Suppl. 35, 419.
Schmitt, J.H.M.M., Pallavicini, R., Monsignori-Fossi, B.C. and Harnden, F.R.Jr.: 1987, Astr. Ap. 179, 197.
Swank, J.H., White, N.E., Holt, S.S. and Becker, R.H.: 1981, Ap. J. 246, 208.
Tagliaferri, G., Giommi, P., Angelini, L., Osborne, J.P. and Pallavicini, R.: 1988, Ap. J. Letters 331, L113.
White, N.E. et al. : 1988, this volume.
White, N.E., Culhane, J.L., Parmar, A.N., Kellet, B.J., Kahn, S., van den Oord, G.H.J. and Kuipers, J.: 1986, Ap. J. 301, 262.
White, N.E. and Peacock, A.: 1988, in X-ray Astronomy with EXOSAT (R. Pallavicini and N.E. White eds.), in press.

X-RAY SPECTRAL SYNTHESIS IN HYDRODYNAMIC FLARE MODELS

S. Serio[1,2], E. Antonucci[3], M.A. Dodero[3], G. Peres[1], and F. Reale[1]

[1]Astronomical Observatory, 90134 Palermo, Italy
[2]Inst. of Interdisciplinary Applications of Physics, 90123 Palermo, Italy
[3]Inst. of General Physics, University of Turin, 10125 Turin, Italy

ABSTRACT. Compact solar flares are triggered by sudden energy release in magnetically confined plasma. This class of flares is well suited to be studied with numerical hydrodynamic models. In particular, one can compare the evolution of observed and synthetic X-ray spectra, computed under various assumptions for the mechanism of impulsive energy deposition, to constrain theoretical models and their parameter space. We discuss recent results on solar flares along this line, non thermal to models of energy depositions by relativistic electron beams. We shall also discuss possible applications of X-ray spectral synthesis to stellar flares.

1. INTRODUCTION

Although the mechanism triggering solar and stellar flares has not yet been clearly identified, it is generally thought that the sudden energy release powering the flare is the result of the violent transformation of energy stored in the magnetic fields shaping the corona. The fact that the initial phase of most solar flares is characterized by emission of hard X-rays has led to the view that the transformation of magnetic into thermal energy can be mediated by beams of relativistic electrons accelerated in a magnetic instability region high in the corona, and impinging on the denser downlying plasma.

Compact solar flares, in which the magnetic topology of the corona does not change appreciably during the flare, are probably best suited to study the detailed flare mechanism. In these flares, in fact, the evolution of the physical plasma parameters is only determined by the impulsive flare mechanism and by the thermodynamics and hydrodynamics of the coronal plasma, and not affected by the hardly observable changes in the detailed topology of the flaring region occurring, for example, in two-ribbon flares. Hydrodynamic numerical models of compact flares can then be built to compare the evolution of observable quantities, with the aim of evidencing observational signatures of different models for the mechanism of impulsive energy release. In these models the corona is described as a rigidly confined single plasma loop.

Such hydrodynamic codes have been applied to model the evolution of compact solar flares (Peres *et al.*, 1987), and of a flare on Prox Cen (Reale *et al.*, 1988). In this last case the hydrodynamic approach allows to constrain the unresolvable dimensions of the flaring region.

2. FLARE HYDRODYNAMICS

The bulk hydrodynamic evolution of a solar flare is essentially similar in different models for the impulsive energy transfer, although some subtle dependence on the model can be evidenced: a sudden energy release in the loop causes, at first, an increase in the coronal plasma temperature and pressure, followed, as soon as the perturbation reaches the lower atmospheric layers, by vigorous evaporation of chromospheric plasma, with velocity of several hundred $km\ s^{-1}$. The relative timing of temperature and velocity surges, as well as their detailed evolution, are instead model dependent. Figure 1 shows the computed evolution of effective Ca XIX temperature and velocity (i.e. averaged over emission in the Ca XIX line at $\lambda = 3.17\ \text{Å}$) for four different models whose parameters have been optimized to best fit the light curves of a flare observed by the *SMM* X-ray Polychromator (flare of Nov. 12 1980 at 17:00 UT; Peres *et al.*, 1987). The four models are characterized, respectively, by thermal energy deposition localized near the top of the loop (a) or near its footpoints (b), and by energy release through beams of electrons with spectral index $\delta = 8$ and low energy cutoff of 10 keV (c) and 25 keV (d).

From the inspection of Fig. 1 we can argue that the detailed diagnostic of flare temperatures and velocities offered by high resolution spectroscopy constitutes a valid tool for discriminating between different impulsive heating mechanisms, with a possible exception between cases a and c.

3. X-RAY SPECTRAL SYNTHESIS

With the aim of obtaining insight into the flare basic mechanism, we are in the process of computing synthetic spectra in the Ca XIX spectral region for different models, and comparing them to observed *SMM* BCS spectra. By comparing the evolution of the synthetic spectra with observations, using models relying on local deposition of thermal energy (e.g. models a and b above), Antonucci *et al.* (1987) have inferred that, in at least 40% of the flares observed by the *SMM* BCS, energy deposition is not occurring near the top of the loops, but more probably near its footpoints. A similar comparison is being carried on for flare models in which energy deposition is provided by electron beams with different amplitude and low energy cutoffs (e.g. models c and d above; Dodero *et al.*, 1988).

Figure 2 shows the computed spectra, integrated over the initial 40 s of evolution, for models a, b, c above. We can easily see that the model relying on local heating near the footpoints (model b) shows a remarkable difference from the others. In cases a and c, in fact, the blue shifted component of the principal component is predominant over the "stationary" component. Since in most observed flares the blue shifted component is typically smaller than the stationary one, we can infer again that localized base heating gives a better description of the observations.

The above result has to be taken with some caution, however, because the analysis is not complete. The next step will address models using electron beams with higher low energy cutoff (e.g. model d above). These models differ from model c because the electrons, being more penetrating, tend to heat more the lower part of the atmosphere (at least before substantial evaporation takes place); therefore we expect that they might behave closer to model b than to model c.

4. STELLAR FLARES

Although observed stellar flares are usually much stronger than compact solar flares, and their energy budget appears, therefore, to be more likely comparable to two-ribbon flares, hydrodynamic modeling has been satisfactorily used to compute observed *Einstein* light curves of a flare on Prox Cen (Reale *et al.*, 1988). Moreover, van den Oord, Mewe and Brinkman (1988) give evidence of a compact flare in the RS CVn binary σ^2 CrB.

Studying stellar flares with hydrodynamic models is appealing because, as shown by Reale *et al.* (1988), it does allow to constrain the *geometry* of the flare, through the dependence of plasma cooling time on its volume. In addition, the exploration of the full range of physical parameters characterizing the coronae of different stars such as solar-like ones, dMe stars and RS CVn systems, can in principle improve our knowledge of the initial phase of the flare.

In order to achieve this goal one might want to use high power spectroscopy, as in *SMM*; however, because of sensitivity, this might be helpful only for a limited number of cases. We wish to point out that even moderate power spectroscopy can give useful diagnostic for stellar flares. To show the potential for moderate X-ray spectroscopy, we draw in figure 3 the effective plasma velocity computed by means of the model of the Prox Cen flare of Reale *et al.* (1988), for two components of the flaring plasma characterized by temperatures above and below 1 keV, respectively. The different behaviour of the two components makes it apparent that by providing a disgnostic over a wide range of temperatures, such as can be given by moderate power spectroscopy, one can fruitfully constrain stellar hydrodynamic flare models.

ACKNOWLEDGMENTS

This work is supported by Ministero della Pubblica Istruzione, Piano Spaziale Nazionale, and Comitato Regionale Ricerche Nucleari e di Struttura della Materia.

REFERENCES

Antonucci, E., Dodero, M. A., Peres, G., Serio, S. and Rosner, R., 1987, *Ap. J.*, **322**, 522.

Dodero, M.A., Antonucci, E., Peres, G., Serio, S., and Reale, F., 1988, in preparation.

Peres, G., Reale, F., Serio, S., and Pallavicini, R., 1987, *Ap. J.*, **312**, 895.

Reale, F., Peres, G., Serio, S., Rosner, R., and Schmitt J.H.M.M., 1988, *Ap. J.*, **328**, 256.

van den Oord, G. H. I., Mewe, R., and Brinkman, A. C., 1988, *Astron. Ap.*, in press.

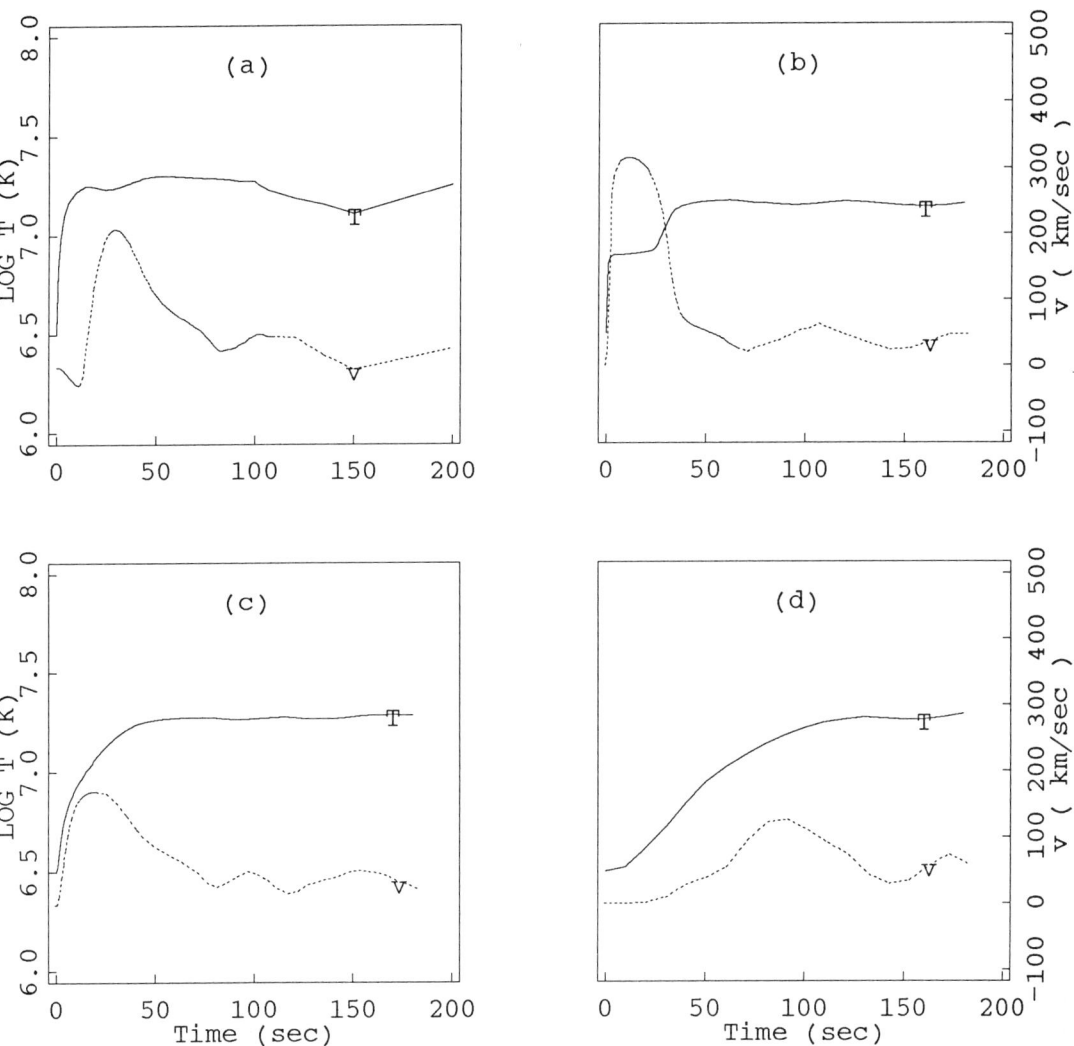

Figure 1 - Evolution of effective Ca XIX temperature and velocity computed for the solar flare of Nov 12 1980 at 17:00 UT (Peres *et al.*, 1987) using two models of localized impulsive energy deposition (a: at the top of the loop; b: near its footpoints) and two models of electron beam heating with spectral index $\delta = 8$ (c: low energy cutoff 10 keV; d: 25 keV).

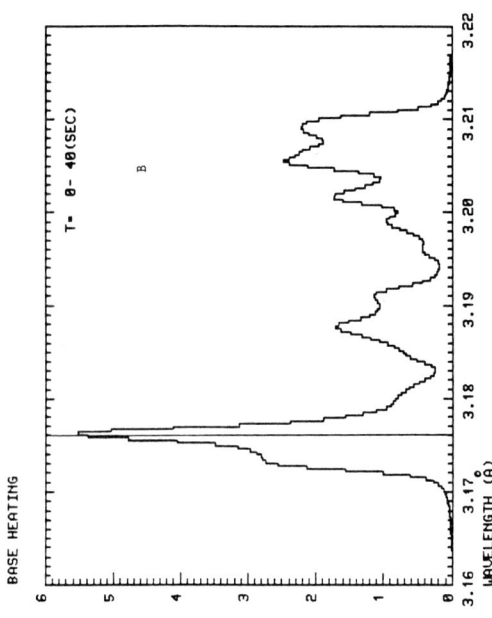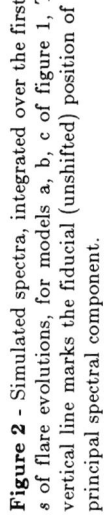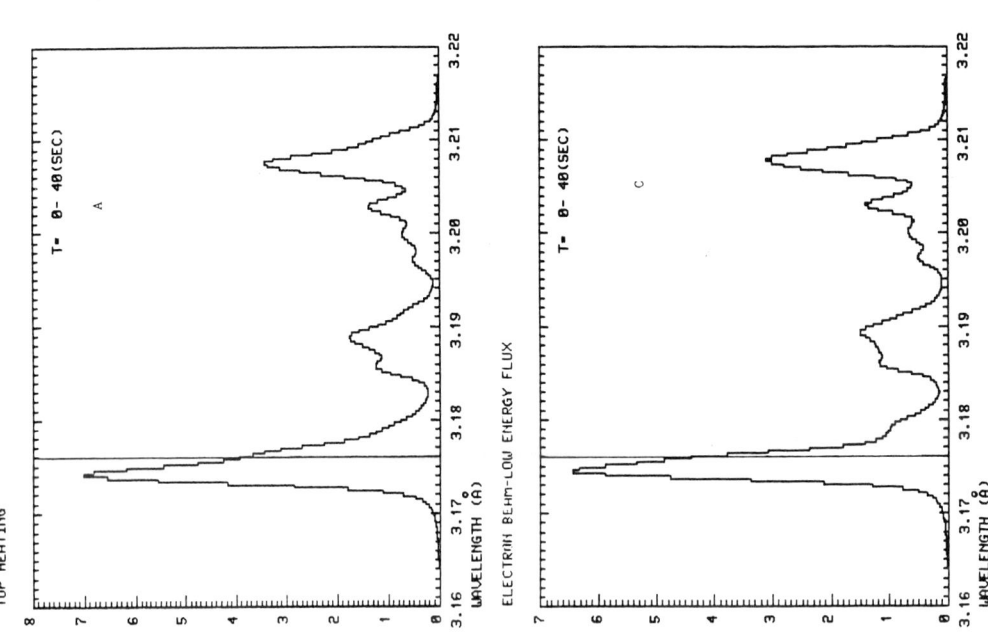

Figure 2 - Simulated spectra, integrated over the first 40 s of flare evolutions, for models a, b, c of figure 1, The vertical line marks the fiducial (unshifted) position of the principal spectral component.

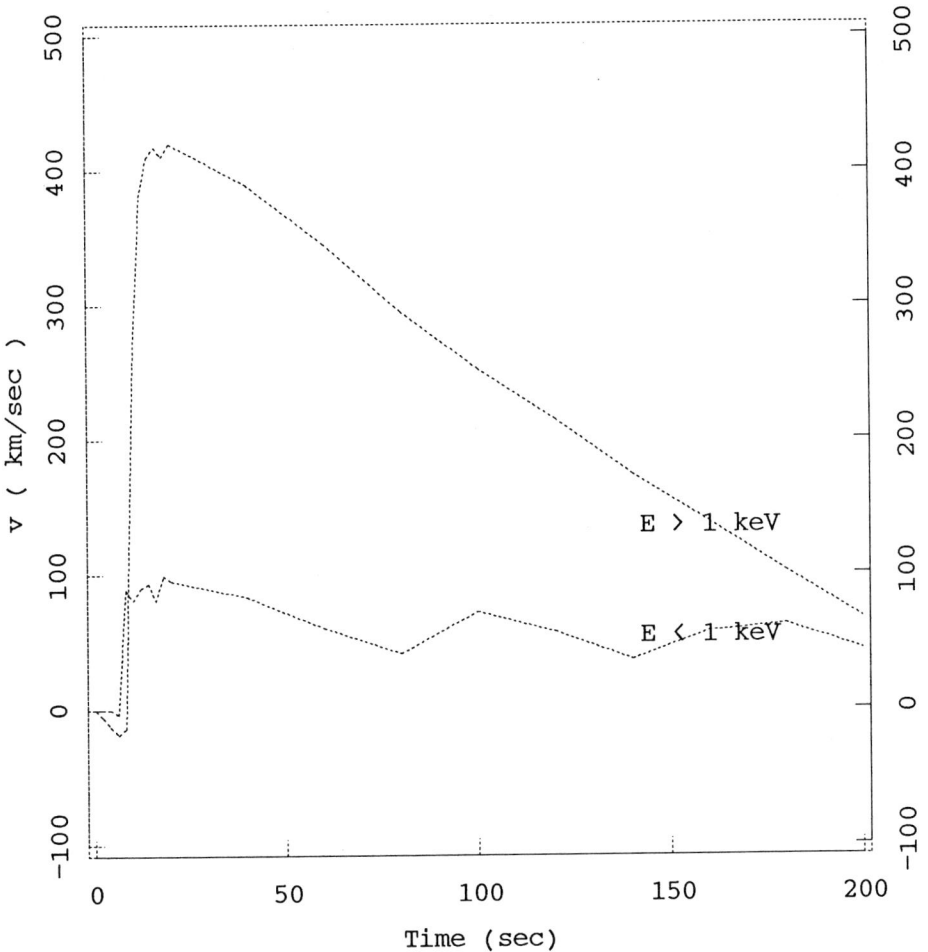

Figure 3 - Evolution of effective velocity in the Prox Cen flare model by Reale *et al.* (1988) for plasma components cooler and warmer than 1 *keV*, respectively.

X-RAY SPECTRAL DIAGNOSTICS FOR CORONAL LOOPS IN THE ACTIVE K DWARF AB DORADUS

O. Vilhu[1] and A. Collier Cameron[2]

[1] Observatory and Astrophysics Laboratory, University of Helsinki, Tähtitorninmäki, SF-00130 Helsinki, Finland

[2] Astronomy Centre, University of Sussex, Falmer, England BN1 9QH

ABSTRACT We discuss theoretical X-ray spectra for coronal loop models in the rapidly rotating young K dwarf AB Doradus (HD 36705), as a typical representative of active X-ray bright stars. The loop models are based on EXOSAT and IUE observations, and further motivated by a possible connection between the observed X-ray flares and co-rotating clouds of neutral hydrogen (a few/day). The resulting synthetic spectra between 0.5 - 7 keV can be approximated by a linear combination of three distinct temperature components. Two components are sufficient between 0.5 - 2.5 keV. Below 0.1 keV the loop spectra deviate significantly from the few component fits. To test some basic assumptions (dynamic vs. static, constant vs. variable cross-sectional area), useful constraints on the DEM(T)-distribution could be obtained with the grating-spectrometer of XMM with 10^3 - 10^4 sec exposure times. The ratio of the He-type (O VII) forbidden and intercombination lines at 0.56 keV will provide sufficient density diagnostics, to distinguish e.g. between compact and large loops. The crystal-spectrometers of XMM and XSPECT could achieve the same but with longer (10^5 sec) observing time. The strong Ly α line of O VIII at 0.65 keV can be observed with the crystals in 10^4 sec, and used even for rotational modulation and flare studies, and giving additional information about flows in flaring loops. At the Iron 6.7 keV lines, where the gratings do not work, the crystals should be used together with low resolution devices, to set constraints on the hottest gas at loop summits. Our discussion applies also to several brighter cool stars but with shorter observing time (like Capella and HR 1099, which are over 3 times X-ray brighter than AB Dor).

1 Loop Models. X-ray Flares and Cool Clouds.

During the last years a picture has emerged in which stellar late type X-ray sources with the greatest X-ray to optical ratios tend to have high axial rotation rates. These are either young objects, associated with regions of star formation, or else are components of close binaries with tidally locked rotation. Their X-ray spectra are usually well fitted by one or two component thin plasma models. Apart from simple scaling laws, not much is known about the structure of the emitting regions, since the heating mechanism is not known.

A good example to study stellar magnetic activity at high levels is the young K dwarf AB Doradus (HD 36705, V=7, P=0.514d). It has been studied by e.g. Vilhu, Gustafsson and Edwardsson (1987), Collier Cameron et al.(1988), and Collier Cameron (1988). Details of the coronal loop models, based on EXOSAT observations, have been described in detail by Collier Cameron (1988). The models are constructed by solving the equations of hydrostatic equilibrium and energy balance along a static loop in a rotating reference frame, so that the effects of centrifugal forces are taken into account. The loop geometry is assumed to be that of a dipole field. The equations are solved subject to the boundary conditions that $Tn_e^2(dT/ds)^{-1} = 3.96 \ 10^{28}$ cm^{-5} in the C IV resonance lines, as derived from the IUE observations, and the total heating rate F(tot) is fixed at $1.6 \ 10^8$ erg cm^{-2} s^{-1}. F(tot) is close to the saturation value (see Vilhu, 1987).

The resulting thermally stable solutions have summit temperatures around $2\ 10^7$ K (compatible with the EXOSAT ME-fits) and summit heights of several stellar radii. For loops of this size the effect of centrifugal forces is important. An interesting scenario for the formation of the observed H α clouds and their connection to large X-ray loops have been developed by Collier Cameron and Robinson (1988a,b) and Collier Cameron (1988). If an X-ray loop on AB Dor grows larger with time, its summit will eventually extend beyond the Keplerian co-rotating radius (2.7 R_*), where the centrifugal acceleration along the field direction will balance the gravity. The plasma density will then increase outwards from this point to the summit, causing the radiative loss rate to increase. When the loss rate exceeds the heating, a temperature minimum develops at the loop summit, and the summit becomes thermally unstable.

If radiative collapse then occurs at the loop summit, we would expect to see material flowing upward along the loop (a few hundreds km/s), following the pressure collapse at the summit. The material in the unstable region will cool until hydrogen recombines, as in solar prominence formation. Actually new clouds of neutral hydrogen, trapped at fixed longitudes, are observed to form at a rate a few/ day. Three strong X-ray flares have been observed in AB Dor (Collier Cameron et al.1988;Pakull,1981;see also Vilhu and Linsky,1987) giving a similar frequency. The long risetimes of flares point to large flaring regions. Once the cloud has formed, it does not dissociate itself from the parent loop immediately. It moves radially outward (in the co-rotating reference frame) and appears to be stretching the confining loop outwards as it goes. We suspect that the cloud finally escapes from the stellar field when the field reconnects below the cloud. The resulting process might resemble a large 2-ribbon flare.

2 Synthetic X-ray Spectra

Each loop model solution specifies the run of temperature and electron density along the loop, whose global shape follows a line of dipole field aligned with the stellar rotation axis. Thermally stable models, with uniform cross-section, have a linear differential emission measure DEM(T) vs. T, and these were used to compute synthetic X-ray spectra over the energy range 0.05 - 7 keV, using the thin plasma code developed by Raymond (1987). The specific model, used in the present discussion, has base-to-summit length $L = 3.1\ 10^{11}$ cm, so that the summit lies at 3 R_* above the photosphere, and the footpoints occupy the entire surface of the star.

The spectra shown give the expected count rates at Earth for two spectral resolutions. These count rates can then be compared with the performances (5σ-detections) of the Grating and Crystal spectrometers in the Report of the Instrument Working Group of XMM (Briel et al.,1987), although the resolution of the crystal is still higher (comparable to the thermal broadenings of 10^7 K plasma: 0.3 eV at 0.57 keV and 2 eV at 6.7 keV). We assume that the OXS-spectrometer of the Danish XSPECT-telescope on board of the future Soviet observatory SPECTRUM-X will have a roughly similar performance than the XMM crystal (Schnopper,1987 and 1988).

Figures 1 - 3 show the synthetic spectra (lower panels) starting from the Oxygen lines at 0.56 keV ,through the Silicon and Sulphur lines (at 1.9 and 2.5 keV, respectively), up to the Iron XXV fluorescence Kα lines at 6.7 keV. A linear combination of three components with temperatures $10^{6.5}$ K, $10^{7.0}$ K and $10^{7.5}$ K, and with relative emission measures 0.4, 1.0 and 0.45, respectively, can fairly well fit the synthetic spectrum in the whole energy interval 0.5 - 6.8 keV. At very low energies (below 0.1 keV) the loop spectrum deviates significantly from the spectrum predicted by the combination of these three single temperatures . A 2T-fit is sufficient for the narrower energy range 0.5 - 2.5 keV, perhaps explaining the success of two component fits with the EINSTEIN and EXOSAT data. The reason is natural since the region between 0.5 keV - 2.5 keV contains lines formed at temperatures between 1 and 10 million Kelvin. The Iron 6.7 keV (He4, He5 and He6) fluorescence lines peak at $T = 10^{7.8}$ K, so that one more higher temperature (somewhat non-unique) must be added to the 2T-model (with EM = 0.45). In the energy interval 0.5 keV - 6.7 keV, even with a moderate resolution (E/dE = 100) one is able to derive useful constraints on loop models. It might not be possible to separate between different static models, but to obtain useful DEM(T)-data to distinguish static vs. dynamic loops and to set important limits for the amount of the hottest gas (loop summits) with the help of the Fe 6.7 keV lines. An important constraint on loop models can be set when a useful estimate of the plasma density becomes available. This is related to the size of the emitting volume; at present only estimates of the total emission measure can be made. The density n_e (about $3 \, 10^{10}$ cm^{-3} at the formation temperature $2 \, 10^6$ K of the Oxygen lines) is directly related to our basic assumpions that the filling factor of the loop footpoints equals to 1 (the whole stellar surface covered by footpoints), and that the CIV emission lines determine the differential emission measure at 10^5 K. These assumptions can be directly tested by observing the Oxygen (O VII) lines at 0.56 keV (Fig.4). The region contains resonance (W), intercombination (X) and forbidden (Z) lines, and in addition a weak OVI-line (Li). The ratio W/Li is sensitive to temperature, while the ratio X/Z is a useful density-indicator.

Figure 4 shows the effect when the loop densities are increased by factors 3 and 10 (the filling factors decreased by 3 and 10, respectively). The ratio X/Z changes by more than a factor of 2, so that reasonably good density -estimates can be expected with the XMM- grating, using 1000 - 10000 sec exposure times. With the crystals, 10^5 sec exposure times are needed. With much better resolution (0.3 eV), the crystals can provide additional important information about flow-velocities (0.5 eV), predicted by dynamic loop models (compare:the thermal broadening at O VII is 0.1 eV). For line profile studies the strong near-by O VIII line at 0.65 keV (19 A) can be observed with the crystals in 10^4 sec,and can be used even for rotational modulation and flare studies. Good density- and velocity-diagnostics can then be made for both the quiescent and flaring states of AB Dor, using one day observation with the crystal at 0.56-0.65 keV.

3 References

Briel U. et al. 1987, The High-Throughput X-Ray Spectroscopy Mission,ESA SP-1092. Collier Cameron A.et.al. 1988,MNRAS 231,131.
Collier Cameron A. 1988,MNRAS 233,235.
Collier Cameron A. and Robinson R.D. 1988a ,MNRAS, in press.
Collier Cameron A. and Robinson R.D. 1988b, in A.K. Dupree and M.T.V. Largo (eds),Formation and Structure of Low Mass Stars, (in press).
Pakull M.W. 1981, Astr.Ap.104,33.
Raymond J.C 1987, private comm.
Schnopper H.W. 1987, Lectures given at the NATO Advanced Study Institute, Cargese,
Schnopper H.W. 1988, X-ray Astronomy: A Status Report, an invited talk given at the Annual Meeting of the Finnish Physical Society,Arkhimedes 2/1988.
Vilhu O. 1987, in J.L.Linky and R.E.Stencel (eds.),Cool Stars, Stellar Systems and the Sun,Lecture Notes in Physics, Springer-Verlag,p.110.
Vilhu O., Gustafsson B. and Edvardsson B. 1987,Ap.J.320,850.
Vilhu O. and Linsky J.L. 1987,PASP 99,1071.

4 Figures

- Figs.1-3 Synthetic loop spectra of AB Dor, together with the few temperature fits.
- Fig.4 The loop spectrum around the O VII triplet at 0.56 keV. The dashed lines correspond to the increase of the density by factors of 3 and 10 (or decreasing the foot-point filling factor by the same amounts).

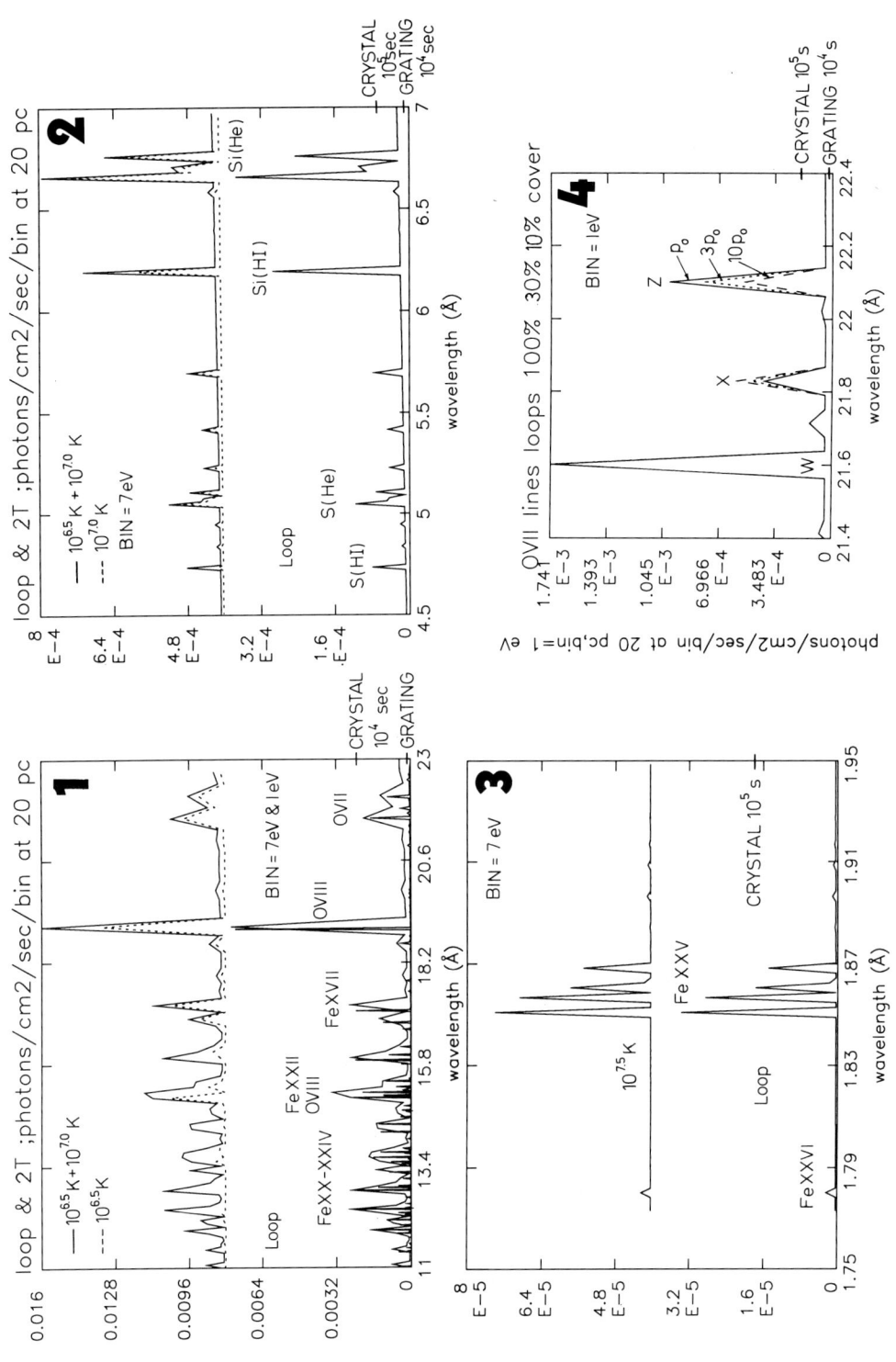

4. Supernova Remnants, Soft X-ray Background

HIGH RESOLUTION SPECTROSCOPY AND PLASMA DIAGNOSTICS OF SUPERNOVA REMNANTS

Claude R. Canizares
Department of Physics and Center for Space Research
Massachusetts Institute of Technology

Abstract. The MIT group has used data from the Focal Plane Crystal Spectrometer on the *Einstein* Observatory to perform plasma diagnostics of four supernova remnants (SNRs), the Cygnus Loop, Puppis A, N132D, and Cas A. The ratio of luminosities of the forbidden line to resonance line of He-like ions of oxygen and neon allow us to show that all four SNRs depart from ionization equilibrium in that they are under-ionized for their electron temperatures. Thus despite the fact that their ages range from 300 yr to 20,000 yr, all four SNRs are still ionizing and, in that sense, are still young. We derive values of ionization time and electron temperature for one or more components in each remnant. The agreement between these values and those deduced by others using entirely different means (e.g. broad-band spectroscopy or imaging) gives us confidence in the reliability of the diagnostics. Two of the SNRs, Puppis A and N132D, show evidence for an overabundance of oxygen by factors of three or more. These results, based on a handful of weak lines, show the great promise of the much more powerful future spectroscopy missions for revealing detailed information about astrophysical plasmas.

I. INTRODUCTION

It has long been traditional to classify supernova remnants (SNRs) according to their ages, because so many of their characteristics are age-dependent. For example, in young remnants it is the shocked ejecta that dominates the emission whereas in old remnants it is the swept up interstellar medium (ISM; e.g. see Hamilton's contribution to this volume). Here I present evidence showing that in some sense *all* SNRs are young. By this I mean that the plasma that dominates the emission, whether ejecta or ISM, is still ionizing and has not reached ionization equilibrium. This idea goes back to Gorenstein, Harnden, and Tucker (1974), Itoh (1977), and others, but the strongest evidence comes from detailed plasma diagnostics of the kind I describe here.

I present results from four SNRs in order of seniority: the Cygnus Loop, Puppis A, N132D and Cas A. All the data come from the Focal Plane Crystal Spectrometer (FPCS) on the *Einstein* Observatory (Giacconi et al. 1979). The high spectral resolution of this instrument allowed us, for the first time, to resolve individual emission lines from other lines and from the continuum.

II. HE-LIKE TRIPLETS AS DIAGNOSTICS OF DISEQUILIBRIUM

Before presenting the data, I will review briefly the use of the relative strengths of lines from He-like ions as diagnostics of departures from ionization equilibrium (see also Dr. Mason's contribution to this meeting). The first excited state of He-like ions can be either a singlet ($1s2p\ ^1P$) or triplet ($1s2p\ ^3P$ or $1s2s\ ^3S$), whereas the ground state is a singlet ($1s^2$ ^1S) (see for example Gabriel and Jordan 1969, Mewe and Schrijver 1978). Decays from the triplet excited state to the singlet ground state are second order transitions and are therefore slower than those from the singlet state, but in plasma at the low densities of SNRs they do eventually occur, giving rise to the forbidden line (from the ^3S state) and the intercombination line (from the ^3P state). Therefore, the relative strengths of these lines compared to that of the resonance line (from the permitted transition between the excited and ground singlet states) depends on the relative populations of the excited states.

For a plasma in ionization equilibrium, the excited states are populated by collisional excitation and by cascades following radiative recombination of the H-like ions. Collisions preferentially populate the singlet state whereas recombination preferentially populates the triplet states (because of their larger statistical weights). This gives rise to line ratios that are somewhat temperature sensitive (e.g. Pradhan 1983). In a plasma that is still ionizing, in other words one that is under-ionized for its electron temperature, the recombination rate is suppressed. This reduces the strengths of the forbidden and intercombination lines relative to the resonance line, giving us the desired diagnostic. The opposite case of an over-ionized, recombining plasma in which the forbidden line is enhanced has been observed in a laboratory plasma (Kallne et al. 1984). This case might be of interest in future studies of stellar flares, for example.

We have computed the expected line ratios as functions of electron temperature and ionization time using the non-equilibrium ionization code of Jack Hughes (Hughes and Helfand 1985) and also using our own calculation of the ionization fractions based on the data of Mewe, Gronenschild, and van den Oord (1985 and references therein) and the collision strengths for the He-like ions of Pradhan (1983 and references therein). Both these computations assume that the electrons are instantaneously heated to the given temperature. The ionization time τ is the product of electron density n_e and time t -- it is proportional to the number of electron collisions which governs the ionization, excitation and recombination processes. Comparing the measured line ratios to the calculated line ratios allows us to constrain the electron temperature T_e and τ. This assumes that the observations correspond to a reasonably homogeneous, isothermal region in the source, an assumption that I will explore further below. In some cases we have also used the computations of Hamilton, Sarazin and Chevalier (1983), which are for an inhomogeneous plasma whose emission measure distribution follows that expected for a plasma heated by an adiabatic Sedov blast wave. Hamilton and Sarazin (1984) have shown that with a modest rescaling of the parameters, these same computations are also valid for a variety of other SNR models (e.g., a reverse shock in the ejecta).

Recently Gabriel et al. (1988) have raised the concern that non-Maxwellian electron distributions could affect the relative ratios of the He-like triplet and disrupt the diagnostic. This is because very energetic electrons will preferentially excite the singlet state, resulting in a stronger resonance-to-forbidden line ratio even if the plasma is in ionization equilibrium. This is a serious objection but, as I will describe in more detail below, it is possible to test for the presence of such supra-thermal electrons. The consistency between our diagnostics and other determinations of the plasma parameters argues against the importance of this effect.

III. THE CYGNUS LOOP

The Cygnus Loop is a rather old remnant, with an estimated age of ~20,000 yr. and a radius of ~20pc at an assumed distance of 700 pc (Ku et al. 1984). The X-ray image shows a nearly circular limb-brightened shell that is several degrees across. We made a series of observations with the FPCS of a 3×30 arc min portion of a bright region in the north (see Vedder et al. 1986). Because of the limited surface brightness of this large diffuse source, we were only able to study three line complexes, the O VII and Ne IX triplets near 570 and 920 eV, respectively, and the O VIII Lyα and O VII 3p-1s lines around 650 eV (see Figure 1). However, even these few lines are sufficient to draw some interesting conclusions about the condition of the source. In particular, Figure 1 shows that the O VII forbidden line is much weaker than the resonance line. This is qualitatively what we expect for an under-ionized plasma.

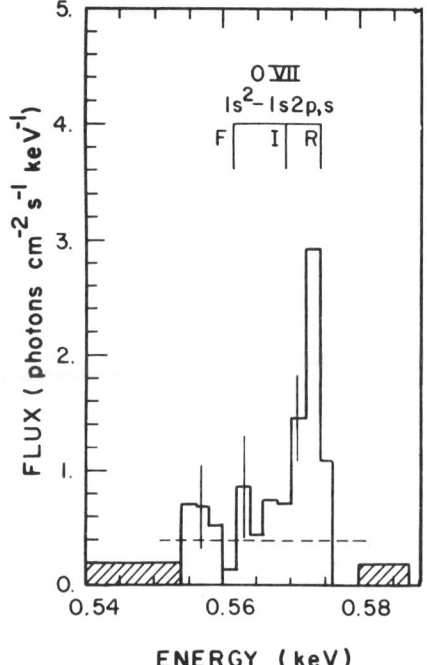

Figure 1. The He-like triplet of O VII from a portion of the Cygnus Loop as observed with the Einstein FPCS (from Vedder et al. 1986). The dashed line shows the level of the non-X-ray background (measured simultaneously in the FPCS imaging detector). The hatched regions of the spectrum were not observed. The energies of the forbidden (F), intercombination (I), and resonance (R) lines are indicated.

Figure 2 shows a plot of the forbidden-to-resonance line ratio vs. the O VIII to O VII density ratio. The grid comes from our computations. It shows the locus of models of impulsively heated plasma at various temperatures as a function of ionization time τ. The rectangle shows the three-sigma confidence region from the data of Figure 1 (the O VIII to O VII density ratio can be deduced from the ratio of the Lα line to the 3p-2s line around 650 eV). It is clear that ionization equilibrium, which corresponds to the upper right hand edge of the grid, is ruled out by the data. Formally we obtain $T_e > 3 \times 10^6$ K and $3{,}000 < \tau < 10{,}000$ yr. cm^{-3}. These values are consistent with those obtained using the grids of Hamilton, Sarazin and Chevalier (1983) for inhomogeneous plasma, so they are relatively model-independent.

Ku et al. (1984) used the continuum shape to measure a temperature of 3×10^6 K, just the value of our lower limit, and they deduced the age to be 18,000 yr. If we adopt their temperature, we find $4{,}000 < \tau < 8{,}000$ yr. cm^{-3}. With their age, this implies that the electron density $n_e \sim 0.2$ to 0.5 cm^{-3}, which is reasonably consistent with their estimate from the surface brightness of 0.65 cm^{-3}. Tsunemi et al. (1988) require a non-equilibrium model with two temperature components to fit the 0.1 - 3 keV spectrum from TENMA and a rocket flight. The parameters of their low temperature component are very similar to those just quoted. The high temperature ($\sim 8 \times 10^6$ K) component would not contribute to the O VII line flux. Furthermore, it is thought to come primarily from the interior of the Loop, outside our aperture.

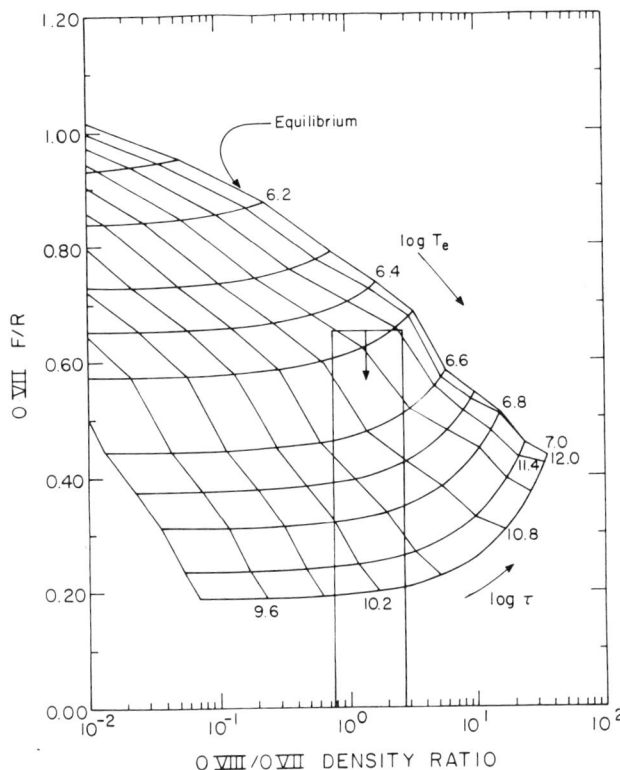

The consistency of our line diagnostics with analyses of broadband spectral and imaging data gives us confidence in the reliability of our conclusion, namely that even this rather aged remnant is in a youthful state of ionization.

Figure 2. *A diagnostic grid of T_e and τ in terms of observed line ratios (see text). The rectangle shows the region allowed by the observations of the Cygnus Loop (three-sigma; see Vedder et al. 1986).*

IV. PUPPIS A

Puppis A is between four and five times younger than the Cygnus Loop; its age is estimated to be ~4000 yr. (Winkler, Tuttle and Kirshner 1988). Puppis A is the brightest X-ray line source in the sky after the sun, and it was the object we studied most carefully with the FPCS. The locations of our aperture during the observations roughly correspond to the interior, the sharp shock front along the northeast, and a bright region in the east that is thought to be a shocked interstellar cloud (see Petre et al. 1981). Much of our data on Puppis A is already in the literature, but some of it is now being studied by Kathryn Fischbach and others in our group (e.g. see the contribution of Fischbach et al. to this meeting). Here I will focus on the question of ionization equilibrium as indicated by the relative strengths of the He-like triplet.

As in the Cygnus Loop, the spectra of Puppis A from all three regions show the large resonance-to-forbidden line ratios indicative of non equilibrium ionization (e.g. see Figure 3). In the case of the interior of Puppis A, we can compare our line strengths not only to the models but also to those seen in the solar corona. This is because the other oxygen line ratios are very similar in the two -- thus if Puppis A were at equilibrium it would have a temperature similar to that of the quiet corona ($2-3 \times 10^6$ K; Winkler et al.

1981a, 1981b, Canizares and Winkler 1981, Canizares *et al.* 1983). However, in the equilibrated corona the forbidden and resonance lines of O VII are nearly equal in strength, whereas in Puppis A their ratio is ~0.3.

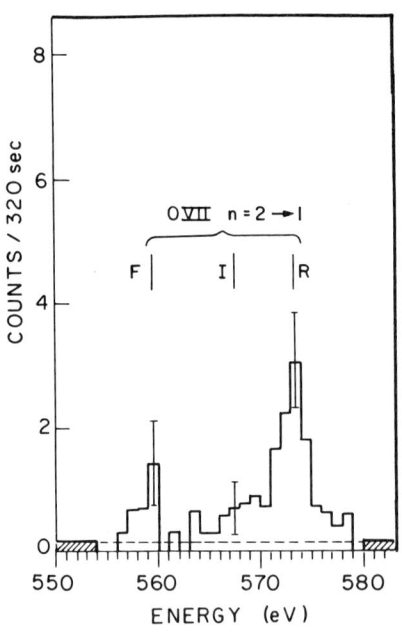

Figure 3. The He-like triplet of O VII from the interior of Puppis A as observed with the Einstein FPCS (from Winkler et al. 1981). See caption to Figure 1 for details.

Fischbach *et al.* (this volume) derive values of τ for the three regions of Puppis A using several line ratios, including the He-like triplet (see their Table 1). The results give a plausible scenario for the remnant: the interior emission comes predominantly from older plasma with densities of 0.1-0.6 cm^{-3}, the shock front is much younger and roughly an order of magnitude denser, and the bright eastern knot is comparably young but still denser by a factor of three or more.

We also have an independent consistency check on the validity of the He-like triplet diagnostic. For the interior of Puppis A, the O VIII Lα to Lβ line ratio taken together with two ratios of O VII to O VIII lines rule out ionization equilibrium without reference to the He-like triplet. This is shown in Figure 2b of Fischbach *et al.* (in this plot equilibrium corresponds to the region of τ vs. T_e space in which the dashed curves become horizontal). These results only depend on oxygen lines, so they are independent of any assumptions about relative elemental abundances.

This consistency between the He-like triplet diagnostic and the other line ratios implies that the forbidden-to-resonance line ratios are not strongly affected by suprathermal electrons (see Section II). A further argument against suprathermals comes from the broad-band spectrum of Puppis A. If several percent of the electrons in Puppis A had characteristic energies of 15 - 20 keV, which is what would be needed to affect significantly the He-like triplet (Gabriel *et al.* 1988), then they will emit *bremsstrahlung* and give rise to a high energy tail. I believe that such a tail would have been detected by experiments such as Ariel V (Zarnecki *et al.* 1978), and probably EXOSAT, TENMA and GINGA can set even tighter limits.

One further concern about our diagnostics comes from the implicit assumption that the material is reasonably homogeneous within each of the three regions. The fact that the deduced densities for the interior, shock front and eastern knot are qualitatively consistent with what one gets from the imaging analysis and the kinematic ages of the various regions implies that inhomogeneities are not grossly distorting our results. However, Jansen (1988) and colleagues have recently analyzed EXOSAT images showing considerable inhomogeneity throughout Puppis A. We are now trying to incorporate these results into our analysis.

Before leaving Puppis A, I will briefly mention the question of the relative elemental abundances in the remnant. Abundances are of great astrophysical importance. To a large extent, the detailed plasma diagnostics of things like ionization equilibrium are useful because they permit us to address such important questions. The spectrum of Puppis A shows much stronger oxygen and neon lines relative to those of Fe XVII than does the spectrum of the solar corona. In our initial analysis of the Puppis A interior, Frank Winkler and I concluded that oxygen and neon were overabundant relative to iron by factors of 3-5 compared to solar values (Canizares and Winkler 1981). Our more recent conclusion regarding the departure from ionization equilibrium opens the possibility that the apparent weakness of the Fe lines are an artifact of disequilibrium -- that the lines are weak because iron is spread among more ionization stages in Puppis A than in the sun (see also Canizares et al. 1983, Szymkowiak 1985). We have performed a preliminary analysis of this by setting limits on the strengths of lines of Fe XVIII and Fe XX. Our conclusion so far is that non-equilibrium effects cannot explain the entire effect, so that Puppis A does indeed contain a significant overabundance of oxygen and neon, although possibly by less than we had originally thought.

Two other observations of Puppis A in different wavebands are also relevant. Winkler, Tuttle and Kirshner (1988) discovered some optically emitting filaments that appear to be almost pure oxygen -- these show very strong O I, O II and O III lines together with Ne III, but no hydrogen lines! These filaments contain only a tiny fraction of the mass which we observe in the X-ray band, but they certainly support the conclusion that the ejecta are enriched with oxygen. The second relevant observation addresses an alternative explanation for our observed relative overabundance of oxygen to Fe, namely that the Fe is trapped in grains (see Canizares and Winkler 1981, where we argued against this possibility on theoretical grounds). Dwek et al. (1987) have analyzed the far infrared data on Puppis A from the IRAS satellite. The ratio of IR to X-ray luminosity is an order of magnitude less than what one would expect if the SNR contained significant amounts of dust, so depletion of iron cannot be a significant effect.

We still feel justified, therefore, in calling Puppis A the oxygen-rich remnant of a Type II supernova in a massive star.

V. N132D

N132D is also a remnant that shows optically emitting filaments that appear to be nearly pure oxygen. It is in the Large Magellanic Cloud, at a distance of about 50 kpc, and its age is roughly a thousand years, making it four times younger than Puppis A (e.g. see Hughes 1987).

We observed several portions of the spectrum of N132D (e.g. see Canizares et al. 1983). The most striking result is the great strength of the oxygen lines -- even more extreme than in Puppis A. Nearly 15% of the X-ray luminosity of the remnant is in the single Lα line of O VIII. In fact, the half-dozen lines that we studied account for nearly 30% of the total luminosity even though they represent only a few tiny slices of the spectrum.

Unfortunately, we have no useful results on the O VII triplet. We did measure the Ne IX forbidden to resonance line ratio, but it does not provide much of a constraint on the ionization time. Our measurements are even consistent with the remnant being at equilibrium. Taken together with the broad-band temperature measured by Hughes (1987), our line ratios give $\tau > 2500$ yr. cm^{-3}. With an age of 1300 yr. (Lasker 1980) this means that $n_e > 2$ cm^{-3}. A value of $n_e \sim 2$ cm^{-3} would be quite consistent with Hughes' value from the imaging analysis.

It is hard to avoid the conclusion that the oxygen abundance is very much enhanced in N132D. The region around 820 eV (15 Å) contains three lines of Fe XVII flanked by the O VIII Lγ and Lδ lines (see Figure 4). In the solar corona, and even in Puppis A, the Fe lines are much stronger than the oxygen lines, whereas in N132D the Fe lines are barely detectable and may even be absent. When we use the line ratio grids of Hamilton, Sarazin and Chevalier (1983) for inhomogeneous, non-equilibrium models we find that a large oxygen overabundance is required. If the oxygen emission comes primarily from the interior of the SNR then the overabundance there is even more extreme (see Hughes 1987).

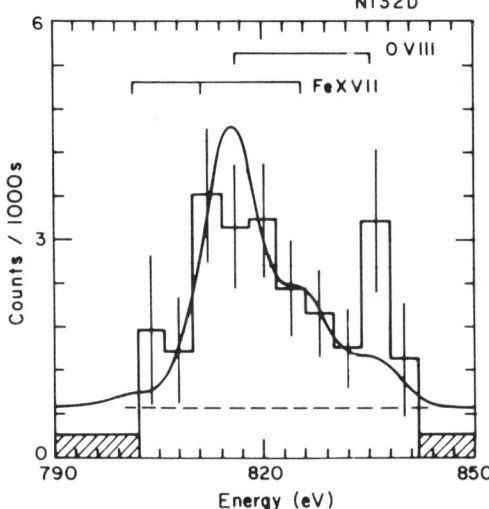
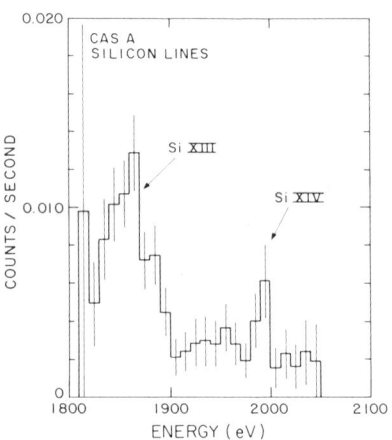

Figure 4. A portion of the spectrum of N132D as observed with the FPCS. The energies of three L lines of F XVII and of the Lβ and Lγ lines of O VIII are indicated. See caption of Figure 1 for details.

Figure 5. A portion of the spectrum of Cas A as observed with the FPCS (Markert et al., 1988). The Si XIII and XIV lines are indicated.

VI. CAS A

Cas A is yet another factor of four younger -- it's age is just over 300 years. Our observations of Cas A yielded surprising kinematic information: we measured a differential Doppler shift across the remnant indicating an asymmetric expansion velocity of ~5000 km s^{-1} (Markert et al. 1983). More recently, Tom Markert and Paula Blizzard have measured the strengths of the hydrogenic and He-like ions of Si, S and Ne as well as a complex of Fe L lines from 850 to 1250 keV (see Figure 5; Markert et al. 1988a,b).

The analysis, being performed with Jack Hughes, shows that the Si and S line ratios together map out an allowed region of T_e vs. τ parameter space. But this region of parameter space is completely incompatible with that required by the Ne lines. We believe that this apparent discrepancy is easily explained by the extreme youth of the SNR. As Andrew Hamilton has described (in this volume) young SNRs contain two distinct emission regions, one behind the primary shock in the ISM and one behind the so-called reverse shock in the expanding stellar ejecta. Our results indicate that the Si and S lines are produced primarily in the hotter ISM while the Ne lines come from the ejecta.

The broadband spectra of Tsunemi *et al.* (1986) from TENMA fit this hypothesis, as do the more recent results from EXOSAT (Jansen *et al.* 1988). The non-equilibrium analysis of Tsunemi finds a value of $T_e = 4 \times 10^7$ K and $\tau = 1500 - 4500$ yr. cm^{-3}, which falls right on the region of parameter space required by the Si and S lines. If we adopt their temperature, our lines constrain the ionization time to be $\tau \sim 2400$ yr. cm^{-3}, giving a pre-shock density of $n_o = 2.0 \pm 0.3$ cm^{-3}, which is quite reasonable for the ISM. We can then deduce the temperature and density of the shocked ejecta by assuming it is in pressure equilibrium with the shocked ISM and that its values of τ and T_e must fall in the region required by the Ne lines. We derive $T_e \sim 2.5 \times 10^6$ K and $n_o \sim 25$ cm^{-3}. These values are also very reasonable. Jansen *et al.* (1988) find a higher temperature for the cooler component, but the discrepancy could easily be due to their assumption of ionization equilibrium.

Of course, the values we deduce are only mean values (weighted by the emission measure), as the imaging analysis of Fabian *et al.* (1980) shows significant variation in the density of the ejecta on the scale of tens of arc seconds. On the other hand, since our values of the density derived from the ionization time (which is linear in density) roughly agree with his based on surface brightness (which depends on the density squared), clumping on very small scales is probably not important. This means that the mass determinations based on the imaging analysis are reasonably correct.

VII. CONCLUSION

I believe that these results demonstrate the great power of plasma diagnostics to reveal the physical conditions of optically thin sources such as SNRs. Even the measurement of only a few well-chosen lines can give remarkably precise information. In each case our deduced parameters agree with those determined by entirely different means, such as broad-band spectral fits or imaging analyses. The diagnostics are also quite model dependent: although I had insufficient time to show this, in most cases the parameters we deduce from our simple homogeneous, single-temperature models are close to those which we derive using the grids of Hamilton, Sarazin and Chevalier (1983) for an inhomogeneous Sedov blast wave.

Therefore, we are now confident in concluding that all four of the SNRs described here are still young, in the sense that they are all still ionizing, despite the fact that they range in age from 300 yr. to ~20,000 yr. Furthermore, our improved understanding of the physical conditions in the plasma gives us greater assurance in our abundance determinations. In particular, we see clear evidence for large oxygen enhancements in Puppis A and N132D.

In some ways these results only serve to whet our appetites. The FPCS was limited in sensitivity and *Einstein* had only a short (albeit happy) life. Future missions will have up to a thousandfold improvement in sensitivity for high resolution spectroscopy, and they will have extended lifetimes that will permit long range observing programs. There is little doubt that they will provide a vast new arena for detailed astrophysical studies.

REFERENCES

Canizares, C.R. and Winkler, P.F. 1981, *Ap. J.*, **246**, L33.
Canizares, C.R., Winkler, P. F., Markert, T. H., and Berg, C. 1983, in Danziger, J. and Gorenstein, P. (eds), *Supernova Remnants and Their X-ray Emission*, (Reidel), 205.
Dwek, E., Petre, R., Szymkowiak, A., and Rice, W. L. 1987, *Ap. J.*, **320**, 27.
Fabian, A. C., Willingale, R., Pye, J. P., Murray, S. S., and Fabbiano, G. 1980, *M. N. R. A. S.*, **193**, 175.
Gabriel, A., Bely-Dubau, F., Faucher, P., and Acton, L. 1988, *Ap. J.* (in press).
Gabriel, A. and Jordan, C. 1969, *M.N.R.A.S.*, **145**, 241.
Giacconi, R. 1979, *Ap. J.*, **230**, 540.
Gorenstein, P., Harnden, R., and Tucker, W. 1974, *Ap. J.*, **192**, 661.
Hamilton, A. and Sarazin, C. 1984, *Ap. J.*, **284**, 60.
Hamilton, A., Sarazin, C., and Chevalier, R. 1983, *Ap. J. Supp.*, **51**, 115.
Hughes, J. 1987, *Ap. J.*, **314**, 103.
Hughes, J. and Helfand, D. 1985, *Ap. J.*, **291**, 544.
Itoh, H., 1977, *Pub. Astron. Soc. Japan*, **29**, 813.
Jansen, F. 1988, *Ph.D. Thesis*, Leiden State University, The Netherlands.
Jansen, F., Smith, A., Bleeker, J. A. M., de Korte, P. A. J., Peacock, A., and White, N. E. 1988, *Ap. J.*, **331**, 949.
Kallne, E. *et al.* 1984, *Phys. Rev. Lett.*, **52**, 2245.
Lasker, B. 1980, *Ap. J.*, **237**, 765.
Markert, T.H., Canizares, C.R., Clark, G.W., and Winkler, P.F. 1983, *Ap. J.*, **268**, 134.
Markert T., Blizzard, P., Canizares, C., and Hughes, J. 1988a, in Roger, R. S. and Landecker, T. L. (eds) *Supernova Remnants and the Interstellar Medium*, (Cambridge Univ. Press), 129.
Markert T., Blizzard, P., Canizares, C., and Hughes, J. 1988b, in preparation.
Mewe, R., Gronenschild, E. H. B. M., and van den Oord, G. H. J. 1985, *Astron. Ap. Supp.*, **62**, 197.
Mewe, R. and Schrijver J. 1978, *Astron. Ap.*, **65**, 99.
Prahdan, A. 1983, *Phys. Rev. A.*, **28**, 2128.
Petre, R., Canizares, C.R., Kriss, G.A., and Winkler, P.F. 1981, *Ap. J.*, **258**, 22.
Szymkowiak, A. E. 1985, *NASA Tech. Memorandum 86169*.
Tsunemi, H., Manabe, M., Yamashita, K., and Koyama, K. 1988, *Publ. Astron. Soc. Japan*, **40**, 449.
Winkler, P.F. Canizares, C.R., Clark, G.W., Markert, T.H., and Petre, R. 1981, *Ap. J.*, **245**, 574.
Winkler, P.F., Tuttle, J.H., Kirshner, R.P. and Irwin, M.J. 1988, in *Supernova Remnants and the Interstellar Medium*, eds. R.S. Roger and T.L. Landecker, (Cambridge University Press, Cambridge 65.)
Zarnecki, J. C., Culhane, J. J., Toor, A., Seward, F. D., and Charles, P. A. 1978, *Ap. J. (Letters)*, **219**, L17.

ACKNOWLEDGEMENTS. I thank many colleagues whose work in progress I have quoted here. In particular I am grateful to Leah Bateman, Paula Blizzard, Kathryn Fischbach, Jack Hughes, Tom Markert, Tim Pfafman, Atul Pradhan, Pablo Saez, and Peter Vedder. I thank Fred Jansen for showing us his work prior to publication, and Elaine Aufiero for her expert preparation of the manuscript.

DISCUSSION-C. Canizares

J. Schmitt: In the case of stellar coronae, one worries very much about multi-thermal plasmas. You don't seem to worry about this much. Can you explain why?

C. Canizares: We do worry about multi-thermal plasmas, but at this stage there is only a limited amount we can do about them. When their effects are large, as in Cas A, we can actually distinguish two components. In most cases, we must make some simplifying assumption, such as adopting an isothermal or Sedov model. Happily, some of the astrophysically important conclusions that we draw are not too dependent on these assumptions, as is shown by the results of Hamilton and Sarazin (1984) and the apparent insensitivity of our conclusions to the model chosen.

P. Winkler: Does your analysis shed any light on the question of post-shock equilibration between ion and electron temperatures – which remains a major source of uncertainty in understanding SNR plasmas?

C. Canizares: Not really. All the results I have quoted assume $T_e = T_i$, which is at least plausible. It is possible that in some cases we will be able to rule out the extreme case of no energy transfer between protons and electrons as requiring implausible parameters (e.g. see Jansen *et al.* 1988), but modest departures from equality will probably be very difficult to discern. In the future, when we have better diagnostics for T_e, we may be able to answer this question.

FEATURES IN THE SOFT X-RAY BACKGROUND

R. ROTHENFLUG
Service d'Astrophysique (DPHG/IRF)
C.E.N. Saclay (France)

ABSTRACT. The soft X-ray background is explained in terms of emission coming from hot gas. Most of these soft X-ray data were obtained by proportional counters with a poor energy resolution. Instruments having the capability to resolve lines were only flown by two groups: a GSPC by a japanese group and a SSD by a french-american collaboration. They both detected the O VII line emission coming from the soft X-ray background and so proved the thermal nature of the emission. The implications of these results on possible models for the local hot medium will be discussed. The same detectors observed part of the North Polar Spur. They detected emission lines coming from different species (O VII, Fe XVII, Ne IX). Spatial variations of line ratios for this object could be due to non-equilibrium ionization effects.

1. INTRODUCTION

The sky has now been almost entirely scanned in the soft X-ray range by experiments using proportional counters. These detectors have a very poor energy resolution (typically $E/\Delta E \sim 2$ for $E < 1$ keV) and the detected X-rays are discriminated by different windows. So their results are given as rates in the following broad bands: Be band (80-110 eV), B band (130-188 eV), C band (160-284 eV) and M bands (M1:440-930 eV; M2:600-1100 eV). In part 2, the broad band results on the soft X-ray background are summarized and we described the spectroscopic observations obtained with detectors having a better energy resolution. These last results are compared to an isothermal model with the additional constraints brought by the broad band rates of the same sky region. The explanation of these spectra as emission coming from an active blast wave is also discussed. In part 3, results obtained with the same kind of detectors are presented for the North Polar Spur. The line intensities of the different species (O VIII, Fe XVII, Ne IX) derived by the two detectors give some indications on the ionization state of the plasma.

2. THE HOT LOCAL BUBBLE

2.1. General characteristics of the soft X-ray background

Maps of the diffuse emission in galactic coordinates for the B and the C bands are shown in McCammon et al(1983). The C map obtained by SAS 3 presents the same general characteristics as the Wisconsin C map (Marshall and Clark, 1984). The most important features of the soft X-ray background are:
   1. The B and C maps appear very similar, with the exception of some features appearing in the C map and which are associated to particular objects (North Polar Spur, Eridanus, etc: see below).

2. The flux in the galactic plane is approximately the same at all longitudes in the B and C bands.

3. Again in the B and C bands, the observed flux is generally higher by a factor of two to three at high latitudes. A large scale anticorrelation with the H I column density exists.

4. The ratio of the Be to B band rates is almost constant over 120° of the sky. This strongly suggests a common origin for the two band emissions. Its constancy means that there is no more than about 5.10^{18} H I atoms/cm² between the observer and the bulk of the Be and B band emissions (Bloch et al,1986). This implies that the soft X-ray emission in these bands must be produced within surely less than ~100 pc.

5. In the M bands (0.5-1.2 keV), the background seems isotropic. In this band, an important contribution of the extragalactic background is expected. One usually assumes that this extragalactic background can be estimated by the extrapolation towards low energies of the following spectrum: $11.E^{-1.4}$ ph/cm².s.keV.sr. with E in keV. This contribution decreases at low galactic latitudes because of the absorption by the interstellar matter. Once this contribution is removed, the remaining M band flux then increases towards the galactic plane, contrary to the emission in the other bands.

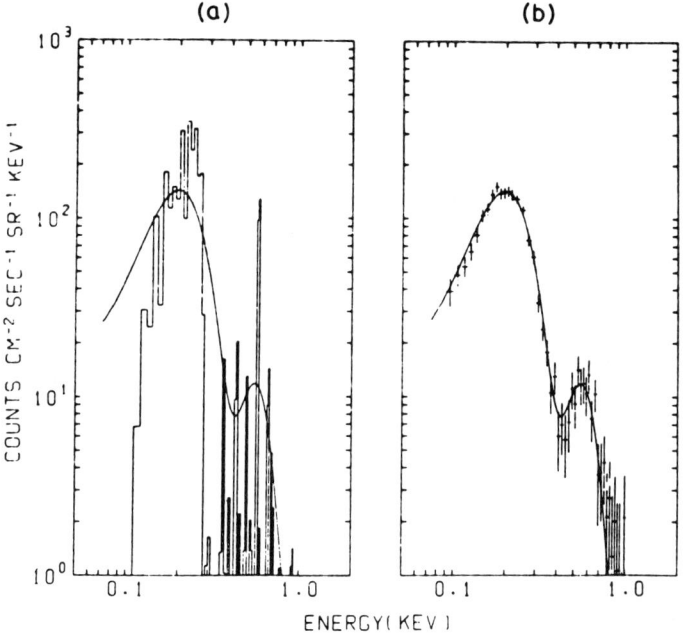

<u>Figure 1</u>: Observation of the Hercules Hole region with a GSPC (Inoue et al, 1979). (a) Thermal spectrum (Kato,1976) for $1.4\ 10^6$ K integrated in 10 eV bins and convolved with the detector energy response (solid line). (b) The convolved spectrum fitted to the observed one.

2.2. Evidence for line emission

Two experiments were performed by two different groups with detectors having $E/\Delta E \sim 4$ at 600 eV, which is better than usual proportional counters. They measured the soft X-ray background spectrum in particular regions of the sky.

1. A Japanese group observed the Hercules Hole region (20°x20° around $l \sim 80°$ and $b \sim 40°$) with a gas scintillating proportional counter (Inoue et al, 1979). This region presents a very low column density in HI ($\sim 1.7 \ 10^{20}$ H atoms/cm²). The observed spectrum is shown in figure 1, the contribution of the extragalactic bakground being subtracted. It revealed a clear peak at about 570 eV which can be identified as the O VII emission line. Emission from continua alone (either power law or thermal bremsstrahlung) can be rejected at better than 99%. The total spectrum is in agreement with a thermal emission at a temperature around $1.4 \ 10^6$ K. The model of Kato (1976) was used for the plasma emission.

2. An experiment observed the soft X-ray background with solid state detectors (Rocchia et al, 1984). It was made in collaboration between the Smithsonian Astronomical Observatory and the Service d'Astrophysique in Saclay. This experiment scanned the region roughly delimited by $b>10°$, $0<l<180°$ with a ~1 ster field of view. In figure 2, we reproduce the spectrum obtained in regions of the sky outside the North Polar Spur, once the contribution of the extragalactic background has been subtracted. The feature around 570 eV is due to the O VII emission line. There is also a strong excess at low energies attributed to a blend of C V (300eV) and C VI (360 eV) emission lines. Their temperature determination was $1.14 \ +/- \ 0.08 \ 10^6$K (at the 90% confidence level), also with the plasma emission model of Kato (1976).

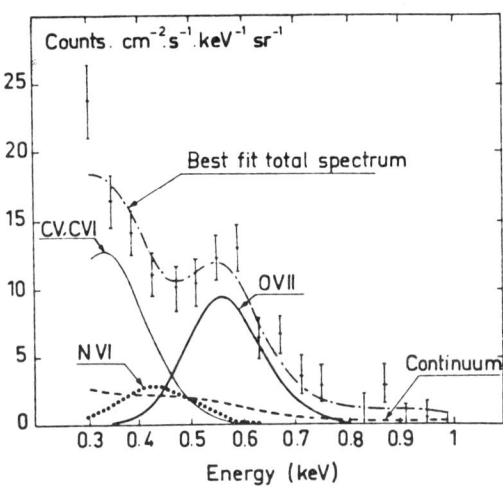

Figure 2: Observation of the soft X-ray backgound with a solid state detector (Rocchia et al, 1984). This concerns a region of the sky outside the North Polar Spur. The dot-dashed line is the total spectrum. The contributions of the different ion species are indicated.

2.3. Comparison with an isothermal model

In summary, the band rate observations revealed that the soft X-ray background is (very probably) produced locally (inside 100 pc from the sun) and the spectroscopic results confirmed that the emission is coming from a hot plasma at a temperature around 10^6 K. In this part, we will investigate if the GSPC and the SSD spectra can both be explained by plasma emissions at the same temperature (see Arnaud and Rothenflug, 1989 for a more complete discussion).

One must also have in mind that the broad bands measure in fact emission coming from different lines : at 10^6 K, iron lines (mainly Fe IX and Fe X lines around 72.5 eV) dominate in the Be band, magnesium lines dominate in the B band and silicon lines in the C band. Thus it could be interesting to take broad band rates as additional constraints. Clearly the Be band rate bring information at very low energy not included in the GSPC and SSD data.

1. The SSD spectrum is expected to be more representative of the whole emission than the GSPC spectrum since it was obtained for a large region of the sky. As a first step, we tried to fit an isothermal plasma to this spectrum together with the band rates. For the B,C,M1 and M2 rates, the mean values over the whole sky were adopted as a crude estimate, the contribution of the extragalactic background being subtracted (McCammon et al,1983). For the Be band, a value of 8 cts/s was thought to be appropriate for this part of the sky from the observations of Bloch et al (1986).

The fits of a thermal model was done using the Arnaud and Rothenflug (1985) ionization model coupled with the Mewe et al (1985,1986) emission model. Normal (undepleted) abundances were taken from Meyer (1985). An isothermal model with T= $1.2 \ 10^6$ K and $N_H = 4.10^{18}$ H I atoms/cm^2 gives an acceptable fit to the SSD spectrum and the Be,B,and C band rates (reduced $\chi^2 = 1.8$) but failed to account for the M1 and M2 rates.

2. The GSPC spectrum contains more information than band rates and we included it into our data set to be compared with an isothermal model. The ratio of the GSPC to SSD experiments was then introduced as a free parameter since the SSD data refer to a wider region of the sky. The best fit of a single temperature model is compared to the whole data set on figure 3.

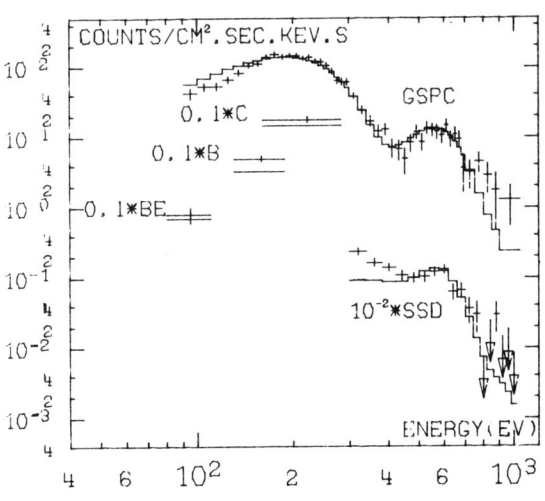

Figure 3: Best fit single temperature model compared to the soft X-ray background spectral observations: GSPC (Inoue et al,1979); SSD (Rocchia et al,1984) Be,B,C: band rates (see text). The M1 and M2 rates are not shown.

The statistical weight of the GSPC data forces the temperature to a value around $1.5 \, 10^6$ K. The Be rate forces the N_H value to $5. \, 10^{18}$ H atoms/cm². The reduced χ^2 is greater than 3. At such temperature, the predicted carbon line emission is much fainter than the observed SSD flux and so this fit fails to reproduce the carbon line intensity. The higher temperature plasma produces more counts in the M bands: the best fit leads to ~10 counts in the M1 band and ~6 in the M2 band.

Such an enhancement of the carbon line could be due to non-equilibrium ionization effects because of the delay in the ionization of helium-like carbon with respect to other ions (Arnaud et al,1984). This kind of effect is expected if the thermal emission comes from an active shock running into a rarefied interstellar medium.

2.4.Comparison with an active blast wave.

The various possible origins of the local hot bubble were recently reviewed by Cox and Reynolds (1987). Here we will only discuss the case of an active blast wave. Cox and Anderson (1982) devised a model of such a blast wawe developing in a medium with finite pressure. In that model, the parameters are the explosion energy, the external density and temperature. Cox and Anderson provided a simple analytical approximation to the hydrodynamical evolution and structure for this SNR type. They computed the ionization structure at the shock and the emissivities of their remnant in the B,C,M bands. They showed that their model is able to explain the mean B and C rates, but failed to account for the M flux.

Arnaud and Rothenflug (1986) made an attempt to compare the spectra produced by such SNRs with the GSPC and SSD results. They used the analytical expressions given by Cox and Anderson for the SNR evolution. They computed the ionization structure behind the shock and the corresponding spectra, using slightly different atomic physics (see part 2.3).

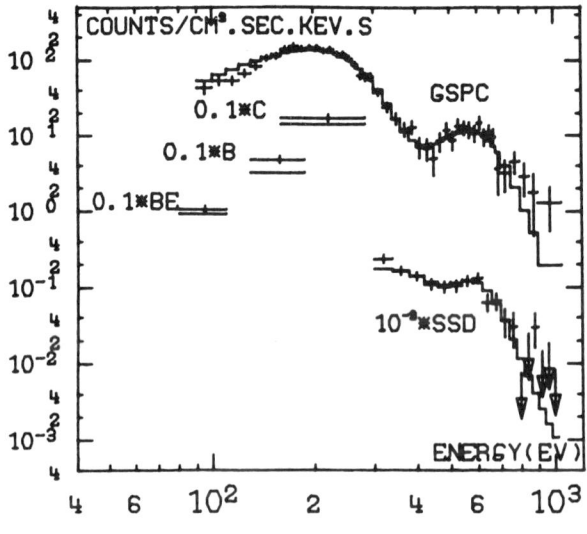

Figure 4: Comparison of spectroscopic and band rate data with the Cox and Anderson model adapted by Arnaud and Rothenflug (1986). Best fit values with depleted abundances:

$N_H = 6.10^{18}$ H atoms/cm²;

$n_{ext} = 1.25 \, 10^{-2}$ el./cm³;

$R_s = 120$ pc;

$E_o = 1.3 \, 10^{51}$ ergs.

They compared them to the GSPC and SSD observed spectra, with the additional constraints brought by the band rates. They showed that SNR models with normal abundances failed to account for the whole data set. An acceptable fit ($\chi^2=1.7$) can be obtained with models allowing depleted abundances. Their best fit is reproduced in figure 4. Possible depletion of the elements in the local hot plasma was also proposed by Bloch et al (1988) to explain the mean energy of the photons observed in the Be band.

In a forthcoming paper, Arnaud and Rothenflug (1989) will extend this discussion to some other possible models of the soft X-ray background. It remains that such SNR models encounter great difficulties to explain why the M flux should increase toward the galactic plane and the B and C fluxes should have an opposite behaviour.

3. THE NORTH POLAR SPUR

Several enhancements show up in the soft X-ray maps. They are explained as emissions from other closeby (d<500 pc) hot cavities, as for instance the Monoceros Loop (Nousek et al,1981). The North Polar Spur (NPS) is the most important of these enhancements.

3.1. The North Polar Spur as SNR.

The NPS forms a part of a 116° diameter circle on the sky called Loop I by radio astronomers: the radius of this shell is estimated to be ~115 pc (Berkuijsen,1973). About 5° outside this radio loop, there is a ridge of neutral hydrogen with an expansion velocity maybe as high as 30 km/s (Heiles et al,1980). Inside the radio emission, the soft X-ray emission is strongly enhanced with a spatial distribution indicating a strong limb brightening (see for instance Davelaar et al,1980 and also the M1 and M2 maps of McCammon et al,1983). The X-ray emission is much harder than outside the spur and its temperature is evaluated around 3.10^6 K or even higher from soft X-ray measurements (Davelaar et al,1980; Iwan,1980).

The most probable scenario involves the formation of the HI shell by the stellar winds of an entire association. Its dimension, its density ($n_{HI} \sim 2$ p/cm^3, Heiles et al,1980) and its expansion agree with the characteristics of such HI bubbles as computed by Bruhweiler et al (1980). The stellar winds have created a low density cavity in which the SNR is now developing. The soft X-ray observations implies an external density of $\sim 10^{-2}$ p/cm^3 and leads to an explosion energy around 3.10^{51} ergs and an age of $\sim 8.10^4$ years (Davelaar et al,1980).

3.3. NPS spectral observations.

The soft X-ray enhancements associated with the NPS was partly observed with a GSPC (Inoue et al,1980) and a SSD (Rocchia et al,1984).

a) The GSPC experiment pointed towards the region roughly centered at l~30°, b~25°, with a 10°x10° FWHM field of view (Inoue et al,1980). The highest energy part of this spectrum is shown on figure 5. At temperatures ~3.10^6 K, the most important lines contributing to this energy range are those of oxygen, iron and neon. With the energy resolution of the GSPC, it becomes possible to sort out the contribution of the most intense lines. This was done by Inoue et al (1980) using a deconvolution method. The result is depicted on figure 5 which shows the lines once convolved by the response of the GSPC.

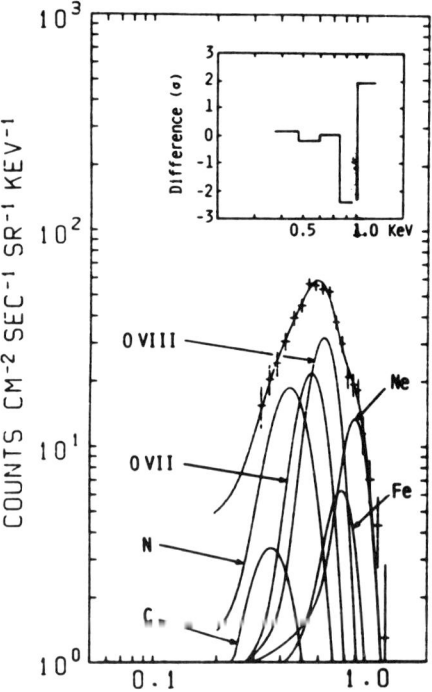

Figure 5: The M band spectrum of the NPS as observed by a GSPC (Inoue et al,1980). Curves indicate the contributions of the six emission-line groups derived by the decomposition and their sum.

b) The SSD experiment had a circular field of view of ~22° FWHM and pointed a region around $l \sim 25°$, $b \sim 25°$, 5° away from the region analysed by the GSPC (Rocchia et al,1984). Their spectrum is also characteristic of a relatively high temperature (5.10^6 K). A deconvolution method was applied to these data in order to estimate the intensities of the different lines (Rothenflug et al,1984). The resulting line contributions are sketched on figure 6.

Let us compare the line intensities derived by the two experiments. The Ne IX line intensity will be used as reference. Once corrected for the detector efficiency, we obtain the following intensities:

        O VIII   GSPC: 2.9 (+/-0.9)   SSD: 1.3 (+/-0.7)

        Fe XVII   GSPC: 0.5 (+/-0.4)   SSD: 2.4 (+/-0.5) (90%confidence)

The O VIII line intensities of the two experiments are marginally compatible. For the Fe XVII line, the two error ranges do not overlap and the intensities measured by the two experiments are quite different.

Let us remark that the GSPC experiment observed a region near the shock (as delimited for instance by soft X-ray maps) and the SSD observed a wider region including it, but also a large NPS part far from the shock. Inoue et al(1980) proposed that the weak Fe XVII line in their data can be explained as a depletion of iron in grains: just behind the shock, grains are present in the gas, but are destroyed far from the shock. Another explanation could be the delay in the ionization of iron with respect to the others elements: a comparison with the computations of Masai(1983) indicates that an ionization parameter of $n_e t \sim 10^{11}$ s/cm³ could be a crude estimate, where n_e is the electronic density and t the time in seconds. Values given above on the external density and age leads to $\sim 8.10^{10}$ s/cm³, not very far from that estimate. A modelling of the NPS taking into account these spectroscopic results would be very useful to constrain its age, its shock temperature and the external density.

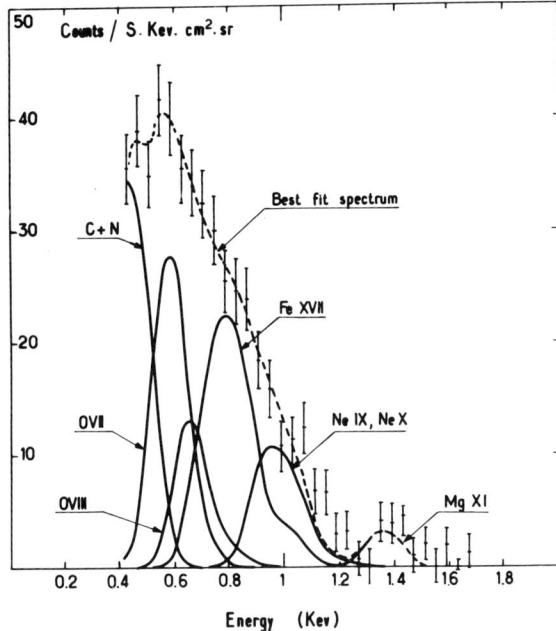

Figure 6: The spectrum of the NPS part observed by the SSD. The contribution of each line blend is drawn together with the total spectrum (Rothenflug et al, 1984).

Acknowledgements

I wish to thank M. Arnaud for her very useful comments and her friendly advices.

REFERENCES

Arnaud,M.,Rothenflug,R.,Rocchia,R.,1984, Physica Scripta, Vol.T7,p.48
Arnaud,M.,Rothenflug,R., 1985, Astron. Astrop. Suppl. Ser. 60, 425
Arnaud,M.,Rothenflug,R., 1986, Adv.Space Res., Vol.6,No.2, p119
Arnaud,M.,Rothenflug,R., 1989, in preparation
Berkuijsen,E.M., 1973, Astr.,Ap., 24, 143
Bloch,J.J.,Jahoda,K.,Juda,M.,McCammon,D.,D.N.,Sanders,W.T.,Snowden,S.L.

1986, Ap.J.(Letters) 308, L59
Bloch,J.J.,Priedhorsky,W.C.,Smith,B.W., 1988, these proceedings
Bruhweiler,F.C.,Gull,T.R.,Kafatos,M.,Sofia,S.,
   1980, Ap.J.(Letters) 238, L27
Cox,D.P.,Anderson,P.R., 1982, Ap.J. 253, 268
Cox,D.P.,Reynolds,R.J., 1987, Ann. Rev. Astron. Astrophys., Vol. 25, 303
Davelaar,J.,Bleeker,J.A.M.,Deerenberg,A.J.M., 1980, Astr.,Ap., 92, 231
Heiles,C.,Chu,Y.H.,Reynolds,R.J.,Yegingil,I.,Troland,T.H., 1980,
   Ap.J., 242, 533
Inoue,H.,Koyama,K.,Matsuoka,M.,Ohashi,T.,Tanaka,Y.,Tsunemi,H.,
   1979, Ap.J., 227, L65
   1980, Ap.J., 238, 886
Iwan,D.A., 1980, Ap.J. 239, 316
Kato,T., 1976, Ap.J.Sup. 30, 397
McCammon,D.,Burrows,D.N.,Sanders,W.T.,Kraushaar,W.L., 1983,
   Ap.J., 269, 107
Marshall,F.J.,Clark,G.W., 1984, Ap.J. 287, 633
Masai,K., 1984, Astrophysics and Space science, 98, 367
Mewe,R.,Gronenschild,E.H.B.M.,Van den Oord,G.H.J., 1985,
   Astron. Astrop. Sup. Ser., 62, 197
Mewe,R.,Lemen,J.R.,Van den Oord,G.H.J., 1986,
   Astron. Astrop. Sup. Ser., 65, 511
Meyer,J.P., 1985, Ap.J.Suppl.Ser., 57, 151
Nousek,J.A.,Cowie,L.L.,Hu,E.,Linblad,C.J.,Garmire,G.P., 1981,
   Ap.J., 248, 152
Nousek,J.A.,Fried,P.M.,Sanders,W.T.,Kraushaar,W.L., 1982, Ap.J. 258, 83
Rocchia,R.,Arnaud,M.,Blondel,C.,Cheron,C.,Christy,J.C.,Rothenflug,R.,
   Schnopper,H.W.,Delvaille,J.P., 1984, Astron. Astrop. 130, 53
Rothenflug,R.,Arnaud,M.,Blondel,C.,Rocchia,R., 1984, Physica Scripta,
   Vol.T7, p.62

DISCUSSION

J. Bloch: The emission mechanism for the M band in the soft X-ray background must be different from the high energy power law and Be,B and C band emissions. This is because the power law does not make up the M band counts and the M band is very isotropic compared to the pole enhanced emission in the Be,B and C bands. How was this other component for the M band taken into account?

P. deKorte: You assume for the fit of spectra, which result in depleted abundances, that the total spectrum is coming from one region. I doubt this very much, especially for part > 0.5 keV.

W.T. Sanders: Your model is able to fit the data from 0.1 to 1. keV in one direction of the sky. The Wisconsin maps show isotropic M band intensity but C variations of a factor of three. How does your model account for this?

Rothenflug: These three questions are related to the emission in the M bands, i.e to the characteristics of the soft X-ray background for energies greater than 0.5 keV. The emission in these M bands contains at least 3 components: the absorbed extragalactic background, a contribution from the local hot medium, and a component with a large scale height (Nousek et al,1982). Our model is able to explain a significative part of the M bands, in a particular direction. The remaining parts are coming from the extragalactic background and from an yet unclear origin. Among the possible candidates, the emission from stars have been proposed.

G. Vaiana: To what extent the detailed conclusions (depleted abundances and non-equilibrium ionization) depend on how well the extragalactic background has been subtracted ? Is the extragalactic background subtracted just the extrapolation of the power law from the high energy part of the spectrum ?

Rothenflug: All these observations assume that the extragalactic background can be estimated by an extrapolation of the part of the spectrum measured above ~2 keV. The same power law is generally used by the different groups (see part 2.1). The contribution of the background is removed taking into account the absorption in the line of sight. For $N_H = 2.10^{20}$ H I atoms/cm², McCammon et al (1983) estimated that the contribution of the extragalactic background is ~3% for the B band, ~11% for the C band, ~30% for the M1 band and 39% for the M2 band. Unless the extragalactic background spectrum presents a sharp cutoff below 2 keV, its contribution will not change dramatically wathever its spectrum shape is at low energies.

DISCUSSION-R. Rothenflug

J. Bloch: The emission mechanism for the M band in the soft X-ray background must be different from the high energy power law and Be, B and C band emissions. This is because the power law does not make up the M band counts and the M band is very isotropic compared to the pole enhanced emission in the Be, B and C bands. How was this other component for the M band taken into account?

P. deKorte: You assume for the fit of spectra, which result in depleted abundances, that the total spectrum is coming from one region. I doubt this very much, especially for part > 0.5 keV.

W. Sanders: Your model is able to fit the data from 0.1 to 1.0 keV in one direction of the sky. The Wisconsin maps show isotropic M band intensity but C variations of a factor of three. How does your model account for this?

R. Rothenflug: These three questions are related to the emission in the M bands, i.e. to the characteristics of the soft X-ray background

NON–EQUILIBRIUM IONISATION IN TYCHO'S SUPERNOVA REMNANT

W. Brinkmann

Max-Planck-Institut für Physik und Astrophysik, Institut für Extraterrestrische Physik, 8046 Garching, F.R.G.

ABSTRACT: We present a detailed numerical study of the X–ray emission of Tycho's supernova remnant. Using Nomoto's W7 model of an exploding white dwarf as intial condition we follow the hydrodynamical and ionisation evolution for different densities of the interstellar medium and compare the emitted spectra with EXOSAT GSPC observations. The results indicate that the density of the ambient medium is $0.5 \lesssim n_0 \lesssim 1.0$ cm^{-3}, the distance to Tycho is about 3 kpc.

1. Introduction

Observations and modelling of the thermal X–ray emission of young supernova remnants provide the most direct tool to study their evolution, their chemical composition, and their interaction with the interstellar medium. Using a nuclear explosion model for the actual supernova event as initial condition in a hydrodynamical – ionisation modelling of a remnant the only free parameters in the calculation are the chemical composition and the density of the interstellar medium swept up by the outgoing shock wave. Matching the shape and intensity of the emitted spectrum with the observed X–ray flux as well as the comparison of the of the measured size of the remnant with the calculated size should result in a selfconsistent determination of the interstellar gas density and the distance of the remnant.

The ideal object for this type of study is the remnant of SN 1572, Tycho. The obtained optical light curves indicate that the event was a type I explosion which are thought to be caused by exploding white dwarfs where a deflagration front, starting at the centre of the white dwarf disrupts the whole star (Nomoto et al., 1984). Detailed numerical modelling of this kind of explosion resulted in well determined spatial profiles for the chemical composition, density and velocity (Thielemann et al., 1986). The X–ray shape of the remnant is nearly perfectly spherical symmetric, indicating that the explosion went off in a largely homogeneous medium and no signs for the existence of a central pulsar have been found. An analysis of the Einstein data (Gorenstein et al., 1983) roughly confirms the "standard" picture, although the mass derived for the progenitor star is larger than the $\simeq 1.4 M_\odot$ expected from the deflagration of a white dwarf (Nomoto et al., 1984).

Recently, detailed spectral observations in the harder X–ray band of $\sim 1.6 - 10$ keV have been made with the GSPC detectors onboard EXOSAT (Smith et al., 1988) and Tenma (Tsunemi et al., 1986). These instruments cover the vital energy range of the higher excited Si, S, Ar, Ca, and Fe lines and provide a good estimate for the temperature of the underlying continuum.

Itoh, Masai and Nomoto (1988) recently compared the Tenma results with hydrodynamical–ionisation calculations using Nomoto's W7 model as intial conditions. As the Tenma data indicate a rather soft X-ray continuum they assumed throughout their calculations that the electrons are heated by Coulomb collisions with the ions only, resulting in a considerably lower electron temperature compared with the ion temperature and, therefore, in a relatively soft X-ray continuum.

We present the results of a detailed numerical modelling of the X-ray emission of Tycho and compare them with the EXOSAT GSPC data. We assume temperature equilibrium between electrons and ions obtained by non–linear plasma processes, as postulated by McKee (1974). The details of the numerical calculations are given in Brinkmann et al. (1988). We are able to fit the observed spectrum quite well, although some marked differences between the simulated and observed spectra pertain indicating that some of the basic underlying model assumptions need readjustment.

2. Results

First we calculated models with the intial abundance distributions as given by Thielemann et al. (1986) and could not reproduce the observed spectra at all. At lower densities of the interstellar medium ($n_0 \lesssim 1$ cm^{-3}) we could not match the enormous flux of the S – Ar – line complex at energies $\lesssim 3$ keV and the iron line at ~ 6.5 keV. Increasing the density of the interstellar material yielded better, but still unacceptable agreement between the measured and calculated spectra and at densities $n_0 > 2$ cm^{-3} the calculated total X-ray emissivity exceeds the observed value.

This problem to model the emission lines is not unexpected. It was encountered by Itoh et al (1988) as well and modelling of the early optical spectra of type I supernovae (Branch et al., 1985) could only be achieved by mixing the material of the ejecta. The physical reason for this mixing is, that in the deflagration models the burning front is convective due to the density inversion across the front and the hydrodynamics has, in principle, to be simulated by a more dimensional hydrocode which would then be able to handle the convective burning and mixing process properly.

Lacking this information we tried to reproduce the measurements by artificially mixing the elemental abundances of several outer zones homogenously over the region considered. As a first result it turned out, that a mixing of the material burned during the passage of the deflagration wave only is insufficient: a mixing of burned matter into the outer, unburned zones of the model is required. This implies, that mixing does not only occur in the convectively burning zones but that, eventually during the later expansion, unburned material is mixed into these zones, perhaps by Rayleigh – Taylor instabilities. Secondly, surprisingly little ^{56}Fe has to be mixed into the outer zones to match the measured iron line at ~ 6.5 keV.

Figure 1 show a spectrum with mixing of matter in the region of incomplete Silicon burning in Nomoto's W7–model which gave in general the best fits to the observed spectra and a density of the interstellar medium, $n_0 = 0.5$ cm^{-3} (for details see Brinkmann et al., 1988). For this amount of mixing only a fraction of $\lesssim 0.05~M_\odot$ of the total ^{56}Fe content of the ejecta participates in the emission which means, that the vast majority of the produced iron remains cold, inside the remnant.

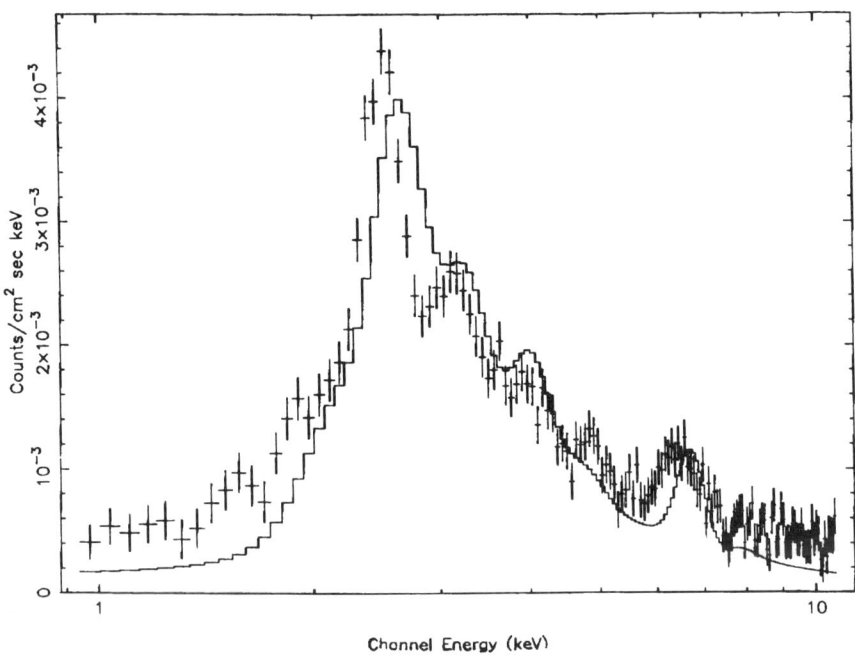

Figure 1: Calculated X–ray spectrum (solid line) convolved with the EXOSAT GSPC detector response in comparison with the observed fluxes (crosses)

Models with densities of the interstellar material in the range of $0.5 \lesssim n_0 \lesssim 1.0~\text{cm}^{-3}$ seem to be in best agreement with the experimental results: For higher densities the continuum flux gets too high putting the remnant too far away. Further, as the material is heated much longer, the slope of the continuum get too flat indicating a higher temperature than obtained by EXOSAT. (It should be noted that for our models the continuum is consistent with the value of ~ 6 keV claimed for the EXOSAT data, not with Tenma's ~ 3 keV). At lower external densities, the total flux as well as the strength of the emission lines is, in general, too low.

A detailed inspection of the modelled spectra shows that the majority of the emission originates from a localized region near the hydrodynamical contact discontinuity. This is a peculiarity of the employed chemical composition of the ejecta: Further out, in the outer shock, the temperature of the gas is too high, only bremsstrahlung is emitted. Near the contact discontinuity temperature and ionisational status are just ideal for line emission and further in, the ionisation time $n_e t$ is still too small to populate the higher ionisation stages effectively. In particular, most of the line emission originates from this region while the high temperature continuum flux is produced in the outer shock. This shows, that both spectral components are mainly produced independently in spatially and physically distinct regions. It further shows that the radius of the remnant, as seen by Einstein in the enegy band ≤ 4 keV, has to be identified with the position of the contact discontinuity and **not** with the position of the outer shock!

An estimate for the distance of the remnant can be obtained by comparing the emitted spectral power in the model calculations with the flux actually observed with EXOSAT. A second, independent measure for the distance is found from comparing the X-ray images of Tycho, taken with the Einstein HRI detector, with the radial brightness distribution resulting from our numerical models. In both cases the distance to Tycho is found selfconsistently to be $\gtrsim 3$ kpc.

3. Conclusions

A modelling of the X-ray emission of Tycho's supernova remnant using Nomoto's W7 model as initial condition results in a total X-ray spectrum which is in reasonable agreement with the observed EXOSAT GSPC spectrum of Tycho. The most severe difference between the observed and modelled spectra is that the centroids of the calculated emission lines of Fe and S are systematically shifted to higher energies, indicating that these elements are over-ionized in the calculations. Possible reasons for this can be our lack of the knowledge of actual microphysical processes involved (state of the electron gas, effects of magnetic fields and relativistic particles) as well as uncertainties in the astrophysical modelling (nuclear explosion model, clumpiness of the matter, global deviations from spherical symmetry, plasma instabilities).

Better, i.e. at least two dimensional explosion models and X-ray spectra with high spectral and spatial resolution together with a detailed hydrodynamic simulation seem to be the only way to get insight into the different physical processes involved and to obtain "physically reliable" parameters for these objects.

References

Branch, D., Doggett, J.B., Nomoto, K., and Thielemann, F.-K. 1985, *Ap.J.* **294**, 619

Brinkmann, W., Fink, H.H., Smith, A., and Haberl, F. 1988, subm. to Astr. Ap.

Gorenstein,P., Seward, F., and Tucker, W. 1983, in *Supernova remnants and their X-Ray Emission*, IAU symposium no. **101**, eds. J. Danziger and P. Gorenstein, D. Reidel, Dordrecht

Itoh, H., Masai, K., and Nomoto, K. 1988, *Ap.J.* in press

McKee, C.F. 1974, *Ap.J.*, **188**, 335

Nomoto,K., Thielemann, F.K., and Yokoi, K. 1984, *Ap.J.*,**286**, 644

Smith, A., Davelaar, J., Peacock, A., Taylor, B.G., Morini, M., and Robba, N.R. 1988, *Ap.J.*, **325**, 288

Thielemann, F.K., Nomoto, K., and Yokoi, K. 1986, *Astron. Astr.*, **158**, 17

Tsunemi, H., Yamashita, K., Masai, K., Hayakawa, S., and Koyama, K. 1986, *Ap.J.*, **306**, 248

Fe VIII, IX, AND X LINE EMISSION FROM THE SOFT X-RAY BACKGROUND: PREVIOUS LIMITS AND A FUTURE MEASUREMENT

J. J. Bloch, W. C. Priedhorsky, and Barham W. Smith
Los Alamos National Laboratory, Los Alamos, NM, USA

ABSTRACT. We discuss pulse height analysis of Be band data in relation to the important 72 eV Fe line cluster emission from the soft X-ray background (SXRB). Pulse height fits to the Be band data suggest that the Fe lines must be suppressed by a factor of ~10 with respect to the rest of the X-ray spectrum. The broad band rates and the mean energy of the pulse height data for the Be band can be brought into agreement by using depleted elemental abundance emission models. A planned measurement of the SXRB Fe lines using the Array of Low Energy X-ray Imaging Sensors (ALEXIS) experiment could resolve this issue.

1. THE Fe LINE EMISSION FROM THE SOFT X-RAY BACKGROUND

The soft X-ray background (SXRB) emission (0.07-0.25 keV) is thought to originate in a ~10^6 K plasma that exists in the local interstellar medium. The SXRB flux has been measured in the Be (0.07-0.111 keV), B (0.130-0.188 keV), and C (0.16-0.284 keV) bands (McCammon *et al.* 1983, Marshall and Clark 1984, Bloch *et al.* 1986, Juda 1988). The relatively constant ratio (see Figure 1) between the Be band and the B band flux from the SXRB has been used as an argument that the SXRB emission source is within 100 pc of the sun and closer than the first ~1×10^{19} cm^{-2} of HI column density (Bloch *et al.* 1986, Juda 1988).

Emission models of a ~10^6 K plasma (Raymond and Smith 1977,1987) with normal cosmic elemental abundances predict that 34% of the atomic cooling power comes from a set of closely-spaced lines around 72 eV from Fe VIII, Fe IX and Fe X (see Figures 2 and 3). These lines should have dominated the observed counts in the Be Band (Bloch *et al.* 1986), but pulse height analysis of these data yields a mean energy which is incompatible with 72 eV (see Figures 4a and 4b, Bloch 1988). The flux from the Fe lines must be reduced by ~10 with respect to the rest of the SXRB spectrum within the Be band to make the model spectra compatible with the observed pulse height distribution. One way that the broad band Be, B and C band rates can be reconciled with the Be band pulse height distribution is to assume an equilibrium plasma emission using standard depleted abundances (Bloch 1988) as observed in higher density regions of the ISM. (see Figures 4c and 4d). Such a situation might arise if the hot plasma was formed by heating cool gas containing dust that has not yet been destroyed by the hot component.

2. THE ALEXIS PROJECT

A direct measurement of the flux in the 72 eV Fe lines would provide an important diagnostic of the hot local interstellar medium. The Earth and Space Sciences division at Los Alamos National laboratory is planning the Array of Low Energy X-ray Imaging Sensors (ALEXIS) experiment (Priedhorsky *et al.* 1988). It consists of 6 normal incidence ultrasoft X-ray telescopes with microchannel plate detectors flown on a mini-satellite. Each telescope will have a 40 degree field of view with one degree resolution and an effective collecting area of 0.56 cm² at the peak of the multilayer response. FWHM spectral resolution of the mirrors will be ~5%. The telescopes will utilize spherical multilayer mirrors with three or four narrow passbands distributed among the six telescopes. One passband will be centered at 72 eV to include the Fe line emission. The flux from these lines in the SXRB could produce a signal of as much as 100 counts per second for ALEXIS telescopes with this passband. With the narrow spectral response of ALEXIS, an unambiguous measurement of the flux from the lines will be possible. ALEXIS' spatial resolution and estimated year-long operating lifetime will allow the generation of all-sky maps of the SXRB in the 72 eV Fe lines with one degree resolution and considerable sensitivity. Comparison of such maps with those at other energies could help resolve remaining questions about the origin of the SXRB.

Several other upcoming missions with capabilities within the EUV regime (e.g., EUVE, ROSAT XUV wide field camera) should provide limits for the flux from these Fe lines. However due to their wider spectral responses or poorer area-solid angle products they will not easily provide true measurements.

Bloch, J. J., Jahoda K., Juda, M., McCammon, D., Sanders, W. T., and Snowden, S. L. 1986, *Ap. J. (Letters)*, **308**, L59.
Bloch, J. J. 1988, Ph. D. thesis, University of Wisconsin-Madison.
Juda, M. 1988, Ph. D. thesis, University of Wisconsin-Madison.
Marshall, F. J., and Clark, G. W. 1984, *Ap. J.*, **287**, 633.
McCammon, D., Burrows, D. N., Sanders, W. T., and Kraushaar, W. L. 1983, *Ap. J.*, **269**, 107.
Priedhorsky, W. C., Bloch, J. J.,Smith, B W., Strobel, K., Ulibarri, M, Chavez, J., Evans, E., Siegmund, O. H. W., Marshall, H., Vallerga, J., and Vedder, P., X-ray Instrumentation in Astronomy, SPIE Proceedings, **982**, 1988.
Raymond, J. C., and Smith, B. W. 1977, *Ap. J. Suppl.*, **35**, 419.
—. 1987, private communication (update to Raymond and Smith 1977).

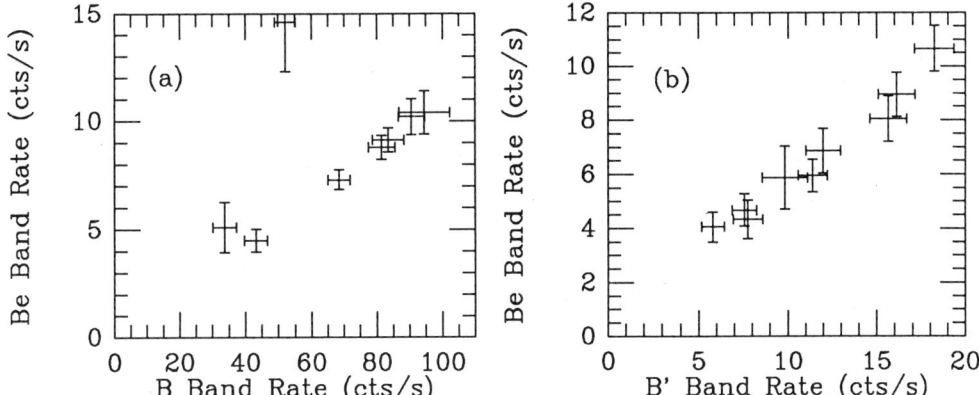

Figure 1. — (a) Be band count rate from the soft X-ray background *vs.* Wisconsin sky survey B band rate (averaged over the 15° field of view of the Be band detector). These results are from the first Be band detector flight reported in Bloch *et al.* 1986. (b) Be band rate vs. B' band rate observed on the second flight of the Be band detector (Juda 1988). The B' band rate was measured with a B band detector on the same rocket flight with the same field of view as the Be Band detector. An optical depth in the Be Band is $\sim 1 \times 10^{19}$ cm^{-2} HI while in the B band it is $\sim 6 \times 10^{19}$ cm^{-2} HI.

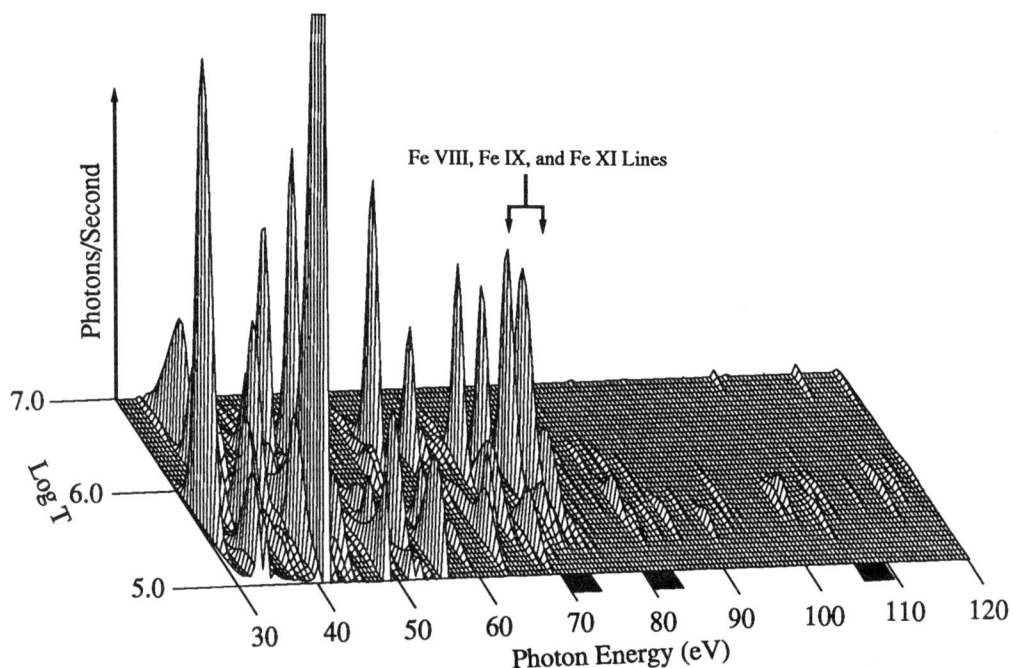

Figure 2.— Surface plot showing the photon spectrum from an optically thin plasma with temperatures ranging from Log T 5.0 to 7.0. The black boxes indicate the three ALEXIS bandpasses.

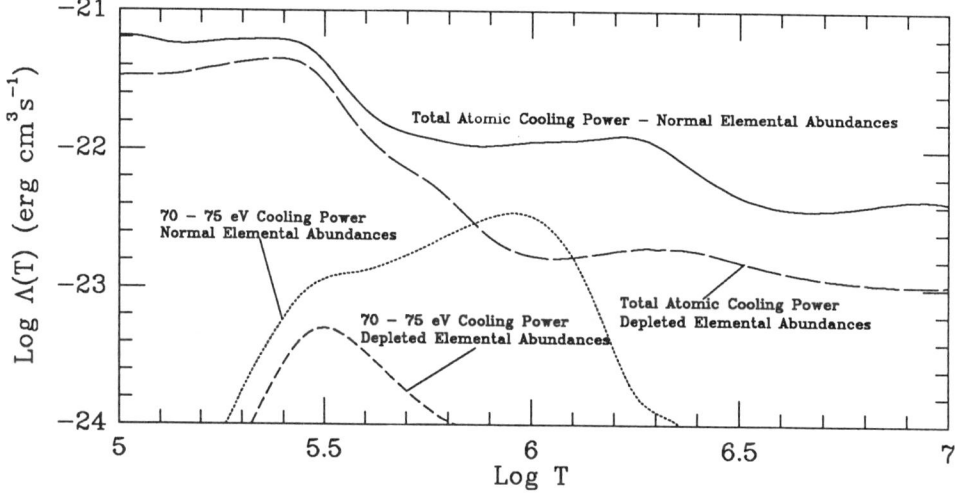

Figure 3.— Atomic cooling power radiated by an optically thin plasma in thermal equilibrium as a function of temperature. Normal and depleted abundance cooling curves are shown for the total power and the power radiated by lines between 70 and 75 eV.

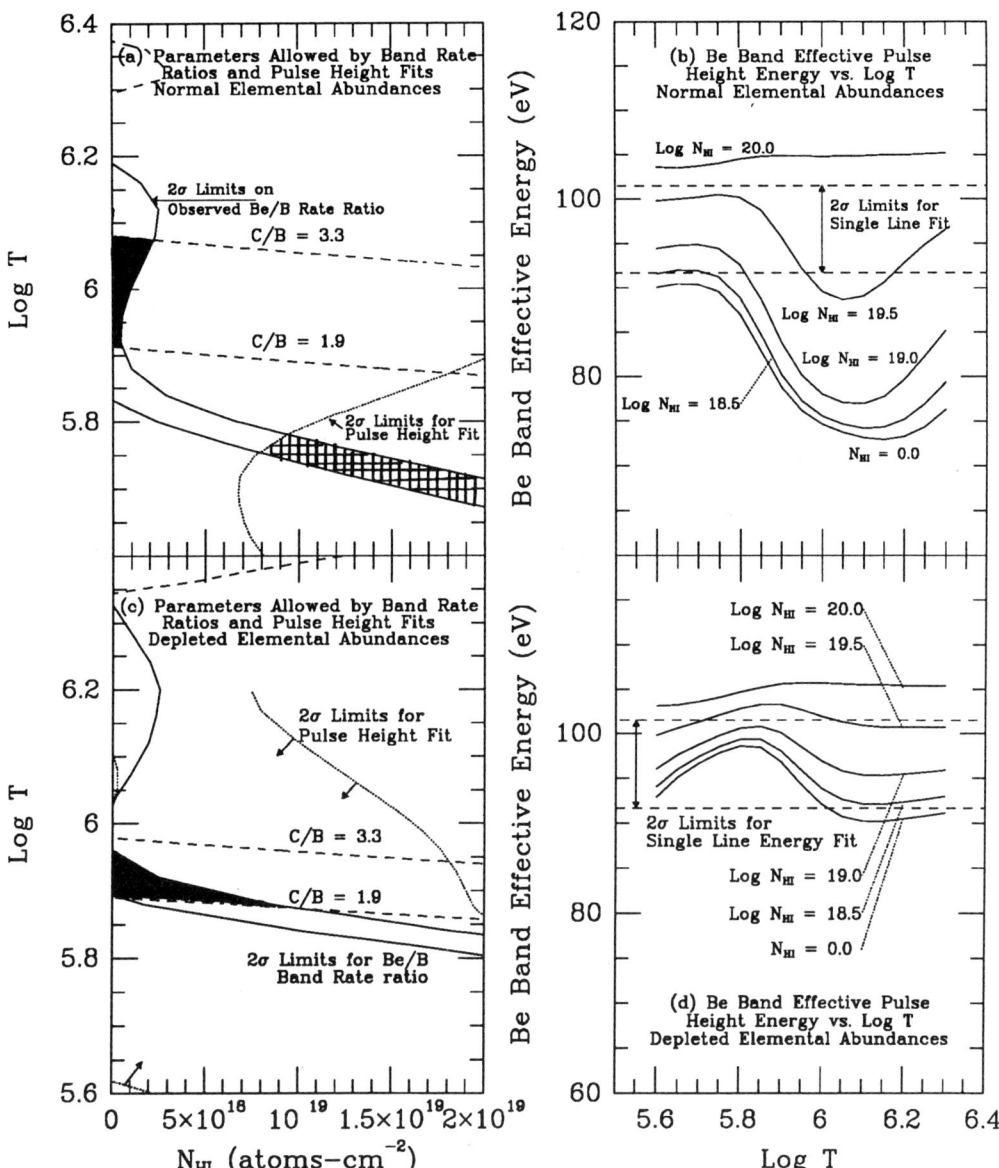

Figure 4. — (a) Temperature *vs.* intervening gas parameter for optically thin plasma emission. The black region is consistent with the 2σ limits on the observed Be/B ratio and the fluctuations in the C/B ratio for the observed regions of the SXRB. The hatched region is consistent with the Be/B ratio and the Be pulse height distribution. This region cannot generate enough C band X-rays and requires more than an optical depth of absorption for the Be band X-rays. (b) Be band effective energy vs. temperature for the same model as (a). (c) Same as (a) except using depleted abundances. The black region is consistent with the pulse height distribution and the Be/B ratio and generates most of the C band emission. (d) Same as (b) except using depleted abundances.

X-RAY EMISSION FROM REVERSE-SHOCKED EJECTA IN SUPERNOVA REMNANTS

Denis F. Cioffi[1][†], and Christopher F. McKee[2]

[1]NASA/Goddard Space Flight Center, Greenbelt, Maryland, U.S.A.
[2]University of California, Berkeley, California, U.S.A.

ABSTRACT. A simple physical model of the dynamics of a young supernova remnant is used to derive a straightforward kinematical description of the reverse shock. With suitable approximations, formulae can then be developed to give the X-ray emission of the reverse-shocked ejecta. The results are found to agree favorably with observations of SN1006.

I. Introduction

Supernova remnants (SNRs) spend most of their lifetimes in stages where the mass of the swept interstellar material, M_s, dominates the mass of the ejecta, M_{ej}, and where the radiation results from the cooling interior and from the actions of the main, forward shock. Before this occurs, however, radiation from reverse-shocked ejecta in young SNRs provides the first evidence of the interaction between a remnant and the medium which surrounds it.

In Cioffi, McKee, and Bertschinger (1988), through a careful comparison with a hydrodynamical simulation, we developed accurate and smooth analytic kinematics of an expanding SNR from soon after the explosion until the remnant merges with the interstellar medium (ISM). In Cioffi and McKee (1988) we indicated how this improved understanding enables a better calculation of the evolving interior luminosity. In Cioffi and McKee (1989; hereafter CM) we shall present a comprehensive analytic treatment of the X-ray luminosity from the interior and from the reverse-shocked ejecta. Here we show some results of our model of the reverse shock.

II. The Dynamics

Immediately after a supernova explosion, the pressure in the initially cold ejecta falls even further due to a rarefraction wave and the adiabatic expansion of the material. The strong pressure which lies behind the high-velocity forward shock drives a compression wave back into the ejecta, and that wave soon steepens into the reverse shock. Until the observations demonstrate the need for a more advanced approach, we rest our simple model (evolved from McKee [1974]) on the basic physics that govern the reverse shock interaction, and treat the problem analytically.

Our first assumption calls for uniform ejecta. Although both hydrodynamical simulations (e.g., Nomoto, Thielemann, and Yokoi 1984) and some supernova theory (e.g., Colgate and McKee 1969) predict an exponential fall-off in the density of

---

† NAS/NRC Resident Research Associate

the ejecta, our apparent success, and that of Hamilton, Sarazin, and Szymkowiak (1986), in matching the X-ray observations of SN1006 suggest that only a small fraction ($\lesssim 10\%$) of the ejecta is found in this "tail" (e.g., Chevalier 1988), and most of the luminosity originates in the dense, uniform part.

We next assume that the pressure which drives the reverse shock remains a fixed fraction φ of the pressure behind the forward shock,

$$\rho_{ej} v_{rs}^2 = \varphi \rho_o v_s^2, \qquad (1)$$

where v_{rs} is the velocity of the reverse shock, v_s is the velocity of the forward shock, ρ_o is the homogeneous mass density surrounding the supernova, and ρ_{ej} is the uniform density of the unshocked ejecta. Hamilton and Sarazin's (1984) numerical simulation shows that the pressure ratio φ varies smoothly with time, but for an average one might choose $\varphi \approx 0.2$. Setting $\varphi = 0.3$, however, produces a better agreement (see the table below) with the dynamics found in Woosley's more sophisticated hydrodynamical model ("CDTG7") used by Hamilton and Fesen (1988).

The Reverse Shock of SN1006

Parameter	Unit	G7	CM
Shocked Fraction F		0.80	0.95
Blast-wave mass M_s	$M_\odot$	4.3	3.9
Distance	kpc	1.7	1.7
Forward Shock Radius	pc	7.4	7.3
Blast-wave shock velocity	km s^{-1}	4000	4195
Reverse-shock velocity	km s^{-1}	4000	3939
Density ratio ρ_{ej}/ρ_o		0.15	0.12

G7: "CDTG7" (Hamilton and Fesen [1988])
CM: Cioffi and McKee (1989)

Our final assumption allows the outer ejecta radius to move with the velocity of the main shock, v_s. For an explosion of energy E, v_s is assumed constant at $v_{ej} = \sqrt{2E/M_{ej}}$ in the ejecta stage of evolution, and then smoothly slows into the Sedov-Taylor (ST) point-blast solution when $M_s/M_{ej} = 2/3$ (CM).

With the definition $F = M_{rs}/M_{ej}$, where M_{rs} is the ejecta mass which has been reverse-shocked, the above approximations eventually lead to

$$\frac{M_s}{M_{ej}} = \frac{9}{4\varphi}\left[1 - (1-F)^{1/3}\right]^2. \qquad (2)$$

When $\varphi = 0.3$, the ST stage begins at $F_{ST} = 0.658$ (CM), so in this formalism SN1006 has entered the ST stage.[1] The above table shows the excellent agreement between our analytic results and Hamilton and Fesen's (1988) hydrodynamical "CDTG7" computation when we use their best SN1006 model values of $M_{ej} = 1.32\,M_\odot$, $E = 1.0 \times 10^{51}$ ergs, an ambient hydrogen number density of $0.069\,\mathrm{cm}^{-3}$, and an age of 980 years.

---

[1] At $F = 1$, $M_s/M_{ej} = 7.5$ and the reverse shock hits the center, with reflected shocks presumably further complicating the interior.

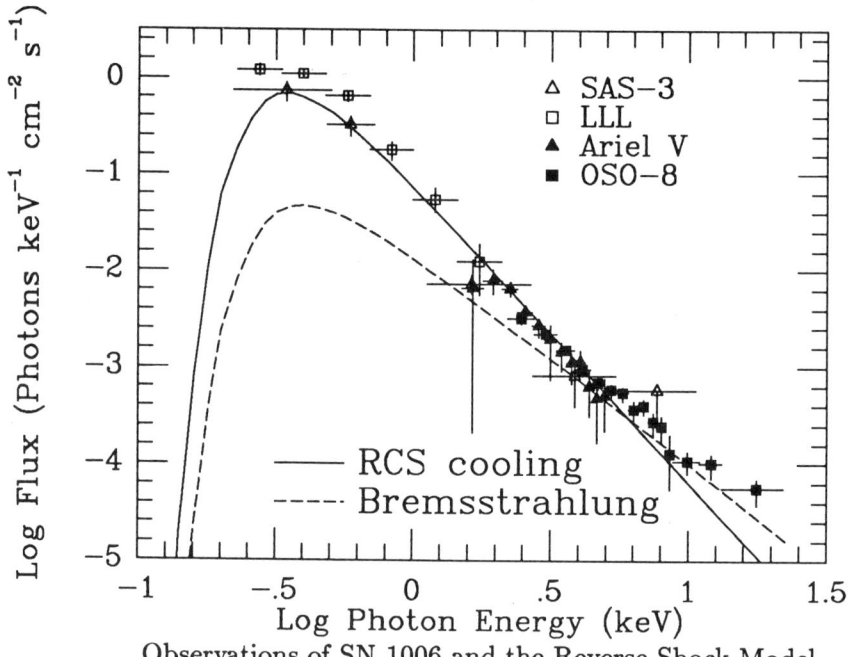

Observations of SN 1006 and the Reverse Shock Model

III. The Luminosity

The cooling of the ejecta produces a negligible decrease in the entropy of the gas during the dynamical timescales of interest, and thus we approximate an adiabatic shock for the luminosity calculation. We can write the total luminosity in terms of some appropriate cooling funcion $\Lambda(T)(\mathrm{ergs\,cm^3\,s^{-1}}) \propto T^m$, defined so that the luminosity per unit volume is given by $n^2\Lambda$, where n is the number density, and m is either $-1/2$ (a fit to Raymond, Cox, and Smith [1976] cooling) or $+1/2$ (bremsstrahulung).

Although non-equilibrium ionization (NEI) typically increases the luminosity of the forward shock by a large factor (e.g., Shull 1982), its specific effect on the reverse-shocked material is less certain. The higher density of the ejecta may make NEI relatively unimportant (e.g., Itoh 1977), especially when compared to the effect of the metal-rich composition on the standard cooling curve, which we do attempt to account for (CM). (For SN1006 we assume pure carbon.)

For a parcel of hot gas at temperature T, the approximate luminosity above (photon) energy ϵ can be obtained from multiplying the total by $\exp[-\epsilon/k_B T(F)]$, where k_B is Boltzmann's constant and $T(F)$ identifies the particular parcel of reverse-shocked material. We integrate through the shocked material in terms of the mass fraction F and obtain

$$L(t,\epsilon) = \int_0^{M_{rs}/M_{ej}} dF \, n\Lambda[T(F)] \frac{M_{ej}}{m_H} \exp[-\epsilon/k_B T(F)], \qquad (3)$$

where m_H is the mass of the hydrogen atom. To integrate the above, we need also assume a uniform pressure inside the reverse-shocked ejecta. For $\bar{P}_{ej}$ we use a fixed multiple of the immediate post-reverse-shock pressure, $\bar{P}_{ej} = f_p(3/4)\varphi\rho_o v_s^2$, where, from an average through Hamilton and Sarazin's (1984) similarity solution, we take $f_p = 1.6$. We differentiate the integral with respect to ϵ to obtain a rough spectrum. The integrals are straightforward, but long, and we present both the complete expressions and useful approximations in CM. For now we display the results through the particular example of SN1006.

The figure displays four sets of X-ray observations of SN1006, from Hamilton, Sarazin, and Szymkowiak (1986). Recent *Tenma* observations (not shown) suggest the possibility that these data are confused with emission from the Lupus region (see Koyama et al. 1987), which would steepen the spectrum above 1.5 keV. The solid line shows the results of the model using a fit to Raymond, Cox, and Smith (1976) cooling, and the dashed line shows our model's bremmstrahlung contribution.

IV. Summary

The reverse-shock in young SNRs reheats the ejected material to X-ray temperatures. We have constructed a model for the kinematics of the reverse shock by using the simple dynamics expected from the supernova's interaction with a homogeneous ISM. With power-law cooling curves, we integrate through the reverse-shocked ejecta to obtain the X-ray luminosity above any energy ϵ as function of time. Our agreement with observations of SN1006 illustrates the validity of this approach and suggests that it provides a useful tool for analyzing the physical conditions in young SNRs.

Acknowledgements. DFC thanks Andrew Hamilton for providing the SN1006 data in a convenient electronic form. CFM's research is supported in part by NSF grant AST 8615177.

References.

Chevalier, R. A. 1988, in *Supernova Remnants and the Interstellar Medium*, eds R. S. Roger and T. L. Landecker (Cambridge: Cambridge University Press), p. 31.

Cioffi, D. F., and McKee, C. F. 1988, in *Supernova Remnants and the Interstellar Medium*, eds R. S. Roger and T. L. Landecker (Cambridge: Cambridge University Press), p. 435.

Cioffi, D. F., and McKee, C. F. 1989, in preparation (CM).
Cioffi, D. F., McKee, C. F., and Bertschinger, E. 1988, *Ap. J.*, **334**, xxx.
Colgate, S. A., and McKee, C. 1969, *Ap. J.*, **157**, 623.
Hamilton, A. J. S., and Fesen, R. A. 1988, *Ap. J.*, **327**, 164.
Hamilton, A. J. S., and Sarazin, C. L. 1984, *Ap. J.*, **281**, 682.
Hamilton, A. J. S., Sarazin, C. L., and Szymkowiak, A. E. 1986, *Ap. J.*, **300**, 698.
Itoh, H. 1977, *Pub. Astr. Soc. Japan*, **29**, 813.
Koyama, K., Tsunemi, H., Becker, R. H., and Hughes, J. P. 1987, *Pub. Astr. Soc. Japan*, **39**, 437.
McKee, C. F. 1974, *Ap. J.*, **188**, 335.
Nomoto, K, Thielemann, F-K., and Yokoi, K. 1984, *Ap. J.*, **286**, 644.
Raymond, J. C., Cox, D. P., and Smith, B. W. 1976, *Ap. J.*, **204**, 290.
Shull, J. M. 1982, *Ap. J.*, **262**, 308.

INTEGRATED X-RAY SURFACE BRIGHTNESSES OF SUPERNOVA REMNANTS AND
COMPARISON WITH RADIO AND INFRARED VALUES

John Dickel[1], Lisa Norton[1], and Paul Gensheimer[1]

[1]Astronomy Department, University of Illinois, Urbana, IL, USA

ABSTRACT. A weak linear correlation is found between the x-ray and radio surface brightnesses of galactic supernova remnants. A flatter slope is found for the infrared-radio correlation.

1. INTRODUCTION

Supernova remnants produce significant amounts of x-ray, radio, and infrared emission. Although the individual emission processes in each wavelength band are different - shock heated gas, synchrotron radiation from relativistic particles trapped in magnetic fields, and collisionally heated dust, respectively - we expect some correlation between the emission at the different wavelengths. For the SNRs in the Large Magellanic Clouds there does appear to be a correlation between the x-ray and radio surface brightnesses (Berkhuijsen 1986). The infrared and x-ray emission also appear to be related (Dwek 1988). To check these correlations more fully for a significant sample of galactic remnants we have compared the integrated fluxes and surface brightnesses of SNRs in the Einstein, IRAS, and radio databases.

2. THE DATA

2.1 X-ray

Imaging data, primarily from the HRI and IPC detectors of the Einstein Observatory (Seward 1988), for SNRs in the Milky Way have been analyzed by numerous authors and by us to give an integrated flux emitted by the source in the energy band from 0.2-4 kev. To do this the source spectrum has been convolved with the instrumental response function to give the power within the chosen wavelength range and a correction has been made for interstellar absorption. Where sufficient IPC counts are present to determine a spectrum, both the temperature and extinction are used to find the correct flux. Various approximations have been used by individual authors and conflicting results exist; they are caused by discrepancies between HRI and IPC measurements, choices between 1- and 2-temperature models, etc. In some cases we have made rather arbitrary decisions but have tried to keep the results as consistent as possible. Where insufficient spectral information exists we have adopted an average temperature of 0.7 kev and a mean column density of hydrogen of 10^{22} cm^{-2}. The total x-ray fluxes are determined

Table 1. X-ray Properties of SNR's

Galactic l	b	Flux (erg/cm^2/s)*10^{-11}	Diameter (')	Temp. (keV)	N_H (atoms/cm^2)*10^{20}
4.5	+6.8	30.	3.3	0.52	28.
6.4	-0.1	18.	42	0.78	100.
11.2	-0.3	1100.	4.2	0.32	400.
21.5	-0.9	2.3	1.1	----	140.
27.4	+0.0	14.	4.3	1.0	200.
29.7	-0.3	7.4	0.5	----	500.
31.9	+0.0	28.	9.9x7.3	0.44	230.
33.7	+0.0	≥12.	9.6	<0.5	>100.
34.6	-0.5	51.	33x1	1.0	100.
39.2	-0.3	13.	8.2	0.78	100.
39.7	-2.0	5.2	93x21	2.2	100.
41.1	-0.3	≥5100.	6.4	<0.25	>500.
43.3	-0.2	9.0	4.5x3.8	2.8	400.
49.2	-0.7	0.14	55x36	0.78	100.
53.6	-2.2	59.	19x26	0.26	100.
65.3	+5.7	≥8.0		0.25	5.5
68.8	+2.6	1.7		1.4	60.
74.3	-8.5	910.	160	0.26	4.0
74.9	+1.2	0.36	10.7	----	130.
78.2	+2.1	≥11.		1.3	100.
82.2	+5.3	~11.		0.78	100.
89.0	+4.7	~16.		0.60	29.
109.2	-1.0	23.	33	0.70	30.
111.7	-2.1	530.	4.6x3.8	0.63	100.
119.5	+10.2	≥18.		0.78	100.
120.1	+1.4	220.	7.7	1.0	100.
130.7	+3.1	1.6	6.3x4.8	----	36.
132.7	+1.3	15.	60	0.60	60.
160.4	+2.8	~51.		0.80	35.
184.6	-5.8	7700.	1.9	----	30.
189.0	+3.0	46.	40	0.80	35.
260.4	-3.4	1300.	42x25	0.70	30.
263.5	-2.7	3800.		0.21	0.59
290.1	-0.8	6.8	19x12	0.78	100.
291.0	-0.1	1.7	29	----	20.
292.0	+1.8	40.	6.7	0.51	38.
296.1	-0.7	120.	20	0.78	60.
296.5	+10.0	27.	71	0.15	140.
315.4	-2.3	18.	38	1.2	2.8
320.3	-1.2	73.	30x23	0.30	90.
326.3	-1.8	96.	34x43	0.43	100.
327.1	-1.1	1.3	13x6	0.78	100.
327.4	+0.4	7.0	19	1.0	150.
327.6	+14.5	95.	34	0.16	6.5
332.4	-0.4	16.	8.7	0.51	100.

( ) means larger than mapped field   (----) means power law assumed

by finding the integrated counts/second from the FITS images provided by F. Seward and using the diagrams in the Einstein User's Manual (Harris and Irwin 1984). All told, reasonable values are available for 40 remnants. By comparison of differing values for the same object we estimate that individual brightnesses should generally be accurate to within a factor of 3. The results are presented in Table 1; specifics of individual sources can be obtained from the authors.

2.2 Radio and Infrared

Radio flux densities at a frequency of 1 GHz and diameters have been taken from Green's (1984) catalog. Most of the sources for which the other data are available are reasonably bright and accurate to ~30%.

Arendt's (1989) catalog of the infrared emission of all SNRs detected by the IRAS satellite presents tables of the infrared emission integrated over the four IRAS bands of 12, 25, 60, and 100 μm. The accuracy is often limited by confusion with the background, particularly in the 12 and 100 μm bands; most values are accurate to ~50%.

3. RESULTS

The surface brightnesses in each wavelength regime and have been correlated against each other and other parameters such as temperature (x-ray and infrared), age guestimates, etc. The best correlations found are shown in figure 1. Although the surface brightnesses do appear to be correlated, there is a very large scatter among the points. The correlation between the infrared and x-ray brightnesses is comparable to that between the infrared and radio.

For expected correlations, we have assumed that most observed remnants are in their point-blast adiabatic phase and integrated the non-equilibrium model spectra calculated by Hamilton et al. (1983) to determine the predicted x-ray emission in the .2-4 kev band. There is some increase in line emission as the continuum decreases so the x-ray brightness does not change much with temperature or time during this evolutionary phase; it is basically dependent only upon the square of the ambient density times the energy of the explosion. In contrast, the radio surface brightness does depend significantly on the shock speed and other time dependent phenomena. Also, the circumstellar medium must be very clumped to provide sites for instabilities which are responsible for particle acceleration and magnetic field amplification (Dickel et al. 1989). The radio brightness is proportional to density$^{17/12}$ · energy$^{7/4}$ · time$^{-1.2}$ (further calculations by Jones and Dickel). The density dependence in the two regimes is similar and is probably largely responsible for the correlation found. The much more significant time dependence of the radio emission than of the x-ray emission does not entirely remove the correlation. The clumps required to explain the observed radio emission may be dense enough to limit the shock speed and x-ray temperature to low values where time evolution caused by further

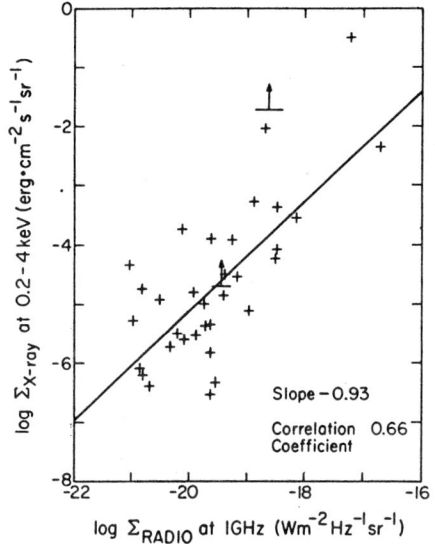
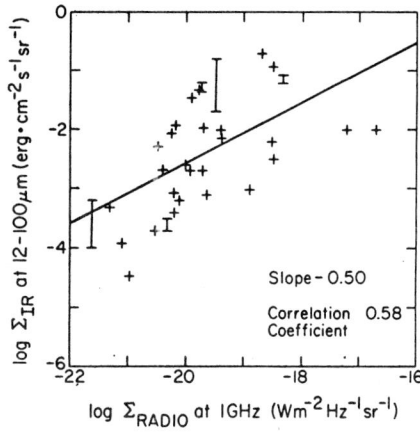

Fig. 1. SNR Brightness Correlations

cooling could have an effect. The true x-ray emission should be characterized by a range of temperatures rather than a single "best-fit" value. The energy dependence is rather different in the two wavelength regimes but the total range of energies is small.

The infrared emission is expected to closely track the x-ray emission from the hot gas responsible for heating the dust (Dwek 1988). However, the slope of the infrared dependence upon the radio (and thus x-ray) emission is flatter than linear. This could be explained by a depletion of dust in the brightest remnants. More collisions caused by the higher density could evaporate dust.

Clearly we need high spatial and spectral resolutions at both x-ray and infrared wavelengths plus multi-temperature models to properly evaluate these effects in real SNR. We eagerly await AXAF and SIRTF.

ACKNOWLEDGEMENTS: F. Seward, R. Arendt, and NASA contract JPL 958014.

REFERENCES

Arendt, R. G. 1989 Ap. J. Suppl., in press.
Berkhuijsen, E. M. 1986 Astron Astrophys., **166**. 257.
Dickel, J. R., Eilek, J. A., Jones, E. M. and Reynolds, S. P. 1989 Ap. J. Suppl., in press.
Dwek, E. 1988 in Supernova Remnants and the Interstellar Medium, ed. by R. Roger and T. Landecker (Cambridge: Cambridge Univ. Press), 363.
Green, D. A. 1984 Mon. Not. Roy. Astron. Soc., **209**, 449.
Hamilton, A., Sarazin, C. and Chevalier, R. 1983 Ap. J. Suppl., **51**, 115.
Harris, D. E. and Irwin, D. 1984 Einstein Observatory Revised Users Manual (Cambridge, MA: Harvard-Smithsonian Center for Astrophysics).
Seward, F. 1988 unpublished FITS tape from Einstein Obs. Databank.

A Comparison of Three Regions of Puppis A

K.F. Fischbach, L.M. Bateman, C.R. Canizares, T.H. Markert, and P.J. Saez

Massachusetts Institute of Technology, Cambridge, Massachusetts, U.S.A.

ABSTRACT High resolution X-ray spectral observations of Puppis A were performed with the FPCS on the *Einstein* Observatory at three regions of the remnant: the shock front, the bright eastern knot, and the interior. Plasma diagnostics of lines from OVII and OVIII constrain the values of electron temperature, ionization timescale, and hydrogen column density. We compare results of the diagnostics for these three regions. A non-equilibrium analysis of previously published fluxes of oxygen lines shows that the interior has not yet reached ionization equilibrium.

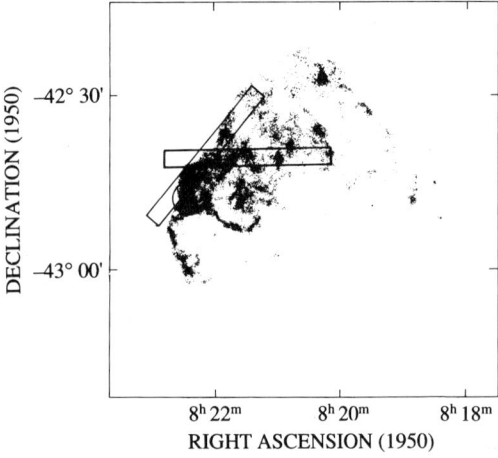

Figure 1 Aperture positions for the bright eastern knot, the shock front and the interior overlaid on an HRI image of Petre, et al. (1981).

Between April, 1979 and July, 1980, the Puppis A supernova remnant was observed using the Focal Plane Crystal Spectrometer (FPCS) on the Einstein Observatory (Canizares, *et al.* 1979). The remnant was observed at several positions; the bright eastern knot, the shock front, and at several closely spaced positions in the interior. A circular (6 arc minute) aperture was used to observe the bright eastern knot; in the other cases a 3 X 30 arc minute rectangular aperture was used to make the observations. The interior was observed with the rectangular aperture in a variety of angular orientations. Figure 1 shows FPCS aperture positions for the three regions, overlaid on the Puppis A image obtained by Petre *et al.* (1981). The X-ray observations consisted primarily of lines of helium-like and hydrogen-like ions of oxygen (O VII and O VIII) and neon (Ne IX and Ne X), and neon-like iron (Fe XVII).

The X-ray emission lines detected by the FPCS provide useful diagnostics of conditions in the line-emitting plasma. For transitions $i \rightarrow g$ and $j \rightarrow k$ of ionization states $+x$ and $+y$ of an element Z, the ratio of X-ray fluxes is given by:

$$\frac{f_{ig}}{f_{jk}} = \frac{\Omega_{ig} \times n_{+x} \times \exp(-\sigma(E_{ig})N_H) \times \exp(-E_{ig}/(kT_e))}{\Omega_{jk} \times n_{+y} \times \exp(-\sigma(E_{jk})N_H) \times \exp(-E_{jk}/(kT_e))}$$

where Ω_{ig} is the effective collision strength for transition $i \rightarrow g$, E_{ig} is the excitation energy, T_e the electron temperature, n_{+x} the density of ion x, σ the cross-section per hydrogen atom for photoelectric absorption at energy E_{ig}, and N_H the hydrogen column density (see Vedder et al. 1986). Since it is difficult to make enough measurements so as to solve for all of the unknowns, the approach we have taken is to measure a few line intensities selected so that as many parameters as possible will cancel. For example, two lines from the same ion define an allowed region in the parameter space of column density N_H and electron temperature T_e. In order to use two different ions from the same atom, it is necessary to perform a non-equilibrium analysis because the relative abundances of the various ions are functions of time since the plasma was shocked. The ionization structure is determined by solving a set of Z+1 simultaneous differential equations. (We have employed the technique of Hughes and Helfand (1985) to solve the ion balance equations.) The resulting ion abundances are then used explicitly in the equations for the line emissivities. The various parameters of the non-equilibrium model are T_e, N_H, and τ. Here τ ($\equiv$ electron density x time since the shock) is the ionization timescale and measures the extent to which ionization equilibrium has been attained. The complete analysis technique is discussed in more detail in Markert et al. (1988).

In order to compare the physical conditions of the three regions, (i.e. interior, bright eastern knot and shock front) we compared the results of the analysis described above (Winkler et al. 1981; Winkler et al. 1983; Fischbach et al. 1988) The derived allowable regions in (N_H, T_e) and (T_e, τ) parameter space are reproduced in Figures 2, 3 and 4 for the respective regions. In order to obtain an estimate of the ionization timescale, τ, of the interior, previously published oxygen line fluxes (Winkler et al. 1981) were applied to the non-equilibrium analysis, from which Figure 2b is obtained. In the analysis N_H was assumed to be 4×10^{21} cm^{-2}, but the results are not particularly sensitive to this assumption.

The deduced constraints on column density N_H, electron temperature T_e, and ionization timescale τ are shown in Table 1 for the three regions.

TABLE 1

PLASMA DIAGNOSTICS

Region	T (10^6 K)	τ (yr cm^{-3})	t (yr) [1]	n_e (cm^{-3}) [2]
Interior	4.3-19 [3]	300-2000	3700	0.1-0.6
Eastern Knot	6-8	800-1200	100	8-12
Shock Front	>7.9	150-400	100	1.5-4

(1) See Winkler et al., 1988; Winkler et al., 1983; Fischbach et al., 1988).
(2) Electron density n_e is estimated by naively assuming $\tau = n_e t$. In fact, the line-emitting region consists of material shocked at different times, so the reported value of n_e is probably a slight underestimate.
(3) Winkler (1981) derives $T_e \geq 1.5 \times 10^6$. These temperature constraints are obtained from the analysis shown in Figure 2b.

The non-equilibrium analysis on oxygen lines from the interior suggests that the plasma departs substantially from ionization equilibrium. Thus, none of the three regions is found to be in ionization equilibrium despite the advanced age (~3700 years) of the remnant.

Acknowledgements We thank Jack Hughes for supplying his computer code for the non-equilibrium analysis. We are grateful to Elaine Aufiero for preparing the manuscript. This work was supported in part by NASA grant NAG 8-494.

REFERENCES

Canizares, C.R., Clark, G.W., Markert, T.H., Berg, C., Smedira, M., Bardas, D., Schnopper, H., and Kalata, K. 1979, *Ap. J. (Letters)*, **234**, L33.
Fischbach, K.F., Canizares, C.R., Markert, T.H., and Coyne, J.M. 1988, in *Supernova Remnants and the Interstellar Medium*, eds. R.S. Roger and T.L. Landecker, (Cambridge University Press, Cambridge) 153.
Hughes, J.P. and Helfand, D.J. 1985, *Ap. J.*, **291**, 544.
Markert, T.H., Blizzard, P.L., Canizares, C.R. and Hughes, J.P. 1988, in *Supernova Remnants and the Interstellar Medium*, eds. R.S. Roger and T.L. Landecker, (Cambridge University Press, Cambridge) 129.
Petre, R., Canizares, C.R., Kriss, G.A. and Winkler, P.F. 1981, *Ap. J.*, **258**, 22.
Vedder, P.W., Canizares, C.R., Markert, T.H. and Pradhan, A.K. 1986, *Ap. J.*, **307**, 269.
Winkler, P.F., Canizares, C.R., Clark, G.W., Markert, T.H., Petre, R. 1981, *Ap. J.*, **245**, 574.
Winkler, P.F., Canizares, C.R. and Bromley, B.C. 1983, in *Supernova Remnants and Their X-Ray Emission*, eds. J. Danziger and P. Gorenstein, (Dordrecht: Reidel), 245.
Winkler, P.F., Tuttle, J.H., Kirshner, R.P. and Irwin, M.J. 1988, in *Supernova Remnants and the Interstellar Medium*, eds. R.S. Roger and T.L. Landecker, (Cambridge University Press, Cambridge 65.}

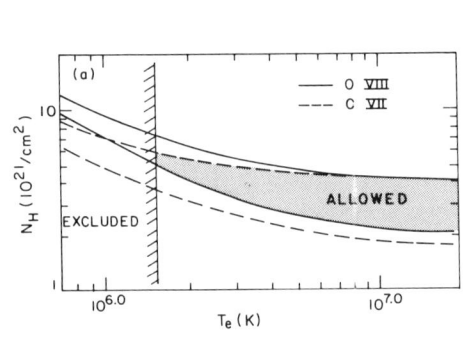
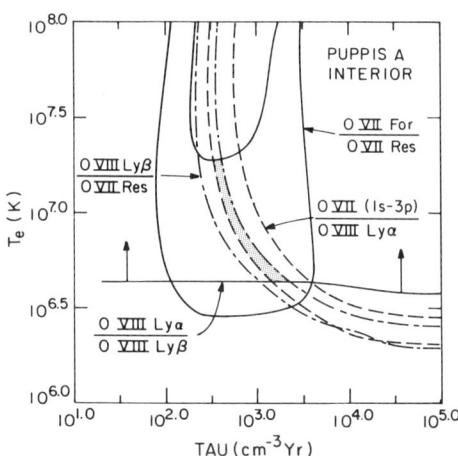

Figures 2a,b   Regions in (N_H, T_e) and (T_e, τ) parameter space allowed by FPCS measurements of the interior. Shaded area is region of overlap consistent with all measurements. Figure 2a is reproduced from Winkler et al. (1981). Figure 2b shows results of a non-equilibrium analysis using oxygen lines as published in Winkler et al. (1981) and assuming $N_H = 4 \times 10^{21}$ cm^{-2}. Each pair of contours in Figure 2b indicates 90% confidence.

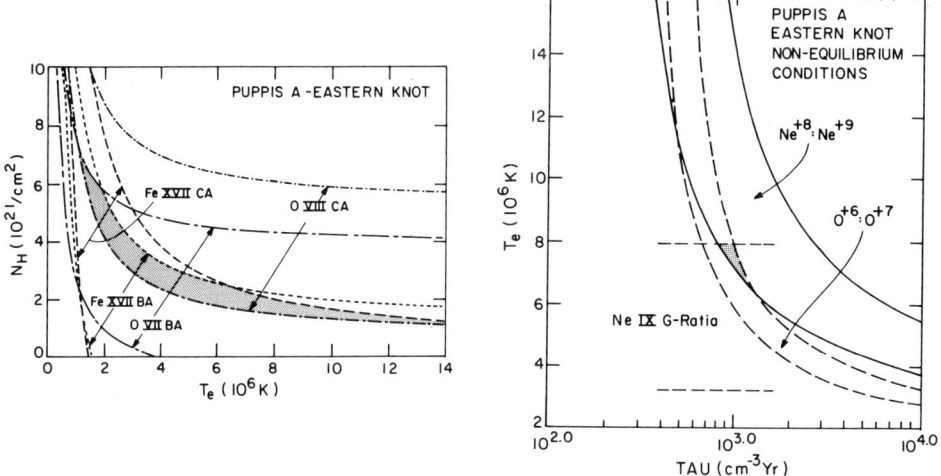

Figures 3a,b Allowed (shaded) regions in (N_H, T_e) and (T_e, τ) parameter space for the bright eastern knot (Winkler et al. 1983).

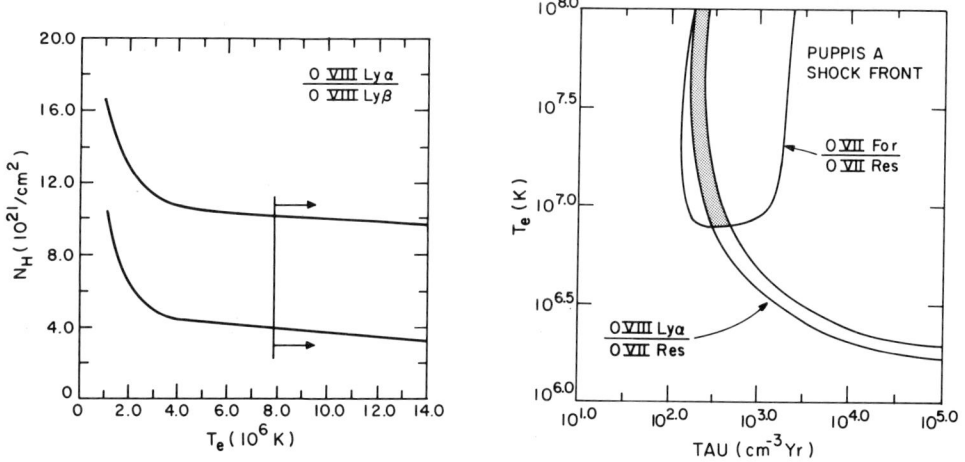

Figures 4a,b Allowed (shaded) regions in (N_H, T_e) and (T_e, τ) parameter space for the shock front (Fischbach et al. 1988). Arrows to the right of the vertical line in Figure 3a indicate allowed region.

5. Compact Binaries

AN EXOSAT OBSERVATION OF SPECTRAL VARIABILITY FROM THE RS CVn BINARY AR LAC

N.E. White[1], R.A. Shafer[1,2], A.N. Parmar[1], K. Horne[3], and J.L. Culhane[4]

[1] EXOSAT Observatory, ESTEC, Noordwijk, The Netherlands
[2] present address: GSFC, Greenbelt, Maryland, USA
[3] Space Telescope Science Institute, Baltimore, Maryland, USA.
[4] Mullard Space Science Laboratory, Holmbury St Mary, Surrey, UK.

ABSTRACT. The eclipsing RS CVn system AR Lac (G2IV+K0IV) has been observed continuously for one 2 day binary cycle with the EXOSAT observatory. Below 1 keV a factor of two intensity modulation is seen with a flat bottomed minimum around the time of primary eclipse and a shallow dip preceeding primary eclipse. Above 1 keV, where only emission with temperatures $> 10^7$ K would be detected, no orbital modulation or eclipse is seen. This suggests that the > 20 million degree emission comes from a large region, comparable in size to the binary seperation. The modulation in the < 1 keV lightcurve has been modelled by χ^2 fitting to X-ray bright spots and by using maximum entropy deconvolution. The lower temperature emission is found to orginate in compact regions with a pressure and temperature similar to that of the flaring sun.

1. INTRODUCTION

The close proximity of the sun allows the luxury of imaging individual magnetic loop structures. For other stars the three dimensional spatial distribution of the X-ray emitting corona must be mapped indirectly by observing through an eclipse of the corona by a binary companion, or by searching for a rotational modulation caused by the self-eclipse of active regions by the underlying star. The RS CVn binaries show extreme coronal activity and are ideal for eclipse and rotational modulation investigations. They typically have orbital periods in the range 1 to 14 day with the hotter component of spectral type F-G and luminosity class IV-V. The cooler, usually more evolved component is typically K0IV. A characteristic feature of these systems is a low-amplitude optical modulation at the orbital period caused by giant star spots on the stellar surface, analogous to sun spots (e.g. Eaton and Hall 1979). Orbital phase dependent features found in rotationally broadened absorption and emission line profiles can be used to locate, respectively, dark starspots and regions of enhanced chromospheric emission (Vogt and Penrod 1983, Walter et al 1987).

A spectral survey of the quiescent spectral properties of seven RS CVn's by Swank et al (1981) using the *Einstein* SSS showed at least two distinct temperatures distributions to be present of $4 - 8 \times 10^6$ K and $\geq 2 \times 10^7$ K. The emission measures of both components are similar and within the range $10^{53} - 10^{54}$ cm^{-3}. Swank et al (1981) suggest that the bimodal temperature distribution indicates two isothermal families of magnetic loops, with possibly, different dimensions and pressures. The spectra of luminous late type stars in the Hyades taken by the *Einstein* imaging proportional counter (IPC) are well fit by a single family of loops, with a continuous temperature distribution and a maximum temperature at the loop apex of $\sim 10^7$ K (Stern, Antiochos, and Harnden 1986). A two temperature model also provides an acceptable fit. This suggested that detectors with strong structure in their energy efficiency (e.g. the carbon absorption edge in the IPC response) and poor energy resolution may artificially create a two temperature distribution (Majer et al 1986). However, EXOSAT transmission grating spectra of two RS CVn systems are only well fit by a two temperature model. The Stern et al (1986) single loop model was excluded. If a solar loop model is relevant then Mewe et al (1986) require at least two families of loops with apex temperatures of $\sim 5\times 10^6$ K and $>10^7$ K. The EXOSAT grating spectra do not show any evidence for emission with a temperature of 1-2 million degrees expected from close to the base of the loops. This requires that the cross section of the loops expand rapidly above the stellar surface. The EXOSAT grating spectra clearly demonstrate that the ability to make unambiguous comparisons between loop models requires observations taken with sufficient spectral resolution to resolve temperature sensitive line complexes.

AR Lac is a 1.98 day period eclipsing RS CVn system containing a G2IV primary and a K0IV secondary with radii of 1.54 $R_\odot$ and 2.81 $R_\odot$ respectively, and a seperation of 9.22 $R_\odot$ (Chambliss 1976). The combination of a close to 90° inclination, relatively short orbital period, nearby distance of 50 pc and high X-ray luminosity of 10^{31} erg s^{-1} make AR Lac a prime candidate for an eclipse/rotational modulation study. Walter, Gibson and Basri (1983) observed over one orbital cycle using the *Einstein* observatory but only with an orbital phase coverage of $\sim$ 17 %. Minima seen around the time of primary and secondary eclipse could be modelled as due to the eclipse of compact regions on each star and one extended region on the K subgiant.

The highly elliptical 90hr earth orbit of the European Space Agency's X-ray astronomy observatory EXOSAT gave for the first time the possibility to obtain uniterrupted X-ray light curves for up to $\sim$76hr (White and Peacock 1988 and refs therein). This paper presents an uninterupted EXOSAT observation over one complete orbital cycle of AR Lac with an orbital phase coverage of close to 100%. We consider in detail the location, height and temperature of the corona in AR Lac.

2. OBSERVATION AND RESULTS

The EXOSAT observation of AR Lac began at UT 1300 on 1984 July 3 and ended at the same time two days later. These times correspond to orbital phases, ϕ, from 310° to 685°, where $\phi = 360°$ is the center of primary eclipse. This assumes an ephemeris for the interval 1983–1985 of JD Hel. 2,445,611.6290 + 1.98316E provided by M. Rodono (1988, Priv. communication). Two instruments were used. The low energy imaging telescope (LEIT) with a channel multiplier array and a 4000Å filter (0.05 - 2keV) and the medium energy (ME) proportional counter array (1 - 30keV). The response of these two instruments are nicely matched to the spectra of the two temperature components found by Swank et al (1981). The ME is sensitive only to the higher temperature component, the LEIT is sensitive to both.

The background subtracted lightcurves from the ME and LEIT are shown in Figure 1. The duration of the two eclipses, from 1st to 4th contact, are indicated by horizontal lines. The most striking feature of the LEIT light curve is a minimum lasting $\sim$12hrs centered on the time of primary eclipse, where a factor of two reduction in count rate occurs. There is also a shallow minimum preceeding the time of secondary eclipse. In the ME the most prominent feature is a flare at the beginning of the observation that lasted for $\sim$2hr. The ME lightcurve shows no evidence for features that might be identified with the eclipses and, in particular, no minimum is seen at the time of primary eclipse corresponding to that seen in the LEIT. There is some low level variability in the ME, with a second smaller flare following secondary eclipse. Both flares are also seen in the LEIT.

Background subtracted spectra from the ME have been accumulated for five different intervals during the observation. The first interval, lasting 2hr, only includes the large flare that occured at the beginning of the observation. The other four spectra have accumulation times of between 8 and 13 hr and cover the remainder of the observation. A significant signal is detected only between 1.5 and 6.0 keV. The flare spectrum has been obtained by subtracting the following source plus background data. This spectrum represents the average over the whole flare which had a very similar rise and decay time of ~ 30 min. The decay overlaps the time of primary eclipse and it is possible that part of the decay was caused by the flaring region being eclipsed by the companion star.

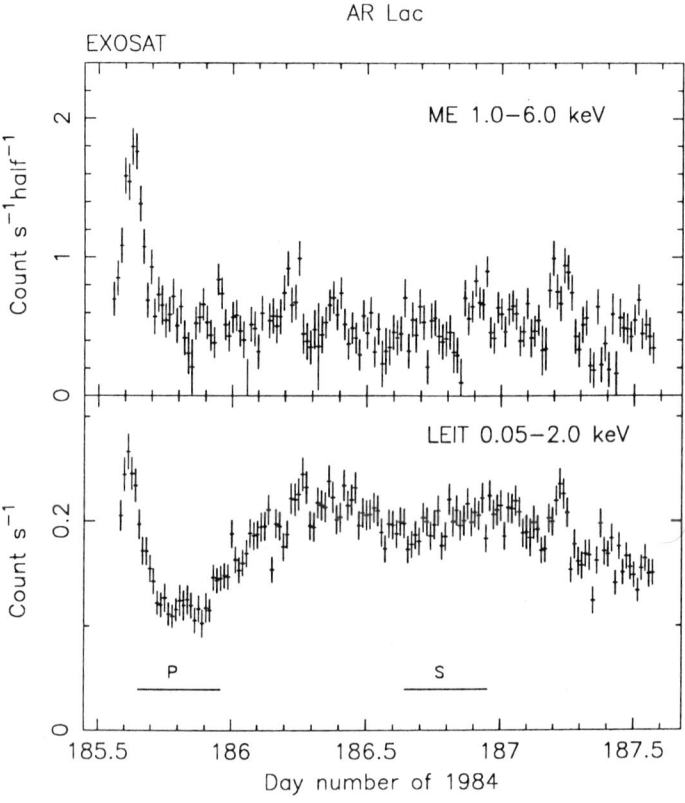

Figure 1. The light curve of AR Lac. The accumulation time is 1200s. The two horizontal lines indicate the time from 1st to 4th contact for primary (P) and secondary (S) eclipses.

The results for the quiescent cases are all similar with a temperature of 1.4×10^7 K and an emission measure of 5×10^{53} cm^{-3}. This is respectively a factor of ~ 3 and ~ 6 lower than the two values reported in Swank et al (1981). The ME flare spectrum has a temperature of 4.5×10^7 K and an emission measure of 4×10^{53} cm^{-3}. The emission measure of the low temperature component and the line of sight interstellar absorption to AR Lac can be constrained by including the LEIT lexan 4000Å filter count rate in the spectral fitting. The temperature of the low temperature component could not be usefully constrained. If it is fixed at $5.0 - 7.0 \times 10^6$ K (Swank et al 1981) then $N_H < 1.3 \times 10^{20}$ H cm^{-2}. The emission measure depends on orbital phase and is $< 5.0 \times 10^{53}$ cm^{-3} and $< 2.5 \times 10^{53}$ cm^{-3} for the maximum and minimum of the modulation. A lower limit to the emission measure comes from the LEIT modulation which requires that the diference between the peak and minimum counting rate comes from a component with a temperature $< 10^7$ K. This corresponds to a minimum emission measure of 2×10^{53} cm^{-3}, which in turn requires N_H to be $> 1 \times 10^{19}$ H cm^{-2}.

3. ECLIPSE MODELLING

The primary objectives are to 1) determine the maximum fraction of the stellar surface covered by an X-ray emitting corona, 2) set limits on the vertical extension above the stellar surface of the X-ray emitting structures, and 3) to locate where the X-ray structures are on the surface of the two stars. We have modelled the LEIT lightcurve both by χ^2 fitting to uniformly emitting structures on one or both of the stars and using a maximum entropy technique. All the non-flaring features in the LEIT lightcurve are be assumed to be the result of either an eclipse by the companion star or a rotational modulation. The X-ray lightcurve recorded by the ME shows two flares and also many small fluctuations, but no overall orbital modulation or eclipses. This precludes any useful eclipse modelling of the ME lightcurve.

The total count rate in the LEIT will be a combination of the 15×10^6 K component seen in ME, plus that from the 7×10^6 K component. The contribution of the hotter component to the LEIT can be estimated from the spectrum measured by the ME, but depends on the interstellar absorption. The range of N_H found in the previous sub-section predicts that the hot component will contribute between 30 and 100% of the LEIT counting rate during primary eclipse. A d.c. component was included as a free parameter to account for the unmodulated contribution to the LEIT counting from the hot component.

Acceptable fits to the LEIT lightcurve are obtained for only two structures, but with multiple solutions caused by ambiguity as to which star they are on. Three different permutations are possible: *Case 1*, the G star is X-ray bright and the K star is dark; *Case 2*, the G star is dark and the K star is bright; *Case 3*, both the G and the K stars are bright. The best fits with two structures require the height of one to be an order of magnitude larger than the other. The significance of this was tested by repeating the fit, but with the height of all the structures the same. In some models a third structure was then required to obtain an acceptable fit. A total of eight different models were tried. The height and the covering fraction are very similar for all eight models, only the locations change. We illustrate two representative fits for case 3 in Figures 2 and 3. Each structure is labeled by the spectral type of the underlying star and its longitude, e.g. G180. Zero longitude corresponds to the central meridians on the G and K stars at primary eclipse and increases in the same sense as increasing orbital phase.

The Maximum Entropy Method (MEM) models the X-ray light curve by distributing the coronal X-ray emission as uniformly as possible over the stellar surfaces. This approach complements the χ^2 fitting because the broad class of all possible positive X-ray maps fall within the scope of MEM and it can be used to investigate if it is possible to fit the data with a larger covering fraction. MEM maps were fitted to the observed X-ray light curve with the aid of the iterative fitting program MEMSYS, which uses an algorithm similar to that described by Skilling and Bryan (1984). MEM maps were generated for each of the models found from the χ^2 fitting. The coronal height and the d.c. background intensity levels were set at their best-fit values. The covering fraction $f(p)$ is defined as the fraction of the stellar surface enclosed by the surface brightness contour that encloses a fraction p of the total X-ray flux. The contour levels used in the Figures are not equally spaced in surface brightness, but instead they enclose the fractions $p = 0.1, 0.3, 0.5, 0.7$ and 0.9 of the total flux. The latter two contour levels are given as dashed lines. Above each map the surface area of each star responsible for 50% and 90% of the total emission are given. We concentrate on the maps made assuming both stars are bright and these are shown in Figures 4 and 5.

The map shown in Figure 4 assumes the coronal height on the K star is small (0.04 $R_\odot$), but that on the G star it is 1.5 $R_\odot$, as given by the χ^2 fitting. The cluster of three bright regions at high

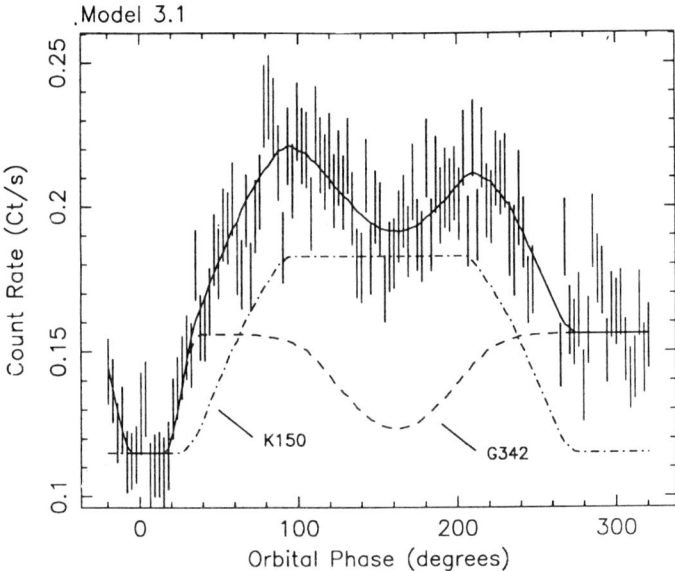

Figure 2. The best fit to the LEIT light curve for two uniformly emitting spots, with one on each star. The G star structure is extended with a height, H, of $1.5R_\odot$ and the K star structure compact with $H < 0.05\ R_\odot$. The dashed lines show the contributions of the individual lightcurves to the total. Two flares have been excluded.

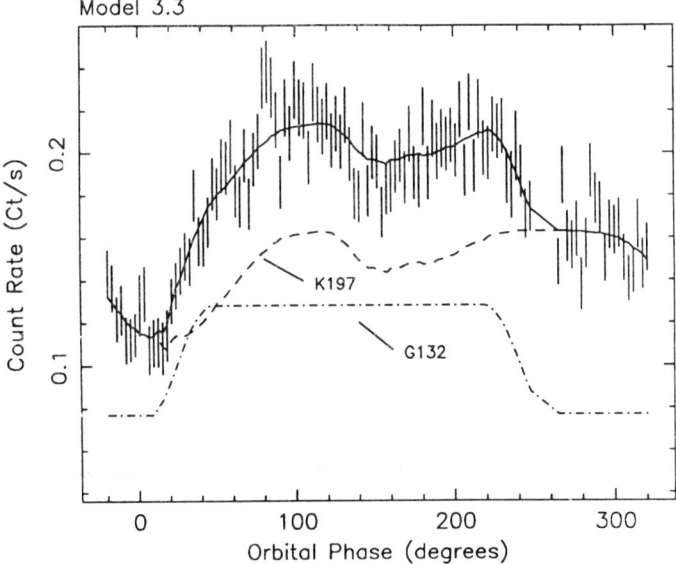

Figure 3. The same as Figure 2, but with the extended structure on the K star with $H = 4.0\ R_\odot$ and the compact structure on the G star with $H < 0.09\ R_\odot$.

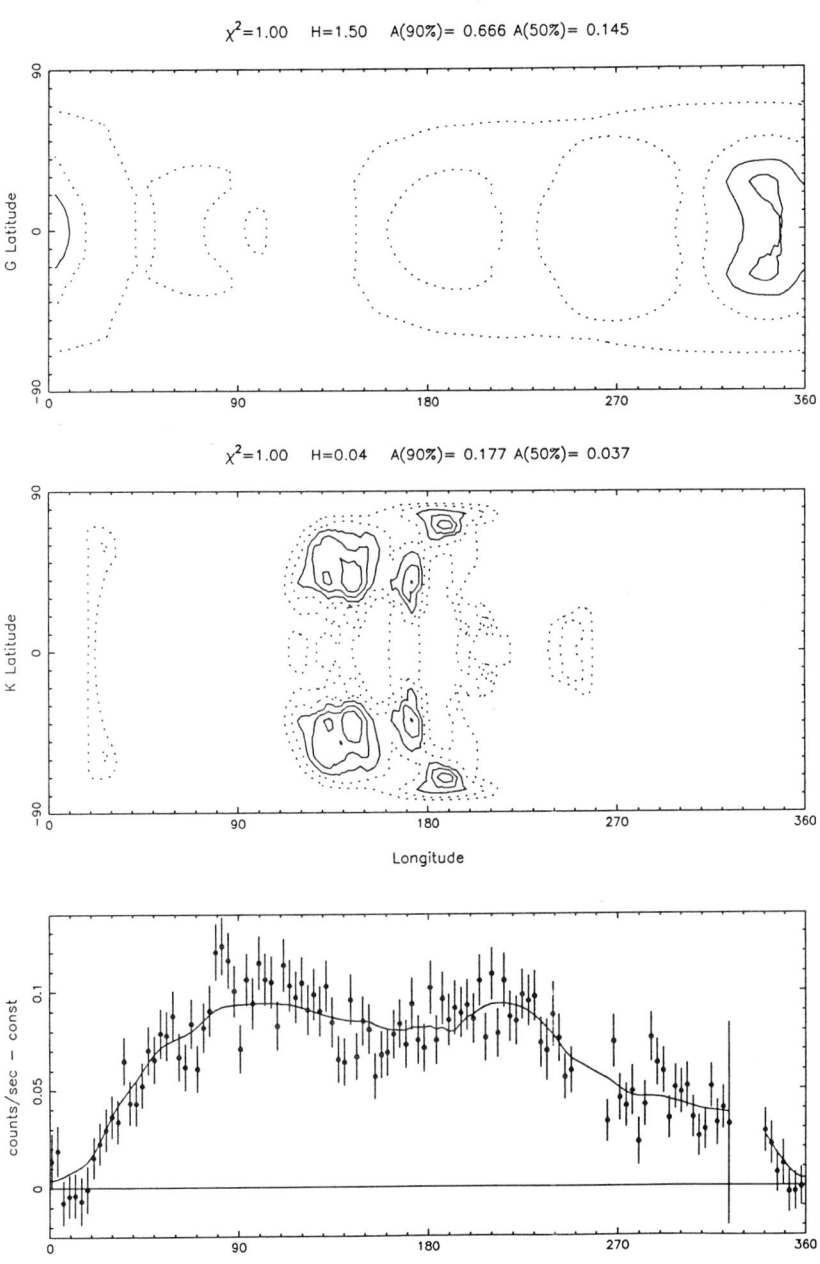

Figure 4. The MEM map generated under the assumption that both stars are X-ray bright. The top panel shows the G star map and the lower panel the K star map. The coronal height on the G star was assumed to be 1.5 $R_\odot$ and on the K star 0.04 $R_\odot$. The contours give pressure isobars of 1.9, 3.4, 5.1, 6.7 and 8.6 dyne cm^{-2} on the K star and 17, 31, 47, 61 and 79 dyne cm^{-2} on the G star. The dashed contours distinguish the two lowest pressures.

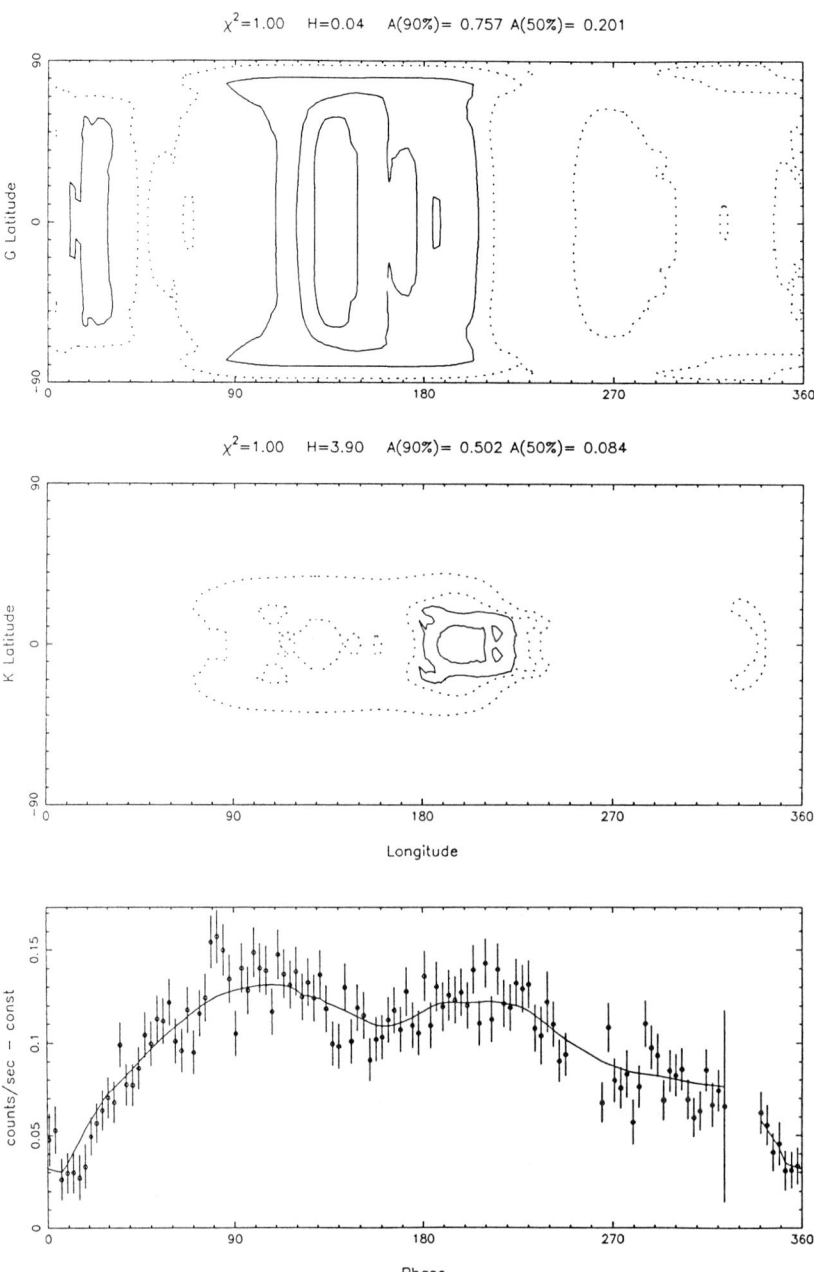

Figure 5. The MEM map generated under the assumption that both stars are X-ray bright and that the coronal height on the K star is 3.9 $R_\odot$ and on the G star 0.04 $R_\odot$. The contours give pressure isobars of 18, 31, 39, 59 and 77 dyne cm^{-2} on the G star and 1.0, 1.8, 2.3, 3.5 and 4.5 dyne cm^{-2} on the K star.

latitudes on the K star between longitudes of 120-200° matches up with K150 in χ^2 modelling. Similarly on the G star the structure near the equator at $\sim 340°$ corresponds nicely with G340. In Figure 5 the K star corona is extended with $H_2 = 3.9$, and the G star corona is compact with $H_1 = 0.04$, the values given by the χ^2 fitting. There is a prominent structure on the equator of the K star at a longitude of 200°. This corresponds to region K215 in Figure 3. This tall spot protrudes so far above the stellar surface that its eclipse by the G star is displaced in phase and so gives the broad minimum centered on phase 160°. The G star flux remains high with the primary eclipse caused by a region at a longitude of 25°. There is also a structure on the G star at a longitude of 140°, which matches well with the region G138.

Because the inclination of AR Lac is 90°, we cannot distinguish between north or south latitudes. In addition, the latitude is usually only constrained to be either high or low. If the structures are on the G star then their latitude is $< 50°$. If the structures are on the K star then they are, with one exception, always between 40° and 80°. The exception is an extended structure at a longitude of $\sim 210°$. This must be at a low latitudes on the K star, otherwise the flux in the d.c. component is less than expected from the extrapolation of the hot component.

All the models require at least one compact X-ray emitting structure on one of the stars. The height of this compact structure is only marginally resolved and because of this we will treat all of the measurements as upper limits. If the structures are all on the G star, or on both the G and the K star, then $H < 0.04$ $R_\odot$ ($< 7 \times 10^9$ cm). This increases to < 0.17 $R_\odot$ ($< 1 \times 10^{10}$ cm) if they are only on the K star. If the height of individual structures are allowed to be different then there is in some of the models evidence for a more extended structure with a height of 1.6 ± 0.8 $R_\odot$ (1×10^{11} cm) if it is on the G star, or $4.1^{+1.6}_{-0.9}$ $R_\odot$ (3×10^{11} cm) if it is on the K star.

In the χ^2 fitting the half-opening angles range between 10 and 40° on the G star and 3 and 40° on the K star. This translates to a maximum covering fraction of $26 \pm 10\%$ for the G star and $\sim 15\%$ for the K star. The MEM maps give the most conservative estimate of the covering fractions. In the worst case 50% of the flux comes from 20% of the surface area of the star, although 10% is more typical.

4. DISCUSSION

The different orbital lightcurves seen above and below 1 keV in the ME and the LEIT provide direct evidence that the two temperature components discovered by Swank et al (1981) come from two physically distinct regions. The lack of any modulation at energies above 1 keV requires that the high temperature emission (> 10 million degrees) be located in such a way as to avoid a deep eclipse by the two stars. In contrast, the factor of two orbital modulation seen below 1 keV in the LEIT requires the presence of compact coronal structures on one or both stars with temperatures of < 10 million degrees. Three types of coronal structure are identified: 1) compact features with $H < 10^{10}$ cm containing the cool component, 2) in some models extended structures with $H \sim 10^{11}$ cm, also containing the cool component, and 3) extended structures containing the hot component with a height that must large compared to that of the underlying stars, and probably pervades the whole binary system.

The measured temperature, T, and emission measure, $EM = n_e^2 V$, combine to give the pressure, $P = 2n_e kT$, of the emitting plasma for a given volume, V. The results of the χ^2 fitting and MEM maps provide the first realistic limits on the volume filled by the low temperature component. The χ^2 fitting gives pressures in the compact regions that range from 100-450 dyne cm^{-2} for both stars. In the extended regions the pressure is a factor of 10-100 lower with a value of ~ 15 dyne cm^{-2} on the G star and ~ 4 dyne cm^{-2} on the K star. The dependence of the pressure on the height of the corona is illustrated in Figure 6 for each star using representative covering fractions of 20% and

3%, and a typical emission measure of 1.0×10^{53} cm^{-3}. These pressure-height curves show how the average pressure increases for smaller covering fraction and height.

The range and distribution of the MEM derived pressures are in good agreement with those from the χ^2 fitting. The maximum entropy technique in general tries to provide images with the smallest range as possible of surface brightness (pressure) and can be considered as lower limits. For the compact emission on either star, the peak pressure of the various models ranged over 60–110 dynes cm^{-2}. The much larger heights of the extended emission components produced smaller peak pressures of 4–7 dynes cm^{-2}.

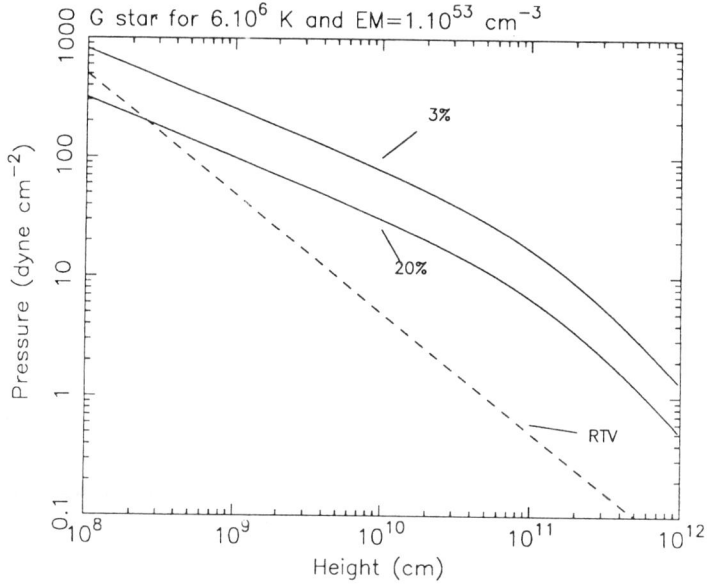

Figure 6. The solid lines indicate the relation between pressure and height for covering fractions of 3% and 20% for an emission measure of 1×10^{53} cm^{-3}. The dashed line shows that predicted by the RTV scaling law. These are for the G star, for the K star the solid lines are a factor of two lower, the dashed lines remain the same.

The limit set on the height of the compact corona of $< 1 \times 10^{10}$ R$_\odot$ is not very restrictive since it includes the dimensions of the entire range of structures found on the sun. However, in all cases the peak pressures for AR Lac stars are a factor of 10-100 higher than the quiescent sun. These pressures and the temperature of the cool component of $\sim 7 \times 10^6$ K are, however, similar to those found in solar flares. For the larger structures found in some of the AR Lac models the heights are comparable to the radius of the underlying star and also to the pressure scale height. While these larger loops are not required by all the models, the pressures are comparable with those found on the sun, but the dimensions are an order of magnitude larger.

The energy balance of a static, thermally insulated magnetic loop heated by an arbitrary energy source gives a scaling law between the half-length of the loop, L, and the peak temperature, T_p, and pressure of the form $T = 1.4 \times 10^3 (PL)^{1/3}$ (Rosner, Tucker and Vianna 1978, hereafter RTV). The EXOSAT results on AR Lac provide an opportunity to test the validity of this formulism in a stellar setting. The RTV relation is plotted on Figure 6 (assuming L= πH/2). The point where the RTV relation intercepts the pressure-height curves is within the upper limits on the height of the compact regions. Taken at face value the solutions require that for the models where the G star is bright the pressure is > 200 dyne cm^{-2} and the height $< 2,000$ km. For the models where the K star is bright the RTV relation gives pressures of > 70 dyne cm^{-2} with a height of $< 7,000$ km. The heights of a few thousand kilometers are reminiscent of X-ray bright points on the sun, although the pressure and temperature implied by the RTV relation are two orders of magnitude larger than in the solar case.

For the large extended structures where the height is resolved to be several solar radii the RTV relation does not intercept the pressure-height curves. The pressures are between 1 and 10 dynes cm^{-2}, comparable to the solar value, but the dimensions are an order of magnitude larger. Such a large discrepancy would require the loop height to be over 10 times the pressure or heating scale height to bring the RTV relation into agreement, which is unlikely (Serio et al 1981). We conclude that either the eclipse models that require these large structures are not correct, or that the energy input and balance for these large structures are very different from those of solar loops.

The radiative cooling times infered for both the compact and extended structures in AR Lac are similar to those found in the flaring sun. The cooling times of less than a 1000 s for the compact structures are much shorter than the assumed lifetime of the structures, and additional heating on a similar or shorter timescale is required to maintain the observed X-ray flux. For the extended structures the cooling time of 10,000 s (i.e. several hours) and hence the corresponding heating timescale required is also long.

We have concluded that the most likely explanation for the hot component is that it originates in large extended magnetic structures that pervade the whole system. The cooling time in such a large volume is of order of a day or longer. A puzzling aspect of the hot component is why the temperature of the plasma should be highest in the largest regions. This requires that the energy input increases with loop length, contrary to that generally assumed in the loop scaling law (e.g. RTV). A clue comes from the fact that the cooling time is of the order of a day or longer, such that a continuous energy source is not required. The only other occasion such high temperatures are seen is during flares where the total energy release can be up to $\sim 10^{35}$ erg (e.g. White et al 1986), a factor of ten more luminous than the two flares reported here. This energy may on occasions be channelled into heating plasma that escapes to a much larger volume. This might occur if during a large flare the plama broke out of its magnetic loop but was then trapped by other closed magnetic field lines further out. This might be expected if the flare was associated with a major reorganization of the magnetic field. Another possibility is that the larger scale magnetic field is constantly changing its configuration and releasing smaller energy impulses.

5. For the Future

The results of the modelling of the X-ray lightcurve of AR Lac have shown the power of eclipse observations to constrain the spatial structure of stellar coronae. The multiple solutions can be eliminated in future observations by making simultaneous high resolution uv observations of line profiles and using the doppler imaging technique to uniquely locate the chromospherically active regions (see Walter et al 1987, Vogt and Penrod 1983). The improvements in spectral resolution and collecting area given by future X-ray observatories such as XMM and AXAF will allow maps, such as those shown in Figures 4 and 5, to be made for individual spectral lines. By combining these with similar maps generated from uv and optical observations, which probe the upper and lower chromosphere, it will be possible to build a complete three dimensional view of the structure of stellar coronae.

6. REFERENCES

Chambliss, C.R. 1976, P.A.S.P., 88, 762.

Eaton, J., and Hall, D. 1979, Ap. J., 227, 907.

Majer, P., Schmitt, J.H.M.M., Golub, L., Harnden Jnr., F.R., and Rosner, R. 1986, Ap. J., 300, 360.

Mewe, R., Gronenschild, E.H.B.M., and van den Oord, G.H.J. 1985, Astr. Ap. Suppl., 62, 197.

Mewe, R., Schrijver, C.J., Lemen, J.R., and Bentley, R.D., 1986, Adv. Space Res., Vol. 6, No. 8, 133.

Rosner, R., Tucker, W. H., and Vaiana, G.S. 1978, Ap.J., 220, 643 (RTV).

Serio, S., Peres, G., Vaiana, G.S., Golub, L., and Rosner, R. 1981, Ap. J., 243, 288.

Skilling, J., and Bryan, R.K., 1984, M.N.R.A.S., 211, 111.

Stern, R.A., Antiochos, S.K. and Harnden, F.R. , 1986, Ap. J., 305, 417.

Swank, J.H., White, N.E., Holt, S.S., and Becker, R.H. 1981, Ap.J., 246, 214.

Vogt, S.S., and Penrod, G.D., 1983, P.A.S.P., 95, 565.

Walter, F.M., Gibson D.M., and and Basri, G.S. 1983, Ap.J., 267, 665 (WGB).

Walter, F.M., Neff, J.E., Gibson, D.M., Linsky, J.L., Rodono, M., Gary, D.E., and Butler, C.J. 1987, Astr. Ap., 186, 241.

White, N.E., Culhane, J.L., Parmar, A.N., Kellett, B.J., Kahn, S., van den Oord, and Kuijpers, J. 1986, ApJ., 301, 262.

White, N.E., and Peacock, A.P. 1988, in *X-ray Astronomy with EXOSAT*, ed. N.E. White and R. Pallivicini, Memoria S.A.It, 59, in press.

THEORY OF ACCRETION DISK CORONAE

T. R. Kallman
NASA/Goddard Space Flight Center, Greenbelt, MD USA

ABSTRACT. Accretion disk coronae are likely to be the dominant site for X-ray absorption and reprocessed emission in low mass X-ray binaries, and may be present in other classes of compact X-ray sources such as active galactic nuclei and cataclysmic variables. In spite of this fact, and in spite of the observational evidence for their existence, there remain many uncertainties about the structure of accretion disk coronae. This paper will discuss the coronal structure and dynamics, their X-ray spectral signatures including coupling to the variability behavior of compact X-ray sources, and the major unsolved theoretical issues surrounding them.

1. INTRODUCTION

Accretion disks are a common occurrence in a variety of astronomical systems, ranging from cataclysmic variables and X-ray binaries to the nuclei of active galaxies. It is likely that the disks that form as the result of accretion share many common features and differ primarily in their scale size. In many situations a large fraction of the energy of the accreted material is released as X-rays near the compact object at the disk center. The illumination of the disk surface by these X-rays produces an accretion disk corona or wind (hereafter collectively reffered to as an ADC) which is the dominant site for the formation of spectral features observable by high resolution X-ray spectrometers.

The idea of accretion disk coronae dates back to early work on accretion disk structure by Shakura and Sunyaev (1973). The most detailed discussions of ADC structure, and those which form the basis for much of the present discussion, are those by Begelman, McKee and Shields (1982) and Begelman and McKee (1982; hereafter referred to collectively as BM), and by White and Holt (1981) and by McClintock, et al. (1982), and others referred to by these authors. Since the theory of accretion disk coronae and their implications for observations have already been discussed in detail by these authors, in this review I will concentrate on the confrontation between theory and the past and future observations of ADC sources using high resolution X-ray spectrometers. In order to do so, it is necessary to summarize some of the past theoretical results; this is done in Section II. Models for ADC X-ray spectra and the comparison with observations are described in Section III, and the prospects for future progress are discussed in Section IV.

2. ACCRETION DISK CORONA BASICS

For the purpose of estimating the properties of accretion disk coronae I will assume that the accretion disk is geometrically "thin" (i.e. vertical thickness much less than radial distance from the X-ray source) and optically thick (to the cooling radiation emitted near the midplane) (see, e.g. Shakura and Sunyaev 1973; the interior structure of the disk is not of crucial importance to the corona structure), that radiation pressure and disk self-gravity are negligilble. I will also assume that the heating of the disk atmosphere is dominated by X-rays from a compact object at the disk center, that sources of mechanical heating such as convection or MHD waves are unimportant, and that the timescales for variability of the X-ray source intensity or the disk structure are long compared to the timescales affecting the corona structure.

The temperature of X-ray heated gas depends only on the shape of the X-ray spectrum and on the ionization parameter, Ξ. This quantity is defined as the ratio of the X-ray flux, F (in energy units over the band 1 - 1000 Ry), to the gas pressure P, according to $\Xi = F/Pc$ (c.f. Krolik, McKee, and Tarter, 1981). For ionization parameter values greater than a critical value, Ξ_c^*, the gas temperature is determined primarily by the effects of Compton scattering. The "Compton temperature" is $T_{IC} = <\epsilon>/4k$, where $<\epsilon>$ is the mean photon energy. For X-ray spectra similar to those observed from LMXRB, the Compton temperature is $T_{IC} \sim 10^7$ K. For ionization parameters less than a critical value Ξ_h^*, the gas temperature is determined by the effects of atomic processes; the temperature in this case is $T_{atomic} \sim 10^4$ K. These critical ionization parameters have values $\Xi_h \simeq 0.3(T_{IC}/10^8 K)^{-3/2}$ and $\Xi_c^* \simeq 10$ (although various effects such as conduction, non-equilibrium effects, strong Compton cooling by the UV radiation from the disk and other sources of heating may modify these estimates). For intermediate ionization parameters, $\Xi_h^* < \Xi < \Xi_c^*$, there exists a third equilibrium temperature which is unstable according to Field's (1965) criterion. This simple Ξ scaling behavior applies provided the gas is optically thin to the X-rays; the validity of this assumption will be discussed later.

The existence of an X-ray heated corona depends on the decrease of gravity with distance from the accretion disk midplane. The conditions necessary for the formation of a corona may be expressed in terms of the hydrostatic condition:

$$\frac{dP}{dz} = -\frac{GMm_H}{R^3 kT} Pz = -\frac{Pz}{z_s^2} \leq 0 \qquad (1)$$

where P is the gas pressure, z is height above the disk midplane, M is the mass of the central X-ray source, R is the distance from the X-ray source and m_H is the mean mass per particle in the gas. The scale height is $z_s = \sqrt{R^3 kT/(GMm_H)} = 8 \times$

$10^7 cm\, R_9^{3/2} T_7^{1/2} (M/M_\odot)^{-1/2}$, and the temperature gradient in the disk atmosphere may be written:

$$\frac{dT}{dz} = \frac{dP}{dz}\frac{dT}{d\Xi}\left(\frac{d\Xi}{dP}\right)_F \qquad (2)$$

Since $(d\Xi/dP)_F > 0$, the necessary condition for a corona to form is that $dT/d\Xi < 0$ which is satisfied for $\Xi_h^* < \Xi < \Xi_c^*$ in X-ray illuminated disks if $T_{IC} \geq 10^7$ K.

Therefore, we expect a corona at $T = T_{IC}$ to exist wherever $\Xi \geq \Xi_h^*$. Near the interface between the corona and the disk photosphere must exist a chromosphere and transition region in which X-ray heating is important, but is not sufficiently strong to heat the gas to temperatures comparable to T_{IC}. These regions must dominate the optical and UV appearance of X-ray illuminated accretion disks. The maximum density of the corona may then be expressed in terms of the incident X-ray flux F according to:

$$n_{max} = \frac{F}{c\Xi_c^* kT_{IC}}$$
$$= 1.9 \times 10^{17} cm^{-2} R_9^{-2} L_{37} f T_{IC7}^{-1} \qquad (3)$$

where R_9 is the radius in units of $10^9 cm$, L_{37} is the X-ray source luminosity in units of $10^{37} ergs^{-1}$, and T_{IC7} is the Compton temperature in units of $10^7 K$. The effects of geometry and the transfer of X-rays through the corona are combined in the factor $f = 4\pi R^2 F/L$. $f = 1$ corresponds to unattenuated X-rays normally incident on the disk. The total coronal column is $\sim n_{max} z_s$ or

$$N_{tot} = \frac{Lf}{4\pi \Xi_c^* c}(RkT_{IC}\mu m_H GM)^{-1/2} E(z_b/z_s)$$
$$= 1.5 \times 10^{25} cm^{-2} L_{37} f R_9^{-1/2} T_{IC7}^{-1/2} E(z_b/z_s) \qquad (4)$$

where z_b is the height of the base of the corona, M is the mass of the compact object, μm_H is the mean mass per particle in the gas and E is a slowly varying function related to the error function. It is clear from this equation that the column density of the corona increases with decreasing distance from the source, and that when $R \geq$

$R_{thick} = 10^9 cm L_{37}^2 f^2 T_{IC7}^{-1}$ the Thompson depth through the corona exceeds unity. This buildup must ultimately be limited by the fact that X-rays will not penetrate to the base of a corona whose Thompson depth, τ_{Th}, exceeds some critical value which we define as $\tau_{Thcrit} \simeq 1 - 10$. Therefore we may treat the requirement $\tau_{Th} \leq \tau_{Thcrit}$ as a second ("optically thick") necessary condition for coronal equilibrium. When this condition applies, the minimum ionization parameter in the corona may be shown to be

$$\Xi_{min} = \frac{Lf\sigma_{Th}}{4\tau_{Thcrit}\pi c(RkT_{IC}\mu m_H GM)^{1/2}} E(z/z_s)$$
$$= 10 L_{37} f R_9^{-1/2} T_{IC7}^{-1/2} \tau_{Thcrit}^{-1} \frac{M}{M_{odot}} E(z/z_s) \qquad (5)$$

where σ_{Th} is the Thompson cross section and z is height at which Thompson depth τ_{Thcrit} occurs ($E(z/z_s)$ is of order unity for our fiducial parameter values and is a very slowly varying function of R). In what follows we assume $\tau_{Thcrit}=1$. Since $\Xi_h^* \simeq 0.1 - 1$ (c.f. Krolik, McKee, and Tarter, 1982) this suggests that in the inner region of the disk where the optically thick criterion applies the mean ionization parameter may be much greater than Ξ_h^*, and that it increases with decreasing distance from the source.

X-ray heating may also have dynamical implications for the corona. These are characterized in terms of the "escape temperature" of disk gas $T_{esc} = GMm_H/(kR) = 1.6 \times 10^9 K R_9^{-1}(M/M_\odot)$, which is the temperature of a gas whose thermal velocity equals the escape velocity from the disk. At radii $R \geq R_{IC} = 1.6 \times 10^{11} cm T_{IC7}^{-1}(M/M_\odot)$ the corona temperature exceeds the escape temperature and the corona is not bound to the disk. In this region, referred to as the "wind region", the mass loss rate from the disk can greatly exceed the mass accretion rate. This can lead to a limit cycle behavior as the disk is depleted by the wind, the X-rays turn off when the depleted zone propogates inward to the X-ray source, and the disk reforms in the absence of X-rays (Shields, et al., 1986). The timescale characterizing this process is the viscous timescale in the disk (e.g. Begelman and McKee, 1982), $\Delta t \sim \Delta t_{viscous} \simeq 85 s \alpha^{-1}(h_d/R_{IC})^{-2}(M/M_\odot)(R/R_{IC}T_{IC8})^{3/2}$

Another process which may affect the disk structure is cooling of the corona by Compton scattering. The flux at the surface of the disk from viscous energy generation has a temperature in the range $10^4 - -10^5 K$ is $F_d = 2.2 \times 10^{16} erg cm^{-2} sec^{-1} R_9^{-3}(M/M_\odot) L_{37}(\eta/0.1)$. The radius at which this equals the flux from the X-ray source is $R_{eq} = 2.7 \times 10^7 cm f^{-1} M/M_\odot (\frac{\eta}{0.1})$; for $R \leq R_{eq}$ the disk will cool the corona to the disk photospheric temperature $T_{disk} = (F_d \sigma)^{1/4} = 1.4 \times 10^5 K R_9^{-3/4}(M/M_\odot)^{1/4} L_{37}^{1/4}(\eta 0.1)^{1/4}$. In the outer region of the disk, where the corona is optically thin, the corona will be cooled by the X-rays which pass through it and are absorbed and re-radiated by the disk photosphere. The temperature of these re-radiated X-rays is $T_{disk} = (F_x/\sigma)^{1/4} = 3.4 \times 10^5 K R_9^{-1/2} f^{1/4} L_{37}^{1/4}$. Since the flux at this temperature will at most equal that of the direct X-rays, this effect will reduce the corona temperature by a factor of 2.

In the optically thin coronal region the emission measure of the corona is

$$EM_{thin} = \left(\frac{fL}{ckT_{IC}\Xi_c^*}\right)^2 \frac{1}{4\pi} \left(\frac{kT_{IC}}{GMm_H}\right)^{1/2} (R_{min}^{-1/2} - R_{max}^{-1/2})$$
$$= 4 \times 10^{61} cm^{-3} \frac{L_{38}^2 f^2}{T_7^{3/2} R_{min9}^{-1/2}} \qquad (6)$$

The results presented so far in this section divide the disk into several qualitatively different zones ordered according to increasing radial distance from the X-ray source: (i) The zone where the coronal temperature is tied to the disk photospheric temperature by Compton cooling; (ii) The optically thick coronal zone, where the vertical column density is limited by radiative transfer effects and where the mean ionization parameter may exceed Ξ_h^*; (iii) The optically thin coronal zone, where the mean ionization parameter is Ξ_h^* and the corona is partially cooled by X-rays reprocessed in the photosphere; and (iv) The wind zone where the corona is not bound to the disk. The distinctions between these regions are displayed schematically in Figure 1.

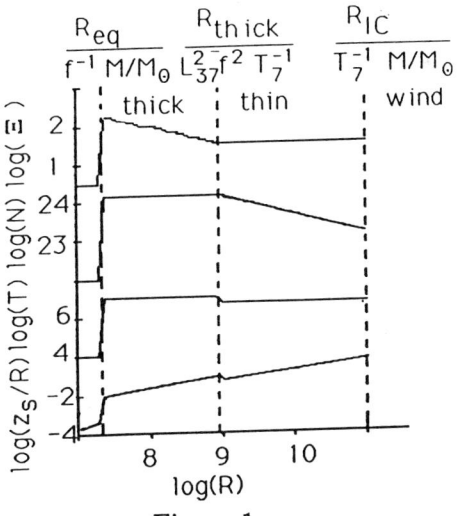

Figure 1.

Inherent in the discussion so far has been the assumption that the flux proportionality constant, f, is known. This factor depends on the detailed geometrical relationship between the compact object and the disk, i.e. the disk flaring angle and the size of the compact object, and on the effects of radiative transfer in the corona. It is possible that the corona may be self-sustaining in the sense that scattered X-rays may be dominant in determining the flux which excites the corona. However, as of this writing there exists no successful calculation of the efficiency of this process. London (1982) showed that in the absence of Compton cooling radiative transfer effects led to a value of $f \simeq 0.01 - 0.1$ for X-ray and disk properties appropriate to low mass X-ray binaries, but was unable to find a self-consistent steady transfer solution when Compton cooling of the corona by the disk was included. In fact, as suggested by Fabian, it is possible that there may exist two distinct states for the corona: a low-f Compton cooled state and a high-f non-cooled state. Given this ambiguity, and given the uncertainties in the disk geometry, we consider the value of f to be an important unknown. As shown in the following section, this quantity can be constrained by observations.

3. OBSERVATIONS

The eclipse light curves of low mass X-ray binaries, revealing a significant ($\sim 10\%$) residual flux during eclipse, provided much of the original motivation for the study of accretion disk coronae. As shown in Figure 1, the ADC scale height can exceed $z_s \simeq R \sim R_{star}$ at large radii, where R_{star} is the radius of the companion star. The eclipse flux depends on the Thompson depth of this material, which may be ~ 0.1 if $f \sim 0.1$. ADC models have been successfully fitted to the orbital light curves of the high-inclination binaries 4U1822-37 (White and Holt, 1982), 4U2129+47 (BM; White and Holt, 1982; McClintock, et al., 1982), and Cyg X-3 (White and Holt, 1982; Molnar, 1986) observed by proportional counter experiments on the *HEAO-1 and Einstein* satellites. As suggested by Kahn, higher spectral resolution ($\lambda/\Delta\lambda \geq 20$) observations of these light curves will test ADC models more severely, since they will allow limits to be set on the existence of photoabsorption edges, and hence on the degree of ionization of the ADC gas.

3.1 Iron K Emission Lines

Another observational test of ADC models is provided by the iron K line emission from low mass X-ray binaries (LMXRB). Such emission has been detected by the gas scintillation proportional counter detectors (GSPC's) on the EXOSAT and Tenma satellites.

A model calculation of the dependence of the emissivities of the K lines from the various stages of iron on radial and vertical position in an ADC, embodying the ADC structure as summarized in the previous section along with detailed calculations of the atomic physics affecting the line emission, are displayed in figure 2(a) (Kallman and White, 1988). Figure 2(b) displays the emitted line flux in the 6.2 - 7 KeV energy band at radii corresponding to those displayed in Figure 2(a). The panels on the left show the spectrum emitted at each radius as it would be viewed by a spectrometer with a resolution $\Delta\epsilon/\epsilon = 0.002$, roughly that attainable by the best X-ray spectrometers contemplated for the future. The panels on the right show the spectrum as it would be viewed by a spectrometer with a resolution approximately that of a GSPC, $\Delta\epsilon \sim 100 eV$. The broadening processes included in the calculations of Figure 2(b) include blending of multiple components from different ionization stages and Compton scattering. The components from various ion stages are apparent as the narrow emission spikes in Figure 4a. Comptonization results in the broader "pedestals" under the narrow components. The narrow unscattered line components all contain more energy than do the broadened Compton scattered components, reflecting the fact that the lines are always emitted at Thompson depths $\tau_{Th} \leq 1$. The ratio of the scattered to unscattered components is greatest at the smallest radius, 10^7 cm. This is consistent with equation (4), showing that the greatest column densities occur at the smallest radii. Broadening due to blending of multiple components results in a line width of approximately 0.5 KeV, and so dominates over Compton broadening for the ADC conditions considered here. The line blends are dominated by emission from ion stages Fe XXV and below, and so have centroid energies in the range 6.2 - 6.7 KeV. In the optically thin coronal region, the line blend is receives a strong contribution from the 6.4 KeV line. These photons are emitted by material at low ionization (Fe XVII and below) in the disk photosphere and chromosphere; such emission is weak or absent in the optically thick coronal region owing to attenuation of the K continuum photons.

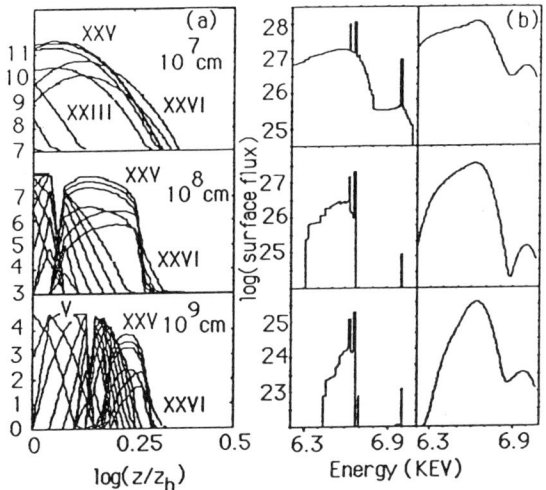

Figure 2

The models shown in Figure 2 may be compared with the results from

EXOSAT compiled by White, et al. (1986) which show that for most objects the line centroid energy is $\simeq 6.7 KeV$, and the width (Full Width Half Maximum) is $FWHM \simeq < \Delta \varepsilon > \simeq 1 KeV$. This suggests that the standard ADC model can account for the observed line centroid energies if the flux incident on the disk is relatively high, $f \simeq 0.1$, and if the lines are emitted at small distances from the compact object, $R \leq 10^8$ cm. If so, the line emission regions must shield the disk at larger radii from X-rays in order to avoid an excess of narrower, lower energy line emission, and the observed width must be due to rotation or some other broadening mechanism. Rotation causes broadening $\Delta \varepsilon / \varepsilon = (GM/Rc^2)^{1/2} = 0.24 KeV (\varepsilon/6.4 KeV)^{-1} R_8^{-1/2}$ for Kepler motion around a 1 $M_\odot$ compact object, so that radii $R \simeq 10^7 cm$ are required to provide the observed broadening.

3.2 Soft X-ray Lines

The wavelength range between 10 and 20 Å contains a large number of emission lines which are useful diagnostics of accretion disk coronae. The high spectral density of lines necessitate spectrometers with higher sensitivity and spectral resolution than can be obtained with proportional counters or GSPC's. Line emission in the 0.5 - 2 KeV energy range due to the L shell transitions of iron and to the K shell transitions of medium-Z elements such as nitrogen and oxygen from several of the brightest low mass X-ray binaries (Kahn, Seward and Chlebowski, 1982; Vrtilek et al., 1986a, b) have been detected by the transmission grating experiments on the Einstein and EXOSAT observatories.

The interpretation of soft X-ray line spectra is considerably more complicated than that of the Iron K lines. Attempts at model fitting (Vrtilek et al. 1986a,b) have shown that neither models of constant temperature mechanically heated gas (Raymond and Smith, 1977) nor models of photoionized constant density clouds (Kallman and McCray, 1982) give adequate fits to all the line strengths. Furthermore, the atomic processes which affect this line emission are subject to considerable uncertainty. A notable example is dielectronic recombination at temperatures much less than the ionization potential of the most abundant ions, which is likely to be important in photoionization-dominated gases (uncertainties in these rates are discussed in more detail elsewhere in this volume).

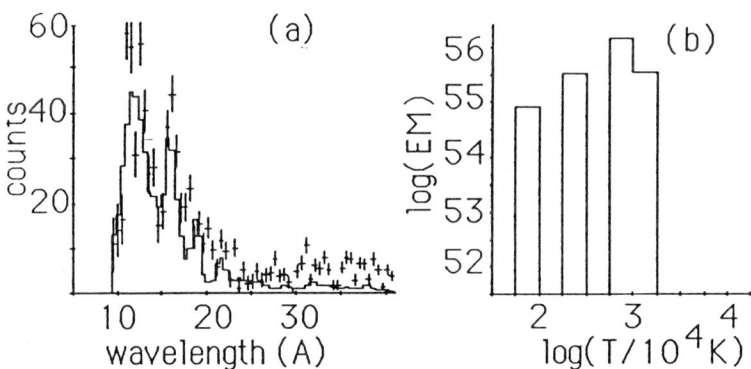

Figure 3

An attempt at fitting such soft X-ray line spectra to a more complicated model is shown in Figure 3(a). The line spectrum of Cyg X-2 as observed by the objective grating spectrograph (OGS) on the *Einstein* observatory satellite ("state G" of Vrtilek et al. ,1986b) is displayed (crosses), together with a model consisting of a multi-temperature optically thin mechanically heated gas. The line spectrum is dominated by approximately four emission features, the 12 - 14 Å features which are likely to be dominated by Fe XVIII - XXII, the 16 Å feature which may be due to O VIII Lβ or to Fe XVII - XVIII, and the 19 Å feature due to O VIII L α. The locations of these strong lines are all matched by the models, and χ^2 per degree of freedom is 0.7 for this model fit shown. In the absence of emission lines the bremsstrahlung continuum alone fits the data with χ^2 per degree of freedom of 1.4, demonstrating that the lines are statistically significant and that the model fit is acceptable. The emission measure distribution, displayed in figure 3(b), ranges from log(T)=6.25 to log(T)=7.25, reflecting the presence in the observed spectrum of emission from ions ranging from Fe XVII to Fe XXIII. Similar results have been obtained from the analysis of the OGS spectra of several other low mass X-ray binaries (Kallman, Vrtilek, and Kahn, 1989).

The emission measures required to fit the Cyg X-2 OGS spectra are much less than those estimated in equation (6), suggesting that only a fraction $\sim 10^{-4}/f$ of the corona will be cool and neutral enough to emit soft X-ray lines. Furthermore, the observed widths of the lines, $\lambda/\Delta\lambda \geq 20$ provide a limit on the rotational broadening and hence on the radius of the emission region, $R \geq 10^8 cm(M/M_\odot)$. Therefore, the soft X-ray emission lines probe a region which is physically distinct from the region responsible for most of the Iron K line emission discussed in the previous subsection.

4. SUMMARY

In spite of the fact that many of the properties of accretion disk coronae can be estimated in a simple way, and that they provide a successful and attractive explanation for the X-ray light curves of many low mass binaries, there remain a number of outstanding problems, including: (i) What is f? The Sco X-1 Iron K line observations suggest $f \geq 0.1$ in the inner disk regions ($R \leq 10^8 cm$) and much smaller values at larger radii, while the calculations of London (1983) suggest $f \simeq 0.01$ at all radii if the Eddington ratio is ~ 0.1. Soft X-ray line emission requires $f \geq 10^{-4}$ for $R \geq 10^8 cm$. The eclipse light curves of high inclination low maxx X-ray binaries require $f \geq 0.1$ at the largest disk radii. (ii) Modelling of the Soft X-ray Emission lines: The full power of the soft X-ray emission line observations can't be utilized until uncertainties concerning the atomic physics affecting recombination line emission (and other processes which are more familiar from the study of mechanically heated gases) are resolved. (iii) Modelling of Wind Flow: The gas flow in the disk wind region may be observationally distinguishable from the bound coronal region, for example owing to the presence of material at $T \leq T_{IC}$ far above the disk plane. (iv) Further Observations of Emission Lines: If the Fe K line widths are due to rotation, they are predicted to have a characteristic double peaked structure which will be resolvable with future high resolution spectrometers. The soft X-ray emission lines are known to vary on timescales less than 10^4 sec; observations with sufficient sensitivity to resolve the fastest such variability will provide a lower limit on the size of the region responsible for this emission.

REFERENCES

Begelman, M., McKee, C. E., and Shields, G. B., 1983, *Ap. J.*, **271**, 70.
Begelman, M. and McKee, C. E., 1983, *Ap. J.*, **271**, 89 (BMS).
Fabian, A., Guilbert, P., and Ross, R.R., 1982, *MNRAS*, **199**, 1045.
Field, G. B., 1965, *Ap. J.*, **142**, 531.
Kahn, S.M., Seward, F.D., and Chlebowski, T., 1984, *Ap. J.*, **283**, 286.
Kallman, T. R., 1984, *Ap. J.*, **280**, 269.
Kallman, T. R., and McCray, R. A., 1982, *Ap. J. Suppl.*, **50**, 263.
Kallman, T. R. and Mushotzky, R., 1985, *Ap. J.*, **292**, 49.
Krolik, J.H., and Kallman, T.R., 1984, *Ap. J.*, **286**, 366.
Krolik, J.H., and Kallman, T.R., 1987, *Ap. J. Lett.*, **320**, L5.
Krolik, J. H., McKee, C. F., and Tarter, C. B., 1981, *Ap. J.*, **249**, 422.
London, R., 1982, *in Cataclysmic Variables and Low Mass X ray Binaries*, J. Patterson and D. Lamb, eds
McClintock, J., London. R.A., Bond. H.E., and Grauer, A.D., 1982, *Ap. J.*, **258**, 245 .
Molnar, L., 1986, *in The Physics of Accretion onto Compact Objects*,, K.O. Mason, M. Watson, and N.E. White, eds. (Berlin: Springer).
Raymond J.C., and Smith, B. H., 1977, *Ap. J. Supp.*, **35**, 419.
Shakura, N. I., and Sunyaev, R.A., 1971, *Astron. and Astrophys.*, **24**, 337.
Shields, G.A., McKee, C.F., Lin, D.C., and Begelman, M. 1986, *Ap. J.*, **306**, 90.
Vrtilek, S.D., Kahn, S.M., Grindlay, J.E., Helfand, D.J., and Seward, F.E., 1985, *Ap.J.*, **307**, 698.
Vrtilek, S.D., Helfand, D.J., Halpern, J.P., Kahn, S.M., and Seward, F.D., 1986, *Ap. J.*, , .
Vrtilek, S. D., Swank, J., and Kallman, T., 1988, *Astrophys. J.*, **326**, 186.
White, N.E., and Holt, S.S., 1982, *Ap. J.*, **257**, 318.
White, N.E. et al., 1986, *M.N.R.A.S.*, **218**, 129.

LINE STRUCTURES IN THE X-RAY SPECTRA OF CYGNUS X-2 OBSERVED WITH EXOSAT

P. E. Freeman[1], S. M. Kahn[1], L. Chiappetti[2], E. G. Tanzi[2], A. Ciapi[3], L. Maraschi[3], A. Treves[3], E. G. Branduardi-Raymont[4], E. N. Ercan[5]

[1] University of California, Berkeley, CA, USA
[2] Institute of Cosmic Physics - CNR, Milan, ITA
[3] University of Milan, Milan, ITA
[4] Mullard Space Science Laboratory, London, GBR
[5] Bogazici University, Bebek, Istanbul, TUR

ABSTRACT. Cygnus X-2 was observed with EXOSAT at five phases of a single orbital cycle in September of 1983. We will summarize the results of spectral fits of the LE + ME (Argon) data in terms of a superposition of thermal bremstrahlung and blackbody components. During the first observation a grating spectrum was obtained and this is described in some detail. The GSPC data are used to investigate the presence of iron features, and their behavior during dips.

1. INTRODUCTION

Cygnus X-2 was observed by EXOSAT in September 1983, at five equispaced phases of a single 9.8d orbital cycle. The aim of the observations was to obtain spectral information in the .1 to 15 KeV band. All the instrumentation onboard the satellite was used toward this end. These included: 1) a medium energy array of proportional counters (ME); 2) a Gas Scintillation Proportional Counter (GSPC); 3) a grazing incidence low energy telescope (LE2), focused on a Channel Multiplier Array and with interposed filters; and 4) a second telescope (LE1) used in conjunction with a 1000 line/mm transmission grating. The grating was used only for the first observation, after which the insertion mechanism jammed.

In a previous paper, light curves and a spectral fit deduced from the ME data were reported (Chiappetti et al. 1987). After the discovery of Quasi-Periodic Oscillations from Cyg X-2 (Hasinger et al. 1986), the short term variability was also reanalyzed (Stella et al. 1986, Chiappetti et al. 1987).

In this paper, we start by giving spectral fits derived from the combination of ME and LE2 data (Section 2), and then concentrate on the line spectra. In Section 3 we report the analysis of GSPC data, while in Section 4 we discuss the analysis of the grating spectrum.

2. X-RAY CONTINUUM

During each observation, the LE2 count rates were measured with four filters- pure polypropylene (PPL), parylene coated with aluminum (Al/Par), a 3000 Å Lexan filter, and a boron filter. This measurement was combined with the output of the ME Ar counter and fitted with a standard χ^2-square minimization procedure. We report the result of a fit with a superposition of thermal bremsstrahlung and blackbody components, plus interstellar absorption modeled after Morrison and McCammon (1984). Results of the fit are reported in Table 1.

Following the convention introduced by Vrtilek et al. (1986), during observations one, two, four, and five the source appeared to be in a high state (L2-10 keV $> 9 \times 10^{37}$ erg/s), while during observation three the source was in low state (L $< 7 \times 10^{37}$ erg/s). The fits give parameters comparable to those obtained with the SSS- Einstein experiment (.5 to 4.5 keV; Vrtilek et al.). The differences are attributable to the intrinsic variation of the source and the different energy range of the observations.

3. GSPC OBSERVATIONS OF CYG X-2

The EXOSAT GSPC is coaligned with the ME array of proportional counters. The GSPC offers the advantage of higher energy resolution (about 10 percent FWHM at 6 keV), which allows the detection and the detailed study of broad spectral features which cannot be resolved with the ME.

We accumulated GSPC spectra of Cyg X-2 over the energy range 2 to 15 keV for the time intervals covered by the ME observations. The spectra were then fitted with a two-component exponential plus blackbody model similiar to that applied to the ME data; the low-energy column density was held constant at the value determined from the combined LE2 and ME data (see Section 2), since the GSPC is insensitive to densities lower than 10^{22} cm^{-2}. The GSPC best-fit parameters for the X-ray continuum are consistent with those derived from the ME data for each observation. Inclusion of an Fe line at 6.7 keV improves the quality of the fits to the GSPC spectra substantially in the case of observations 1 and 3, and marginally for observations 4 and 5 (quality is determined by χ^2 values: e.g. in observation 1, the inclusion of the Fe line decreases the reduced χ^2 value from 1.50 to 1.12 for 203 and 201 degrees of freedom respectively).

Table 2 lists the Fe line energy and equivalent width (EW), along with the 90 percent confidence errors for each of the five observations.

For observations 1 and 2 the parameters listed in Table 2 correspond to time intervals outside the dips. We have investigated the behavior of the Fe line during the dips: the best fits to the GSPC spectra accumulated during the dip intervals do not require inclusion of an Fe line, with a 90 percent confidence upper limit to the EW of 41 and 33 eV respectively. Although these upper limits are not very stringent when compared with the EW measured in observations 3 to 5, they are significantly different from the values measured in observations 1 and 2 outside of the dips (60 and 55 eV), and indicate a change in the behavior of the source during the dips.

TABLE 1
SUMMARY OF ME + LE2 SPECTRAL AND TEMPORAL ANALYSIS

Obs	N_H (10^{21} cm^{-2})	min max T_{TH} (keV)	min max T_{BB} (keV)	min max	L_{TH} (10^{37} erg/s)	L_{BB} (10^{37} erg/s)	χ^2 (28 dof)
1	1.92	1.86 2.03	5.1 5.2	4.9	1.12 1.15	1.10	8.5 3.1	3.7 0.09	85.3
2	2.09	2.05 2.13	4.7	4.5 4.9	1.13 1.16	1.10 3.9	9.3	3.5 0.08	77.8
3	2.18	2.03 2.18	7.5	7.2 7.9	1.44 1.60	1.30 2.3	7.7	0.68 0.008	30.4
4	2.40	2.38 2.50	7.1	6.7 7.3	1.30 1.36	1.27 3.1	10.	2.7 0.05	68.2
5	2.19	2.13 2.25	8.2	7.8 8.5	1.30 1.36	1.27 2.3	9.3	2.7 0.05	38.8

For the luminosities, the upper figure represents the value in the 1 to 16 keV band, while the bottom figure shows the value in the 0.2 to 1 keV band.

TABLE 2
OBSERVATIONS OF THE DISCRETE IRON FEATURE (GSPC)

Observation	Phase	Fe line energy (keV)	Fe line EW (eV)	Comment
1	0.86	6.67 ± .07	60 ± 8	dip during observation
2	0.05	6.67 ± .10	55 ± 7	dip during observation
3	0.33	6.86 ± .14	34 ± 12	
4	0.55	6.69 ± .20	14 ± 9	
5	0.75	6.57 ± .23	18 ± 8	

4. GRATING OBSERVATION OF CYG X-2

The 1000-line Objective Grating Spectrometer (OGS) was used during the first observation, in conjunction with the LE1 telescope and a Channel Multiplier Array (CMA).

The OGS spectrum was fit using a standard χ^2-minimization procedure, with parameter constraints introduced by the results of the ME + LE2 data fit. The wider band of this latter spectrum serves to make it more sensitive to temperature and normalization, so these numbers were held constant. In the case of the normalization, an overall fudge factor was introduced, to account for uncertainties in relative effective area calibration between the various instruments. This factor was within ten percent of unity, consistent with the expected uncertainty. Since the grating data are sensitive to absorption, the column density and the abundances in the absorbing medium were allowed to float freely.

Inclusion of all data in the fitting process led to an inability to derive continuum parameters consistent with the ME + LE2 results. Refering to the residuals of the fit between 10 and 19 Å, in Figure 2, we note a discrete excess of emission. If this region of the spectrum is excluded, then the continuum model deduced from the ME + LE2 data can also be fitted to the remaining OGS data. The excess is most likely associated with a complex of L-shell emission lines from several multiply-ionized species of iron. The structure of this emission complex is similiar to that found by Vrtilek et al. for Cyg X-2 using the Einstein OGS, though the lower resolution of the EXOSAT OGS precludes definite line identifications in our case (see Table 3). The unknown discrete feature found by Vrtilek et al. at 19.7 Å does not appear in the EXOSAT spectrum.

Also evident in the spectrum is oxygen K-shell absorption edge at 23.3 Å. Our spectral fits suggest an abundance 30 percent lower than the cosmic value given by Morrison and McCammon. The abundances of nitrogen and carbon were found to be normal, although the constraints are much less restrictive for these elements.

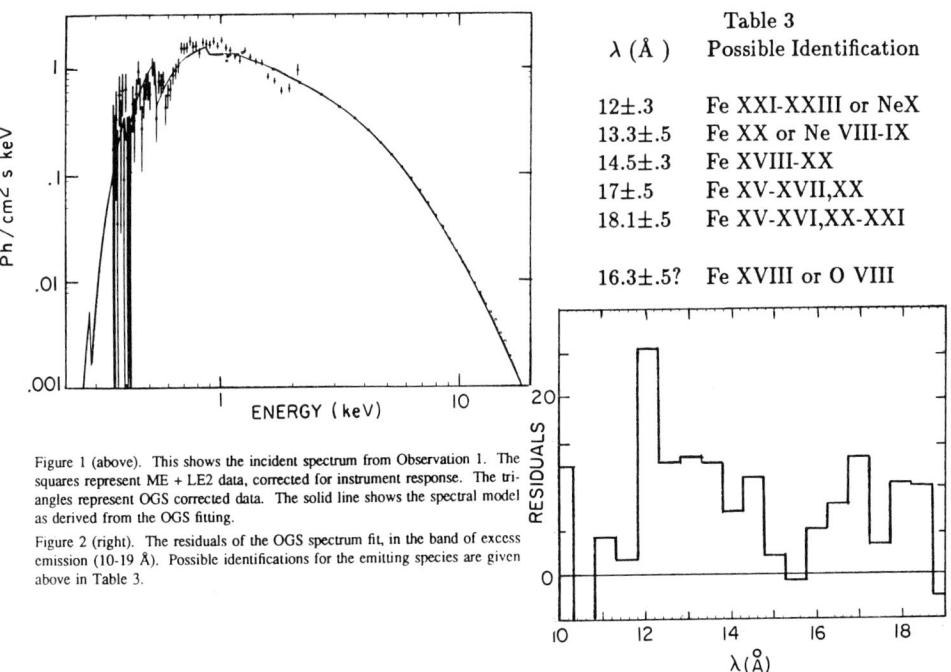

Table 3

λ (Å)	Possible Identification
12±.3	Fe XXI-XXIII or Ne X
13.3±.5	Fe XX or Ne VIII-IX
14.5±.3	Fe XVIII-XX
17±.5	Fe XV-XVII,XX
18.1±.5	Fe XV-XVI,XX-XXI
16.3±.5?	Fe XVIII or O VIII

Figure 1 (above). This shows the incident spectrum from Observation 1. The squares represent ME + LE2 data, corrected for instrument response. The triangles represent OGS corrected data. The solid line shows the spectral model as derived from the OGS fitting.

Figure 2 (right). The residuals of the OGS spectrum fit, in the band of excess emission (10-19 Å). Possible identifications for the emitting species are given above in Table 3.

5. DISCUSSION

Our observations of Cygnus X-2 using a variety of instruments and covering a wide energy band yield results which are encouragingly consistent with previous studies of this source. In particular, we find that the continuum is well-described by a model involving a superposition of thermal bremsstrahlung and blackbody components.

The GSPC spectral fits explicitly require the presence of a resolved line feature at 6.7 keV which may be attributed to helium-like Fe K-shell emission broadened by multiple Compton scattering in an optically thick emitting region. The derived equivalent width of this feature is similar to that detected for other low mass X-ray binaries. Our data suggest that the equivalent width may anticorrelate with the appearance of "dips" in the X-ray light curve. In the conventional interpretation, these dips are associated with obscuration of the central emitting regions by "bulges" in the accretion flow at the edge of the disk. If this is correct, then the observed decrease in the line equivalent width during dips must indicate that the Fe K emitting region is more compact than the continuum emitting region. Presumeably then, the continuum arises in an extended accretion disk corona, whereas the line is produced in a *cooler* region closer to the compact object. It will be interesting to see if this phenomenon is observed for other sources as well.

The appearance of discrete excesses in the soft X-ray spectrum in the range 10 - 15 Å confirms the detections of similar features for Cyg X-2 and other sources using the *Einstein* objective grating spectrometer. These excesses are most likely associated with a complex of Fe L-shell emission lines produced in a moderate density plasma ($n_c \sim 10^{11}$ cm^{-3}) located near the edge of the accreting flow. Such material will be photoionized by the strong continuum X-ray flux from the central source. The lines are mostly excited by radiative cascades following recombination. This process is quite complex so that modest changes in the photoionizing flux or in the geometry and density of the accreting flow can produce qualitative variations in the details of the discrete spectral structure. The observed EXOSAT features are similar but not identical to those observed with *Einstein*. Eventually, detailed comparisons of the variations in these features may yield interesting constraints, however, at present the limited spectral resolution of the grating spectrometer precludes a more sophisticated analysis.

This work was supported in part by a grant from NASA in connection with the Space Astrophysics Data Analysis Program.

REFERENCES

Chiappetti, L., Maraschi, L., Tanzi, E.G., and Treves, A. 1983, *Ap. J.*, 265, 354.

Chiappetti, L., Ciapi, A.L., Maraschi, L., Stella, L., Tanzi, E.G., and Treves, A. 1987, *Ast. and Sp. Sci.*, 131, 691.

Hasinger, G., Langmeier, A., Sztajno, M., Trumper, J., Lewin, W.H.G., and White, N.E. 1986, *Nature*, 319, 469.

Hirano, T., Hayakawa, S., Nagase, F., and Tawara, Y. 1986, *Ast. and Sp. Sci.*, 119, 77.

Stella, L., Chiappetti, L., Ciapi, A.L., Maraschi, L., Tanzi, E.G., Treves, A., "Quasi-Periodic Oscillations in Cyg X-2" in *Proceedings of the Fourth Marcel Grossman Meeting on General Relativity*, Rome, June 1985, Ed. Ruffini, R.

Vrtilek, S.D., Kahn, S.M., Grindlay, J.E., Helfand, D.J., and Seward, F.D. 1986, *Ap. J.*, 307, 698.

White, N.E., Peacock, A., Hasinger, G., Mason, K.O., Manzo, G., Taylor, B.G., and Branduardi-Raymont, G. 1986, *Mon. Not. R. Astr. Soc.*, 218, 129.

X-RAY ABSORPTION BY IONIZED GAS IN EXOSAT SPECTRA FROM THE BINARY SYSTEM 4U1700-37/HD153919

F. Haberl[1], T.R. Kallman[2], and N.E. White[1]

[1]EXOSAT Observatory, SSD, ESTEC, ESA, Noordwijk, The Netherlands
[2]NASA Goddard Space Flight Center, Greenbelt, Md., USA

ABSTRACT. We observed the 3.41 day eclipsing, massive binary system 4U1700-37/HD153919 with EXOSAT for more than one complete binary period to investigate the spectral variations during the orbital cycle of the neutron star. The spectra show a low energy excess below $\sim$ 3 keV when modelled by a powerlaw spectrum attenuated by photoelectric absorption by neutral gas, suggesting partial ionization of the absorbing gas. The column density derived from spectra above 3 keV shows an asymmetric distribution around orbital phase 0.5 with higher absorption before eclipse ingress. We approximated the with distance to the X-ray source gradually decreasing ionization of the wind by two zones. One of higher ionized wind around the X-ray source for which X-ray opacities of a gas in photoionization equilibrium were used and a zone of neutral gas further away from the X-ray source. We find that our spectra below 3 keV can be well fitted by a powerlaw which is attenuated first by photoelectric absorption of ionized gas and then by neutral gas. Since around phase 0.5 the major contribution of the wind column density along the line of sight arises from the ionized part we found that the total column density can be higher up to a factor of about 4 taking ionization into account.

1. INTRODUCTION

An accreting compact object orbiting an OB supergiant can be used to investigate the properties of the wind from the OB star. The X-ray emission from the compact object will suffer photo-electric absorption by the stellar wind and X-ray spectroscopy can be used to investigate the density and ionization structure of the wind. The driving force behind the mass loss from the OB star is the transfer of momentum from the supergiant's radiation field to the wind by scattering of radiation in UV spectral lines (see e.g. Lucy and Solomon 1970). X-ray photo-ionization of the wind will reduce the X-ray absorption and also can inhibit or enhance the velocity of the flow by changing the UV line transitions available to accelerate the wind (MacGregor and Vitello 1982). The X-ray spectra of the massive X-ray binaries (MXRB's) show variable absorption with increases in absorption around the times of eclipse ingress and egress, caused by the neutron star passing behind the dense innermost regions of the wind near to the supergiant (e.g. Branduardi, Mason and Sanford 1978, hereafter BMS). Strong photoelectric absorption is evident in the X-ray spectra of 4U1700-37 at lower energies with an equivalent hydrogen column density (n_H) varying between $1 \cdot 10^{22} - 4 \cdot 10^{23}$ H cm^{-2}. (Jones et $al.$ 1973; Mason, Branduardi, and Sanford 1976; BMS; White, Kallman, and Swank 1983, hereafter WKS). An increase in photoelectric absorption is evident at orbital phases greater than 0.5 as well as close to eclipse (see e.g. BMS). Possible explanations are a slow moving wind trailing the X-ray source due to X-ray ionization disrupting the radial acceleration of the wind (Fransson and Fabian 1980, MacGregor and Vitello 1982), a dense gas stream from the primary that trails behind the X-ray source (Fahlman and Walker 1980), or a bow shock in the wind caused by the supersonic passage of the neutron star (Jackson 1975).

2. RESULTS

2.1 The Observation

The EXOSAT observation of 4U1700-37 reported here began on 1985 April 4 at 04.30 UT and ended at 01.00 UT on 1985 April 8 with the only major interruption on April 4 between 09.00 and 20.00 UT by the perigee passage of EXOSAT. The principle instrument used in this study is the medium energy proportional counter array (ME, Turner, Smith, and Zimmermann 1981). The background subtracted X-ray light curve from the ME in the 2-10 keV band is shown in Figure 1 (upper panel) with a time resolution of ~8 min. The X-ray eclipse starts at ~10.00 UT on April 5 and ends at ~06.14 UT on April 6. The source undergoes constant flaring activity on time scales of minutes to hours with intensity variations up to a factor of 100 with several extended low states of exceptionally low intensity.

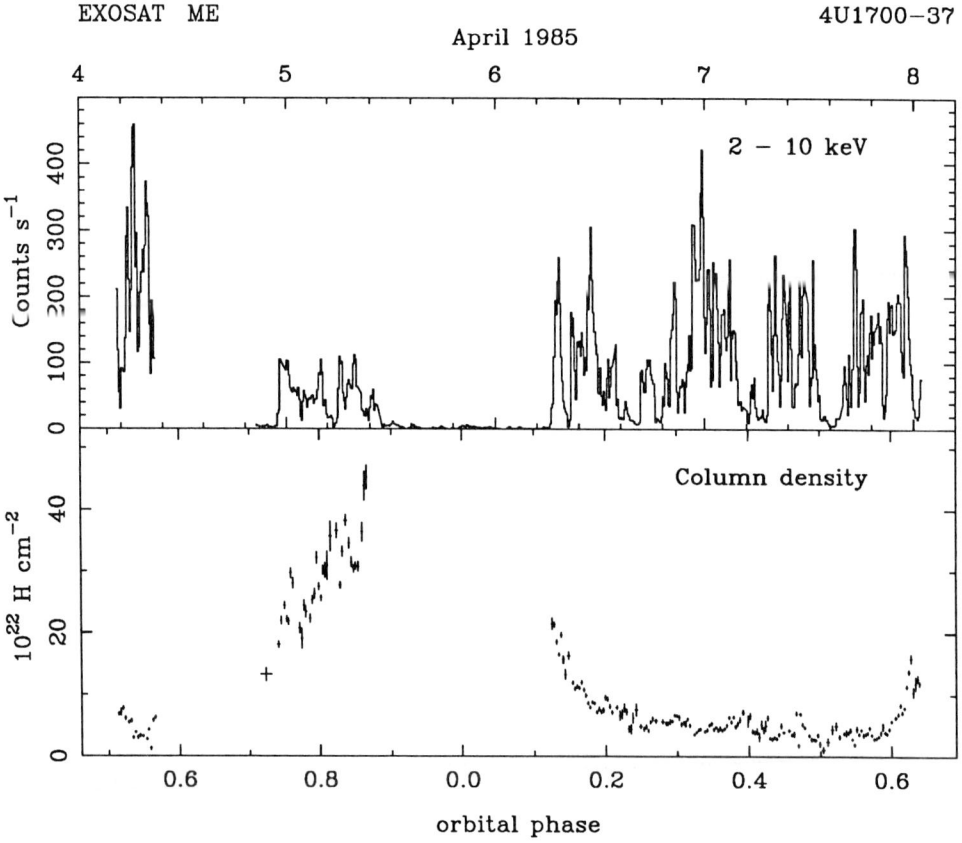

Figure 1. The background subtracted 2 - 10 keV light curve of 4U1700-37 obtained from the ME on 1985 April 4 to 8 (upper panel), with only one major interuption from about phase 0.6 to 0.7 caused by a perigee passage of EXOSAT. The count rate is in units of counts per second per half detector array with a time resolution of about 8 min. The lower panel shows the equivalent hydrogen column density versus orbital phase, derived from ME spectra above 3 keV using opacities for neutral gas.

2.2 The Spectra above 3 keV

Pulse height analysed, PHA, spectra were obtained from the argon chamber of the ME in the energy range 1-20 keV. The spectra have been background subtracted and summed into ~15 min intervals. In the 3-20 keV band the spectra are well represented by a powerlaw with a high energy cutoff of the form $\exp[(E_c - E)/E_f]$ above a break energy E_c, with photoelectric absorption by cold material as predicted for a cosmic abundance by Morrison and McCammon (1983) and a narrow iron emission line around 6.4 keV. The best determination of the unabsorbed continuum parameters was obtained by taking a spectrum from the ME over a 4.4 hr interval around phase 0.4 starting at 22.00 on April 6 where the absorption was lowest. This gives an energy index α of 0.13 ± 0.10, E_c of 6.6 ± 0.7 keV and E_f of $21.1 ^{+4.3}_{-3.3}$ with a χ^2_r of 0.5. All given errors indicate the 90% confidence region. To better constrain n_H the parameters E_c and E_f were fixed at 6.6 and 21.1 keV. The column density as a function of orbital phase derived from fitting to the 3-20 keV spectra is shown in the lower half of Figure 1. Apart from the expected increase in n_H around the time of eclipse ingress and egress, a sharp rise in n_H around binary phase 1.6 is visible, which also occurs on the previous cycle (between phase 0.74 and eclipse ingress) where it causes an additional absorption of $\sim 20 \cdot 10^{22}$ H cm^{-2} compared to eclipse egress.

2.3 An Excess In The Spectrum Below 3 keV

In the first spectral fitting we excluded energies below 3 keV because the spectra show an excess over the model prediction in this energy range. This excess increases with higher X-ray intensity and varies with orbital phase.

The increasing deviation in the energy spectrum from the model using opacities for neutral gas with increasing source intensity is most noticable in three ME spectra taken across an intensive flare that occured on April 4 around binary phase ~ 0.54. In Figure 2 a, b, and c the fits to the PHA spectra accumulated respectively before the flare, during the flare rise and at the flare maximum are shown. The average count rate during each is 46 cts s^{-1}, 381 cts s^{-1} and 605 cts s^{-1} respectively. The spectrum at the lowest intensity gives a good fit to the model used in paragraph 2.2 with χ^2_r of 1.1, but with increasing intensity the fit becomes progressively worse with e.g. for the highest intensity spectrum a χ^2_r of 5.5. This poor fit is caused by a deficiency in the observed spectrum around 3.5 keV and a low energy excess below 2.5 keV. This effect is most obvious around superior conjunction of the X-ray source and is less detectable in spectra taken close to eclipse egress. Excluding data below 3 keV gives acceptable χ^2_r between 1.0 and 1.2 for the three spectra.

We fitted our spectra to a "two zone model" where a powerlaw is attenuated by two different absorbing media in series, one with the opacities for a gas in photoionization equilibrium (Krolik and Kallman 1984) and one with opacities for neutral gas. This two zone model is an approximation for the wind near the X-ray source where the gas is ionized and far away from the X-ray source where the gas is neutral. The ionization parameter and column density in the ionized zone and the column density in the neutral zone are free parameters. The total absorption is given by the sum of the two column densities. In reality the degree of ionization around the X-ray source will decrease smoothly with distance, but a two zone approximation is good enough for this data. The powerlaw index was fixed at the value 0.13 obtained from the fits that excluded the data below 3 keV. The two zone model gives good fits to the flare spectra with χ^2_r between 1.0 and 1.2. The ionization parameter increases from $\log(\xi)$ of 1.56 to 1.65 during the flare which is still within the errors of typically 0.1. For the neutral column density only an upper limit of $2.5 \cdot 10^{22}$ H cm^{-2} is obtained while the ionized column density lies at $18 \pm 3 \cdot 10^{22}$ H cm^{-2}. Both column densities are constant within the errors and give a total of $\sim 20 \cdot 10^{22}$ H cm^{-2} while the fits with only neutral absorption give $\sim 5 \cdot 10^{22}$ H cm^{-2}.

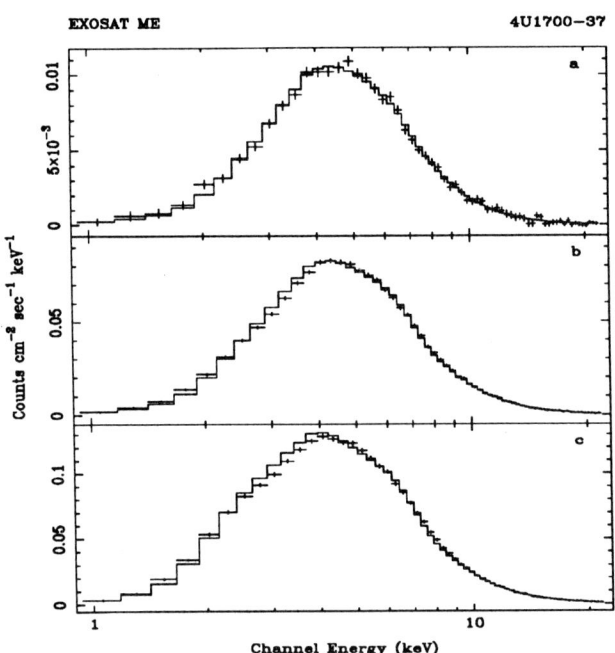

Figure 2.
Three ME spectra obtained during a flare on 1985 April 4. The spectrum on panel (a) was obtained before the flare, (b) during flare rise and (c) during flare maximum. Deviations from a powerlaw attenuated by neutral gas absorption (plotted as histogram) increase with source intensity.

3. DISCUSSION

The spectra during flare rise deviate more from a powerlaw attenuated by neutral gas absorption the higher the X-ray intensity was. One possible explanation is that X-rays change the ionization state of the wind in the vicinity of the X-ray source. The highly ionized gas has lower absorption cross sections for low energetic X-rays which leads to an excess in the spectra at low energies compared to spectra which are attenuated by absorption from neutral gas. Fits to flare spectra using a model in which a zone of ionized gas and a zone of neutral gas in series absorb the X-rays, gave acceptable χ^2. Due to the lower opacities of ionized gas the total column density is increased. This effect is most important around phase 0.5 where the line of sight does not go through the dense nearly neutral regions near the primary star. Hence it can change the column density profile and one has to be careful in modelling the wind using the column densities.

REFERENCES
Branduardi, G., Mason, K.O., and Sanford, P.W. 1978, *M.N.R.A.S.*, **185**, 137.
Fahlman, G.G.and Walker, G.A.H. 1980, *Ap.J.*, **240**, 169.
Fransson, C., and Fabian, A.A. 1980, *Astr. Ap.*, **87**, 102.
Jackson, J.C. 1975, *M.N.R.A.S.*, **172**, 483.
Jones, C., Forman, W., Tananbaum, H., Schreier, E., Gursky, H., Kellog, E., and Giacconi, R. 1973, *Ap.J.(Letters)*, **181**, L43.
Krolik, J.H. and Kallman, T.R. 1984, *Ap.J.*, **286**, 366.
Lucy, L.B. and Solomon,,P. 1970, *Ap.J.*, **159**, 879.
MacGregor, K.B., and Vitello, P.A.J. 1982, *Ap.J.*, **259**, 267.
Mason, K.O., Branduardi, G., and Sanford, P.W. 1976, *Ap.J.(Letters)*, **203**, L29.
Turner, M.J.L., Smith, A., and Zimmermann, H.U. 1981, *Space Sci. Rev.*, **30**, 513.
White, N.E., Kallman, T.R., and Swank,J.H. 1983a, *Ap.J.*, **269**, 264.

X-RAY SPECTROSCOPY OF THE ULTRA-SOFT TRANSIENT 4U1543-47

H. van der Woerd[1], N.E. White[1], and S.M. Kahn[2]

[1] EXOSAT Observatory, Astrophysics Division, Space Science Department of ESA, ESTEC, Postbus 299, 2200 AG Noordwijk, The Netherlands.

[2] Department of Physics, University of California, Berkeley, CA 94720, U.S.A.

ABSTRACT. The X-ray transient 4U1543-47 was observed in 1983 by the EXOSAT observatory near the maximum of an outburst. The X-ray spectrum was measured using a gas scintillation proportional counter (GSPC) and a transmission grating spectrometer (TGS). Two emission line features are resolved. A broad (FWHM~2.7 keV) line at 5.9 keV is detected in the GSPC, which we interpret as a redshifted and broadened iron Kα line. The line broadening and redshift may arise from either Compton scattering in a cool plasma with small optical depth ($\tau \approx 5$), or from Doppler and relativistic effects in the vicinity of a compact object. The spectrum below 2 keV, obtained with the TGS, shows evidence for a broad emission line feature at 0.74 keV, which may be an iron L-transition complex. However, we find that such an emission feature could be an artifact caused by an anomalously low interstellar absorption by neutral Oxygen. The continuum emission is extremely soft and is well described by an unsaturated Comptonized spectrum from a very cool plasma (kT = 0.84 keV) with large scattering depth ($\tau = 27$). The continuum spectrum is strikingly similar to that of black hole candidate LMC X-3.

1. INTRODUCTION AND OBSERVATIONS

X-ray transients show dramatic outbursts in the X-ray band with typical recurrence times between 1 and 60 years and durations lasting several months. Their high luminosity (10^{37} - 10^{38} erg s^{-1}), optical counterparts, spectral characteristics and time variability during outburst are similar to the persistent bright X-ray sources; close binary systems powered by accretion of matter onto a neutron star or black hole. White, Kaluzienski and Swank (1984) suggested that transients which show an "ultra-soft" (kT $\lesssim$ 3 keV) spectrum are good candidates for containing a black hole. The *persistent* X-ray sources which show these ultra-soft spectra include the black hole candidates Cyg X-1 (in its high state) and LMC X-3 (White and Marshall 1984). The X-ray source 4U1543-47 was identified as an ultra-soft transient when it was detected in a high state in 1971 (Li, Sprott, and Clark 1976). Twelve years later another outburst from 4U1543-47 was discovered in 1983 August by the TENMA satellite (Tanaka 1983, Kitamoto et al. 1984). We report the results from high-resolution X-ray spectra taken with the European Space Agency's X-ray observatory EXOSAT (White and Peacock 1988). EXOSAT observed 4U1543-47 for 4 hours on 1983 August 28. The Gas Scintillation Proportional Counter (GSPC) detected the source to be bright with a counting rate (2.0 to 13.0 keV) at constant value of 335 counts s^{-1}, or about one eigth the counting rate seen from Sco X-1. Spectra below 2 keV were collected with the Transmission Grating Spectrometers (TGS) of the Low Energy imaging telecopes. We present the spectrum obtained with the LE1 spectrometer and thin Lexan filter, which has the best spectral resolution ($\Delta E/E \approx 10$ (E/1 keV) %) and photon statistics.

2. SPECTRAL RESULTS

Various trial model spectra, folded through each detector response, were simultaneously fit to the GSPC spectrum and the negative and positive orders of the LE1 spectrum. For all combinations of many one and two component trial continuum models a broad emission line at ~ 5.9 keV in the GSPC was required. In the TGS an excess at ~ 0.7 keV was found. The continuum model with the least number of free parameters that gives an acceptable fit is a Comptonized spectrum as prescribed by Sunyaev and Titarchuk (1980), that includes absorption by neutral material with cosmic abundances (Morrison and McCammon 1983). The best fitting parameters (which also included two emission lines discussed below) give an electron temperature (kT_e) of 0.843 ± 0.004 keV, an optical depth (τ), of 26.6 ± 0.6, and an equivalent hydrogen column density, $N_H = (4.26 \pm 0.15) \times 10^{21}$ H atoms cm^{-2} with a χ^2 of 449 for 348 degrees of freedom (d.o.f.). Fig. 1 shows the model fit to both the GSPC and LE1 grating spectrum. For clarity only the LE1 negative orders are shown. The source luminosity in the band 0.4 to 13.0 keV, corrected for absorption, is 2.0×10^{37} erg s^{-1}, for an arbitrary distance of 1 kpc. The line in the GSPC was modelled using a Gausian profile and was centered at 5.93 ± 0.24 keV with a full width half maximum (FWHM) of 2.71 ± 0.47 keV and an equivalent width (EW) of 115 eV. The line profile is shown in Fig 1b, and is clearly a very broad feature. The structure in the residuals between 4 and 5 keV arises from uncertainties in the calibration, but does not strongly influence the line parameters. The emission line in the TGS is at 0.742 ± 0.027 keV with FWHM= 0.204 ± 0.068 keV, and EW = 86 eV. The excess flux near 0.74 keV can, however, also be interpreted as the result of an overestimation of the continuum absorption at this energy. The fit is sensitive to the depth of the oxygen K-edge at 0.53 keV (Ride and Walker 1977), and an acceptable fit can be obtained by reducing the abundance of oxygen by 35%, as compared to the solar abundance (χ^2 of 496 for 350 d.o.f.).

3. DISCUSSION

The broad feature around 0.7 keV is either due to an emission line complex at 0.74 keV, or due to depletion of the number of neutral oxygen atoms in the line of sight. In the former we find the line to be broad with a Gaussian FWHM of 0.2 keV. Since the energy resolution is only a factor of four better than this, the broad feature may be caused by the smearing together of an unresolved line complex, centered on 0.74 keV. The best candidates for such a complex would be the L-transitions of a wide spectrum of iron ions. Similar Fe L-shell complexes have been observed in other LMXB with the TGSs flown on EINSTEIN and EXOSAT (Kahn, Seward and Chlebowski 1984 (KSC), Brinkman et al. 1985, Vrtilek et al. 1986a,b). These lines are most likely formed by radiative recombination in a plasma which is photoionized by the strong continuum emission from the central source (KSC). An equally acceptable fit to the data is obtained without these lines, however, if oxygen is allowed to be underabundant by 65% from the cosmic value. It is notable that KSC also find in addition to the emission line features also an underabundance of oxygen for Sco X-1. It is possible that we are observing a combination of reduced abundance and an emission complex at 0.74 keV. Part of the observed absorption could be intrinsic to the system, in which case the apparent underabundance would reflect a real underabundance of oxygen within the accreting matter. Another possibility is that the observed absorption is interstellar and that oxygen is depleted due to grain formation (Fireman 1974, Ride and Walker 1977). When the grains are big enough they become opaque for soft X-rays and part of the grain is effectively shielded from the radiation. The average atom in the grain only absorbs 65% of the X-rays compared to a purely gaseous medium. This infers a typical grain size of 0.37 μ in the direction of 4U1543-47 (cf. Fireman 1974).

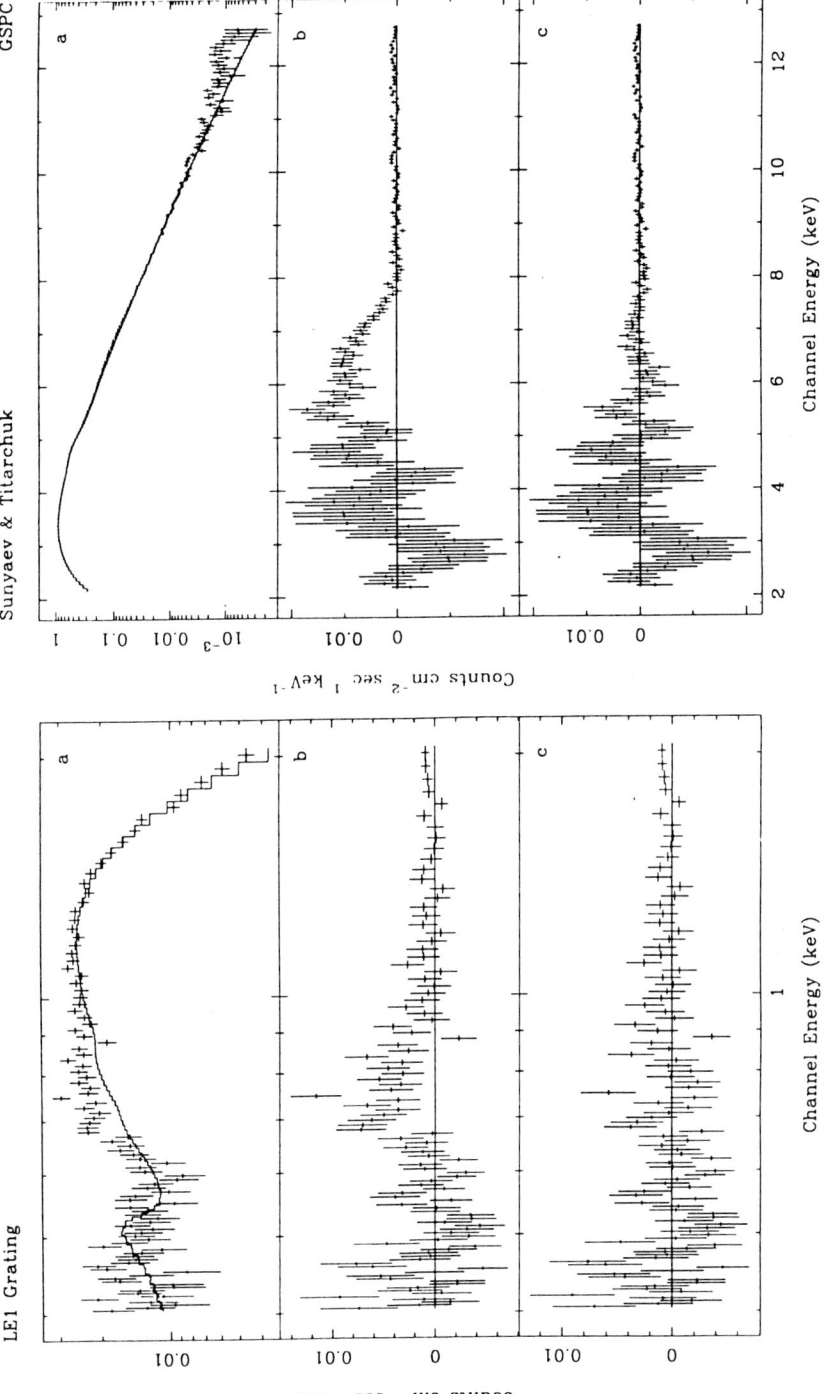

FIG. 1. The negative spectral orders of the LE1 grating spectrum and the GSPC spectrum of the soft X-ray transient 4U1543-47 near outburst maximum, fitted by an unsaturated comptonized spectrum (a). The (b) panels show the residuals, where the strengths of the emission feature at 0.74 keV and of the iron line at 5.9 keV was set to zero. Panels (c) shows the residuals for the fit with these emission features included.

The broad emission feature seen in the GSPC centered on 5.9 keV seems most likely associated with the iron K-line complex between 6.4 and 6.7 keV. Broad iron K-lines, typically with a FWHM of 1 keV and line centroid at 6.4-6.7 keV, have been seen from several other LMXB (White et al. 1986, Hirano et al. 1987). An exception to this is the broad iron line discovered from the blackhole candidate Cyg X-1 by Barr, White and Page (1985) which is at 6.2 keV with a FWHM of 1.2 keV. The broadening may arise from either Comptonization in an accretion disk corona (White et al. 1986, Hirano et al. 1987, Kallman and White 1989) or from the effects of rotation and general relativistic effects in the inner regions of an accretion disk, close to the compact object (Fabian et al. 1989). Comptonization in an optically thick accretion disk corona can explain the broadening and red shift, although Kallman and White (1989) have shown this requires a greater optical depth than that expected from a purely X-ray heated corona. If we consider only the recoil effect, which is certainly correct if the iron line is generated and broadened in a plasma with approximately the continuum temperature, we find that both the redshift and broadening are consistent with an optical depth of $\tau \approx 5$ (cf. Sunyaev and Titarchuk 1980).

The Comptonisation model (Titarchuk and Sunyaev 1980) found to give a good fit to the continuum spectrum of 4U1543-47, is the same as that found for the spectrum of the blackhole candidate LMC X-3. The best fitting parameters are also very similar for 4U1543-47 with $kT_e = 0.84$ keV and $\tau = 27$, which corresponds to a Comptonisation y parameter of 4.7 ($y = 4kT_e\tau^2/m_ec^2$) and for LMC X-3 $kT_e \sim 1$ keV, $\tau \sim 23$ and $y \sim 4.1$ (Treves et al. 1988). This correspondence between the spectra of 4U1543-47 and LMC X-3 underlines the suggestion (White and Marshall 1984; White, Kaluzinski and Swank 1984) that ultrasoft transients should be considered good black holes candidates.

acknowledgements. We thank L. Chiappetti, R. Blissett, W. Craig and R. Rogers for contributions during an earlier phase of this work. JMK acknowledge support from a NASA Astrophysics Data Program. The EXOSAT Observatory team are thanked for their support. NEW thanks L. Stella and A. Fabian for discussions.

REFERENCES

Barr, P., White, N. E., and Page, C. G. 1985, *M.N.R.A.S.*, **216**, 65p.
Brinkman, A. C., et al. 1985, *Space Sci. Rev.*, **40**, 201.
Fabian, A.C., Rees, M.J., Stella, L., and White, N.E. 1989, *M.N.R.A.S.*, in press.
Fireman, E. L. 1974, *Ap. J.*, **187**, 57.
Hirano, T., et al. 1987, *Publ. Astron. Soc. Japan*, **39**, 619.
Kahn, S.M., Seward, F.D., and Chlebowski, T. 1984, *Ap. J.*, **283**, 286 (KSC).
Kallman, T.R., and White, N.E. 1989, *Ap. J.*, in press.
Kitamoto, S., et al. 1984, . *Publ. Astron. Soc. Japan*, **36**, 799.
Li, F. K., Sprott, G. F., and Clark, G. W. 1976, *Ap. J.*, **203**, 187.
Morrison, R., and McCammon, D. 1983, *Ap. J.*, **270**, 119.
Ride, S. K., and Walker, A. B. C. 1977, *Astr. Ap.*, **61**, 339.
Sunyaev, R. A., and Titarchuk, L. G. 1980, *Astr. Ap.*, **86**, 121.
Tanaka, Y. 1983, IAU Circ. No. 3854.
Treves, A. et al. 1988, *Ap. J.*, **325**, 119.
Vrtilek, S.D., et al. 1986a, *Ap. J.*, **307**, 698.
Vrtilek, S.D., et al. 1986b, *Ap. J.*, **308**, 644.
White, N. E., and Marshall, F. E. 1984, *Ap. J.*, **281**, 354.
White, N. E., and Peacock, A. 1988, *EXOSAT Preprint* No. 75.
White, N. E., Kaluzienski, J. L., and Swank, J. H. 1984, in *High Energy Transients in Astrophysics*, ed. S. E. Woosley, AIP Conf. Proc. **115**, p. 31.
White, N. E., et al. 1986, *M.N.R.A.S.*, **218**, 129.

6. Clusters of Galaxies, Cooling Flows

X–RAY SPECTRA OF CLUSTERS OF GALAXIES

Craig L. Sarazin
Department of Astronomy
University of Virginia
Charlottesville, Virginia U.S.A.

ABSTRACT. X-ray line observations of clusters of galaxies have shown that the X-ray emission in clusters is mainly thermal emission from hot diffuse gas, and that much of this gas has come out of stars, probably having been ejected from galaxies in the cluster. Future high resolution observations should allow us to determine the physical state of the gas. X-ray line measurements and abundance determinations can lead to strong constraints on the origin of the intracluster gas, and on the chemical evolution and history of galaxies. Some of the stronger resonant X-ray lines may be observable as absorption lines against a background quasar. Such X-ray absorption line measurement can be used to directly derive distances to clusters, using a technique similar to (and possibly complementary to) that the well–known method using the Zel'dovich–Syunyaev effect.

1. INTRODUCTION

The study of the X-ray emission from clusters of galaxies is an area of astrophysics where X-ray spectroscopy has already played an essential role. In the first case, the discovery of iron K lines in the spectrum of the Perseus cluster (Mitchell et al., 1976), followed by their discovery in the Coma and Virgo clusters (Serlemitsos et al., 1977), established immediately that the emission mechanism of clusters was thermal emission. This implied that all of the apparently empty space between the galaxies in clusters was actually filled by diffuse, hot intracluster gas with typical temperatures and electron densities of $T \sim 7 \times 10^7$ K and $n_e \sim 10^{-3}$ cm^{-3}. The total mass of this gas is found to somewhat exceed the total mass of all of the galaxies in the clusters. The strength of the iron K lines indicated that the iron abundances in these three clusters were about half of the solar value. Since the only known sources for the production of heavy elements are in the interiors of stars, and the only significant populations of stars are located in galaxies, this indicated that much of the intracluster gas had come out of stars and may have been been ejected from galaxies. Since the present rates of stellar mass loss in galaxies are generally thought to be at least two orders of magnitude too small to have produced this gas, this suggested that galaxies may have had much more star formation and mass loss in the past.

A second case where X-ray line spectroscopy has played an essential role in clusters was the discovery of low ionization X-ray lines emitted at the centers of some clusters of galaxies (Canizares et al, 1988; Mushotzky 1984). This discovery, made with the *Einstein* X-ray Observatory, indicated that there are considerable amounts of cooler gas ($T \sim 10^6 - 10^7$ K) at the cluster centers. It is thought that this gas originates in "cooling flows" of gas moving in from the intracluster medium. While there is also considerable evidence supporting the cooling flow picture from X-ray images of clusters, the X-ray spectroscopy is still the most difficult piece of evidence for alternative theories. Although cooling flows provide some of the most interesting applications of X-ray spectroscopy to clusters, they are reviewed by Fabian in this volume and I will not discuss them further. I will concentrate on the applications of X-ray spectroscopy to the hotter gas filling the bulk of the volume in clusters.

In this volume, there are several excellent reviews of the atomic processes by which X-ray lines are emitted, and I will not review this subject. The gas density is very low and the radiation field is very weak in clusters, and as a result, there are no direct spectral diagnostics for the density of the gas. The likely time scales for changes in the physical conditions in clusters are very long, and one therefore expects the gas to be in collisional ionization equilibrium. The emission lines are certainly not very optically thick, although resonant scattering may be somewhat important and can produce observable absorption lines toward background quasars (see §4 below). Under these conditions, the X-ray spectrum of clusters depends only on the electron temperature and abundances in the gas. At the temperatures found in clusters, the strongest emission lines are expected (and observed) to be the iron K lines (from Fe XXV, Fe XXVI, and satellites from lower ions) at photon energies of about 7 keV, the iron L lines (from Fe XVII to Fe XXIV) at about 1 keV, and the K lines of the lighter elements O, Ne, Mg, Si, S, Ar, and Ca (Sarazin and Bahcall 1977).

2. EXISTING OBSERVATIONS

Because the *Einstein* X-ray Observatory was not sensitive to 7 keV photons, all of the observations of the Fe K lines have been made with proportional counters or gas scintillation proportional counters with poorer spatial resolution. The Fe Kα line ($n = 2$ to $n = 1$, where n is the principal quantum number) has been observed in about 30 clusters. Major surveys include the OSO–8 survey of Mushotzky et al. (1978), the *Ariel*–5 survey of Mitchell et al. (1979), and the HEAO A–1 survey of Mushotzky (1984) and Henriksen and Mushotzky (1985). Many of the results have been compiled by Rothenflug and Arnaud (1985), and reviewed by Mushotzky (1984,1988). The Fe Kα observations all indicate iron abundances in the intracluster gas of about 1/2 of the solar value (Mushotzky 1984,1988; Henriksen and Mushotzky 1985; Rothenflug and Arnaud 1985). In the present data, there is no significant evidence that the iron abundance depends on any cluster property, such as richness and X-ray luminosity. Of course, the number of clusters observed is small enough that it is difficult to rule out any such variation. Henriksen and Mushotzky (1985) found a slight tendency for the iron abundance to be higher in lower temperature clusters, although this effect was not seen in the compilation of Rothenflug and Arnaud (1985), and might be due to errors in the atomic physics used to calculate the line emission.

Recently, gas scintillation proportional counters on EXOSAT and *Tenma* have been used to observe the spectra of clusters of galaxies (Hughes et al., 1988a,b; Singh et al., 1986; Okumura et al. 1988). These instruments have about twice the spectral resolution of normal proportional counters. This higher resolution is particularly useful for the study of the iron Kβ line ($n = 3$ to $n = 1$). The presence of this line confirms the thermal excitation of the iron lines, and the ratio of Fe Kβ/Kα should be useful as a diagnostic for the temperature in the gas. The best observations of this ratio would appear to be those of Okumura et al. (1988) for the Coma, Perseus, and Ophiuchus clusters. However, these observations pose a serious problem, since the Kβ/Kα ratio measured for Coma and Ophiuchus is higher than that which can be produced at *any* temperature assuming thermal excitation and collisional ionization equilibrium.

Observations of K lines from intermediate elements such as Mg, Si, and S and of the L lines of Fe have been made with the Solid State Spectrometer (SSS) on the *Einstein* X–ray Observatory (Lea *et al.*, 1982; Mushotzky 1984,1988). At the SSS resolution, it is somewhat difficult to resolve these lines from the continuum, and they have been observed mainly in clusters with cooling flows where a range of gas temperatures contribute to the emission. As a result, the abundances of the intermediate elements are very uncertain, but are probably in the range of 1/3 to 3 times solar (Mushotzky 1984).

The L Lines from O VIII and Fe XVII–XXIV have been observed at higher resolution with the Focal Plane Crystal Spectrometer (FPCS) on the *Einstein* X-ray Observatory (Canizares *et al.*, 1988). Only a few lines in the nearest clusters were strong enough to be detected. Because these lines are mainly produced by relatively cool gas, only clusters with cooling flows were detected. In these clusters, a range of gas temperatures contribute to the line emission, making the abundance determination more difficult. Nonetheless, the intensities of oxygen and iron lines in M87/Virgo and Perseus clusters strongly suggest that the ratio of oxygen to iron abundances in these clusters is about three times the solar value (Canizares *et al.*, 1988).

One important question is whether the heavy elements in the intracluster gas are spread uniformly throughout the cluster, or are concentrated in the center of the cluster. All of the abundances quoted above assume that the composition of the intracluster gas is uniform. If the heavy elements were concentrated in the center of the cluster, their emission per unit mass would be increased because of the higher electron density, and the required abundances might be decreased by as much as a factor of $\sim$20. There are at least two reasons why the heavy elements might concentrate in the cluster core. First, the heavy elements would eventually settle at the bottom of the cluster gravitation potential unless their diffusion were impeded or the gas were kept well mixed (Fabian and Pringle 1977; Rephaeli 1978; Abramopoulos *et al.*, 1981). Alternatively, the heavy element enriched gas may come only from galaxies, and may be deposited primarily in the center of the cluster because the higher ram pressure there is more effective in removing the gas from galaxies (Nepveu 1981). Although the present X-ray line observations lack the spatial resolution to map out the heavy element abundances in any detail, there are now a number of cases where the observations are not consistent with any very strong heavy element abundance gradients (Lea *et al.*, 1982; Ulmer *et al.* 1987; Hughes *et al.*, 1988b).

In the future, it should be possible to determine the temperature of the intracluster gas from line ratio diagnostics, preferably from two or more lines from the same ion. At present, the gas temperature in clusters is usually determined from the shape of the continuum X-ray spectrum. Because the gas in clusters is hot and because the *Einstein* and EXOSAT telescopes were insensitive to hard X-rays, most of the temperature data come from low spatial resolution proportional counters. These observations mainly provide global or average gas temperatures. Such global temperature measurement can be used to assess the importance of heating processes in the gas in clusters. Except for the small cooling flow regions at the centers of some clusters, the cooling time in the intracluster gas is generally longer than the Hubble time. Thus, the intracluster gas can retain any heat it acquires, either during its formation or subsequently. If the intracluster gas either falls into the cluster, or is ejected without much extra energy from galaxies, it will initially have a thermal energy per unit mass which is comparable to the depth of the gravitation potential well in the cluster (see, for example, Sarazin 1988a). Then, one would expect the gas temperature to satisfy $T \approx \mu m_p \sigma^2/k \propto \sigma^2$, where μ is the mean mass per particle in the gas in units of m_p and σ is the cluster velocity dispersion. Mushotzky (1988) has shown that the observed relation between T and σ has a flatter slope than $T \propto \sigma^2$, albeit with a great deal of scatter. Although part of the scatter and flattening might be due to errors in the determination of cluster velocity dispersions, if the trend found by Mushotzky is real, it suggests that other heating processes might play a role. Such processes include heating by supernovae during the ejection of gas from galaxies, heating by nonthermal particles associated with cluster radio sources, and heating by drag on the galaxy motions. In any case, if heating occurs it is apparently most important in the poorer clusters with smaller velocity dispersions.

In addition to average temperatures, it is very important to determine the spatial variation of the gas temperatures in clusters. Since the intracluster gas is hydrostatic, one can derive the total mass and dark matter distribution in clusters from the radial variation of the gas temperature and gas density (Fabricant and Gorenstein 1983). The radial variation of the temperature also tells us to what extent thermal conduction has modified the temperature distribution in clusters. Theoretically, one expects that if conduction is not suppressed by magnetic fields, it will make the central ~ 1 Mpc of the cluster gas isothermal, but will not affect the temperatures of the outermost gas. Observational limits on the temperature variation in the Coma cluster by Hughes *et al.* (1988b) are consistent with this sort of variation.

3. ABUNDANCES AND THE EVOLUTION OF THE INTRACLUSTER GAS

The heavy element abundances in the intracluster gas provide many interesting constraints on the origin and evolution of this gas, and on the history of star formation and the chemical evolution of galaxies. The intracluster gas shows the integrated effect of the large number of galaxies in the clusters, and thus gives a global picture of galactic evolution which may compliment the more detailed information acquired by studying individual galaxies.

Several of the most obvious lessons from the intracluster gas abundances were mentioned in the introduction. A significant amount of the intracluster gas must have come out of stars. Now, at present the only significant population of stars we observe are located in galaxies; this suggests that a significant part of the intracluster gas came out of galaxies. Another possible source might be a generation of pregalactic stars ("Population III"), but since the intracluster gas abundances are higher than those of many galactic stars, it is unlikely that such pregalactic stars contributed much to the heavy elements in the intracluster medium. Since stellar mass loss rates from the present stellar population in the elliptical and S0 galaxies are rather low, this also suggests that these galaxies had many more massive stars and more star formation in the past.

One simple way to parameterize the effect of stellar and galactic ejection is to consider the intracluster gas to be a mixture of unprocessed, primordial gas and of processed gas ejected from galaxies. A number of calculations of the evolution of the stellar populations of elliptical galaxies suggest that they will produce gas with heavy element abundances of several times solar (Larson and Dinerstein 1975). This suggests that the intracluster medium may be composed of comparable amounts of primordial and ejected gas.

The total mass of iron in the intracluster gas in a typical rich cluster is about

$M_{Fe} \gtrsim 10^{11}\ M_\odot$, which is very considerable when compared to the total mass of stars in all of the galaxies in a typical cluster. This confirms the notion that these galaxies had an earlier generation which included many massive stars; it is possible that the remnants of these earlier stars contribute to the "dark matter" in galaxies and clusters. The comparison of the abundances of different heavy elements in intracluster gas can also provide useful information on the stars which produced the heavy elements. At present, the only useful information on this question comes from the observation that the oxygen to iron ratio is several times solar in the M87/Virgo and Perseus clusters (Canizares et al., 1988). It is believed that oxygen is produced primarily in Type II supernovae from short–lived massive stars, while much of the iron production may occur in Type I supernovae involving older, lower mass stars. The high O/Fe ratio in clusters provides another indication that high mass stars were more common in the past and contributed significantly to the enrichment of the intracluster gas.

From the observed mass of iron in the intracluster gas one can estimate the nuclear energy released in the fusion leading to the iron; this gives $E_{nucl} \gtrsim 10^{63}$ ergs, which is comparable to the gravitational binding energy of all of the intracluster gas. For massive stars, part of this energy appears as optical radiation, and part appears as kinetic energy in stellar winds and supernovae. Unless the overwhelming majority of this energy is radiated away, the heating due to supernovae and winds from the massive stars which produced the heavy elements in clusters will be very significant. This heating will be most important in smaller systems (galaxies and poorer groups), which have shallower gravitational potential wells. There should be enough energy to easily remove most of the gas from galaxies, and to remove part of the gas from poorer groups. Thus, one would expect to find a higher ratio of gas to stellar mass in rich clusters than in poor clusters or in individual elliptical galaxies. In fact, Rothenflug and Arnaud (1985) and Mushotzky (1988) find that the ratio of M_{gas} to M_{stars} does increase with M_{star} for clusters. Moreover, the remaining gas in galaxies or poor clusters should have been significantly heated. One would expect to find that the gas in poorer groups and individual elliptical galaxies is hotter relative to the velocity dispersion than in richer clusters. In §2, I noted that this appears to be the case for poorer clusters, and the ratio of T/σ^2 is even higher for ellipticals. As a result of this, one would expect the gas in elliptical galaxies and in poorer clusters to have a flatter density distribution (slower decrease with radius) than in rich cluster.

The picture presented above represents one very simple scenario for the chemical evolution of galaxies and clusters. In this volume, Forman, Jones, and David present a somewhat different model in which the variation in the ratios of M_{gas} to M_{stars} represents a variation in the efficiency of galaxy formation with environment, with rich clusters being less efficient at galaxy formation. Future high resolution X-ray line spectroscopy can distinguish between these two models. For example, the Forman–Jones–David models predicts that the heavy element abundances should be greater in poor clusters and in elliptical galaxies. In general, one would like to measure the abundances of several different elements in a wide range of clusters and in individual elliptical galaxies. For at least a sample of nearby clusters, it is important to map out the spatial variation of the line emission to determine whether the heavy elements are uniformly distributed throughout the gas. Finally, it is very important to extend these spectral observations to clusters at significant redshifts. Such observations should constrain the time at which the heavy elements were produced.

4. X–RAY ABSORPTION LINES AND THE DISTANCE TO CLUSTERS

Several authors have noted that the strongest resonance X-ray lines from clusters may be slightly optically thick (Shapiro and Bahcall 1980; Basko et al., 1981; Gil'fanov et al., 1987). Basko et al. suggested that these resonance lines might be detectable in absorption in the X-ray spectrum of a background quasar behind a cluster of galaxies. Detailed calculations show that some ions should have column densities of of at least 10^{16} cm^2, leading absorption line equivalent widths of several eV (Basko et al. 1981; Sarazin 1988b). The strongest lines are generally the Fe XXVI $1s - 2p$ and Fe XXV $1s^2\,{}^1S - 1s2p\,{}^1P$ lines. Gil'fanov et al. suggested that the lines would be somewhat optically thick. The optical depth of the lines is more difficult to predict than the absorption equivalent width, since the optical depth depends inversely on the line width. The minimum line width is that due only to thermal broadening. Because these X-ray lines are produced by heavy elements, the thermal line widths are quite narrow (a small fraction of the sound speed). However, it is likely that the turbulent line broadening exceeds thermal line broadening. The expected optical depths of the strongest lines are in the range 0.01–10, with the larger values applying if the lines are only thermally broadened. Gil'fanov et al. argue for the larger values of the line optical depth, and suggest that resonant scattering might affect the line profiles and the surface brightness distributions of the lines.

If an X-ray absorption line due to intracluster gas could be detected in the spectrum of a background quasar and if the same line could be observed in emission from the cluster, this would allow a direct determination of the distance to the cluster. The reason this leads to a distance measurement is simple; absorption is independent of distance, as long as the absorber lies between the observer and the background light source. The observed flux due to emission decreases with distance squared, and so comparing absorption and emission leads to a distance estimate. The arguments leading to the distance estimator for cluster X-ray absorption lines are very similar to those which have been applied to the Zel'dovich–Syunyaev effect in clusters (Zel'dovich and Syunyaev 1969; Cavaliere et al., 1977,1979; Gunn 1978; Silk and White 1978). I learned at this conference that the same idea of using X-ray absorption lines to determine distances has been developed independently by Krolik and Raymond (see the article by Krolik in this volume).

The basic idea can be described easily; a more rigorous treatment is given in Sarazin (1988b). Let us assume that an X-ray-bright quasar is found which is located behind an X-ray emitting cluster of galaxies. We assume that the gas in the cluster is homogeneous (not highly clumped) and spherically symmetric. If the gas is not spherically symmetric, it would be necessary to observe a sample of clusters and average the results. Using a spectroscopic detector with a small aperture, the X-ray spectrum of the quasar is observed and is found to contain an absorption feature from a resonance line from an ion located in the hot gas of the cluster. Let W be the equivalent width of the line in the quasar spectrum. Let us assume that the optical depth of the line is not large. Then, the equivalent width determines the column density of the ion N_{ion} as

$$\left(\frac{N_{ion}}{10^{16}\ \text{cm}^{-2}}\right) = 0.911 f^{-1} (1+z) \left(\frac{W}{\text{eV}}\right), \qquad (1)$$

where z is the redshift of the cluster and f is the absorption oscillator strength of the line. The column density is given by

$$N_{ion} \equiv \int n_{ion} dl \qquad (2)$$

where n_{ion} is the number density of the ion and l is the path length through the cluster along the line–of–sight to the quasar.

Let us consider the emission by the cluster gas in the *same* resonance line as the absorption. The emissivity of the cluster gas in such a line can be written as

$$\epsilon_{line} = n_{ion} n_e \Lambda_{line}(T), \qquad (3)$$

(see, for example, Sarazin and Bahcall 1977), where n_e is the electron density and Λ_{line} is a function of the electron temperature T which is determined by atomic physics. The surface brightness of the cluster in this line at the same position where the absorption line is measured is given by

$$I_{line} = (1+z)^{-4} \int n_{ion} n_e \Lambda_{line}(T) dl. \qquad (4)$$

Then, comparing equations (2) and (3), we see that the average electron density along that line–of–sight is roughly

$$\langle n_e \rangle \approx \frac{I_{line}(1+z)^4}{N_{ion} \langle \Lambda_{line} \rangle}, \qquad (5)$$

where $\langle \Lambda_{line} \rangle$ is an average value along the line–of–sight.

The total X-ray emissivity (in any band) in the gas can be written as

$$\epsilon_X = n_e^2 \Lambda_X(T); \qquad (6)$$

again, Λ_X is a function of the electron temperature T which is determined by atomic physics. Then, the total flux of X-ray emission in this band as observed at the Earth is

$$F_X = \frac{1}{4\pi d_A^2 (1+z)^4} \int n_e^2 \Lambda_X(T) dV, \qquad (7)$$

where d_A is the angular–diameter distance to the cluster (Weinberg 1972), and V is the volume of the cluster. Let us assume that the X-ray emission extends out to a characteristic radius r_X, and that the volume associated with the emission is roughly $4\pi r_X^3/3$. The quantity we might actually observe is the angular size of the X-ray emitting region $\theta_X \equiv r_X/d_A$. The flux can be written as

$$F_X \approx \frac{1}{3} \langle n_e \rangle^2 \langle \Lambda_X \rangle \theta_X^3 d_A (1+z)^{-4}. \qquad (8)$$

Then, the distance to the cluster is given approximately by

$$d_A \approx \frac{3 N_{ion}^2 \langle \Lambda_{line} \rangle^2 F_X}{(1+z)^4 I_{line}^2 \langle \Lambda_X \rangle \theta_X^3}. \qquad (9)$$

This is expression is approximate because the correct definitions of $\langle n_e \rangle$, $\langle \Lambda_{line} \rangle$, $\langle \Lambda_X \rangle$, and θ_X involve conflicting averages over the gas distribution in a cluster. Exact expressions in which the correct averages are determined directly from the X-ray surface brightness profile of the cluster are given in Sarazin (1988b). However, equation (9) is useful in that it shows fairly directly the dependence of the distance estimate on the observed quantities. For example, one notes (sadly) that distance estimate depends on the *square* of the quantities which are most difficult to measure (N_{ion}, and I_{line}), and only linearly on the easiest quantity to measure (F_X). A similar misfortune befalls distance estimates based on the Zel'dovich–Syunyaev effect (Silk and White 1978).

It will obviously be difficult to measure the required quantities (particularly W and I_{line}) with sufficient accuracy to improve upon the present determinations of the Hubble constant H_o or to determine the deceleration parameter q_o. However, this method involves many of the same measurements as the distance determinations based on the Zel'dovich–Syunyaev effect, and there is a chance that the X-ray absorption line measurements can be done more reliably than the microwave diminution measurements required for the other method. Although I have considered the application of this method to a cluster with a background quasar, some clusters (for example, M87/Virgo and Perseus) have AGNs with significant nonthermal X-ray luminosities at their centers. These clusters often have cooling flows, in which the lower gas temperatures can greatly increase the equivalent widths of the X-ray absorption lines. The measurements are likely to be much easier in these cases, although the interpretation of the measurements is likely to be more difficult.

5. CONCLUSIONS

The observations of X-ray emission lines from clusters of galaxies can provide a wealth of information concerning the physical state of the intracluster gas. From the abundances in the intracluster gas we should be able to determine the origin of the intracluster gas and to place strong constraints on the chemical evolution and star formation history of galaxies. Because clusters are spatially extended, these observations require only moderate spatial resolution (a fraction of a minute of arc). Depending on the particular observation, moderate-to-high spectral resolution is needed.

It should also be possible to detect X-ray absorption lines due to the intracluster gas in the spectra of background quasars. By comparing the equivalent width of such an absorption line and the flux from the same line seen in emission, one can determine directly the distance to the cluster. If this could be done with sufficient accuracy for a sample of clusters, such X-ray absorption line distance measurement could be used to determine H_o and possibly q_o. Although the required X-ray observations are difficult, they are very similar to the observations which must be made to implement the Zel'dovich–Syunyaev distance estimator, and may complement that effort.

Finally, I should note that nearly everything that I have said applies as well to the hot gas in individual elliptical galaxies. In fact, since this gas is apparently cooler than that in clusters, the X-ray emission lines (and possibly, the absorption lines as well) should be even stronger than in clusters.

ACKNOWLEDGMENTS

I would like to thank Jack Hughes, Julian Krolik, and Richard Mushotzky for useful comments, and Paul Gorenstein and Martin Zombeck for the effort they made to organize this very useful conference. This work was supported in part by NASA Astrophysical Theory Program Grant NAGW-764.

REFERENCES

Abramopoulos, F., Chanan, G., and Ku, W. 1981, *Ap. J.*, **248**, 429.
Basko, M. M., Komberg, B. V., and Moskalenko, E. I. 1981, *Sov. Astr.*, **25**, 402.
Canizares, C. R., Markert, T. H., and Donahue, M. E. 1988, in *Cooling Flows in Clusters and Galaxies*, ed. A. C. Fabian (Kluwer: Dordrecht), p. 63.
Cavaliere, A., Danese, L., and deZotti, G. 1977, *Ap. J.*, **217**, 6.
Cavaliere, A., Danese, L., and deZotti, G. 1979, *Astr. Ap.*, **75**, 322.
Fabian, A. C., and Pringle, J. E. 1977, *M.N.R.A.S.*, **181**, 5p.
Fabricant, D., and Gorenstein, P. 1983, *Ap. J.*, **267**, 535.
Gil'fanov, M. R., Syunyaev, R. A., and Churazov, E. M. 1987, *Sov. Astr. Lett.*, **13**, 3.
Gunn, J. E. 1978, in *Observational Cosmology*, ed. by A. Maeder, L. Martinet, and G. Tammann (Geneva Obs.: Geneva).
Henriksen, M. J., and Mushotzky, R. F. 1985, *Ap. J.*, **292**, 441.
Hughes, J. P., Yamashita, K., Okumura, Y., Tsunemi, H., and Matsuoka, M. 1988a, *Ap. J.*, **327**, 615.
Hughes, J. P., Gorenstein, P., and Fabricant, D. 1988b, *Ap. J.*, **329**, 82.
Larson, R. B., and Dinerstein, H. L. 1975, *P.A.S.P.*, **87**, 911.
Lea, S. M., Mushotzky, R., and Holt, S. 1982, *Ap. J.*, **262**, 24.
Mitchell, R. J., Culhane, J. L., Davison, P. J., and Ives, J. C. 1976, *M.N.R.A.S.*, **175**, 29p.
Mitchell, R. J., Dickens, R. J., Bell-Burnell, S. J., and Culhane, J. L. 1979, *M.N.R.A.S.*, **189**, 329.
Mushotzky, R. F. 1984, *Phys. Scripta*, **T7**, 157.
Mushotzky, R. F. 1988, in *Proc. NATO Summer School on Hot Astrophysical Plasmas*, ed. by R. Pallavicini (Kluwer: Dordrecht), in press.
Mushotzky, R. F., Serlemitsos, P. J., Smith, B. W., Boldt, E. A., and Holt, S. S. 1978, *Ap. J.*, **225**, 21.
Nepveu, M. 1981, *Astr. Ap.*, **101**, 362.
Okumura, Y., Tsunemi, H., Yamashita, K., Matsuoka, M., Koyama, K., Hayakawa, S., Masai, K., and Hughes, J. 1988, preprint.
Rephaeli, Y. 1978, *Ap. J.*, **225**, 335.
Rothenflug, R., and Arnaud, M. 1985, *Astr. Ap.*, **144**, 431.
Sarazin, C. L. 1988a, *X-ray Emission from Clusters of Galaxies*, (Cambridge Univ.: Cambridge).
Sarazin, C. L. 1988b, preprint.
Sarazin, C. L., and Bahcall, J. N. 1977, *Ap. J. Suppl.*, **34**, 451.
Shapiro, P. R., and Bahcall, J. N. 1980, *Ap. J.*, **241**, 1.
Serlemitsos, P. J., Smith, B. W., Boldt, E. A., Holt, S. S., and Swank, J. H. 1977, *Ap. J. (Lett.)*, **211**, L63.
Silk, J., and White, S. D. M. 1978, *Ap. J. (Lett.)*, **226**, L103.

Singh, K. P., Westergaard, N. J., and Schnopper, H. W. 1986, *Ap. J. (Lett.)*, **308**, L51.
Ulmer, M. P., Cruddace, R. G., Fenimore, E. E., Fritz, G. G., and Snyder, W. A. 1987, *Ap. J.*, **319**, 118.
Weinberg, S. 1972, *Gravitation and Cosmology: Principles and Applications of the General Theory of Relativity*, (New York: Wiley), pp. 407–468.
Zel'dovich, Y. B., and Syunyaev, R. A. 1969, *Ap. Sp. Sci.*, **4**, 301.

SIGNATURES OF COOLING FLOWS

A.C. Fabian
Institute of Astronomy
Madingley Road
Cambridge CB3 0HA
U.K.

ABSTRACT. The evidence for cooling flows in clusters of galaxies is discussed. Peaked X-ray surface brightness profiles and the presence of soft X-ray spectral components are characteristic signatures of cooling flows. The best available data are consistent with gas cooling at a rate of $\sim 100\,M_\odot\,\mathrm{yr}^{-1}$ and depositing cooled matter over a radius of $100-200\,\mathrm{kpc}$. The same rate is obtained from observations of gas cooling over a range of temperatures and thus a range of cooling times covering less than $3\times 10^7\,\mathrm{yr}$ to more than $3\times 10^9\,\mathrm{yr}$. It appears that cooling flows are both steady and long-lived. Some indirect optical evidence for distant cooling flows ($0.5 < z < 1$) is presented.

1. INTRODUCTION

Cooling flows appear to be relatively common in the centres of the hot gaseous atmospheres of clusters and groups of galaxies. They also occur in many early-type galaxies. The gas is densest in the core of a cluster and its cooling time, t_{cool}, due to the emission of X-rays such as those observed, is shortest there. A cooling flow is formed when t_{cool} is less than the age of the system, $t_a (\sim H^{-1})$.

In the cases considered here, t_{cool} exceeds the gravitational free-fall time, t_{grav}, within the cluster (except perhaps in some very small region at the centre), so, for a cooling flow,

$$t_a > t_{cool} > t_{grav}.$$

The flow takes place because the gas density has to rise to support the weight of the overlying gas.

If that is not immediately clear, consider the gaseous atmosphere trapped in the gravitational potential well of the cluster or galaxy to be divided into two parts at the radius, r_{cool}, where $t_{cool} = t_a$. The gas pressure at r_{cool} is determined by the weight of the overlying gas, in which cooling is not important. Within r_{cool}, cooling is tending to reduce the gas temperature and so the gas density must rise in order to maintain the pressure at r_{cool}. The only way for the density to rise (ignoring matter sources within r_{cool}, which is a safe assumption in a cluster of galaxies) is for the gas to flow inward. This is the cooling flow.

If the initial gas temperature exceeds the virial temperature of the central galaxy (this is generally the case for rich clusters but not for poor ones or individual galaxies) then the gas does cool as it flows in. When the temperature has dropped to the virial temperature of the central galaxy, the gas then heats up as it flows further in due to the release of gravitational energy. The gas temperature eventually drops catastrophically in the core of the galaxy if its gravitational potential flattens there. All this means that the gas within r_{cool} radiates its thermal energy, plus PdV work and gravitational energy released in the flow.

This is how an idealized, homogeneous cooling flow, in which the gas has a unique temperature and density at each radius, will behave. Observations of real cooling flows shows that they are inhomogeneous and must consist of a mixture of temperatures and densities at each radius. The homogeneous flow is, however, still a fair approximation of the mean flow.

In this review, I concentrate on the X-ray evidence for cooling flows and their implications for central cluster galaxies. A brief discussion of the optical appearance of cooling flows is provided and used to suggest that they were widespread at redshifts of 0.5 and more. Apart from a short comment, I ignore any continuous heat source capable of disrupting a flow since the X-ray data show that there are no such sources. An alternative view is given in the following paper by W. Tucker (this Volume).

2. X-RAY EVIDENCE FOR COOLING FLOWS

2a X-ray Images

A sharply-peaked X-ray surface brightness distribution is often indicative of a cooling flow. It shows that the gas density is rising steeply towards the centre of the cluster or group which means either that the gas pressure is rising steeply or, if the pressure is roughly constant, that the gas temperature is falling. The equation of hydrostatic equilibrium

$$\frac{dP_{gas}}{dr} = -\rho_{gas}g$$

shows that the first option requires a large gravitational field with a virial temperature for that region which exceeds that of the gas. This is not the case within the core of a cluster of galaxies where the potential is flattening off *. The constant pressure option is the simplest one consistent with the data (Fabian et al. 1981; Stewart et al. 1984b).

Most of the images have been obtained with the *Einstein Observatory* and with EXOSAT, although the peaks were anticipated with data from the Copernicus satellite (Fabian et al. 1974; Mitchell et al. 1975), from rocket-borne telescopes (Gorenstein et al. 1977) and with the modulation collimators on SAS 3 (Helmken et al. 1978).

Deprojection, or modelling, of the X-ray images shows that $t_{cool} < H_0^{-1}$ within the central region of more than 30 to 50 per cent of the clusters well-detected with the Einstein Observatory (Stewart et al. 1984b; Arnaud 1988). Whether H_0^{-1} should be used for t_a is debatable (but see §2b), and it is not obvious how to extrapolate from 'well-detected' clusters to all clusters. Inspection of the results shows however that reducing t_a by 2, say, does not much change the fraction of clusters which contain cooling flows, and it must be remembered that the spatial resolution of the commonly-used IPC was not sufficient to resolve the central regions of many clusters. In those cases, the measured t_{cool} is just an upper limit. The overall picture is that the prime criterion for a cooling flow, $t_{cool} < 10^{10}$ yr, is satisfied in a large fraction of clusters. It is also satisfied in a number of poor clusters

---

* Jones & Forman (1984) modelled the emission as due to gas at constant temperature in the potential well of the cluster. They found that a 'central excess' of emission is required in the cooling flow cases.

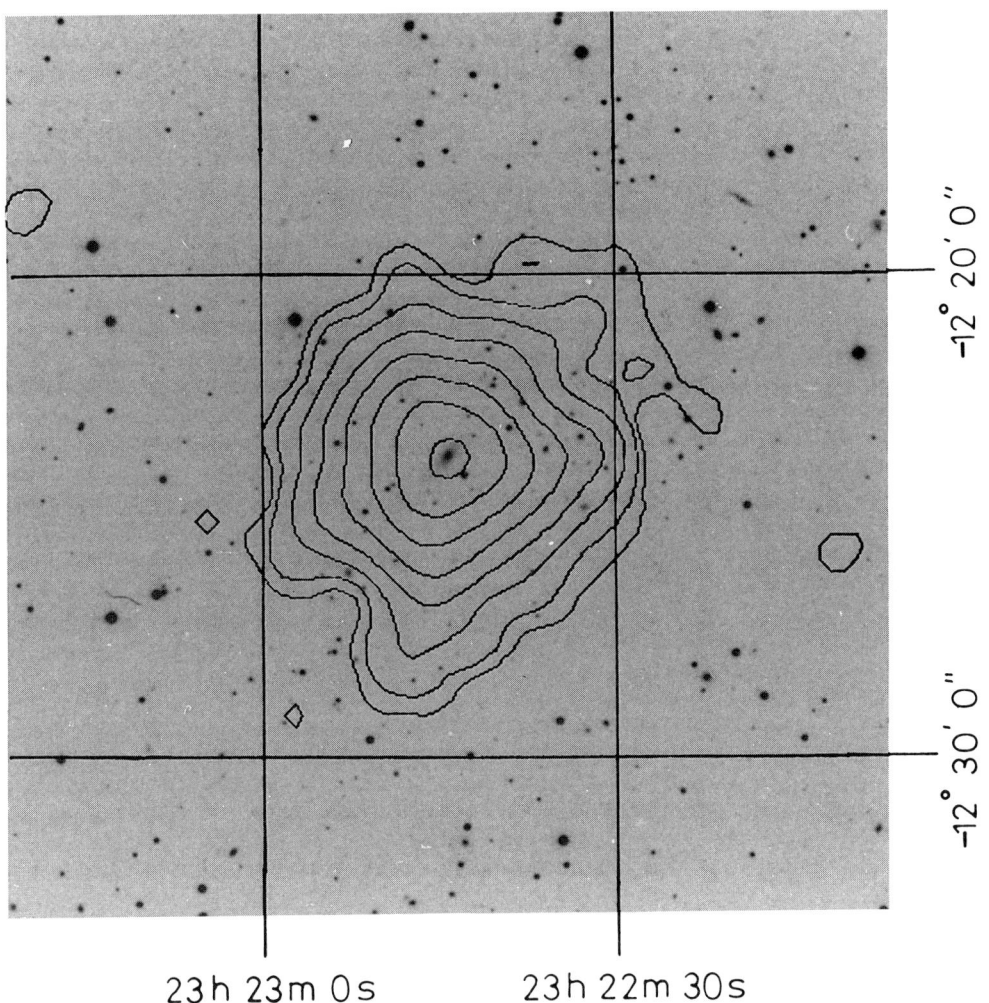

Figure 1. IPC X-ray surface brightness contours of A2597 superimposed on an optical image of the cluster. Note that the contours peak onto the central cluster galaxy. The mass deposition rate in A2597 is $\sim 370 \, M_\odot \, yr^{-1}$ (Crawford et al. 1989).

and groups (Schwartz, Schwarz & Tucker 1980; Canizares, Stewart & Fabian 1983; Singh, Westergaard & Schnopper 1986). The implication of these statistics is that cooling flows are both common and long-lived.

The mass deposition rate, $\dot{M}$, due to cooling (i.e. the accretion rate, although I resist using that term since most of the gas does not change its radius very much) can be estimated from the X-ray images by using the luminosity associated with the cooling region (i.e. L_{cool} within r_{cool}) and assuming that it is all due to the radiation of the thermal

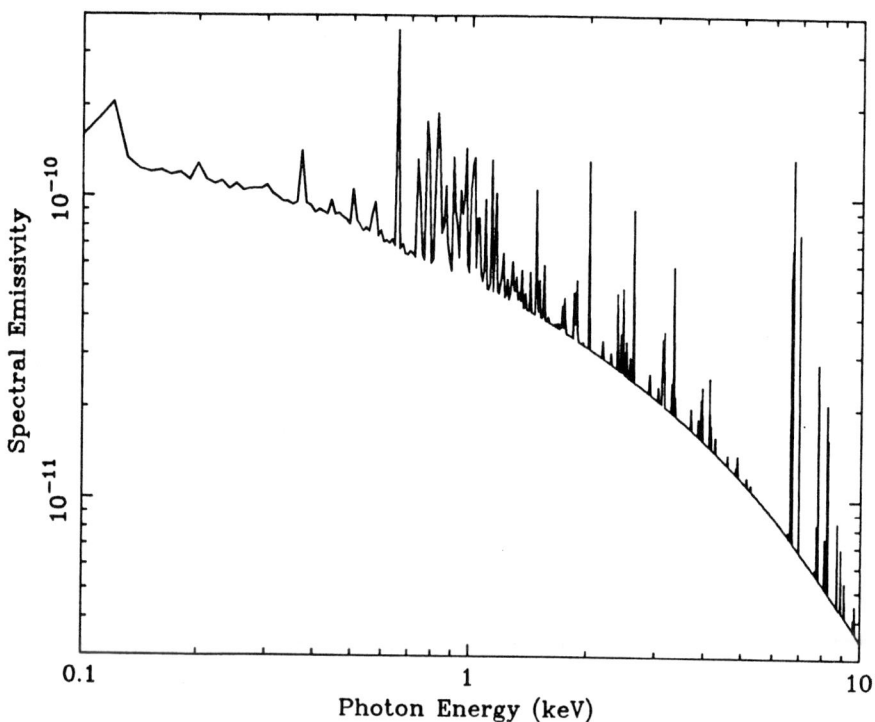

Figure 2. X-ray spectrum at 10 eV resolution of gas cooling from 6 keV.

energy of the gas, plus the *PdV* work done.

$$L_{cool} = \frac{5}{2}\frac{\dot{M}}{\mu m}kT,$$

where T is the temperature of the gas at r_{cool}. Values of $\dot{M} = 50 - 100 \, M_\odot \, \text{yr}^{-1}$ are fairly typical for cluster cooling flows. Some clusters show $\dot{M} \sim 500 \, M_\odot \, \text{yr}^{-1}$ (*e.g.* PKS0745, A1795, A2597 and Hydra A).

Since we often measure a surface brightness profile for the cluster core (where it is well-resolved), we have $L_{cool}(r)$ which can be turned into $\dot{M}(r)$, the mass deposition rate within radius r. Generally, we find

$$\dot{M}(r) \propto r.$$

The surface brightness profiles are less peaked than they would be if all the gas were to flow to the centre. This means that the gas must be inhomogeneous, so that some of the gas cools out of the flow at large radii and some continues to flow in. The actual computation of $\dot{M}(r)$ is in detail complicated, since we need to take into account how the gas cools and

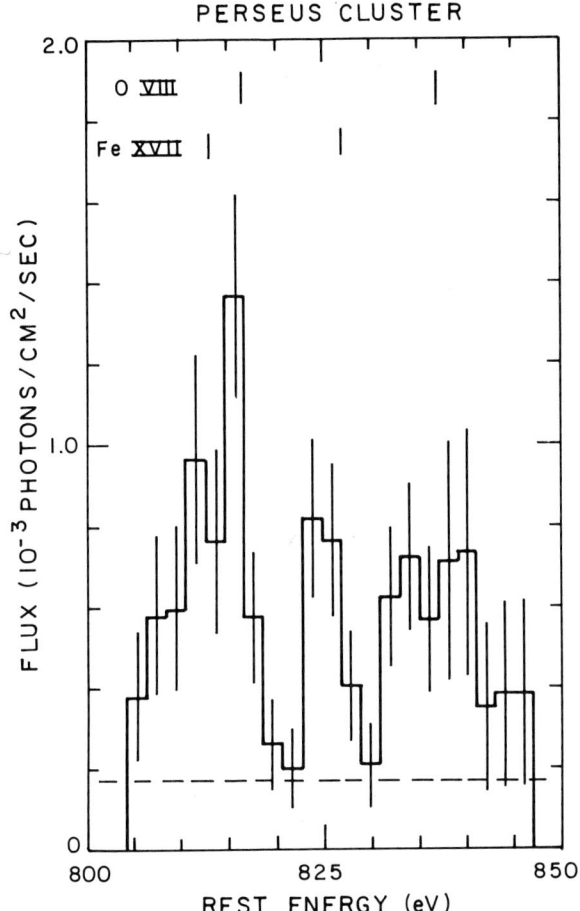

Figure 3. Part of FPCS spectrum of the Perseus cluster. Note the prominent emission lines of OVIII and Fe XVII (from Canizares, Markert & Donahue 1988). The emission measure of the gas producing the Fe XVII lines is consistent with about $200 \, M_\odot \, yr^{-1}$ of gas cooling through $\sim 5.10^6 - 10^6 \, K$. This is the $\dot{M}$ found from imaging and SSS studies, which are most sensitive to higher temperature gas (typically $5.10^6 - 3.10^7 \, K$).

any gravitational work done, but since plain cooling dominates, a simple analysis gives a fair approximation to the profile (see Fabian, Arnaud & Thomas 1986; Thomas, Fabian & Nulsen 1987; White & Sarazin 1987).

2b X-ray Spectra

Key evidence that the gas does actually cool is given by moderate to high resolution spectra of the cluster cores. Canizares et al. (1979, 1982), Canizares (1981); Mushotzky et al. (1981) and Lea et al. (1983) used the Focal Plane Crystal Spectrometer (FPCS) and the Solid State Spectrometer (SSS) on the *Einstein Observatory* to show that there are low temperature components in the Perseus and Virgo clusters, consistent with the existence of cooling flows. Detailed examination of the line fluxes and of the emission measures of the cooler gas by Canizares, Markert & Donahue (1988) and Mushotzky & Szymkowiak (1988) shows that, in the case of the Perseus cluster, the gas loses at least 90 per cent of its thermal energy and that the mass deposition rates are in agreement with those obtained from the images. Good agreement is obtained also in several other clusters. The SSS results show that the emission measures vary with temperature in the manner expected from a cooling gas. *The importance of these data cannot be overemphasized since they show that the gas does cool.* Any 'alternative interpretation' of the images must confront this spectroscopic evidence successfully.

The cooling time of the gas in the Perseus cluster which emits the FeXVII line ($T < 5 \times 10^6$ K) is less than 3×10^7 yr. Since the emission measure of this gas agrees with that inferred from the gas cooling at the higher temperatures which dominate the images and the SSS result, we must conclude that the flow is steady (Nulsen 1988). The shape of the continuum and line spectrum observed with the SSS is consistent with the same mass deposition rate at all X-ray temperatures (as expected) so we must again conclude that the flow is long-lived. It cannot be some intermittent or transient phenomenon only a billion years old.

2c Summary

The overwhelming evidence of the images and spectra shows that cooling does occur at a steady rate over long times (at least several billion years). Since mass is then cooling out of the hot phase at rates of hundreds of solar masses per year an inflow must occur. We do not expect yet to have direct evidence of any inward flow since the velocity is highly subsonic at $\sim 10 \,\mathrm{km\,s^{-1}}$.

Cooling flows are common and all of the nearest clusters (Virgo, Centaurus, Fornax, Perseus, Ophiuchus...) contain one apart from the Coma cluster. Even that could contain a disrupted flow (Fabian, Nulsen & Canizares 1984). Many flows are observed out to a redshift of 0.1, with more distant ones being in 3C295 (Henry & Henriksen 1987) and 1E0839 (Wolter et al., this Volume).

The values of $\dot M$ are probably good to a factor of 2 (Arnaud 1988) and could be higher if there are denser blobs beyond r_{cool} (Thomas et al. 1987).

3. The Fate of the Cooled Gas

The accumulated mass of cooled gas can be considerable;

$$\dot M H_0^{-1} = 2 \times 10^{12} \left(\frac{\dot M}{100 \,\mathrm{M_\odot\,yr^{-1}}} \right) \,\mathrm{M_\odot}.$$

This is a significant fraction of the mass of the central galaxy. It suggests that we are witnessing the continued formation of that galaxy, which is typically one of the largest galaxies known.

If the gas forms stars, then cooling flows are the largest and strongest regions of star formation in our part of the Universe. Even a casual comparison of a central cluster galaxy and a spiral galaxy such as our own, which is thought to be forming stars at a rate of $3 - 10 \,\mathrm{M_\odot\,yr^{-1}}$, shows that cooling flows must form low-mass stars (Fabian, Nulsen

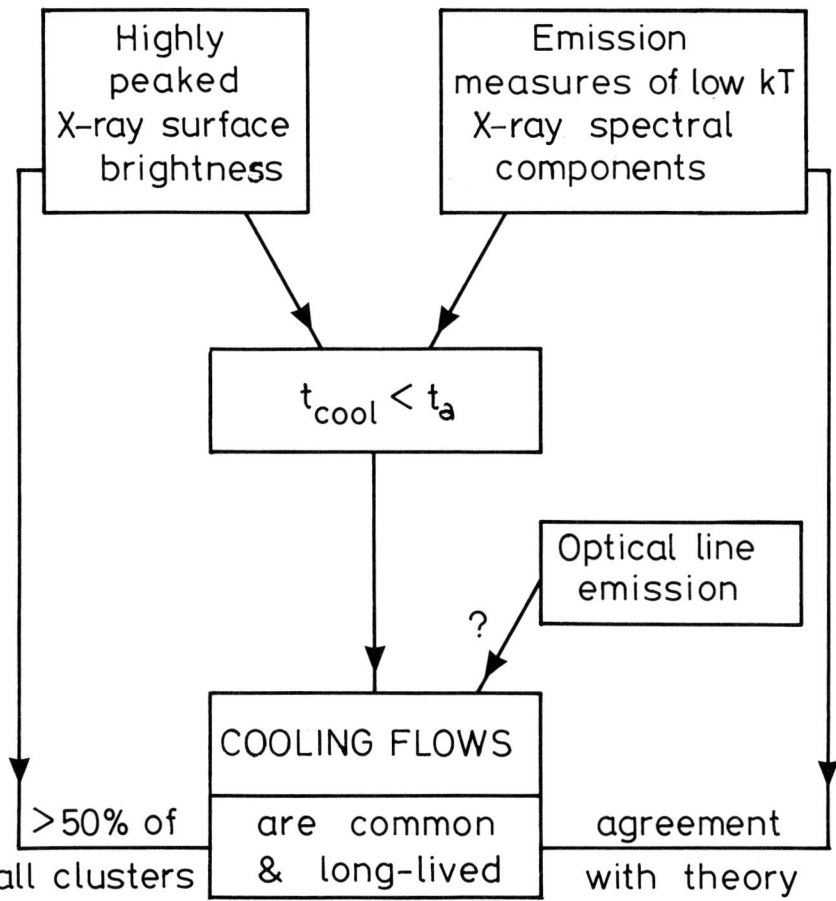

Figure 4. The evidence for Cooling Flows.

& Canizares 1982; Sarazin & O'Connell 1983). Massive stars would make central cluster galaxies much bluer than they are.

It should be stressed that the cooling gas is not directly detected once it has cooled below about 3×10^6 K. If it recombines and forms low-mass stars ($\langle M_* \rangle < 0.5\,M_\odot$) in a distributed manner ($M(r) \propto r$) then there is no reason for it to have been seen. *The optical line luminosity from the gas and the continuum from the stars are undetectable with current means.*

There is, however, plenty of evidence for dark matter in clusters and low-mass stars are one plausible form of dark matter. The manner of the mass deposition with radius, $\dot{M}(r) \propto r$, leads to an isothermal halo which is consistent with the dark matter distribution around large galaxies (*e.g.* M87, Stewart *et al.* 1984a; Mould *et al.* 1987). Cooling flows are a source of baryonic dark matter.

3b Heating

Since the implied star formation rates are so large and there is little sign of it optically, there have been a number of studies suggesting that the rates have been grossly over-estimated. Some heat source that balances the cooling is the obvious solution. Cosmic rays Tucker & Rosner 1982), conduction (Bertschinger & Meiksin 1986), supernovae (Silk et al. 1986) and galaxy motions (Miller 1986) have all been invoked as heat sources. Unfortunately for these models, the X-ray spectra indicate cooling without heating. None of the models proposed so far is able (or even attempts!) to account for the X-ray line emission. There are other problems with these heat sources as well (see Fabian 1988; Bregman & David 1988).

The total level of heating necessary is very large, $\sim 10^{62}$ erg for a large flow over t_a and so if some heat source is found that can accommodate the X-ray spectral measurements successfully, it must be one of the major (unseen!) energy flows in the Universe! Whilst the luminosity of a cooling flow may be only 10 per cent of the total cluster X-ray luminosity and the mass lost through cooling a negligible drain on the enormous outer atmosphere, the cooling luminosity is a major loss of energy from the cluster core. Whatever is eventually decided about cooling flows, they cannot be an insignificant process.

3b Star Formation in Cooling Flows

As already mentioned, the average mass of a star formed in a cooling flow must be considerably smaller than in our Galaxy. In particular, the fraction of the mass turned into massive OB stars must be very small, since there is little ultraviolet light seen with the IUE (*e.g.* Fabian, Nulsen & Arnaud 1984). (The shortest cooling times for the X-ray emitting gas ($\sim 3 \times 10^7$ yr) are comparable to the lifetime of B stars, so intermittency of the flow cannot be important here.) In understanding why there are these differences, it would be helpful to have a predictive theory of star formation for our Galaxy. As we do not, we must look for differences. Some are a) the lack of dust in gas that has cooled from $T > 10^7$ K (Draine & Salpeter 1979), which presumably means that there are no molecular clouds such as give birth to massive stars in our Galaxy; b) the thermal pressure of the gas is 100 – 1000 times higher than in our interstellar medium and that affects the Jeans mass; c) differential motions, cloud masses, angular momentum and magnetic field strengths may be different.

The general statement about the necessity for low mass stars applies to the bulk of the cooled gas. In the centre, there are often seen optical emission line blobs or filaments, which may be atypical of most of the flow. These blobs may give rise to higher mass stars. There is some excess blue light observed at the centres of many cooling flows and it does correlate in strength with the mass deposition rate (Johnstone, Fabian & Nulsen 1987). Spectral fits of the blue light together with upper limits from IUE spectra show that the upper mass limit for stars must be around $1.5 - 2 \, M_\odot$ there. Some F and early G stars are seen. Of course, the best place to look for the bulk of the cooled gas is at large radii where the underlying stellar light of the galaxy is least, so the contrast is highest.

3c The Behaviour of Gas Blobs

The distributed manner of the mass deposition shows that the cooling flow is inhomogeneous. This means that it contains blobs of gas that are denser than the surrounding gas. How these blobs behave is ill-understood. Malagoli et al. (1987) and Balbus (1988) have shown that cooling flows are not expected to be thermally unstable and so cannot generate sizable blobs from initially infinitesimal perturbations. A region that is slightly overdense with respect to its surroundings will fall ahead of the flow under gravity and join a region of similar properties to itself.

Nulsen (1986) has pointed out that the gravitationally-induced motions of a blob relative to its surroundings will cause it to break up and so increase its area-to-mass ratio and have a lower terminal velocity. Magnetic fields can then help to 'pin' a blob to the flow so that it can become unstable.

Another likely property of a flow is turbulence, or at least chaotic motions (Loewenstein & Fabian 1988; Pringle 1988). This can stir the gas around and reduce the tendency of dense blobs to flow inward. The motion of cluster galaxies, of subcluster infall and of the flow itself can all promote chaotic motions. Turbulence of the hot gas can also explain the large velocity spread seen in the optical line-emitting filaments and blobs common at the centres of flows. It can also help to heat the cold blobs. Future high-resolution X-ray spectroscopy can test this idea.

4. Optical Evidence for Cooling Flows

The presence of optical emission-line filaments and blobs in many cooling flows (note that some, *e.g.* A2029, do not have any detectable emission) can be used as a diagnostic of the conditions in those flows. We can also use them to identify candidate cooling flows where X-ray observations are not available or at higher redshifts if there is some resemblance to nearby flows.

One property that can be obtained from the optical spectra is the gas pressure, in particular from the [SII] lines. In the Perseus cluster, these lines change their ratio indicating high density and thus high pressure within 5kpc of the nucleus (Johnstone & Fabian 1988). Since the pressure has risen above the X-ray inferred pressure (from the X-ray surface brightness) at 20 kpc, the mean gas temperature must be down to the virial temperature of the central galaxy, NGC1275, of about 10^7 K. This is further confirmation that the gas has cooled there below the outer temperature of $\sim 8 \times 10^7$ K. Since the gas pressure is so high there, the magnetic pressure cannot be more than about twice the gas pressure (and is probably less than that).

The velocity spread of the optical lines does also indicate large chaotic motions in the hot gas, as discussed earlier. The origin of the optical line emission remains a problem. As already mentioned in §3, the cooling gas by itself cannot produce much detectable optical emission; some of the gas must be held at $\sim 10^4$ K in order that we do see the optical lines. This means that there is a distributed, weak heat source at the centres of many cooling flows (see *e.g.* Johnstone & Fabian 1988). The optical emission is typically less than 1% of L_{cool} (and always less than 10%) and so this heat source cannot significantly affect the cooling flow itself. It is likely that the turbulent energy of the cooling gas is the heat source.

The optical line emission of nearby flows shows some resemblances to the nebulosities surrounding many distant (3CR) radio galaxies and radio-loud quasars (Fabian *et al.* 1986, Hintzen & Romanishin 1986). One way to check whether they are indeed embedded in hot cooling gas is to measure the gas pressure. If the emission-line gas is very extended and has a pressure, $nT > 10^5$ cm^{-3} K, then it is likely to be confined, or it would rapidly disperse. The confining hot gas is then part of a cooling flow unless its temperature exceeds 5×10^7 K. The arguments against either unconfined or gravitationally confined gas are discussed in detail in Fabian *et al.* (1987).

The pressure of gas around quasars cannot easily be measured from the [SII] lines, which are faint and only change where the pressure is very high close to the centre. We have pursued a photoionization method which assumes that the ionization state of, say, oxygen is due to the competition between photoionization by UV radiation from the quasar

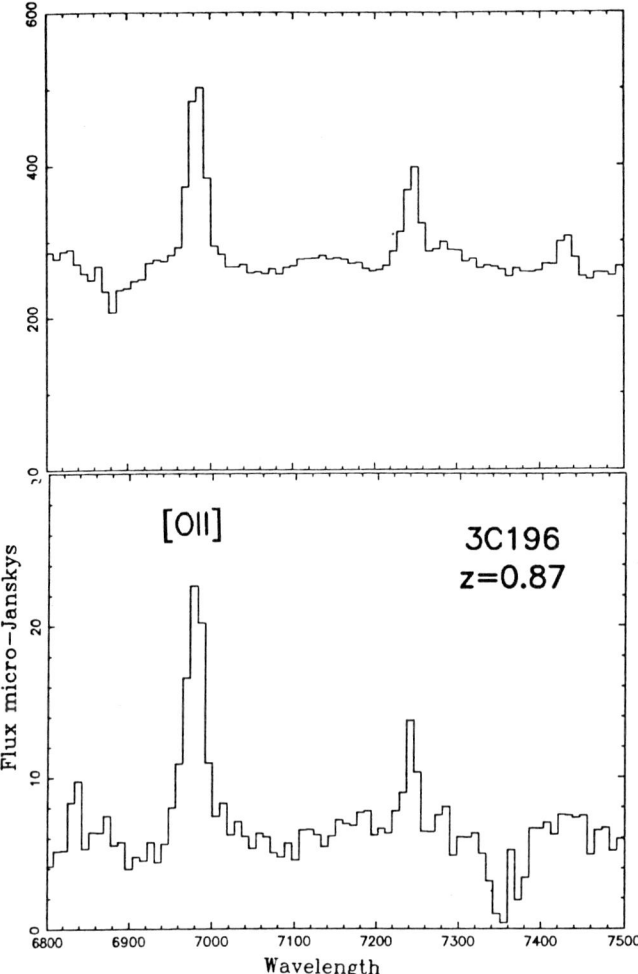

Figure 5. Optical spectra of the radio-loud quasar 3C196 on-nucleus (top panel) and 3 arcsec off-nucleus (bottom panel) showing the increased equivalent width of the [OII] line off-nucleus. Comparison of the [OIII] and [OII] emission shows that the surroundings of 3C196 are at high gas pressure, consistent with a $\sim 100\,{\rm M}_\odot\,{\rm yr}^{-1}$ cooling flow (Crawford 1988).

nucleus and recombination. The higher the gas pressure and so density, the more rapid will be the recombination and so the lower will be the ionization state. Basically, if we use the forbidden oxygen lines to measure the ionization state of the gas,

$$\frac{[{\rm OIII}]}{[{\rm OII}]} \propto f\left(\frac{L_{ion}}{nR^2}\right),$$

where the function f is obtained by computation (using G. Ferland's code), L_{ion} is the ionizing luminosity of the quasar nucleus, obtained from an interpolation between the observed IUE and X-ray emission, R is the radial distance of the nebulosity where the intensity ratio of [OIII] to [OII] line is measured and n is the gas density. The gas pressure is then $\sim 10^4 n \, \mathrm{cm^{-3} \, K}$. We have found that 3C48 at redshift $z = 0.37$ and several other radio-loud quasars out to $z = 0.87$ have pressures indicating surrounding cooling flows with $\dot{M} \sim 100 \, \mathrm{M_\odot \, yr^{-1}}$. There is other evidence that radio-loud quasars lie in clusters and groups from galaxy counts around the quasars (Yee & Green 1984).

This work suggests that cooling flows can have been more common in the past and possibly more massive, if the high line luminosities of the distant 3CR galaxies (*e.g.* Spinrad & Djorgovski 1984) may be scaled up from NGC1275. The cooling flows surrounded the dominant galaxies in groups and poor clusters, which have since merged with other groups to form the present day clusters (Fabian 1988). Some of the evolution of radio-loud quasars and of radio sources may be due to the evolution of their surrounding cooling flows.

5. What Next?

There is much left to do. On the theoretical side there are many unsolved problems on the behaviour of blobs and on how the flow became, or becomes, inhomogeneous. What are the length scales involved? How does the star formation take place? How is conduction suppressed? On the observational side we clearly need to find the cooled gas in some form, hopefully as stars. This can be achieved with sensitive searches for an r^{-1} surface brightness profile at near IR wavelengths.

Of greatest need are more X-ray images and more X-ray spectra. ROSAT, BBXRT, ASTRO-D and AXAF will supply them. Spatially-resolved, high-resolution spectroscopy is the ultimate goal. Then we can tell whether cooling flows really do have something to do with galaxy formation in general.

Acknowledgements

I thank Roderick Johnstone for help and the Royal Society for supporting my work.

References

Arnaud, K.A., 1988. In *Cooling Flows in Clusters and Galaxies*, ed. A.C.Fabian, Reidel, 31.
Balbus, S., 1988. *Astrophys. J.*, **328**, 395.
Bertschinger, E. & Meiksin, A., 1986. *Astrophys.J*, **306**, L1.
Bregman, J.D. & David L.P., 1988. *Astrophys. J.*, **326**, 639.
Canizares, C.R., Clark, G.W., Markert, T.H., Berg, C., Smedira, M., Bardas, D., Schnopper, H. & Kalata, K., 1979, *Astrophys. J.*, **234**, L33.
Canizares, C.R., 1981. In *X-ray Astronomy with the Einstein Satellite* ed. R. Giacconi, Reidel, 215.
Canizares, C.R., Clark, G.W., Jernigan, J,G. & Markert, T.H., 1982. *Astrophys. J.*, **262**, L33.
Canizares, C.R., Stewart, G.C. & Fabian A.C., 1983. *Astrophys. J.*, **272**, 449.
Canizares, C.R., Markert, T.H. & Donahue, M.E., 1988. In *Cooling Flows in Clusters and Galaxies*, ed. A.C.Fabian, Reidel, 63.

Crawford, C.S., 1988. PhD Thesis, University of Cambridge.
Crawford, C.S., Arnaud, K.A., Fabian, A.C. & Johnstone, R.M., 1989. *Mon. Not. R. astr. Soc.*, **236**, 277.
Fabian, A.C. et al., 1984. *Astrophys. J.*, **189**, L59.
Fabian, A.C., Hu, E.M., Cowie, L.L & Grindlay, J.,1981. *Astrophys. J.*, **248**, 47.
Fabian, A.C., Nulsen, P.E.J. & Canizares, C.R., 1982. *Mon. Not. R. astr. Soc.*, **201**, 933.
Fabian, A.C., Nulsen, P.E.J. & Canizares, C.R., 1984. *Nature*, **311**, 733.
Fabian, A.C., Nulsen, P.E.J. & Arnaud, K.A., 1984. *Mon. Not. R. astr. Soc.*, **208**, 179.
Fabian, A.C., Arnaud, K.A., Nulsen, P.E.J. & Mushotzky, R.F., 1986. *Astrophys. J.*, **305**, 9.
Fabian, A.C., Arnaud, K.A. & Thomas, P.A., In *Dark Matter in the Universe*, eds. J. Kormendy & G.R. Knapp, Reidel, 201.
Fabian, A.C., Crawford, C.S., Johnstone, R.M. & Thomas, P.A., 1987. *Mon. Not. R. astr. Soc.*, **228**, 963.
Fabian, A.C., 1988. In *Cooling Flows in Clusters and Galaxies*, ed. A.C.Fabian, Reidel, 315.
Fabian, A.C., 1988. In *Hot Thin Plasmas in Astrophysics*, ed. R. Pallavicini, Reidel, 293.
Gorenstein, P., Fabricant, D., Topka, K., Tucker, W. & Harnden, F.R., 1977. *Astrophys. J.*, **216**, L95.
Helmken, H., Delvaille, J.P., Epstein, A., Geller, M.J., Schnopper, H.W. & Jernigan, J.G., 1978. *Astrophys. J.*, **221**, L43.
Hintzen, P. & Romanishin, W., 1986. *Astrophys. J.*, **311**, L11.
Henry, J.P & Henriksen, M.J., 1986. *Astrophys. J.*, **301**, 689.
Johnstone, R.M., Fabian, A.C. & Nulsen, P.E.J., 1987. *Mon. Not. R. astr. Soc.*, **224**, 75.
Johnstone, R.M. & Fabian, A.C., 1987. *Mon. Not. R. astr. Soc.*, **233**, 581.
Jones, C. & Forman, W., 1984. *Astrophys. J.*, **276**, 38.
Lea, S.M., Mushotzky, R.F. & Holt, S.S., 1982. *Astrophys. J.*, **262**, 24.
Loewenstein, M. & Fabian, A.C., 1988. Preprint.
Malagoli, A., Rosner, R. & Bodo, G., 1987. *Astrophys. J.*, **319**, 632.
Miller, L., 1986. *Mon. Not. R. astr. Soc.*, **220**, 713.
Mitchell, R.J., Charles, P.A., Culhane, J.L., Davison, P.J.N. & Fabian, A.C., 1975. *Astrophys. J.*, **200**, L5.
Mould, J.R., Oke, J.B. & Nemec, J.M., 1987. *Astr. J.*, **92**, 53.
Mushotzky, R.F., Holt, S.S, Smith, B.W., Boldt, E.A. & Serlemitsos, P.J., 1981. *Astrophys. J.*, **244**, L47.
Mushotzky, R.F. & Szymkowiak, A.E., 1987b. In *Cooling Flows in Clusters and Galaxies*, ed. A.C.Fabian, Reidel, 47.
Nulsen, P.E.J., 1986. *Mon. Not. R. astr. Soc.*, **221**, 377.
Nulsen, P.E.J., 1988.In *Cooling Flows in Clusters and Galaxies*, ed. A.C.Fabian, Reidel, 378.
Pringle, J.E., 1988. Preprint.
Sarazin, C.L. & O'Connell, R.W., 1983. *Astrophys. J.*, **258**, 552.
Schwartz, D.A., Schwarz, J. & Tucker, W.H., 1980. *Astrophys. J.*, **238**, L59.
Silk, J., Djorgovski, G., Wyse, R.F.G. & Bruzual, G.A., 1986. *Astrophys. J.*, **307**, 415.
Singh, K.P., Westergaard, N.J. & Schnopper, H.W., 1986. *Astrophys. J.*, **308**, L51.
Spinrad, H. & Djorgovski, G., 1984. *Astrophys. J.*, **280**, L9.
Stewart, G.C., Canizares, C.R., Fabian, A.C. & Nulsen, P.E.J., 1984a. *Astrophys. J.*, **278**, 536.
Stewart, G.C., Fabian, A.C., Jones, C. & Forman, W., 1984b. *Astrophys. J.*, **285**, 1.
Thomas, P.A., Fabian, A.C. & Nulsen, P.E.J., 1987. *Mon. Not. R. astr. Soc.*, **228**, 973.
Tucker, W.H. & Rosner, R., 1982. *Astrophys. J.*, **267**, 547.

White, R.E. & Sarazin, C.L., 1987. *Astrophys. J.*, **318**, 612, 621, 629.
Yee, H.K.C. & Green, R.F., 1984. *Astrophys. J.*, **280**, 79.

DISCUSSION.

A. Hamilton. How important is a finite metallicity to a cooling flow? Is a zero metallicity flow modified in any interesting way?
A.C. Fabian. Not particularly, most of the cooling is by bremsstrahlung above 10^7 K.
S.A. Balbus. You mentioned that some 50% of clusters show cooling flows, according to Arnaud's sample. What was the selction criterion used for picking the sample members?
A.C. Fabian. They are the better observed clusters with more than a few hundred counts. They tend to be round and peaked. Christine Jones has studied a wider sample and suggests a value nearer 30%.
G.S. Bisnovatyi-Kogan. How can the enrichment of intergalactic gas in the clusters occur in the presence of cooling flows? Is it necessary to have the flow switched-off by a very luminous object, observed at higher redshift?
A.C. Fabian. Much of the enrichment in nearby clusters presumably occurred long ago and from the cluster galaxies, not just the central one. For galaxy formation, the cooling flow may well be a mess, although enrichment may still occur first (see Fabian et al., *Astrophys. J.*, 1986).
J.Grindlay. For 3C196, at z=0.87, if the inferred cooling flow was $\dot{M} \sim 100\,{\rm M}_\odot\,{\rm yr}^{-1}$, the responsible galaxy cluster (supplying the flow) must be large enough to be detected in deep optical images. Is there evidence for this?
A.C. Fabian. I'm not aware that anyone has looked. The cluster may be poor (cf. MKW 3s). Yee & Green (1984, 87) find that radio-loud quasars are in groups and clusters at $z < 0.6$, but that they are difficult to detect at higher z.
F. Winkler. How severely should observations with the present IR arrays be able to constrain the Mass Function for the posited stars resulting from cooling flows?
A.C. Fabian. Interesting results are possible since the stars from the cooling flow have $\rho_* \propto r^{-2}$, whereas the visible galaxy halo drops off more steeply. We hope to take suitable IR images early next year at UKIRT.
H. Tananbaum. You stated that cooling flows may be able to account for the dark matter in clusters (or cluster cores) in a baryonic form (*i.e.* low-mass stars). At present rates the cooling flows would only be able to account for a small fraction of the 'observed' (or measured) dark matter. Do you think that cooling flow rates were higher in the past, or that 2 forms/types of dark matter exist, or both?
A.C. Fabian. The simplest implication from the present formation of some dark matter by cooling flows is that cooling flows were stronger in the past. I'm not yet sure that non-baryonic matter is absolutely required.

ALTERNATIVES TO THE EXISTENCE OF LARGE COOLING FLOWS

Wallace Tucker
Harvard-Smithsonian Center for Astrophysics, Cambridge, MA, USA

ABSTRACT. Arguments against the existence of large scale cooling flows in clusters of galaxies are presented. The evidence for cooling flows is all circumstantial, consisting of observations of cool gas or hot gas with a radiative cooling time less than the Hubble time, or a central peak in the x-ray surface brightness profile. There is no evidence for large quantities (several tens to several hundreds of solar masses per year) of matter actually flowing anywhere. On the contrary, several lines of evidence — stellar dynamics, observations of the amount of star formation, x-ray surface brightness observations, theoretical calculations of the growth of thermal instabilities, the amount of cold gas — suggest that cooling flows, if they exist, must be suppressed by one to two orders of magnitude from the values implied by simple estimates based on the radiative cooling time of the x-ray emitting gas. Two heat sources which might accomplish this — thermal conduction and relativistic particles, are considered and an alternative to the standard model for cooling flows is presented: an accretion flow with feedback wherein the accretion of gas into a massive black hole in the central galaxy generates high energy particles that heat the gas and act to limit the accretion.

> "... because our descriptions of the 'real' world are metaphors based on limited abstractions from a more complex reality, it is possible to arrive at quite different, even contradictory concepts of the 'thing' which is being observed."
>
> *Hanbury Brown, The Wisdom of Science*

This quotation from Hanbury Brown's wise book should be posted in every scientist's office as a constant reminder that we are dealing with metaphors based on limited abstractions from limited data on a very complex reality. Cooling flows are a case in point. From the same data it is possible, by changing what is assumed about the unknown, to arrive at almost contradictory pictures of what is being observed. Hopefully in the not too distance future high resolution x-ray spectroscopy will identify the set of unknowns so that the seemingly contradictory pictures or metaphors can be revised and reconciled into one that more accurately reflects the richness and complexity of the underlying reality.

1. COOLING FLOWS — THE STANDARD METAPHOR

The standard cooling flow metaphor has been well reviewed by Andy Fabian (1988). The key observations are high resolution x-ray surface brightness profiles and spectral measurements which demonstrate conclusively that the radiative cooling time inside a certain critical radius, called the cooling radius, in the central regions of clusters of galaxies is shorter that the assumed cluster lifetime of 10 to 20 billion years. Under the assumption that no significant heat sources are present, the observations can be taken to mean that radiative cooling is driving mass accretion onto a stationary supergiant galaxy at a rate given by

$$\dot{M}_o = \frac{4\pi r^2 n^2 \Lambda(t)}{\frac{d}{dr}(\frac{5}{2}\frac{kT}{\mu m_p} + \phi)} \approx \frac{2\mu L_x}{5(k/m_p)T} \text{ gm/sec} \qquad (1)$$

where r is the radius, n the electron density, $\Lambda(T)$ is the plasma emissivity at temperature T, k is Boltzmann's constant, μ is the mean molecular weight, m_p is the proton mass, $\phi(r)$ is the gravitational potential and L_x is the x-ray luminosity. From this equation, values of the accretion rate ranging from several to several hundred solar masses per year have been derived for a wide variety of cluster types (Sarazin 1986, Mushotzky and Szymkowiak 1988, Arnaud and Fabian 1988).

2. PROBLEMS WITH THE STANDARD METAPHOR

These large accretion rates pose a vexing problem: what has become of the tens of billions to trillions of solar masses of accreted gas implied by these estimates? It has been argued that the optical observation of H-alpha filamentary structures is evidence in favor of these large cooling rates. However, neither the mass in these filaments, nor the H-alpha emissivity is consistent with the predictions of the standard cooling flow model. The mass is rather small, on the order of a million solar masses and the filaments require an additional heat source to explain their emissivity. The connection of the optical filaments to the cooling x-ray gas is indirect at best. Their existence certainly cannot be taken as evidence that massive cooling flows are occurring.

The cooling gas does not show up in radio observations of HI at 21 cm, so it cannot be hidden in cold gas either. Nor can it be in black holes in the nuclei of the central galaxies, for it would be detectable both through its effects on the stellar dynamics of the galaxy (Mathews 1988) and through the energy released by accretion of several solar masses of gas per year into the black hole:

$$L_{acc} \approx 6 \times 10^{45} \dot{m}_0 \text{ erg/s} \qquad (2)$$

where $\dot{m}_0$ is the accretion rate in solar masses per year.

Finally, the accreting mass cannot have gone into the formation of a normal stellar population. Otherwise, the rate of star formation would be inconsistent with the observed colors of the central galaxies. O'Connell and McNamara (1988) have shown that evidence for star formation does exist in about 50% of the central galaxies in cooling flow clusters, but a rate that is less than 10% and more often less than 1% of the level implied by the accretion rates derived in the standard model. The implication of their work is that, if the accreted mass goes into the formation of stars, it must be an invisible population with an average mass $<< 0.7$ solar masses.

Several authors have suggested that the accreted mass is going into just such a population (see, e.g., Sarazin and O'Connell 1983, Fabian et al. 1982). They argue that the formation of loss mass stars is expected in cooling flows because the high pressures imply small values of the Jeans mass (proportional to $T^2 p^{-1/2}$ for a protostellar cloud of temperature T and pressure p). The observation that $\dot{m}_0$ is not constant, but increases with radius roughly as $\dot{m}_0 \propto r$, has been taken as evidence that matter is dropping out of the flow (Fabian 1988a,b), presumably to form a population of very low mass stars.

However, several theoretical treatments of the problem of matter condensing out of the flow indicate that it cannot occur in such a way as to explain the observed $\dot{m}_0$ versus r dependence unless ad hoc assumptions are made concerning the existence and distribution of large, pre-existing clumps of matter in the cooling flow (Balbus and Soker 1988, Malagoli, Rosner, and Bodo 1987, Nulsen 1988). Further, it has been shown the pressures in accretion flows are comparable to those in dense molecular clouds, which are thought to be some of the most prolific sites of star formation in our galaxy, and stars of a solar mass or higher have no difficulty forming there (Scalo and Miller 1979). The possibility of hiding large amounts of cooling matter from accretion flows in low mass stars remains speculative, with no empirical basis.

3. AN ALTERNATE METAPHOR — AN ACCRETION FLOW WITH A HEATING FEEDBACK MECHANISM

In view of the difficulties encountered with the standard model, it seems worthwhile to consider an alternate model in which we drop the assumption that no heat sources exist (Tucker and Rosner 1983, Rosner and Tucker 1989). In the presence of a heat source q_H, the accretion rate is

$$\dot{m}_a = \dot{m}_0 \{1 - [q_H / n^2 \wedge (T)]\} \tag{3}$$

Where $\dot{m}_a$ is the actual accretion rate, as contrasted with $\dot{m}_0$, calculated from equation (1), which is the apparent heating rate, derived on the assumption that no heat sources are present. This equation shows that the accretion rate can be arbitrarily small for an arbitrarily fine balance between heating and radiative losses. This provides a simple solution to the problem of hiding excessive amounts of accreted mass: very little mass is actually accreted, so there is little or no mass to hide.

The existence of a heat source can also resolve the paradox posed by the observation that $\dot{m}_0$ varies with radius, whereas the theoretical calculations indicate that it should show little or no variation. If equation (3) is re-written in the form

$$\dot{m}_0 = \dot{m}_a/[1 - (q_H/n^2 \wedge (T))] \qquad (3')$$

it is apparent that, if $q_H/n^2 \wedge (T)$ is a function of radius, then the apparent accretion rate $\dot{m}_0$ will be a function of radius, even though the actual accretion rate $\dot{m}_a$ is constant.

Of course, this resolution of the difficulties of the standard model is academic if no acceptable heat sources exist. On the one hand, Fabian (1988b) has maintained at this symposium that this is the case. On the other hand, I contend that sufficiently strong heat sources do exist in many if not all of the cooling flow clusters and present evidence below to support this contention.

4. ESTIMATES OF THE IMPORTANCE OF HEATING BY THERMAL CONDUCTION

The importance of thermal conduction as a heat source in clusters of galaxies has been recognized by a number of authors (Rephaeli 1977, Tucker and Rosner 1983, Bertschinger and Meiksen 1986, Friaca 1986, Volkov 1985). We can estimate the strength of thermal conduction relative to radiative losses by taking the ratio of the divergence of the conductive heat flux with the radiative energy losses at the cooling radius, that is, the radius at which the radiative cooling time = 10^{10} years. This ratio, which I call C, is given by

$$C = q_{cond}/n^2 \wedge (T) = 0.5 \ T_7^2 b/r_{100}^z \qquad (4)$$

where r_{100} is the cooling radius in units of 100 kpc, T_7 is the temperature in units of 10^7 degrees, and b represents the factor by which thermal conduction is reduced by magnetic field effects. Rosner and Tucker have given arguments that this factor should not be smaller than about 0.2, so I take $b = 0.2$.

The accompanying table shows the computed values for C for the cooling flow clusters listed in the catalog of Arnaud and Fabian (1989). The values of the temperature and cooling radius derived by Arnaud and Fabian from the observations are used in the computation, so C is an empirical, or at least, semi-empirical estimate of the importance of thermal conduction. Of the 45 clusters in the table, 32 have $C > 1$, indicating that thermal conduction should be considered as a major heat source for the majority of cooling flow clusters. The effect of $C > 1$ is to shift the cooling radius inward from the value predicted by the standard model, to change the temperature profile, making it flatter at smaller radii, and to reduce the accretion rate. The temperature profiles are not sufficiently well known to determine observationally just how important thermal conduction is. This is an area where high-resolution x-ray spectroscopy can make a crucial contribution.

One interesting suggestion, which literally turns the standard model inside-out, is that in Centaurus and possibly other clusters, we observing, not a cooling inflow, but an evaporative outflow of gas (de Jong et al. 1988). If this were the case, then the gas supply would have to be replenished at an average rate of several solar masses per year or more. de Jong et al. suggest that this could be accomplished by collisions and stripping of galactic halos in the central regions of rich clusters.

5. ESTIMATES OF THE IMPORTANCE OF HEATING BY RELATIVISTIC PARTICLES

Another potentially important heat source in cooling flow clusters is relativistic particles (Sofia 1973, Lea and Holman 1978, Rephaeli 1979, Scott et al 1980, Tucker and Rosner 1983, Pedlar et al. 1988, Miller 1988). Using the expression derived by Scott et al. for heating by relativistic plasma processes, together with the estimates of relativistic electron density given by Odea and Baum (1986) and the data on the hot gas from Arnaud and Fabian (1989), the ratio H of heating by relativistic electrons to radiative cooling of the hot gas has been estimated for 21 sources studied by Odea and Baum (1986). The results, shown in Table 1, indicate that relativistic electron heating is important in 12 of these sources, is probably not important in 5 sources, and the data is inconclusive in 4 sources.

In only one source is the sum $C + H < 1$: MKW 4, a poor cluster with a central dominant galaxy. MKW 4 is the best candidate to be standard cooling flows in which no heating is occurring. Detailed x-ray spectroscopic observations of this source would be especially interesting. In particular it would be interesting to know if the accretion rate is constant with radius, as the standard model in its simplest form predicts.

6. COOLING FLOWS WITH FEEDBACK

The data in Table 1 indicate that we must take seriously the possibility that heating by conduction and relativistic particles modifies the standard cooling flow model. If thermal conduction is sufficiently strong, it will make the cluster gas virtually isothermal or it may even drive an evaporative wind. In many cases, conduction will not shut off the accretion rate, completely, but will move the cooling radius inward and reduce the accretion rate by a factor of two or three. In these cases, a feedback mechanism involving accretion and heating by relativistic particle may come into play, reducing by accretion rate by a large factor.

An alternative scenario for cooling flows might go as follows: radiative cooling in the central regions lead to pressure gradients and an accretion flow. A portion

TABLE 1
Estimates of the Importance of Heating by Thermal Conduction and Relativistic Particles

Source	C $(= q_{cond}/n^2 \wedge (T))$	H $(= q^{rel}/n^2 \wedge (T))$
SC0004	3.0	?
A85	4.8	7.2
A133	1.1	?
A262	0.2	1.2
A278	4.3	?
A376	1.1	?
AWM 7	1.4	< 0.3
A401	18.0	?
A426 Perseus	3.3	8.5
A478	0.8	?
A496	1.	< 0.6
A539	2.2	?
A576	5.0	< 6.7
A592	3.0	?
PKS0745	2.5	1123
A644	0.5	?
A665	1.4	?
A1060	0.5	< 0.9
MKW 4	0.2	< 0.02
Centaurus	0.4	?
A1644	2.5	?
A1689	2.5	?
A1767	1.4	?
A1795	1.1	113
A1877	1.9	?
3C295	1.8	< 25
A1983	9.8	?
A1991	1.2	14
A2029	1.5	6
A2052	0.3	?
A2063	1.0	?
MKW 3s	0.2	?
A2142	3.3	< 0.
A2151	1.3	?
AWM 4	0.6	137
A2199	0.5	208
Her A	0.1	?
A2244	1.2	?
SC184	27.8	?
Cyg A	1.5	?
A2415	2.2	< 615

of the accreted matter falls into a black hole in the nucleus of the central galaxy, thereby generating nuclear activity, some of which appears in the form of relativistic particles. These relativistic particles stream out into the gaseous halo, heating it and reducing the accretion rate, which after a certain time delay τ reduces the relativistic particle production.

The equation for the accretion rate then becomes

$$\dot{m}_a(t) = \dot{m}_0 - \beta \dot{m}(t-\tau) \tag{5}$$

where $\dot{m}_0$ is given by equation (1),

$$\beta = \alpha m_p c^2 / kT \approx 10^6 \alpha / T_7 \tag{6}$$

and α is a factor that takes into account the efficiency of the accretion process in producing relativistic particles. The steady state solution to equation (5) is

$$\dot{m}_{ss} = \dot{m}_0 / (1+\beta) \tag{7}$$

It may be that the system does not reach a steady state, but exhibits limit cycle behavior, oscillating between relatively long-lived states of no accretion and short-lived states of accretion rates near $\dot{m}_0$, so that the time-averaged accretion rate is close to the value given by equation (7).

Finally, another byproduct of the model is that it may be ale to explain the optical filaments. As discussed by Bohringer and Morfill (1988), a large relativistic electron population can lead to a large non-thermal pressure, which can in turn lead to Rayleigh-Taylor instabilities. These instabilities could be the generators of the optical filaments, and the relativistic particles could supply the missing heat source.

ACKNOWLEDGEMENTS

I thank K. Arnaud and A. Fabian for making the data in their cooling flow catalog available prior to publication. I also thank Carolyn Stern for her computational assistance and W. Forman, C. Jones and R. Rosner for helpful discussions. This work was supported by NAS8-3075.

REFERENCES

Arnaud, K. and Fabian, A. 1988, in preparation.
Balbus, S. and Soker, N. 1988, this symposium.
Bertschinger, E. and Meiksen, A. 1986, *Ap.J.*, **306**, L1.
Bohringer, H. and Morfill, G. 1988, *Ap.J.*, **330**, 609.
de Jong, T., Norgaard-Nielsen, H., Jorgensen, H., and Hansen, L. 1988, preprint.
Fabian, A. 1988a, in *Cooling Flows in Clusters and Galaxies*, ed. A. Fabian (Kluwer Academic Publishers, Dordrecht).

Fabian, A. 1988b, this symposium.
Fabian, A., Nulsen, P. and Canizares, C. 1982, *MNRAS*, **201**, 933.
Friaca, A. 1986, *Astron.Ap.*, **164**, 6.
Lea, S. and Holman, G. 1978, *Ap.J.*, **222**, 29.
Malogoli, A., Rosner, R. and Bodo, G. 1987, *Ap.J.*, **319**, 622.
Mathews, W. 1988, in *Cooling Flows in Clusters and Galaxies*, ed. A. Fabian (Kluwer Academic Publishers, Dordrecht).
Miller, L. 1988, in *Cooling Flows in Clusters and Galaxies*, ed. A. Fabian (Kluwer Academic Publishers, Dordrecht).
Mushotzky, R. and Szymkowiak, A. 1988, in *Cooling Flows in Clusters and Galaxies*, ed. A. Fabian (Kluwer Academic Publishers, Dordrecht).
Nelsen P. 1988, in *Cooling Flows in Clusters and Galaxies*, ed. A. Fabian (Kluwer Academic Publishers, Dordrecht).
O'Connell, R. and McNamara B. 1988, in *Cooling Flows in Clusters and Galaxies*, ed. A. Fabian (Kluwer Academic Publishers, Dordrecht).

DISCUSSION-W. Tucker

C. CANIZARES: Andy Fabian has already noted that the SSS and FPCS have detected x-ray emission lines from approximately 10 clusters. The values for the accretion rate deduced from these lines agree with those derived from the x-ray images, but leaving that aside, the lines indicate the presence of large quantities of relatively cool gas. These are crucial observations that must be explained by any model. The thermal conduction models fail to do this (at least those in the literature). How do the heating models do?

W. TUCKER: In the relativistic particles heating model described here, the balance between heating and cooling requires, roughly, that the temperature $T \propto 1/p^{rel}$, where p^{rel} is the relativistic particle pressure. So, for example, if $p^{rel} \propto 1/r$, then $T \propto r$, inside the cooling radius, in rough agreement with the observations.

J. CULHANE: Can you estimate a characteristic time for the cycle of cooling flow followed by a heat input due to particles? Presumably there will be radio emission while the particle source is on, so this might have observational consequences.

W. TUCKER: The cycle will be composed of at least three parts:
    (1) cooling and accretion on a time scale of $10^9 - 10^{10}$ years;
    (2) production and diffusion of electrons throughout the source on a time scale of about 10^8 years;
    (3) decay of the electron population on a time scale of $10^8 - 10^9$ years.

EVOLUTION OF THE CORONAE IN EARLY-TYPE GALAXIES

L.P. David, W. Forman, and C. Jones
Harvard-Smithsonian Center for Astrophysics, Cambridge, MA, USA

ABSTRACT. We present numerical simulations of the gaseous coronae in elliptical galaxies. These models consist of a modified King profile for the luminous portion of the galaxy and an isothermal dark halo. We include evolving stellar mass loss from planetary nebulae, and type I and II supernovae. Our models show that elliptical galaxies are likely to produce strong galactic winds at early times with x-ray luminosities of $10^{42} - 10^{44}$ ergs s^{-1} and temperatures of 10 keV. Galaxies can lose approximately 10-30% of their initial luminous mass in the wind which has an oxygen-to-iron ratio twice the solar value. Since elliptical galaxies are a principle component of rich clusters and compact groups this early wind phase affects the metallicity and temperature of the intracluster medium.

1. INTRODUCTION

Elliptical galaxies were thought to be nearly devoid of gas prior to the launch of the Einstein observatory. Constraints on the amount of gas in elliptical galaxies through optical and radio observations indicated that the gas mass was well below that which should accumulate through normal stellar evolutionary mass loss over the life time of a galaxy (Faber and Gallagher 1976). Due to the lack of cool gas in elliptical galaxies, most early theoretical studies were concerned with mechanisms for removing the stellar mass loss. Mathews and Baker (1971) showed that galactic winds may currently exist in elliptical galaxies which can efficiently remove all stellar mass loss. Observations with the Einstein satellite led to the discovery that x-ray emission from hot ($T \approx 10^7$ K) extended coronae is nearly ubiquitous among early-type galaxies (Forman, Jones, and Tucker 1985). The present x-ray luminosity of elliptical galaxies ranges from $10^{39} - 10^{42}$ ergs s^{-1} in the 0.5 - 4.5 keV band pass and is correlated with the optical luminosity (Forman, Jones, and Tucker 1985; Canizares, Fabbiano, and Trinchieri 1987). The observed x-ray luminosities are four orders of magnitude greater than the galactic wind models, indicating that the hot coronae must be in approximate hydrostatic equilibrium with the gravitational potential of the galaxy. The radiative cooling time of the gas within the hot coronae is typically less than a Hubble time. If heating is unable to balance cooling, this gas must cool as it settles into the center of the galaxy, and establish a cooling flow (Fabian and Nulsen 1977; Cowie and Binney 1977).

There have been a number of steady-state galaxy cooling flow models presented in the literature (White and Chevalier 1984; Sarazin and White 1987; Vedder, Trester, and Canizares 1988). These papers have extensively examined the effect of heavy halos, supernova rate, mass deposition, and environment on the present x-ray properties of elliptical galaxies. Recently, Loewenstein and Mathews (1987) have developed models of the evolution of the hot coronae in elliptical galaxies and find that accretion flows are common at late times. These models begin at a time corresponding to the main sequence life time of a $5M_\odot$ star and hence do not include the consequences of type II supernovae at early epochs.

In this paper we present models for the evolution of the coronae in elliptical galaxies which begin shortly after the formation of the stellar component of the galaxy. Our models calculate the evolving rate of stellar mass loss from type I and II supernovae, and planetary nebulae. We find that the high rate of type II supernovae at early times generates hot, x-ray luminous galactic winds that enrich the intracluster medium with heavy metals. We discuss the consequences of these early winds in clusters and the overall x-ray luminosity evolution of the coronae.

2. METHOD

In this section we briefly outline our method for simulating the evolution of the coronae in elliptical galaxies (see David, Forman, and Jones (1989), hereafter DFJ, for a more detailed development). We present the results of three different galaxy models with present optical luminosities of 10^9, 10^{10}, and $10^{11} L_\odot$. Each galaxy model consists of a modified King profile for the luminous portion of the galaxy along with an isothermal dark halo which comprises 90% of the total mass of the galaxy. The stellar population is assumed to form instantaneously without any residual gas remaining. The stars are assumed to form with a Salpeter initial mass function with an upper mass cutoff of $40M_\odot$. Stellar mass loss occurs instantaneously as stars evolve off the main sequence. All stars more massive than $8M_\odot$ develop into type II supernovae and leave behind $1.4M_\odot$ remnants. Stars less massive than this lose mass through a planetary nebula and leave behind a remnant with a mass given by the expression in Tinsley (1976). The evolving rate of type I supernovae is determined using the method discussed in Matteucci and Greggio (1986) and is normalized to the rate given in van den Bergh *et al.* (1987) at the present epoch. All of this information is then incorporated into a time-dependent, spherically symmetric, hydrodynamics code in order to determine the evolving dynamical state of the interstellar medium in elliptical galaxies.

3. X-RAY LUMINOSITY EVOLUTION

The evolution of the corona in an elliptical galaxy is completely governed by the evolving stellar mass loss rates, and the mass-averaged temperature of the injected gas. At early times, gas is injected into a galaxy with a temperature of $\sim 10^8$ K due to the high rate of type II supernovae (see DJF). Once all stars more massive than $8M_\odot$ have evolved off the main sequence, the temperature of the injected gas drops precipitously. At late times the temperature of the injected gas increases gradually since the mass injection rate from planetary nebulae decreases more rapidly than the mass injection rate from type I supernovae. We find that the interstellar medium evolves through three dynamically distinct stages in these models. These are: 1) an initial galactic wind phase generated by the high rate of type II supernovae, 2) a transition phase during which the interstellar medium evolves from a pure wind to a pure accretion flow as the rate of type II supernovae diminishes, and 3a) a cooling flow phase which endures for a Hubble time in optically luminous galaxies, or 3b) a second wind phase which commences at late times as the mass-averaged temperature of the injected gas surpasses the escape temperature in low luminosity models ($\lesssim 10^9 L_\odot$).

The evolution of the x-ray luminosity of a corona is primarily determined by the central gas density since the x-ray emissivity is proportional to $\rho^2 T^{1/2}$ and the gas temperature only varies by a factor of 10 over a Hubble time. Figure 1 shows that the x-ray luminosity increases at first as gas is injected into the galaxy, and attains a maximum value when the mass injection rate is balanced by the mass loss rate in the wind. The x-ray luminosity during the wind phase is 100 - 1000 times greater than the present epoch x-ray luminosity of elliptical galaxies and the wind temperature is 10 keV. Extending the formation of the stellar population over 10^8 years reduces the peak x-ray luminosity during the wind phase by approximately an order of magnitude (see DJF). As the wind continues to evacuate the galaxy, the central density and x-ray luminosity decrease, reaching a minimum value at the start of the transition phase. As the gas begins accreting into the center of the galaxy, the x-ray luminosity increases reaching a second maximum at the start of the cooling flow phase. At late times L_x decreases due to the continually decreasing stellar mass loss rate. The sharp drop in the x-ray luminosity in the $10^9 L_\odot$ model (see figure) is due to the development of a galactic wind at late times. In models which maintain a cooling flow, the x-ray luminosity at late times can be approximated by $L_x \propto t^{-0.9}$.

4. CONSEQUENCES OF THE EARLY WIND PHASE

The observation of X-ray emission lines from cluster gas requires that a significant amount of elements synthesized in stellar interiors are injected into the intracluster medium. The enrichment of the intracluster medium can result from both galactic winds and ram pressure stripping. Assuming that all galaxies within a cluster form with a Salpeter IMF and develop winds which remove all of the metal enriched type II supernovae ejecta, we can estimate the metallicity of the cluster gas from the results given in DFJ. All galaxies lose approximately 10-30% of their initial luminous mass during the wind phase. The mass fraction yield of Fe produced by type II supernovae in a galaxy with a Salpeter IMF is 5×10^{-4} (DFJ). Using these values, the iron abundance of the cluster gas relative to the solar value will be $0.30 M_{stars}/M_{gas}$ after the cessation of the early wind phase, where M_{stars} is the initial stellar mass of the cluster and M_{gas} is the initial gas mass in the cluster (the residual gas remaining after galaxy formation ends). Significant amounts of oxygen are produced in type II supernovae which result in an oxygen-to-iron abundance ratio after the wind phase which is approximately twice the solar value. Using a

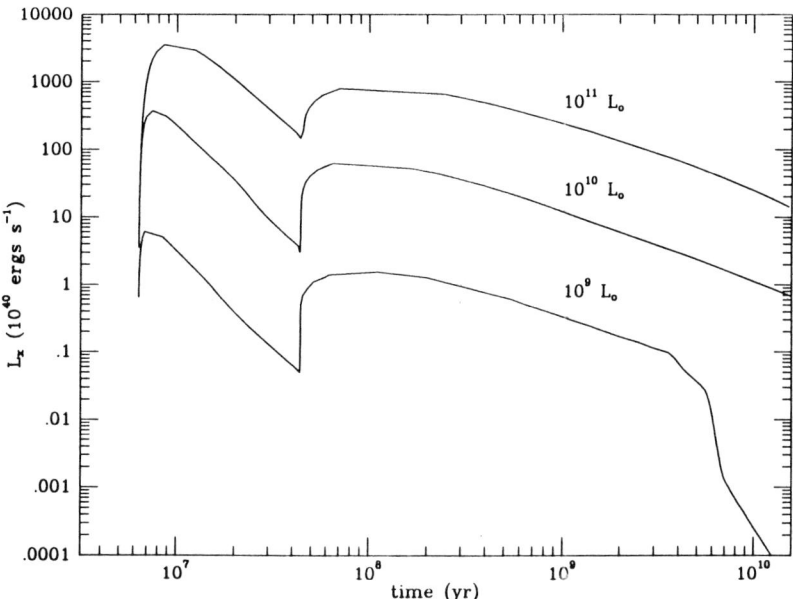

Fig 1. X-ray luminosity evolution of three galaxy models with different optical luminosities at the present epoch.

present supernovae rate equal to the rate given in van den Bergh *et al.* (1987), the mass fraction yield of Fe from type I supernovae is 6×10^{-4}. Assuming that ram pressure stripping is one hundred percent efficient in the removal of stellar mass loss, the resulting Fe abundance of the cluster gas after a Hubble time will be $0.66 M_{stars}/M_{gas}$. Very little oxygen is produced in type I supernovae relative to type II so that ram pressuring stripping reduces the oxygen-to-iron abundance ratio to approximately the solar value. By determining the oxygen and iron abundances for clusters at different epochs with observations from future x-ray telescopes, the IMF of the first generation of stars and the efficiency of ram pressure stripping can be determined. Also, due to the observed decrease of M_{stars}/M_{gas} with cluster richness (David *et al.* 1989; Jones *et al.*, this proceedings), our models predict that the metallicity of the intracluster gas should decrease from poor to rich clusters.

A second important effect of galactic winds is to inject energy into the intracluster medium. The density distribution of the hot gas in many clusters of galaxies is well fit using isothermal hydrostatic β models, where β is the ratio of the energy per unit mass in galaxies to the specific internal energy of the gas. The best fitting values of β range from 1/2 for groups and poor clusters to 2/3 for rich clusters. If the temperature of the gas and galaxies resulted from gravitational collapse in a cluster, these temperatures should be more nearly equal. One explanation for this discrepancy is heating of the cluster gas during the early galactic wind phase. From our models presented in DFJ, the mass-averaged temperature of the gas ejected in the wind is 1.6×10^8 K. If the initial temperature of the cluster gas is assumed equal to the galaxy temperature, the net heating of the cluster gas can be determined. Using $\beta = T_{gal}/T_{gas}$, where T_{gal} is the galaxy temperature, and T_{gas} is the initial gas temperature, the fractional change in the temperature of the intracluster gas can be written as

$$\beta = 1 - 0.45 \left(\frac{M_{stars}}{M_{gas}} \right)$$

This expression indicates that ample heating of the cluster gas can be provided by early galactic winds. Since M_{stars}/M_{gas} decreases systematically from poor to rich clusters (Jones *et al.*, this proceedings), our models predict that heating by galactic winds should produce a direct correlation between β and cluster richness (see Forman *et al.* this proceedings for further discussion).

5. CONCLUSIONS

We find that the gas within elliptical galaxies evolves through three dynamically distinct stages. Our models indicate that a substantial portion of the heavy elements observed in the intracluster medium can be supplied during the early wind phase, and that the energy injected can alter the ratio of energy per unit mass in galaxies to that in gas. The high x-ray luminosity and temperature during the early wind phase indicate that x-rays are a prime region of the spectrum to search for primeval galaxies. If galaxies form at redshifts of 4 to 5, the observed emission will be redshifted to 2 keV, an energy at which future x-ray observatories will be very sensitive.

REFERENCES

Canizares, C.R., Fabbiano, G., and Trinchieri, G. 1987, Ap.J., 312, 503.
Cowie, L.L., and Binney, J. 1977, Ap.J., 215, 723.
David, L.P., Arnaud, K.A., Forman, W., and Jones, C. 1989 (preprint).
David, L.P., Forman, W., and Jones, C. 1989 (preprint).
Faber, S., and Gallagher, J. 1976, Ap.J., 204, 365.
Fabian, A.C., and Nulsen, P.E.J. 1977, MNRAS, 180, 479.
Forman, W., Jones, C., and Tucker, W. 1985, Ap.J., 293, 102.
Loewenstein, M., and Mathews, W. G. 1987, Ap.J., 319, 614.
Mathews, W.G., and Baker, J. 1971, Ap.J., 170, 241.
Matteucci, F., and Greggio, L. 1986, A.A., 154, 279.
Sarazin, C.S., and White ,R.E. 1987, Ap.J., 320, 32.
White, R.E., and Chevalier, R.A. 1984, Ap.J., 280, 561.
Tinsley, B.M. 1976, Ap.J., 208, 797.
van den Bergh, S., McClure, R., and Evans, R. 1987, Ap.J., 323, 44.
Vedder, P.W., Trester, J.J., and Canizares, C.R. 1988, Ap.J., 332, 725.

IMPLICATIONS OF ABUNDANCE MEASUREMENTS OF THE INTRACLUSTER MEDIUM

W. Forman, C. Jones and L. David
Smithsonian Astrophysical Observatory, Cambridge, MA, USA

ABSTRACT

The correlation of $M_{gas}/M_{stellar}$ with the intracluster gas temperature in clusters of galaxies has implications for the efficiency of galaxy formation and predicts trends for the heavy element abundances in these systems.

1.0 INTRODUCTION

The discovery of a correlation between the ratio of the gas mass to stellar mass, $M_{gas}/M_{stellar}$, and the gas temperature (T_{gas}, a measure of the depth of the potential) in groups and clusters of galaxies has important implications for some models of the formation of large scale structure in the Universe (David *et al.* 1989a). In this contribution we describe the importance of determining the heavy element (iron) abundances of groups and clusters with gas temperatures, that span the range of the correlation of $M_{gas}/M_{stellar}$ vs. T_{gas}.

We will confine our discussion to the cold dark matter scenario for the formation of structure in the Universe which is described by Blumenthal *et al.* (1984). In this scenario, it is assumed that the baryonic mass in the Universe is just that component of the mass that is luminous. Thus, in groups and clusters of galaxies the predominant baryonic (luminous) components are the gas mass, M_{gas}, and the stellar mass (in galaxies), $M_{stellar}$. Thus, $M_{luminous} = M_{gas} + M_{stellar}$ and the remaining mass, $M_{virial} - M_{luminous}$, consists of cold, dark matter and is not baryonic. What can we learn from abundance determinations of the hot gas in groups and clusters in a cold, dark matter scenario?

2.0 IMPLICATIONS OF THE CORRELATION OF $M_{gas}/M_{stellar}$ WITH T_{gas}

As described by Jones *et al.* (1989; especially Figure 1), the ratio of the gas to stellar mass, $M_{gas}/M_{stellar}$, increases from unity in groups and poor clusters with low temperatures (~ 2 keV) to values of 3-6 in systems with high gas temperatures (6-10 keV). This correlation, combined with an understanding of the production of heavy elements, predicts a correlation of heavy element abundance with gas temperature.

The groups which are luminous x-ray sources are dense systems and have stellar populations comparable to rich clusters (Morgan *et al.* 1975). Also, the correlation of galaxy population with local density (Dressler 1978, and Postman and Geller

1984) supports the similarity of the galaxy populations in the groups and clusters. Therefore, the production of heavy elements should be directly proportional to the stellar light, or equivalently stellar mass, since comparable populations will have similar mass-to-light ratios. Thus, the larger the ratio of gas mass to stellar mass, the more dilute the stellar products like iron. Since $M_{gas}/M_{stellar}$ increases with increasing T_{gas}, we predict that hotter clusters (those with larger $M_{gas}/M_{stellar}$) will have lower iron abundances than cooler clusters. This prediction assumes that the clusters and groups are closed systems, i.e., no gas is expelled or accreted.

Figure 1 shows the quantitative prediction for the correlation of iron abundance with gas temperature. The two solid curves are derived by taking a simple parameterization for the dependence of $M_{gas}/M_{stellar}$ on T_{gas} and assuming that enriched material is expelled from galaxies only during an early wind phase during which Type II supernovae can readily drive a galactic wind (see David et al. 1989b). The two curves assume different initial mass functions (the upper curve has a power law exponent $\alpha = 2$ and the lower curve has $\alpha = 2.5$). Note that an amount of enriched material equal to that expelled in the wind is produced by stellar evolution and could be liberated by ram pressure stripping. The present estimates of supernova

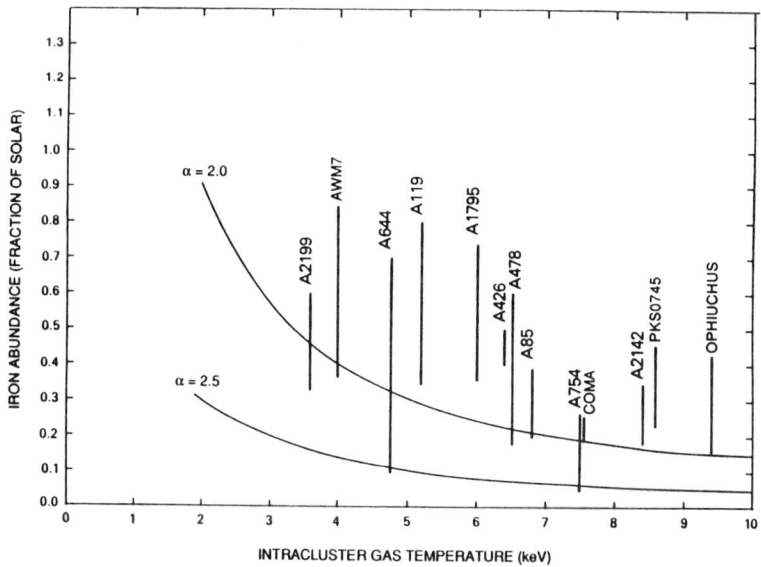

Figure 1. The iron abundance (as a fraction of the solar value) is plotted against gas temperature. The data are taken from Henriksen (1985), Hughes et al. (1988) and Arnaud et al. (1987). The smooth curves are predictions based on a parameterization of the relation between $M_{gas}/M_{stellar}$ and T_{gas} as well as a model for the evolution of stars with two different initial mass functions.

yields can explain the observed heavy element abundances in the intracluster gas as Figure 1 shows. The ejected gas is extremely enriched and is diluted to the observed values by mixing with the predominantly primordial component of the intracluster medium.

The previous discussion assumes that the groups and clusters are closed systems, i.e., gas is not expelled or accreted in significant quantities. This can be tested by observing clusters with progressively lower temperatures. If ejection becomes important below some temperature, T_{crit}, then one would observe an increasing heavy element abundance from the hottest clusters down to those with temperatures equal to T_{crit}. Below T_{crit}, the winds would serve to expel enriched material and the abundance would decline (or remain constant) as the gas temperature decreases further.

3.0 IMPORTANCE OF FUTURE ABUNDANCE DETERMINATIONS

Future experiments will provide greatly enhanced spectroscopic capabilities and will open new avenues for studies of the intracluster gas. The ability to obtain spatially resolved spectra is most important for studies of the intracluster gas since it provides a means to determine abundance and temperature gradients. Below we mention two important observations and their implications.

First, the measurement of temperature profiles in clusters allows the direct determination of the underlying gravitational mass (luminous and dark matter). Assuming hydrostatic equilibrium and spherical symmetry, the gravitating mass is given by

$$M_{grav}(r) \propto \left(\frac{d\ln\rho}{d\ln r} + \frac{d\ln T_{gas}}{d\ln r}\right) T_{gas}(r)\, r$$

where ρ is the gas density and r is the radial distance from the cluster center. Only M87 and Centaurus (NGC4696) have measured density and temperature gradients (Fabricant and Gorenstein 1983, Matilsky et al. 1985). For most clusters $T_{gas}(r)$ remains unknown (or very uncertain). Thus, ASTRO-D, AXAF, and XMM can add immeasurably to our present knowledge.

Second, abundance measurements of clusters will have implications for a variety of problems. Average abundances, in the context of the previous discussion of the changing ratio of gas to stellar mass, can place constraints on models for galaxy formation (e.g., biasing) in some scenarios. Radial gradient determinations will provide information on the origin of the heavy elements. Since the galaxies and gas in clusters have different distributions, the radial variation of abundances could have implications for the existence of Population III objects. Finally, abundance measurements at different redshifts will tell us about the enrichment mechanisms of the intracluster medium and the evolution of the stellar populations producing the enriched material.

Acknowledgements

We would like to thank M. Fonseca and K. Modestino for their help in preparing this manuscript. This work was supported by the Smithsonian Institution Scholarly Studies Program and by NASA contract NASA-30751.

REFERENCES

Arnaud, K. A., Johnstone, R. M., Fabian, A. C., Crawford, C. S., Nulsen, P. E. J., Shafer, R. A., and Mushotzky, R. F. 1987, *MNRAS*, **227**, 241.

Blumenthal, G. R., Faber, S. M., Primack, J. R., and Rees, M. J. 1984, *Nature*, **311**, 517.

David, L., Arnaud, K., Forman, W., and Jones, C. 1989a, preprint.

David, L., Forman, W., and Jones, C. 1989b, this volume.

Dressler, A. 1978, *Ap. J.*, **226**, 55.

Fabricant, D. and Gorenstein, P. 1983, *Ap. J.*, **267**, 535.

Henriksen, M. 1985, *Ph.D. Thesis* (University of Maryland).

Hughes, J. P., Yamashita, K., Okumura, Y., Tsunemi, H., and Matsuoka, M. 1988, *Ap. J.*, **327**, 615.

Jones, C., David, L., Forman, W. 1989, this volume.

Matilsky, T., Jones, C., and Forman, W. 1985, *Ap. J.*, **291**, 621.

Morgan, W. W., Kayser, S., and White, R. A. 1975, *Ap. J.*, **199**, 545.

Postman, M. and Geller, M. 1984, *Ap. J.*, **281**, 95.

CLUSTERS OF GALAXIES AND THE HOT INTRACLUSTER MEDIUM

C. Jones, W. Forman and L. David
Harvard/Smithsonian Center for Astrophysics, 60 Garden St., Cambridge, MA

ABSTRACT

The luminous material in clusters of galaxies falls primarily into two forms — the visible galaxies and the X-ray emitting intracluster medium. The hot intracluster medium is the major observed baryonic component of clusters with a mass equal to or greater than that of the stellar matter. In this paper we discuss changes in the efficiency of galaxy formation for different clusters and the origin of the intracluster medium.

THE ORIGIN OF THE INTRACLUSTER MEDIUM AND THE EFFICIENCY OF GALAXY FORMATION

The ICM is a major component of the cluster being equal or greater in mass than the stellar matter. It is of particular importance to determine the origin of such a large fraction of the known baryonic mass of the cluster. Equally fundamental and related problems to address are the effects of the ICM on the morphology of galaxies and the efficiency of galaxy formation. To begin to address the questions of the origin of the ICM and the efficiency of galaxy formation in different environments, it is useful to compare the ratio of gas mass to stellar mass in groups of galaxies and rich clusters. It is well known that the mass-to-light ratio increases with the size of the system. However, from poor to rich clusters the fraction of X-ray emitting gas to virial mass remains relatively constant ($\sim 10\%$ within the central five core radii) (e.g. Abramopolilos and Ku, 1983). Therefore, the ratio of gas mass to stellar mass should increase from the poor to rich clusters. As shown in Figure 1, in groups of galaxies the gas mass is approximately equal to the stellar mass, while in very rich clusters, the gas mass exceeds the stellar mass by as much as a factor of six (David et al. 1988).

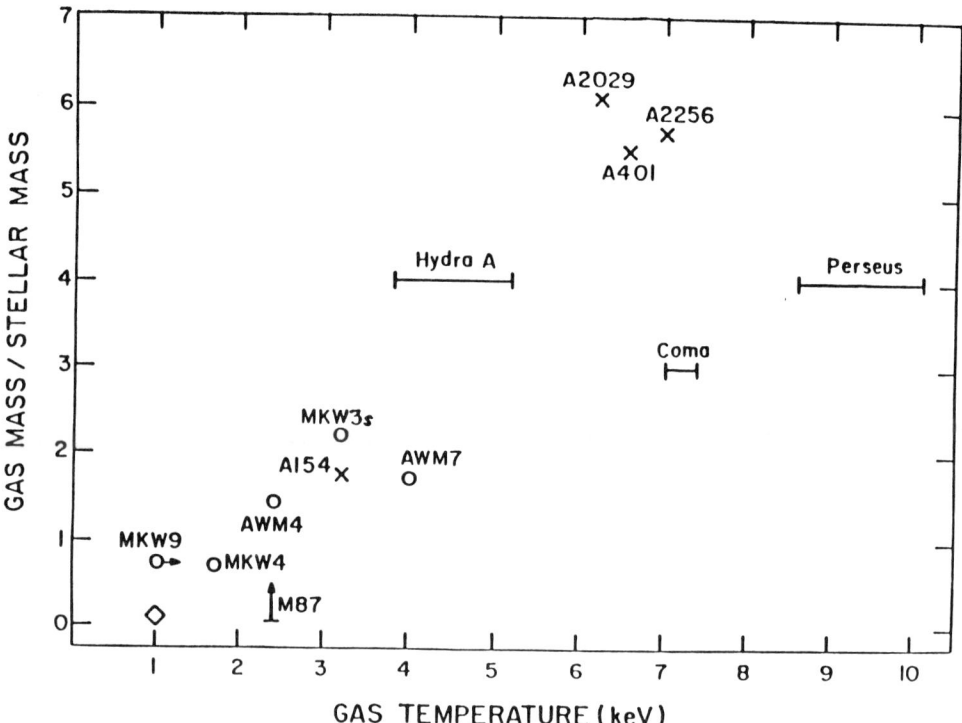

Fig. 1. The ratio of the gas mass as measured by the X-ray observations to the stellar mass is plotted against the temperature of the gas. The gas and stellar masses are evaluated within five core radii. The temperature of the gas reflects the gravitational potential of the system. The increasing ratio of gas mass to stellar mass with increasing gas temperature suggests a decreasing efficiency of galaxy formation (conversion of gas to luminous stellar matter) between groups and rich clusters.

The discovery of heavy elements in the ICM (Mitchell et al 1976, Serlemitsos et al. 1977) revealed an entirely new aspect to the study of the ICM — one crucial in determining its origin. Since heavy elements can be produced only through thermonuclear reactions in stars or by supernovae, the discovery that the intracluster medium was enriched in heavy elements required that material processed through stars be ejected into the ICM. The near solar abundance of the ICM measured by early X-ray experiments led to the suggestion that a large fraction of the material in the ICM was ejected from galaxies (e.g. DeYoung, 1978).

While the enriched material must come from the galaxies (in the absence of a Population III component), more recent studies have suggested that the bulk of the ICM of a rich cluster could not have originated within the galaxies because its mass is several times larger than the mass of the galactic stellar component. Thus, the bulk of the ICM in rich clusters must be "left over" from the formation of the galaxies. In particular, numerical modelling of the hot gaseous coronae around elliptical galaxies shows that over a Hubble time these galaxies can contribute only a fraction of their stellar mass to the IGM (David, Forman and Jones 1989).

While such an analysis constrains the contribution from present epoch galaxies to the ICM, if we assume that the groups and clusters are "closed" systems, we can use the ratios of the stellar mass to gas mass in groups and clusters to limit both the mass loss from present epoch galaxies to the ICM as well as any contribution from early, population III stars. Specifically, so long as the IMF's and the population III component of groups and clusters are similar and no gas is lost or gained by the system, then we would expect the stellar contribution to the ICM per unit stellar mass to be the same in all groups and clusters. The approximate equality of gas mass and stellar mass in the low X-ray luminosity Morgan groups (MKW4, MKW9, and AWM4) limits the contribution to the ICM by all stars to no more than the present stellar mass. Thus, in the richest clusters where the gas mass is three to six times the stellar mass, only a small fraction of the ICM could have been produced in stars. In the rich clusters, most of the gas in the intracluster medium must be primordial.

The ratio of the gas mass to the stellar mass, $M_{gas}/M_{stellar}$, shown in Figure 1, can be related to the efficiency of star formation. We assume a scenario in which the luminous matter (stars and the ICM) form the bulk of the baryonic material and the remainder of the virial mass is in the form of hot or cold dark matter. Then the efficiency of galaxy formation, the conversion of baryons from gas to stars in galaxies, can be written as

$$\epsilon = M_{stellar}/M_{lum} \tag{2}$$

where $M_{lum} = M_{stellar} + M_{gas}$, or equivalently as

$$\epsilon = (1 + M_{gas}/M_{stellar})^{-1} \tag{3}$$

(assuming the expelled gas from galaxies is small and can be neglected). Thus by measuring $M_{gas}/M_{stellar}$ we can study the efficiency of star formation in systems ranging from groups to rich clusters. Our analysis shows that the star (and galaxy) formation efficiency ranges from 50% for groups to as little as ≈15% for rich clusters. If all the ICM in groups is instead gas ejected from galaxies, and we use this injection rate for all clusters, then while the galaxy formation efficiency would be 100% for groups, a lower efficiency is still found for rich clusters (as low as ≈17% for $M_{gas}/M_{stellar} = 6$ to as high as ≈50% for $M_{gas}/M_{stellar} = 3$). Although the amount of luminous material (gas+stars) remains relatively constant for all clusters (Blumenthal et al. 1984), the efficiency of galaxy formation decreases as one moves to richer systems. In other words, although the richest systems obviously produced more galaxies, their efficiency of galaxy formation was lower.

In such a "closed" cluster system, as long as the galaxies in different clusters have similar initial mass functions so that the amount of enriched material is directly related to the stellar mass, then the decrease in efficiency of galaxy formation between groups and rich clusters implies that the intracluster medium in poor clusters should show a greater abundance of heavy elements than that in rich clusters. Since the conclusion concerning different enrichments of the ICM hinges on clusters and groups having similar IMF's, it is important to note that Morgan, Kayser and White (1975) found the galaxy populations of many of the cD dominated groups to be similar to those of rich Abell clusters of Bautz-Morgan Type I. There is no evidence that the galaxies in rich clusters are significantly different from the galaxies in these dense groups.

Since it is likely that the galaxies in rich and poor clusters formed with similar IMF's, the iron (and other heavy elements) produced during the evolution of the component stars would yield a mass of iron directly proportional to the present epoch stellar light of the galaxies:

$$M_{Fe} = \eta L_{galaxies} \qquad (4)$$

where η is determined by the IMF. Let us assume that the fraction of material ejected from the galaxies into the ICM is a constant fraction of the material liberated by the constituent stars. This is supported observationally by the similar colors of Morgan group galaxies and cluster galaxies implying comparable metal abundances. Further support comes from the similarity in the effectiveness of gas removal mechanisms. Internal mechanisms (e.g. supernova driven galactic winds) would be indifferent to the environment and ram pressure stripping (proportional to ρv^2) of the gas from galaxies is comparable in groups and clusters (the gas temperature, a measure of v^2, increases by a factor of 3-4 from groups to rich clusters and the central gas densities are perhaps slightly higher in groups and hence partially offsetting). Thus, in the simplest scenario we can predict a trend of iron abundance with the depth of the potential of the system — increasing potentials, correlated with lower galaxy formation efficiencies, imply decreasing iron abundances.

The present measurements of iron abundances are too inaccurate to verify the above model or test possibilities for the origin of the ICM. Mushotzky (1984) and Henriksen (1985) summarize present results. For rich clusters, Henriksen (1985) reports a possible correlation of decreasing iron abundance with increasing gas temperature, as predicted, but the data are not sufficiently precise to yield quantitative results. Those observations were for only quite luminous clusters ($L_x > 2 \times 10^{44}$ ergs sec^{-1}) while in general we expect the abundances to be highest in the low luminosity clusters. Hughes et al. (1988) have measured a precise iron abundance of 22% of the solar value for the rich Coma cluster. To adequately test for differences in abundances, it is particularly important to obtain comparable measurements for low temperature (high galaxy formation efficiency) systems. By determining accurate values of the heavy-element abundances of the ICM in both poor and rich clusters, one could better investigate the properties of the IMF (e.g. exponent), the efficiency of galaxy formation, and the origin and enrichment of the ICM. A precise determination of the heavy element abundance of the intracluster medium for a sample of clusters ranging from groups to rich clusters has implications for the amount of material in the ICM that must be primordial. In particular, determining a high solar abundance for the ICM in Morgan groups, as suggested by the arguments above, would confirm that the origin of most of the hot gas in rich clusters must be primordial.

The changing ratio of gas mass to stellar mass also will affect the energy (or temperature) of the ICM. By measuring the surface brightness profiles and independently by measuring the ratio of the velocity dispersion to the gas temperature, one can estimate the energy per unit mass of the galaxies compared to that of the gas. From the surface brightness profiles, this value for rich clusters is generally $\sim 2/3$. The values calculated from the measured velocity dispersions and gas temperatures have a wider range (but see Flanagan et al. (1989) who suggest a resolution for the Perseus discrepancy). By comparison to rich clusters, the surface brightness profiles for hot gas around single dominant cluster galaxies such as M87 and the cD groups such as AWM7 yield a value $\sim 1/2$ (Fabricant and Gorenstein 1983, Jones and Forman 1989). This implies that the groups and individual central galaxies have more energy per unit mass in gas compared to the constituent galaxies than do rich clusters. For the groups and poor clusters where the stellar mass is comparable to the gas mass, there may be significant heating of the ICM by the ejected material which may account for the observed difference between the groups and the clusters.

Acknowledgements

We thank Karen Modestino and Donna Wyatt for their excellent preparation of this manuscript. This work was supported through NASA Contract NAS8-30751.

REFERENCES

Abramopoulos, F. and Ku, W. 1983, *Ap.J.*, 271, 446.
Blumenthal, G.R., Faber, S.M., Primack, J.R., and Rees, M.J. 1984, *Nature*, 311, 517.
David, L., Forman, W., and Jones, C. 1989, in preparation.
DeYoung, D.S. 1978, *Ap.J.*, 223, 47.
Fabricant, D. and Gorenstein, P. 1983, *Ap.J.*, 267, 535.
Flanagan, J., Hughes, J., Arnaud, K., Forman, W., and Jones, C. 1989, in preparation.
Henriksen, M. 1985, Ph.D. Thesis (University of Maryland).
Hughes, J.P., Yamashita, K., Okumura, Y., Tsunemi, H., and Matsuoka, M. 1988, *Ap.J.*, 327, 615.
Jones, C. and Forman, W. 1989, in preparation.
Morgan, W.W., Kayser, S., and White, R.A. 1975, *Ap.J.*, 199, 545.
Mushotzky, R. 1984, *Phys. Sci.*, T7, 157.
Serlemitsos, P.J., Smith, B.W., Boldt, E.A., Holt, S.S., and Swank, J.H. 1977, *Ap.J.*, 211, L63.

THE EVOLUTION OF COOLING FLOW AND THE MASS DEPOSITION PROCESS

Makoto Hattori and Asao Habe

Department of Physics, Hokkaido University, Sapporo 060, Japan

ABSTRACT.
  We perform the numerical hydrodynamical calculation of cluster cooling flow including the mass deposition process and examine evolution of cooling flow. We take two mass deposition models. In one model, mass deposition occurs in large extent. In other model, mass deposition occurs only in central region. In any model, diagnostics of X-ray surface brightness do not agree with observation in any evolutional phases.

1. INTRODUCTION
  Studies of observational results of X-ray surface brightness of cluster cooling flows have indicated that mass flux of accretion flow has to decrease with decreasing radius from cluster center (Stewart *et al.* 1984; Thomas *et al.* 1987). Mass deposition from the cooling flow in large extent due to thermal instability have been discussed by many authors as one possible explanation for decreasing mass flux (Matthews and Bregman 1978; Nulsen 1986; Thomas *et al.* 1987). Effects of the mass deposition process on cluster cooling flow are studied by White & Sarazin (1987a, c) using steady state solutions. They proposed prescriptional mass deposition rate. Chevalier (1987) examines the calculation of cooling flow using self-similar solutions. The numerical calculation of evolution of cluster cooling flow is important subject for two main reasons. First, solutions of evolutional calculation could show the entire structure of cluster cooling flow. If one makes steady assumption, that study would be restricted to the inner parts of the cooling flow (Chevalier 1987). Second, numerical calculation could apply to more realistic situations. The approach taken self-similar solutions is restricted in the variety of physical paramaters and because of the mass distribution of cluster of galaxies have core radius self-similarity of the system is destroyed by this radius in the case of cluster cooling flows. In this paper, we perform numerical hydrodynamical calculation of the cluster cooling flow including mass deposition process and examine evolution of cooling flow.

2. MODELS
  In this study, we use two mass deposition models. In one model, we assume that inside the cooling radius, where cooling time equals the age of cluster, mass is deposited from cooling flow in the White & Sarazin's (1987a) 'thermal instability' mass deposition rate. We call this the inhomogeneous model. In the other model, we assume that when the temperature is below 10^6 K and cooling time is below 10^8 yr, mass is deposited at the rate which equals the hot gas density divided by isobaric cooling time, where isobaric cooling time equals the thermal energy divided by cooling rate. We call this the homogeneous model. We construct mass distribution of cluster of galaxies, stellar component and dark halo of central galaxy corresponding to Virgo cental galaxy M87 and a total gravitational velocity potential of these models is consistent with the

spectroscopical velocity dispersion (Sargent et al. 1978) and with the range of masses determined from the X-ray data (Fabricant & Gorenstein 1983) of Virgo central galaxy M87. We ignore self gravity of hot gas and gravity of deposited gas. We take the radiative cooling function given by Raymond, Cox and Smith (1976). Initially, isothermal gas with temperature of 3×10^7 K and central electron density of 1×10^{-3} cm is placed in hydrostatic equilibrium in the total gravitational field. Isobaric cooling time of initial gas is 2×10^7 yr at the centre of the cluster of galaxies. We consider the stellar mass loss (Gisler 1976), supernova heating (Tamman 1974; Macdonald & Bailey 1981; Hattori, Habe & Ikeuchi 1987) from central galaxy and heating by stellar random motions in central galaxy (Sarazin and White 1987). The numerical method is the spherical symmetric MacCormack scheme.

3. RESULTS OF EVOLUTIONAL CALCULATIONS

The entire cluster gas evolves to nearly steady inflow after time comparable to the isobar cooling time of initial intracluster medium at the centre of cluster of galaxies. We summarize our results at 2×10^{10} yr when cooling flow evolves to nearly steady state, below. In inhomogeneous model with $q_i = 0.5$, mass deposition occurs inside of 95 kpc and this radius expands very slowly. Total amount of deposited gas is 2.7×10^{11} $M_\odot$. Mass deposition rate is 30 $M_\odot$ yr^{-1}. X-ray luminosity from inside of 100 kpc, between 0.5–8 keV band, is 2.5×10^{43} erg s^{-1}. In this case to calculate X-ray luminosity we assume the emissivity due to cooling condensations dropping out of cooling flows same as White & Sarazin (1987b). In homogeneous model, mass deposition occurs only inside 0.8 kpc and mass deposition rate is 36 $M_\odot$ yr^{-1}. Total amount of deposited gas is 1.7×10^{11} $M_\odot$. X-ray luminosity is 3.27×10^{43} erg s^{-1}. In both models, until flow evolves to nearly steady state, mass accretion rate to central galaxy is larger than mass deposition rate and gas is stored as hot halo around central galaxy. Fig. 1 shows the flow structures of our results when cooling flow evolves to nearly steady state. The density distribution of inhomogeneous model with $q_i = 0.5$ is flatter than the results of homogeneous model. Temperature gradient is positive in all models and $q_i = 0.5$ model's result is smaller than the result of homogeneous model. These tendencies of flow structures of cluster gas inside of cooling radius are very similar to the steady solution of White and Sarazin (1987c). In each model, velocity is highly subsonic and has negative sign in most of the region. Especially in $q_i = 0.5$ model, in entire region mach number is less than 1. Because of our spatial resolution at the centre is 200 pc, if flow becomes transonic inside of 200 pc, we would not resolve the transonic region. Then we calculated evolution of cooling flow of $q_i = 0.5$ model using high resolution mesh system that spatial resolution at the centre is 20 pc, transonic region appeared inside 200 pc, but flow structure and properties of evolution did not change from previous calculation.

4. DISCUSSION

Fig. 2 shows the evolution of X-ray surface brightness in 0.5–8 keV band. Dotted circles are the observational results for M87 (Fabricant and Gorenstein 1983). The X-ray surface brightness of homogeneous model is centrally too peaked at the centre in the steady state. In the evolutional phase, X-ray emission from inner part of cooling flow is mainly originated from emission of stellar ejecta. If the spatial distribution of metallicity in cooling flow could be observed in detail, the possibility that cooling flow is unsteady could be checked. In inhomogeneous model, X-ray surface brightness becomes flatter distribution than homogeneous model and central peak is suppressed. There is enhancement in X-ray surface brightness, in the mass deposition occurring region which is inside of cooling radius. In both models, diagnostics of X-ray surface brightness do not agree with observation in any evolutional stages (Fig. 2). It is necessary to reexamine the standard cooling flow model.

ACKNOWLEDGEMENTS

We are very grateful to Dr. C.L. Sarazin for his valuable discussions and constructive comments. We also thank Professor S. Sakashita and Professor S. Ikeuchi for their continuous encouragement.

REFERENCES
Chevalier, R.A. 1987, Ap.J., 318, 66.
Fabricant, D. and Gorenstein, P. 1983, A;.J., 267, 535.
Gisler, G.R. 1976, Ast. Ap., 51, 137.
Hattori, M., Habe, A. and Ikeuchi, S. 1987, Prog. Theor. Phys., 78, 1099.
Innanen, K.A. 1973, Ap. Space Sci., 22, 393.
King, R.I. 1972, Ap. J., 174, L123.
Macdonald, J., and Bailey, M.E. 1981, M.N.R.A.S., 197, 995.
Matthews, W.G., and Bregman, J.N. 1978, Ap. J., 224, 308.
Nulsen, P.E.J. 1986, M.N.R.A.S., 221, 377.
Raymond, J.C., Cox, D.P., and Smith, B.W. 1976, Ap. J., 204, 290.
Sarazin, C.L. 1986, Rev. Modern Phys., 58, 1.
Sarazin, C.L., and White, R.E., III. 1987, Ap. J., 320, 32.
Sargent, W.L.W., Young, P.J., Boksenberg, A., Shortridge, K., Lynds, C.R., and Hartwick, F.D.A. 1978, Ap. J., 221, 731.
Stewart, G.C., Canizares, C.R., Fabian, A.C., and Nulsen, P.E.J. 1984, Ap. J., 278, 536.
Thomas, P.A., Fabian, A.C., and Nulsen, P.E.J. 1987, M.N.R.A.S., 228, 973.
Tammann, G.A. 1974, in Supernovae: A Survey of Current Research, ed. Comovici, C.B., Reidel, S., pp 371.
White, R.E., III, and Sarazin, C.L. 1987a, Ap. J., 318, 612.
White, R.E., III, and Sarazin, C.L. 1987b, Ap. J., 318, 621.
White, R.E., III, and Sarazin, C.L. 1987c, Ap. J., 318, 629.

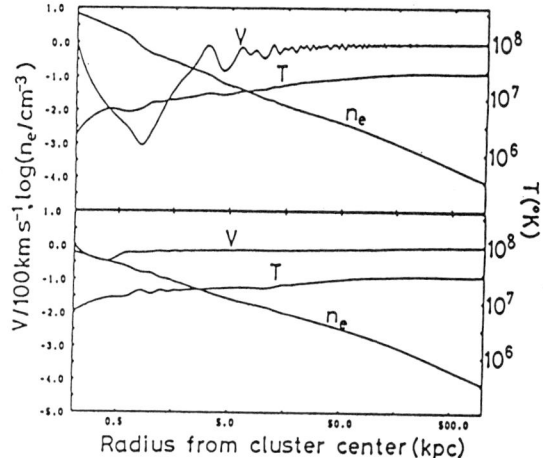

Fig. 1.—The distributions of temperature T, electron density n and velocity v of our results when cooling flow evolve to nearly steady state. Upper panel is the result of homogeneous cooling flow model at 2.03×10^{10} yr. Lower panel is the result of inhomogeneous cooling flow model with $q_i = 0.5$ at 2.1×10^{10} yr.

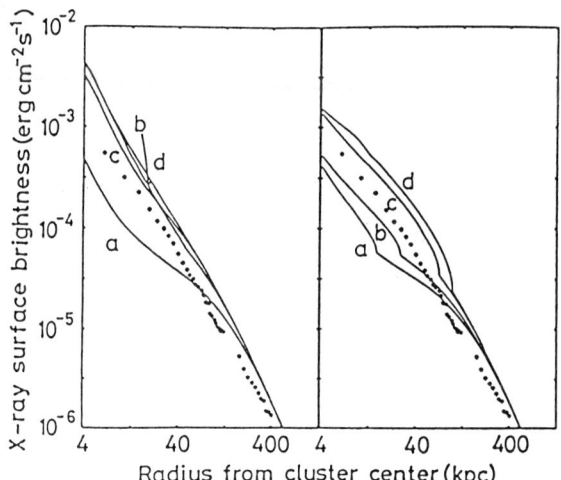

Fig. 2.—The evolution of X-ray surface brightness distributions of our results. Left panel is the results of homogeneous model and a, b c and d are results at 9.1×10^9 yr, 1.78×10^{10} yr, 2.03×10^{10} yr and 2.1×10^{10} yr, respectively. Right panel is the results of inhomogeneous model and a, b, c and d are results at 7.0×10^9 yr, 1.07×10^{10} yr, 1.67×10^{10} yr and 2.1×10^{10} yr, respectively. Dotted-circles are the observational results for M87 (Fabricant and Gorenstein 1983).

A COOLING FLOW CLUSTER AT REDSHIFT Z=0.2

Anna Wolter[1], I.M. Gioia[1,2], T. Maccacaro[1,2], S.L. Morris[3], R. Nesci[4], G.C. Perola[4], and R. Schild[1].

1) Harvard/Smithsonian Center for Astrophysics, Cambridge, MA
2) Istituto di Radioastronomia del CNR, Bologna, Italy
3) Mount Wilson and Las Campanas Observatories, Pasadena, CA
4) Istituto Astronomico dell'Universita', Roma, Italy

Abstract

In this paper we discuss the cluster of galaxies 1E0839.9+2938, serendipitously discovered at x-ray wavelengths. The collected data imply that this is an example of cooling flow cluster.

Introduction

Most of the candidates for the study of cooling flows in clusters of galaxies are optically selected clusters: they are typically at low redshifts ($z < 0.1$). To define the onset of the cooling flow phenomenon, it is necessary to include more distant clusters. We show that the x-ray selection is a suitable tool to identify candidate cooling flow clusters at higher redshifts. In this paper we present a serendipitous x-ray source, 1E0839+2938, identified with a cluster at redshift $z=0.195$. This cluster is a bright x-ray emitter (luminosity: 3.9×10^{44} erg s^{-1}; $H_0=50$ km s^{-1} and $q_0=0$ are adopted throughout the paper.). Its central galaxy is a radio emitter (1.1×10^{24} WHz^{-1}), and presents strong optical emission lines. Furthermore, the emission line region of the galaxy (at Hα and [OII]) is not confined to the nucleus, but is extended. While this is only a summary of the major results, a more detailed presentation of the data and interpretation of the results is given in Nesci et al., 1988.

The data

The Extended Medium Sensitivity Survey (EMSS, Gioia et al., 1988) is a complete sample of serendipitous x-ray sources, and provides a large sample of high-redshift clusters (up to $z\sim0.5$). One of these sources is found at $08^h39^m53.9^s$ $+29^\circ38'48''$. The source is close to the window supporting structure of the IPC (see Fig.1), and it has a relatively small number of net counts (~100), so an analysis of the x-ray surface brightness distribution is not feasible.

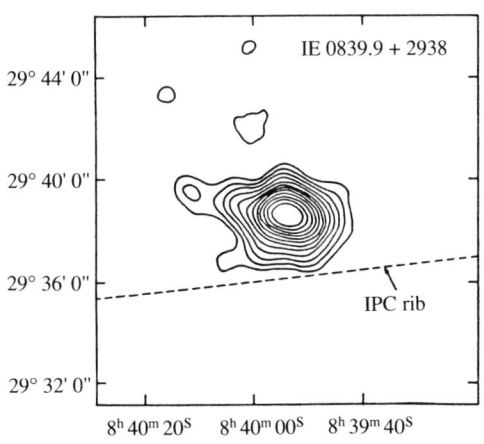

Fig. 1: X-ray contours for 1E 0839+2938. The first contour corresponds to the 3σ level.

The 0.3-3.5 keV flux is $(2.23 \pm 0.24) \times 10^{-12}$ erg cm^{-2} s^{-1}, computed assuming a Raymond-Smith thermal bremsstrahlung model with a temperature of 6 keV, 50% of solar abundance and a hydrogen column density corresponding to the galactic absorption in the direction of the source ($N_H = 4.2 \times 10^{20}$ cm^{-2}; Stark et al., 1989).

A radio source (flux S=5.6 mJy) has been detected with the VLA (6 cm, C configuration) at the optical position of the brightest galaxy in the cluster. At the source redshift this flux corresponds to a luminosity of 1.1×10^{24} WHz^{-1}, a typical value for compact radio sources in cooling flow clusters (O'Dea and Baum 1986; Jones and Forman 1984).

Fig. 2: A CCD image of the field of 1E0839.9+2938, taken with the 24" telescope at the Whipple Observatory.

Photometry of the cluster was performed with the RCA CCD camera at the Whipple Observatory 24" telescope; a frame is is shown in Fig. 2. Identification spectra and follow-ups for the brightest galaxy, G1, and other cluster members were taken at the Multiple Mirror Telescope with the FOGS (0.42"/pixel along the slit, 4-5 Å/pixel resolution). Absorption features typical for an elliptical galaxy are observed in G1, as well as Hα, [OI], [OIII], and [OII] in emission. A redshift of 0.193±0.001 is derived. From these long-slit spectra a spatial extent was suggested, but no firm results were obtained because of poor seeing during the observations. Further observations were made with the University of Hawaii 88" telescope at Mauna Kea (0.58"/pixel along the slit; 26 Å resolution) with two slit orientations (P.A. 0° and 90°). A spectrum is shown in Fig. 3.

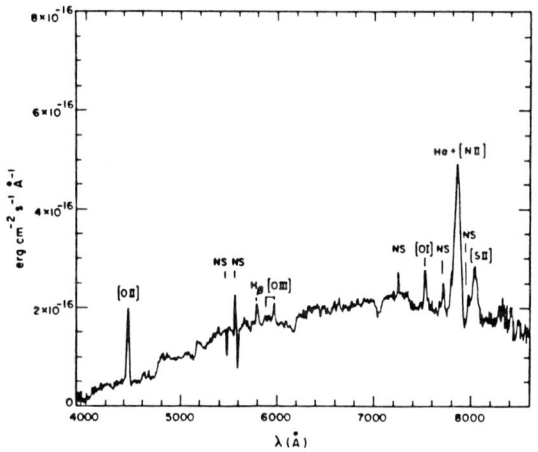

Fig. 3: Optical spectrum, flux calibrated, of galaxy G1 obtained at the University of Hawaii 88" telescope (P.A.=0°) plotted against observed wavelength.

In Table 1, equivalent widths from the MMT and Hawaii spectra for the stronger emission lines are compared with an average value for cD galaxies in cooling flow x-ray clusters (Hu et al., 1985, Johnstone et al., 1987, Kent and Sargent, 1978). The line ratios observed in galaxy G1 are within the observed range for cD galaxies with cooling flows. The net line profiles for the strongest lines (Hα+[NII] and [OII]) are shown in Fig. 4.

Table 1: Equivalent widths, line ratios relative to Hα and typical values for cooling flows.

Emission line	E.W.(MMT) Å	E.W.(UH) Å	line ratio	cooling flows average ratio
(1)	(2)	(3)	(4)	(5)
[OII]3727/29	100.	93	0.53	1.5 (±1.0)
Hβ	9.	8.	0.14	0.31(±0.12)
[OIII]5007	9.	7.	0.13	0.26(±0.12)
[OI]6300	15.	10.	0.24	0.33(±0.18)
Hα	46. ⎫		1.00	1.00
[NII]6583	71. ⎭	114.	1.54	1.43(±0.37)
[SII]6717/30	24.	20.	0.47	1.02(±0.37)

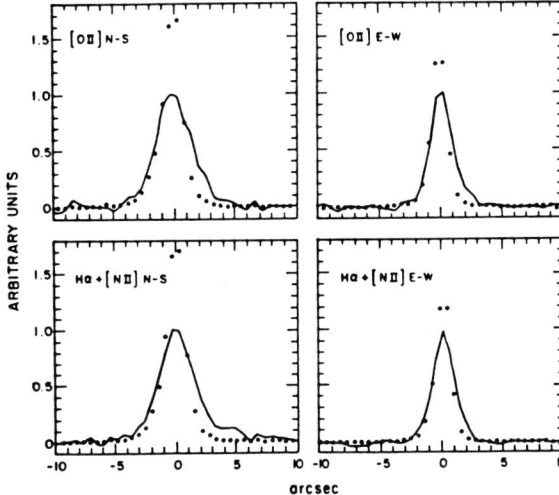

Fig. 4: Spatial profile at Hα+[NII] and [OII]3727 at two position angles (0° and 90°). The solid line is the galaxy profile, and the dotted line is a stellar profile for comparison.

The net line profile is obtained by interpolating the galaxy continuum profile at wavelengths redward and blueward of the emission line. The extent of the emission region (at least in the N-S direction) corresponds, at the cluster redshift, to 13 kpc, comparable to the size of A1795 (FWHM~12 kpc; Cowie et al. 1983).

Some relevant parameters for the dominant galaxy are summarized in Table 2.

Table 2: Relevant data for galaxy G1

Total V magnitude	16.24
V-R	0.82
angle of major axis (from North to East)	75°
eccentricity	0.54
Absolute V magnitude	-24.65
X-ray luminosity (0.3-3.5 keV)	3.9×10^{44} ergs^{-1}
Radio luminosity (6 cm)	1.1×10^{24} WHz^{-1}

Conclusions

Based on the presence of a dominant galaxy at the peak of the x-ray emission, the presence of strong emission lines in the optical spectrum of the dominant galaxy, the spatial extent and asymmetry of the line emitting region, and the detection of G1 at 6 cm, we suggest that 1E0839+2938 is a cooling flow cluster, the second most distant known after 3C295 (z=0.461, see Henry at al., 1986). Due to the small number of net counts, the x-ray data do not allow us to determine the x-ray luminosity distribution, and thus the cooling time of the intracluster gas. This study will require future x-ray observations, which will be possible with ROSAT and AXAF. A more detailed study of the spatial extent of the optically emitting region will be feasible with the Hubble Space Telescope.

Acknowledgments

The observations at the UH 88" telescope were obtained in collaboration with Dr. J.P. Henry. This work has received partial financial support from the Smithsonian Scholarly Studies Grant SS88-3-87, and NASA contract NAS8-30751. G.C.P. and R.N. acknowledge financial support from the Italian MPI and CNR/GIFCO.

References

Cowie, L.L., Hu, E.M., Jenkins, E.B., and York, D.G., 1983, Ap.J., 272, 29.
Gioia, I.M., Maccacaro, T., Morris, S.L., Schild, R.E., Stocke, J.T., and Wolter, A., 1988, in "Large scale surveys of the sky", J.J. Condon and J.F. Lockmann eds., in press.
Henry, J. P., and Henricksen, M. J., 1986, Ap. J., 301, 689.
Hu, E.H., Cowie, L.L., Zhong Wang, 1985, Ap.J. Suppl., 59, 447.
Johnstone, R.M., Fabian, A.C., Nulsen, P.E.J., 1987, MNRAS 224, 75.
Jones, C., and Forman, W., 1984, Ap.J. 276, 38
Kent, S.M., Sargent, W.L.W., 1979, Ap.J. 230, 667.
Nesci, R., Gioia, I.M., Maccacaro, T., Morris, S.L., Perola, G.C., Schild, R.E., and Wolter, A., 1988, Ap.J., submitted.
O'Dea, C.P., and Baum, S.A., 1986 in Radio Continuum processes in clusters of galaxies, NRAO Workshop n.16, O'Dea and Uson Edts, pag. 141.
Stark, A.A., 1989, in preparation.

7. Active Galactic Nuclei, Cosmic X-ray Background

HIGH RESOLUTION X-RAY SPECTROSCOPY OF ACTIVE GALACTIC NUCLEI

Julian H. Krolik

Johns Hopkins University, Baltimore MD, USA

ABSTRACT. High-resolution X-ray spectroscopy has the potential to reveal a number of interesting features of active galactic nuclei, primarily, though not exclusively, through the measurement of absorption lines. After a brief review of the principal problems of AGN research, selected potential high-resolution observations are discussed with a view toward assessing their scientific value and the degree of resolution they will require. Two classes of observations pertaining directly to AGNs are discussed: Fe Kα spectroscopy relevant to the dynamical and thermal character of the emission line zones; and measurement of resonance line absorption by highly-ionized species in BL Lac objects, which should tell us about entrainment of interstellar material by relativistic jets. A third class of potentially important observations uses AGNs as background light sources in order to directly measure the distance to clusters of galaxies.

1. OVERVIEW OF ACTIVE GALACTIC NUCLEI

1. 1. Salient Observational Features

The single most striking attribute of AGNs is the extraordinary luminosity they are capable of generating. The current record is 10^{48} erg s^{-1}(Kühr, *et al.* 1983), *i.e.*, $\sim 10^5 \times$ the luminosity of a typical galaxy. Although in many objects there is a local peak in the luminosity per logarithmic bandwidth near 2.5×10^{15} Hz, this peak generally accounts for only $\simeq 20\%$ of the total luminosity, the rest being spread almost evenly in $\log \nu$ from $\sim 10^{11}$ to $\sim 10^{20}$ Hz (*e.g.*, Ward, *et al.* 1987). Moreover, substantial variations in this large power output are seen, sometimes over periods as short as hours or days, and often over longer timescales. The combination of extremely broad-band emission and very large power produced in a very small volume makes us quite confident that something other than an assembly of stars is responsible for this phenomenon.

In addition to these attention-grabbers, two other properties attract our interest: their strong cosmological evolution, and the presence of very high velocity gas at relatively low temperature. In the last year or so, enough survey data has been accumulated that we now have a reasonably clear picture of the evolution of the brighter AGNs back to $z \simeq 2.5$ (Boyle, et al. 1987; Koo and Kron 1988): the most luminous AGNs were many orders of magnitude more common (in co-moving density) then than now. Somehow the adolescence of galaxies was especially conducive to nuclear activity.

The most prominent features of the optical and UV spectra of AGNs are the strong emission lines: these frequently double the specific flux, and extend over as much as 10,000 km s^{-1}. Studies of their relative strengths have shown that the temperature of the emitting gas is in the range $1 - 2 \times 10^4$ K, so that the Mach number of the gas's motion is $\sim 10^3$! How such quantities of cool gas are accelerated to such high speeds is still quite mysterious (*e.g.* as reviewed by Mathews and Capriotti 1985).

1. 2. Consensus Theoretical Cartoon

Some twenty-five years of work has led to a consensus view of AGN structure which I caricature in Fig. 1. At the center lies a massive black hole ($M \sim 10^8 (L_E/L) L_{46} M_\odot$, where the luminosity is referred both to the Eddington luminosity and to the convenient unit 10^{46} erg s^{-1}) into which material drops at the rate $\dot{M} \sim 1 L_{46}$ M$_\odot$ yr^{-1} if the efficiency of transforming fuel into radiated power is ~ 0.1, as might be expected from material accreting with non-zero angular momentum. The extent of the accretion disk is very hard to determine, which is why its shading in the figure is shown trailing off with increasing radius. Photoionization modelling (Davidson and Netzer 1979; Kwan and Krolik 1981; Ferland and Mushotzky 1982) argues that the emission lines with the largest velocities are made $\sim 1 L_{46}^{1/2}$ pc from the source of the ionizing continuum (this is the "broad line region") while the lines of smaller width ($\sim 300 - 500$ km s^{-1}) are made in a region $\sim 100 - 1000\times$ farther away (the "narrow line region"); variability monitoring of one object, NGC 4151, has recently (weakly) confirmed this distance estimate for the high velocity gas (Clavel, et al. 1987; Edelson and Krolik 1988).

The only change to this picture made in recent years is the discovery of the "intermediate zone". In a *tour de force* of spectropolarimetry, Antonucci and Miller (1985) demonstrated that in the archetype (NGC 1068) of a particular sub-class of AGN (type 2 Seyfert galaxies) an extremely optically thick torus blocks our line of sight to the central continuum source and the high-velocity emission line gas, so that the only way we can detect the true nucleus is through the polarization imparted to the light reflected off warm plasma which fills the hole of the torus and extends some ways above its top. More recent work reported by Miller (1988) suggests that this structure is indeed generic.

Given these properties and supposed structure, several questions immediately demand answers. First, is the energy source truly accretion, and if so, how does the

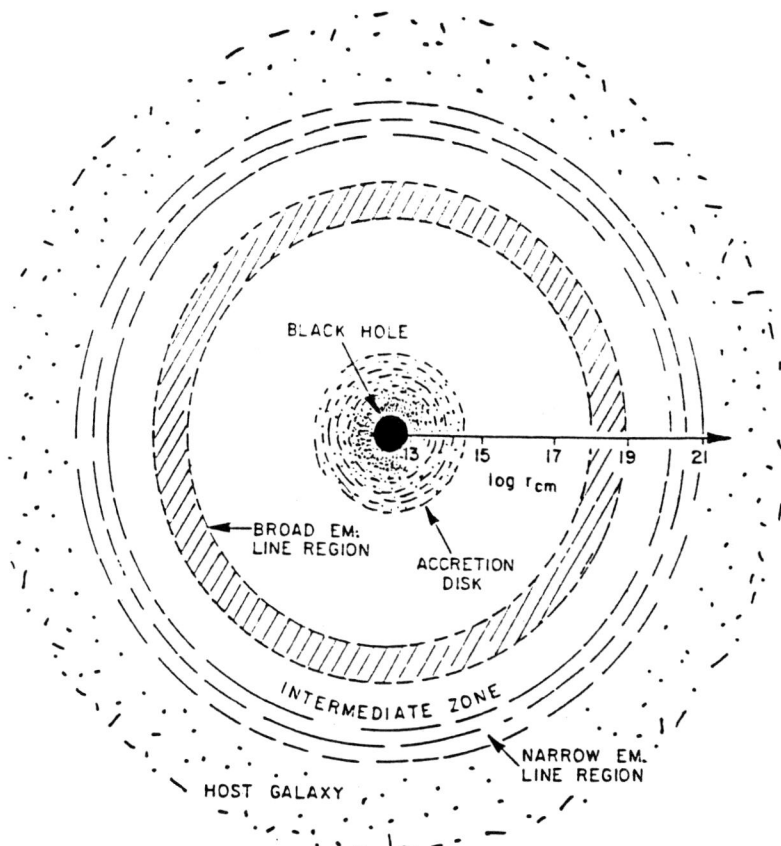

Figure 1. A sketch of the shell structure of AGN as it is envisaged by most workers in the field.

material travel to the black hole? Although 1 $M_\odot$ yr^{-1} is not a huge amount of mass loss for all the stars in a galaxy to toss off, 100 $M_\odot$ yr^{-1} is much harder to envisage, and in practise one might expect much greater requirements when all the likely inefficiencies are taken into account. The most important of the likely inefficiencies is, of course, the difficulty of removing all but 10^{-5} of the gas's angular momentum. We would very much like to understand how that happens. A prerequisite of that inquiry is, of course, to infer the physical conditions in which the accretion flow takes place.

Second, how are the photons generated? The simplest plausible picture of an accretion flow is that of an optically thick, geometrically thin disk (Lynden-Bell and Pringle 1974; Shakura and Sunyaev 1973) in which the state of thermodynamic equilibrium obviates any necessity to isolate the specific mechanisms of photon production. In fact, the notorious "blue bump" found in so many AGNs lends support to this picture, for its characteristic energy (~ 10 eV) is right in line with what might be expected *a priori* (Sun and Malkan 1988; Wandel and Petrosian 1988).

Unfortunately, fits to the spectra of some objects suggest physical inconsistencies (most notably, sometimes L/L_E becomes large enough to make the thin disk approximation inapplicable), and the majority of the light, as remarked above, is *not* contained in the blue bump, and appears to be radiated by some genuinely non-thermal mechanism. Our usual repertory of non-thermal mechanisms (synchrotron radiation, inverse Compton scattering) generally requires relativistic electrons; what is their source?

Third, just what is the connection to young galaxies? The growing evidence for a correlation (albeit imperfect) between unusually high rates of star formation and nuclear activity in galaxies of the present epoch suggests that young may mean both chronologically and evolutionarily. For example, it may be that having large quantities of gas is the important element of "youth", or it may also be that recent dynamical disturbance is the important thing.

Fourth, what is the origin of variability? When the variability seems especially violent and rapid (those AGNs exhibiting this sort of behavior are often labelled "blazars"), relativistic jets directed toward us are often invoked. Is there any way to study these jets directly?

Finally, though it is certainly not the last question one could ask, what are the forces which accelerate the cool gas to such high speeds? Radiation pressure, gravity, and drag against an external fluid medium have all been suggested, but none has clearly won the day.

By citing just a few examples in the next several pages, I hope to demonstrate that high-resolution X-ray spectroscopy has something to contribute towards answering all these questions.

2. NARROW FEATURES INTRINSIC TO X-RAY PRODUCTION

The strong suggestion that most of the continuum (outside the UV) is generated non-thermally also implies that—at least in the simplest picture—there is probably little fine-scale structure intrinsic to the X-ray continuum. If relativistic electrons (and maybe positrons) really are responsible for generating the X-ray photons, the only feature one might expect would be a very broad one associated with $m_e c^2 = 511$ keV. Even if narrow features are somehow present in the spectrum emitted from individual small regions, the large velocities associated with the neighborhood of a black hole are likely to spread out these featuresd over a range $\delta\nu/\nu \sim 0.1$ in the composite spectrum. Thus, if any narrow X-ray features exist, the simplest theoretical prediction is that they are likely to be created in regions of smaller velocity farther away from the center of activity, and their interpretation will provide relatively little guidance in understanding X-ray production.

However, the arguments of the preceding paragraph are far from water-tight. Some have argued that thermal material should always be present, even in regions of predominantly non-thermal radiation (Guilbert and Rees 1988). If that is the case, absorption by relatively cool material may provide diagnostics of the pressure and velocity in the region of X-ray radiation. In Saha equilibrium, for example, temperatures as high as ~ 100 eV $\sim 10^6$ K still permit some H-like Si and S, while Fe still retains half a dozen electrons. The thermal widths of lines in these conditions are extremely small—fractions of 1 eV—so that the line width is a direct indicator of dynamics in the absorbing region, while the specific ionic stage responsible gives information about the thermal state of the material.

3. Fe Kα FROM EMISSION LINE REGIONS

3. 1. The Broad Line Region

There are several places within the various emission line regions likely to produce narrow Fe Kα features. I will discuss them starting on the inside (the broad line region) and moving out. A fair amount is known about the material which radiates the broad lines: most of its Fe is singly-ionized, and the total column is at least $\sim 10^{23}$ H cm^{-2} on average. We can therefore expect hard X-rays striking it to produce fluorescent Kα photons spread from 6.38 to 6.42 keV by the same motions that broaden the optical and UV lines. The equivalent width of this feature is proportional to the covering factor of this material over the X-ray source; photoionization models which predict optical/UV line emissivities yield values of C for different objects ranging from 0.01 up to nearly unity, and so we might expect in a typical source to see equivalent widths anywhere from 6 to 600 eV. In fact, turning this around, measuring the equivalent width of this cold Fe fluorescent Kα line would be very useful, for it would provide an estimate of C which would be much more nearly model-independent. A resolved line observation would be that much more specific, giving dC/dv_r where v_r is the velocity projected on the line of sight.

Even when $C \sim 1$, the volume filling factor of the line-emitting gas is small. This fact has led many people to imagine that the line-emitting gas is contained within a large number of small clouds, surrounded and confined by a much hotter, much less dense medium. This putative intercloud medium is thought to have a temperature of at least 10^8 K and a column density of $\sim 10^{24} L_{46}^{1/2} T_8^{-1/2}$ H cm^{-2} (Krolik, McKee, and Tarter 1981), but we have no direct indication of its existence, or, if it does exist, of its state of motion. Kallman and Mushotzky (1985) pointed out that if T is not too much greater than 10^8 K, the intercloud medium should produce a P Cygni (or reverse P Cygni) profile in scattered Fe XXVI Kα. For example, if the intercloud medium moves at a speed comparable to the clouds' speeds, there would be a feature of optical depth order unity stretching from $\simeq 6.7$ to $\simeq 7.1$ keV.

Detection of such a feature would teach us a great deal. Even an upper limit would give a constraint on its pressure formed by an upper bound on its density and a lower bound on its temperature. Detection would, of course, confirm the existence of the intercloud medium, and provide a fairly good description of its ionization state and density. But beyond that, we would also learn both the sign and magnitude of its velocity. If its speed matches that of the line-emitting material, then we will instantly understand how the cool gas came to move so fast. If it turns out to be flowing inward, there would be strong grounds for identifying it with the accretion flow, and for concluding that accretion is quite inefficient, for the likely magnitude of the mass flow it carries is $\sim 1000 L_{46} T_8^{-1} (v/10^4 \mathrm{km\ s^{-1}}) M_\odot\ \mathrm{yr}^{-1}$.

3. 2. The Intermediate Zone

The second emission line zone from which an interesting Fe K feature can be expected is the intermediate zone, where the host galaxy and the active nucleus have their dynamical interface. It is reasonable to suppose that the accretion mechanics are controlled by whichever system dominates the local gravity; the boundary between the central black hole and the host galaxy occurs between the broad and the narrow emission line regions over a wide range of L/L_E and L. This domain is also precisely the zone in which the obscuring torus and reflecting plasma described in §1 can be found. In fact, theoretical analysis (Krolik and Begelman 1988) has shown that the dynamics of both the torus and the reflecting plasma probably play an important role in regulating the accretion into the nucleus. Because the torus is probably composed of a large number of dusty molecular clouds which sift inward on a timescale not greatly longer than the orbital period, its very existence depends on having an adequate reservoir of molecular gas farther out in the host galaxy, and on having a sufficient level of dynamical disturbance in that reservoir to bring in the gas necessary to resupply the torus. The considerable geometrical thickness of the torus (Seyfert galaxy statistics require its height to be almost as large as its cylindrical radius) then implies that these clouds have a significant vertical velocity dispersion. Whatever mechanism stirs the torus material hard enough to explain its geometrical thickness is also likely to create a distribution of specific angular momentum broad enough to allow the material with the least angular momentum to be captured by the nucleus. Moreover, unless a fraction of the mass inflow through the torus does contribute to the accretion luminosity, the position of the torus's inner edge is unstable. Krolik and Begelman (1986) showed that the reflecting plasma is created by photoionization evaporation of the inner edge of the torus, so the capture fraction depends on its dynamics.

Fe K spectroscopy may help us understand this regulation of accretion because the reflecting plasma should produce sizable features both in fluorescent Kα and in the K-edge (Krolik and Kallman 1987). From estimates of the fraction of the nuclear light reflected and the total luminosity of the nucleus, it is possible to predict the ionization state of the plasma: Fe XVII – Fe XXIII. The ratio of K-edge to Thomson opacity is nearly independent of physical conditions, so the equivalent width of the fluorescent Kα from this gas, when viewed in the equatorial plane of the torus, is $\simeq 0.5[Fe/H]/[Fe/H]_\odot$ keV, essentially independent of all variables but the Fe abundance. In fact, just such a line has now been detected by *Ginga* (Koyama 1988). With sufficient resolution, this line should break up into half a dozen components with central energies ranging from 6.50 to 6.70 keV, each betraying the abundance of a particular ionization stage, and each offset and broadened by a few eV due to bulk and thermal speeds ~ 100 km s^{-1}. The same galaxies viewed along the toroidal axis would produce a Kα line of roughly ten times smaller equivalent width, and a K-edge of optical depth ~ 0.1.

Spectra which resolved these features would allow a very precise determination of the temperature and density of the reflecting plasma, as well as its velocity. From these quantities, much about the capture of accretion fuel by the central engine could be learned.

4. ABSORPTION LINES IN BLAZARS

We already know of one example of a narrow absorption feature appearing in a blazar: using the OGS on the *Einstein Observatory*, Canizares and Kruper (1984) found an absorption line of optical depth $\simeq 3$ stretching from $\simeq 650$ to $\simeq 750$ eV (in the rest-frame) in the BL Lac object PKS 2155-304. In four other BL Lac objects, there are hints of similar absorption in the coarser-resolution SSS data, but they were too faint to observe with the OGS (Urry, Mushotzky, and Holt 1986). No information exists on "optically violently variable" quasars, which, along with BL Lac objects, comprise the class of blazars. Clearly, greater sensitivity in high resolution spectroscopy is a priority for this class of objects.

The interpretation of this absorption is quite controversial. The observed energy matches OVIII Lα quite nicely if we are only seeing material moving towards us. The mass loss rate and energy associated with such an outflow are dauntingly large if it is spherically symmetric: $\dot{M} \simeq 2700 M_\odot$ yr^{-1} and $L_{kin} \simeq 10^{48}$ erg s^{-1}(Krolik, et al. 1985). On the other hand, beaming might well be expected in BL Lac objects, and we would only see them when the beam is directed towards us. The problem then is in explaining why this material has a maximum velocity $\sim 0.1c$ when the velocity of the beamed material producing the photon luminosity should be very near c in order to successfully explain the extreme variability of these objects. Perhaps the absorption is due to ordinary interstellar matter entrained along the outer edge of the jet? If so, can its study teach us about jet mechanics?

More high resolution measurements of features like these would be very helpful. We need to learn, for a start, whether this really is a general phenomenon. Detection of absorption in other species would also be an important clue, for the mass loss estimates derived by Krolik, et al. (1985) depend on the ionization balance of the gas. It would be particularly interesting, for example, if absorption by other ions were detected, to see whether they are present over the same range of velocities as OVIII. Unfortunately, identification of lines could be complicated if there is evidence for substantial differences in their velocities; Doppler shifts are automatically substantial when the motions are even as close to relativistic as these.

5. ABSORPTION LINES IN GALAXY CLUSTERS

As Sarazin discusses at greater length in this volume, clusters of galaxies in general contain substantial quantities (column density $N \sim 10^{22}$ H cm^{-2}) of hot ($T \sim 10^7 - 10^8$ K) gas. Most elements are totally stripped of electrons in such an environment, but several high-Z species, such as Si, S, Ca, Ar, and Fe, still retain a few. Given the typical column density, resonance transitions in these ions should produce absorption lines of equivalent width $\sim 1 - 20$ eV in the spectrum of any X-ray continuum source, e.g. a background quasar, which happens to be lined up behind the cluster. Depending on the temperature and actual column density through a given cluster, anywhere from two to a dozen lines should have significant equivalent widths. These have considerable intrinsic interest for the study of elemental abundances, temperature structure within the gas, and dynamics of both the gas and galaxies in the cluster, but I wish to focus attention here on how they may be used for another purpose: to measure the angular diameter distance from us to the cluster, and thereby calibrate the Hubble ratio and measure q_o (Krolik and Raymond 1988).

The basic idea is very similar to a decade-old proposal (Gunn 1978; Silk and White 1978; Cavaliere, et al. 1979) to measure the distances of galaxy clusters, but removes the principal stumbling block which has hindered its implementation. Stripped to its bare essentials, the scheme is to obtain a column density nl by measurement of an absorption feature, while measurement of an optically thin emission feature gives an emission measure $n^2 l$. If the line of sight lengthscale l is the same as the transverse lengthscale, the angular diameter distance is:

$$D_A \simeq \frac{(nl)^2}{(n^2 l)\theta},$$

where θ is the angular size of the hot gas in the cluster.

In the old version, the absorption measurement was the reduction in intensity of the microwave background in the direction of the cluster due to the Sunyaev-Zel'dovich effect. Its expected magnitude is $\Delta I/I \sim nl\sigma_T(kT/m_e c^2) \sim 10^{-4}$. The temperature weighting in $\Delta I/I$ can be removed by fitting temperature-dependent emissivity models to the total flux spectrum of the hot gas. The emission measure was to be derived from the X-ray continuum surface brightness. The central problem which has prevented the implementation of this scheme is the difficulty of measuring $\Delta I/I$ (see, e.g., Birkinshaw, Gull, and Hardebeck 1984).

In the new version, the absorption feature is any of the X-ray absorption lines described at the beginning of this section, while the emission measure can be taken from either the continuum surface brightness, or the surface brightness of these same lines as they appear in emission away from the background quasar. To get from the column density in a particular ion to the hydrogen column density, one uses the abundance in that ion defined by the strength of its forbidden or subsidiary emission lines. For each resonance line, one then has two independent distance determinations.

This new method has many advantages over the old version. Provided one has an instrument capable of resolving the lines, they are easy to detect, in contrast to the Sunyaev-Zel'dovich effect whose intrinsic smallness makes it difficult to reliably detect independent of instrument. Furthermore, the large number of lines which should be measurable makes it possible to severely control the error of the distance determination: in a typical case, one is likely to have $\sim O(10)$ *independent* measurements of the distance to a given cluster!

The only drawback to this scheme is that it depends on having an aligned quasar bright enough to measure accurately behind any cluster whose distance is desired. While it is true that even with the sensitivity of AXAF, such a quasar can be found in only a small minority of clusters, there should still be enough clusters with aligned bright quasars—probably something like several dozen—to permit independent measurements of both H_o and q_o.

6. CONCLUSIONS

Although I have presented only a small sample of the sorts of observations which could be done with high-resolution X-ray instrumentation, I hope I have demonstrated their substantial interest. Finding narrow features intrinsic to the X-ray continuum is something of a long-shot, but if they are present, they could be extremely helpful in understanding the physical conditions prevalent in the region of continuum radiation. Fe K spectroscopy, however, is virtually a sure bet, and has ramifications ranging from the dynamics of the broad emission line region to the regulation of the accretion flow and the evolution of nuclear activity. Beyond these matters of intrinsic interest to AGNs, high resolution X-ray spectroscopy applied to active galactic nuclei can also be used to make measurements of the fundamental geometric quantities of the Universe.

ACKNOWLEDGMENTS. I am grateful to Craig Sarazin for a conversation at this meeting which helped clarify the role emission line measurements play in defining the heavy element abundances in intra-cluster gas. This work was partially supported by NASA Grant NAGW-1017.

REFERENCES

Antonucci, R.R.J. and Miller, J.S., 1985, *Ap. J.* **297**, 621.
Birkinshaw, M., Gull, S.F., and Hardebeck, H., 1984, *Nature* **309**, 34.
Boyle, B.J., Fong, R., Shanks, T., and Peterson, B.A., 1987, *M.N.R.A.S.* **227**, 717.
Canizares, C. and Kruper, J., 1984, *Ap. J. Lett.* **278**, L99.
Cavaliere, A., Danese, L., and De Zotti, G., 1979, *Astr. Ap.* **75**, 322.
Clavel, J., al., 1987, *Ap. J.* **321**, 251.
Davidson, K. and Netzer, H., 1979, *Rev. Mod. Phys.* **61**, 715.
Edelson, R.A. and Krolik, J.H., 1988, *Ap. J.* **333**, 646.
Ferland, G. and Mushotzky, R.F., 1982, *Ap. J.* **262**, 564.
Guilbert, P.W. and Rees, M.J., 1988, *M.N.R.A.S.* **233**, 475.
Gunn, J.E., 1978, in *Observational Cosmology*, A. Maeder, L. Martinet, and G. Tammann, eds. (Geneva: Geneva Observatory).
Kallman, T.R. and Mushotzky, R.F., 1985, *Ap. J.* **292**, 49.
Koo, D. and Kron, R., 1988, *Ap. J.* **325**, 92.
Koyama, K., 1989, in *IAU Symposium 134: Active Galactic Nuclei*, D. Osterbrock and J.S. Miller, eds. (Reidel: Dordrecht)
Krolik, J.H. and Begelman, M.C., 1986, *Ap. J. Lett.*, **308**, L55.
Krolik, J.H. and Begelman, M.C., 1988, *Ap. J.* in press.
Krolik, J.H. and Kallman, T.R., 1987, *Ap. J. Lett.* **320**, L5.
Krolik, J.H., Kallman, T.R., Fabian, A.C., and Rees, M.J., 1985, *Ap. J.* **295**, 104.
Krolik, J.H., McKee, C.F., and Tarter, C.B., 1981, *Ap. J.* **249**, 422.
Krolik, J.H. and Raymond, J.C., 1988, *Ap. J. Lett.* in press.
Kühr, H., Liebert, J.W., Strittmater, P.A., Schmidt, O.D., Mackay, C., 1983, *Ap. J. Lett.* **275**, L33.
Kwan, J.Y. and Krolik, J.H., 1981, *Ap. J.* **250**, 478.
Lynden-Bell, D. and Pringle, J., 1974, *M.N.R.A.S.* **168**, 603.
Mathews, W.G. and Capriotti, E.R., 1985, in *Astrophysics of Active Galaxies and Quasi-Stellar Objects*, ed. J.S. Miller (University Science Books: Mill Valley CA)
Miller, J.S., 1988, in *Active Galactic Nuclei*, H.R. Miller and P.J. Wiita, eds. (Reidel: Dordrecht)
Shakura, N.I. and Sunyaev, R.A., 1973, *Astr. Ap.* **24**, 337.
Silk, J. and White, S.D.M., 1978, *Ap. J. Lett.* **226**, L103.
Sun, W.-H. and Malkan, M.A., 1988, preprint
Urry, C.M., Mushotzky, R.F., and Holt, S.S., 1986, *Ap. J.* **305**, 369.
Wandel, A. and Petrosian, V., 1988, *Ap. J. Lett.* **329**, L11.
Ward, M., Elvis, M., Fabbiano, G., Carleton, N., Willner, S., and Lawrence, A., 1987, *Ap. J.* **315**, 74.
Zdziarski, A.A., 1986, *Ap. J.* **305**, 45.

DISCUSSION-J. Krolik

S. Kahn: Isn't OVIII Lα also a good candidate for absorption line studies through clusters? I would have expected that it would have appreciable optical depth.

J. Krolik: At temperatures above 10^7 K O is pretty thoroughly stripped. Probably you are thinking of the very center of the hypothetical cooling flow, where the temperature is low enough to permit some O to recombine. Although the emission measure in this region can be substantial, it doesn't occupy enough solid angle to be a likely target for absorption line studies–except for cases like M87 where an x-ray continuum source sits right in the middle of a cooling flow.

C. Sarazin: I would think that Broad Absorption Line Quasars would be ideal candidates for X-ray absorption lines. Any comments?

J. Krolik: You're absolutely right. Although their K-edge opacity should produce a feature of only small contrast, the species observed in them are highly-enough ionized (Li-like) that Kα lines should be present and quite optically thick. Unfortunately, I don't see that the information we gain by observing them will add qualitatively from what we already have from the UV resonance lines.

COSMIC X-RAY BACKGROUND FROM EARLY ACTIVE GALACTIC NUCLEI

Andrzej A. Zdziarski

Space Telescope Science Institute, Baltimore, MD 21218, USA

A model for the origin of the cosmic X-ray background (hereafter XRB) is presented. The component of the background left after subtraction of the known classes of sources is explained by emission from a population of black hole sources at the redshift of $z \sim 4$–5. The model is presented in more detail elsewhere (Zdziarski 1988). Here, we summarize its most important results.

The cosmic X-ray background after subtraction of the estimated contributions of Seyfert galaxies and quasars yields a residual spectrum in the 3–50 keV range,

$$\frac{dF}{dE} \propto E^{-\alpha_x} \exp(-E/E_0), \qquad (1)$$

with 23 keV $\leq E_0 \leq$ 30 keV and $|\alpha_x| \lesssim 0.2$ (Leiter and Boldt 1982; Setti 1985).

We study here a hypothesis in which the spectrum (1) is due to the contribution from a population of presently unresolved discrete sources (e.g., Leiter and Boldt 1982). The existence of such a population is suggested by Hamilton and Helfand (1987), who found that the quasar spatial density is too small to account for the low amplitude fluctuations in the X-ray counts of the *Einstein* Deep Survey. The spectrum of an individual source contributing to the residual XRB should follow eq. (1) with $E_0 = (23$–$30)(1+z)$ keV in the rest frame of the source.

In the model, the emission of an individual source is due to a turbulent spherical accretion flow (Mészáros 1975). The gravitational energy is used to heat thermal electrons in the flow and to generate magnetic fields in equipartition with the flow energy density. The emitted spectrum, which is self-consistently solved for, is due to Compton upscattering by the thermal electrons of cyclo-synchrotron photons emitted by the same electrons and partially self-absorbed. The X-ray spectrum is found to have an approximately constant spectral index, $\alpha_x \simeq 0$, in the wide range of the dimensionless mass accretion rate of $0.5 \gtrsim \dot{m} \gtrsim 20$. Here $\dot{m} \equiv \dot{M}c^2/L_{\rm E}$, $\dot{M}$ is the mass accretion rate, and $L_{\rm E}$ is the Eddington luminosity. The spectra are cut off above $\sim 2kT \sim$ const ~ 150 keV in the range of $\dot{m}$ specified above. Such spectra closely resemble that of the residual cosmic XRB if $z \sim 4$.

This stability of the spectral parameters with the changing accretion rate is critical for our model. The approximate constancy of $kT \sim 10^2$ keV is explained in part by the thermostatic effect of thermal e^+e^- pair production (e.g., Leiter and Boldt 1982; Svensson 1984). The spectral index of ~ 0 is characteristic of saturated Comptonization. The Comptonization process is saturated because of the relative deficiency of the self-absorbed thermal cyclo-synchrotron photons.

The model requires that the sources no longer radiate the $\alpha_x \sim 0$ spectrum at $z \lesssim 4$. This change of the character of the sources can be associated, for example, with the accretion rate becoming sub-Eddington ($\dot{m} \lesssim 1$), which results in an onset of nonthermal processes (Cavaliere and Morrison 1980), which then steepen the X-ray spectra. In the process of spherical accretion, the physical mass accretion rate is not Eddington-limited, as the super-Eddington part of the photon flux is advected to the black hole (Begelman 1979). If the rate of matter supply, $\dot{M}$, was constant for some initial period, we find that $\dot{m} = t_E/t$, where $t_E \simeq 5 \times 10^8$ yrs is the Eddington time. Thus, $\dot{m}$ becomes ≤ 1 after one Eddington time and the onset of nonthermal processes is expected then. If the sources were formed at $z \gg 1$, $t(z=4) \simeq t_E$ at $\Omega = 1$ and $H_0 = 100$ km s^{-1} Mpc^{-1}. Thus, those cosmological parameters are most consistent with our model.

In a detailed cosmological evolution model of the turbulent accreting sources formed at $z_0 = 6$, we found $\alpha_x = 0.2 \pm 0.1$ for $2.6 \lesssim z \lesssim 5.6$. The observed spectral break energy, E_0, was found to be between 20 and 40 keV for $3.9 \lesssim z \lesssim 5.6$ (see Fig. 4 in Zdziarski 1988). Thus, emission coming from a relatively large range of the redshift could give rise to the residual XRB spectrum of eq. (1).

If the sky angular density of the sources contributing to the XRB is ≥ 5000 deg^{-2} (Hamilton and Helfand 1987), the comoving source density is $\geq$ the estimated density of Seyfert I&II galaxies (Persic et al. 1988). If $H_0 = 100$ km s^{-1} Mpc^{-1}, the accumulated mass at $z = 4$ is $\leq 5 \times 10^7 M_\odot$, and $L_x \leq 5 \times 10^{43}$ erg sec. Such low-luminosity sources would be at the sensitivity limit of AXAF.

REFERENCES

Begelman, M. C. 1979, *M. N. R. A. S.*, **187**, 237.
Cavaliere, A., and Morrison, P. 1980, *Ap. J.*, **238**, L63.
Hamilton, T. T., and Helfand, D. J. 1987, *Ap. J.*, **318**, 93.
Leiter, D., and Boldt, E. 1982, *Ap. J.*, **260**, 1.
Mészáros, P. 1975, *Nature*, **258**, 583.
Persic, M. *et al.* 1988, *Ap. J.*, in press.
Setti, G. 1985, in *Non-Thermal and Very High Temperature Phenomena in X-Ray Astronomy*, ed. G. C. Perola and M. Salvati, (Rome), p. 159.
Svensson, R. 1984, *M. N. R. A. S.*, **209**, 175.
Zdziarski, A. A. 1988, *M. N. R. A. S.*, **233**, 739.

DISCUSSION

Dan Schwartz: I challenge the residual spectrum that you fit in your model. I believe it is an artifact of subtracting an exact power law over many decades.

Andrzej Zdziarski: The exact form of the residual spectrum is inded not known at present (cf. the uncertainties in the parameters α_x and E_0 of eq. [1]). The model of this work predicts spectra with α_x and E_0 adjustable to certain extent by its free parameters.

Jerzy Madej: Could you explain in more detail the way you computed X-ray spectra of black hole sources?

Andrzej Zdziarski: A one zone approximation has been used. The part of the source where photons are not advected into the central black hole is assumed to be a homogeneous cloud, in which the supplied power equals the total luminosity. Comptonization is treated using a diffusion approximation. See Zdziarski (1988) for more detail.

COSMIC X-RAY BACKGROUND FROM SELF-ABSORBED LOW LUMINOSITY AGNs

Jonathan E. Grindlay and Michael Luke
Harvard/Smithsonian Center for Astrophysics
Cambridge, MA U.S.A.

ABSTRACT. A model is proposed for the origin of the diffuse cosmic x-ray background whereby it is primarily due to the contribution of low luminosity active galactic nuclei which are increasingly self-absorbed at low luminosities. Strong self-absorption for low luminosity objects allows the observed background spectrum to be flatter in the ~2-20 keV band than the asymptotic spectrum, assumed to have mean index $\gamma \sim$ -0.7 up to a high energy cutoff at ~125 keV. The model can account for the spectral shape and intensity of the background spectrum, as well as its possible fluctuations, if the AGN undergo modest density evolution and the bulk of the CXB arises from AGNs at redshifts $z \sim$ 1-3. The model can be tested with AXAF observations of low luminosity AGNs at 5-10 keV and sensitive new hard x-ray observations.

1. INTRODUCTION

The problem of the origin of the diffuse cosmic x-ray background (CXB) continues to be one of the major problems of high energy astrophysics. Prompted by the Einstein results for the spectra of AGN and the apparent correlation between x-ray luminosity and low energy absorption (e.g. Lawrence and Elvis 1982), as well as the constraints on the spectrum of the CXB summarized by Boldt (1987), we have developed models for the contribution from low luminosity active galaxies. Although low luminosity AGN have been considered for the CXB (e.g. Giacconi and Zamorani 1987 and references therein), their probable increasingly important self-absorption does not seem to have been considered.

2. DESCRIPTION OF MODEL

The basic assumptions of the model are two: that the AGN x-ray spectra are all described by a power law with a luminosity-dependent low energy cutoff and a fixed high energy cutoff, and that the number density of AGN evolve at some rate with cosmic redshift z. A third assumption implicit in applying an assumed density evolution is that the form of x-ray luminosity function, which increases (in amplitude) with z, is just the locally derived value; here we used the parameterization of the luminosity function given by "model a" of Tucker and Schwartz (1986). We note that this luminosity function extends with a power law slope of 1.75 down to a low luminosity break (flattening) at $L_x \sim 10^{40.5}$ erg/s (cf. also Elvis et al 1984); at luminosities above 10^{43} erg/s, however, it has a steeper slope 3.25 up to an assumed cutoff of 10^{48} erg/s.

The spectral cutoff for low energy absorption (as well as Compton scattering) is assumed to be correlated with x-ray luminosity (scaled to $\sim 10^{43}$ erg/sec) as E_a = const/ L_x^β, where E_a is the cutoff energy related to the absorption and scattering column density by the energy and abundance- dependent relation given in Zombeck (1982), L_x is the luminosity in the Einstein band (0.5-4 keV) and β is a parameter of the model. The high energy cutoff is assumed to be fixed at 125 keV (although several trial values in the range 100–300 keV were investigated), as suggested by Schwartz and Tucker (1988) if the cutoff is due to Comptonization of a pair plasma spectrum. The power law index γ between the low and high energy cutoffs was taken to be -0.7 in accordance with the mean value found for a broad range of AGN luminosities (e.g. Halpern 1982, Turner and Pounds 1989), although in fact a range of spectral index values (0.5 - 0.8) were explored in our models. The form of the assumed spectrum and low energy absorption is shown in Figure 1 for the case β = 0.5 and γ = -0.6 together (cf. inset) with the data presented by Lawrence and Elvis (1982).

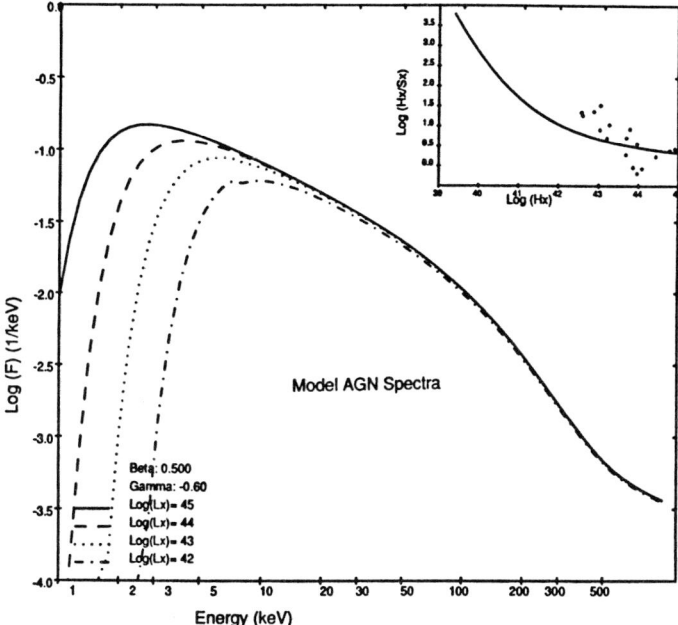

Figure 1: Assumed form of model AGN spectra and cutoffs.

Our key assumption that the low energy cutoff is inversely proportional to the x-ray luminosity is motivated by the results of Lawrence and Elvis (1982) as well as the recent body of work on Seyfert IIs (e.g. Antonucci and Miller 1985, Elvis and Lawrence 1988) showing that they contain heavily obscured AGN sources. It is evident from Figure 1 that with an increasingly large contribution from low luminosity AGN, their integrated spectrum will flatten below $\sim$10-20 keV.

The assumed density evolution is included in the normalizing constant of the x-ray luminosity function $N_{o,z} = N_o \cdot e^{(\alpha \cdot \tau)}$, where α is the second parameter of the model and τ is the 'lookback' time and is $\tau = 0.67(1 - [1+z]^{-3/2})$.

3. RESULTS

Given these assumptions, we fit the residual CXB spectrum of Boldt (1987), which is the spectrum after the contributions from galaxy clusters and bright QSOs are subtracted (which together account for less than 30% of the flux), over the range 3-60 keV. The results are shown in Figure 2. The inset in the Figure shows the percentage contribution (out to an assumed maximum redshift $z = 3.5$) to the background at two observed energies; most of the background comes from low luminosity AGN beyond $z = 1$.

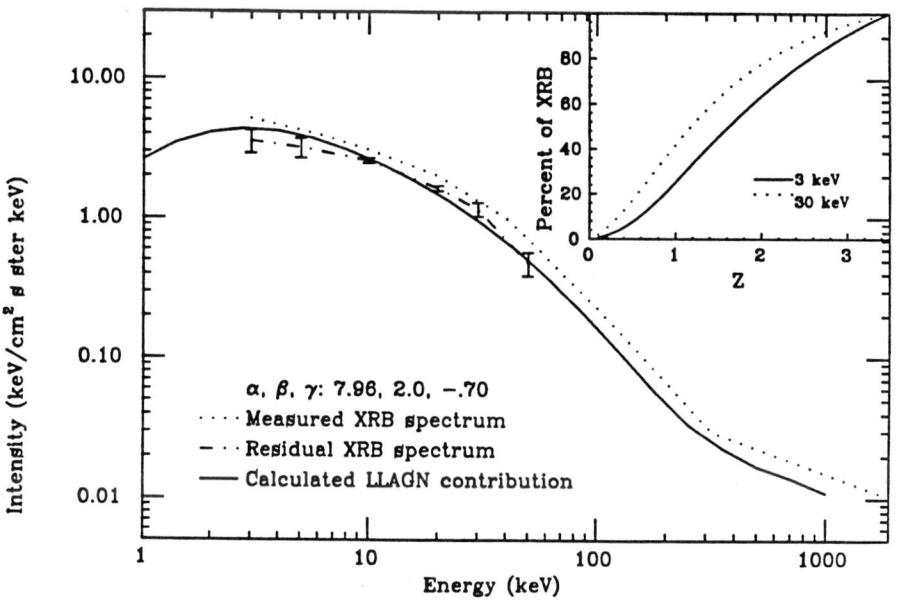

Figure 2: Fit of model to the CXB spectra.

The overall spectral shape and normalization of the CXB spectrum can be fit rather well in the x-ray band (2-100 keV). This model for the CXB can connect to the high energy background spectrum (≥ 100 keV) if we also assume (as in Figure 1) that the intrinsic AGN spectra contain a non-thermal power law component with the same spectral slope (e.g. -0.7) but with only 0.1 the flux and a gamma ray cutoff at ~ 1 MeV. This component could naturally arise from the inverse Compton or SSC models of Band and Grindlay (1986).

The "best-fit" values of the parameters $\alpha \simeq 7\pm1$ and $\beta \simeq 2\pm2$ are such that the required evolution (α) is modest and the required low energy absorption (β) is consistent with individual spectra. In Figure 3 we show the contours derived from χ^2 fits between the models and the CXB spectrum (Boldt 1987) for the allowed ranges of the parameters α and β for three different choices of the spectral index γ.

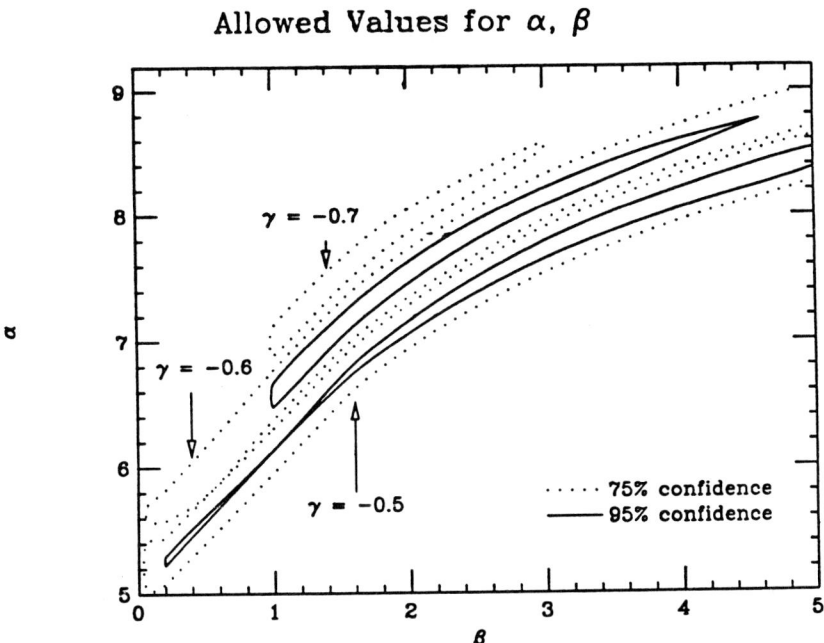

Figure 3: Allowed values for model parameters α and β.

We note that the range of β values indicated would differ for a different choice of scaling (other than the assumed 10^{43} erg/s) in our assumed E_a vs. L_x relation. This could be determined from observations but will require complete samples with good x-ray spectra.

The fact that the CXB is primarily due to low luminosity AGN ($< 10^{43}$ erg/sec) in this model also means that the fluctuation limits (Hamilton and Helfand 1987, Barcons and Fabian 1988) of ~ 3000 sources degree^{-2} are easily satisfied.

4. CONCLUSIONS

The CXB and possibly also diffuse γ-ray background spectrum can be fit satisfactorily by the integrated effects of self-absorbed, low luminosity AGN. A clear cut test of this model can be made by measuring the spectra of low luminosity AGNs with AXAF. The greatly increased sensitivity and spectral coverage up to nearly 10 keV will allow a search for the expected low energy absorption (≤ 10 keV) and power law spectra for a large sample of the lowest luminosity ($\sim 10^{41}$ erg/sec) AGN. The spectral resolution with AXAF will also allow the expected low energy absorption vs. scattering effects to be measured. Another test would be observations of many AGNs out to ≥ 100 keV to search for the expected high energy break (which is redshifted down to the observed ~ 25 keV break). This is a principal objective of coded aperture imaging hard x-ray telescopes, such as EXITE (Grindlay et al 1986), currently under development.

ACKNOWLEDGEMENTS

We thank M. Elvis and D. Schwartz for comments. This work was supported in part by grants NAGW-624 and NAS 8-30751.

REFERENCES

Antonucci, R. and Miller J. 1985, *Ap. J.*, **297**, 621.
Barcons, X. and Fabian, A. 1988, *MNRAS*, *230*, 189.
Band, D. and Grindlay, J. 1986, *Ap. J.*, **308**, 576.
Boldt, E. 1987, *Physics Reports*, **146**, No. 4.
Elvis, M., Soltan, A. and Keel, W. 1984, *Ap.J.*, **283**,479.
Elvis, M. and Lawrence, A. 1988, *Ap. J.*, **331**, 161.
Giacconi, R. and Zamorani, G. 1987, *Ap.J.*, **312**, 503.
Grindlay, J., Garcia, M., Burg, R. and Murray, S. 1986, *Trans. Nucl. Sci.*, **33**, No. 1, 750.
Halpern, J. 1982,*Ph. D. Thesis*, Harvard University.
Hamilton, T. and Helfand, D. 1987, *Ap. J.*, **313**, 20.
Lawrence, A. and Elvis, M. 1982, *Ap. J.*, **256**, 410.
Schwartz, D. and W. Tucker, W. 1988, *Ap. J.*, **332**, 157.
Turner, M. and Pounds, K., 1989, *MNRAS*, in press.
Tucker, W. and Schwartz, D. 1986, *Ap. J.*, **308**, 53.
Zombeck, M. 1982, *Handbook of High Energy Astrophysics*, Cambridge University Press, Cambridge.

8. Future X-ray Observatories

COMMENTS ON THE FUTURE OBSERVATORIES AND THEIR X-RAY SPECTROSCOPY CAPABILITY

J L Culhane

Mullard Space Science Laboratory, Department of Physics and Astronomy, University College London

ABSTRACT. Following a brief discussion of the possibilities offered by X-ray spectroscopy at various values of resolving power (E/ΔE), the spectroscopic capabilities of a number of future missions are presented. The possible roles of Charge Coupled Devices (CCD's), Microcalorimeters, Diffraction Grating and Bragg Crystal Spectrometers are then evaluated.

1. INTRODUCTION

The power of high resolution X-ray spectroscopy has been amply demonstrated for studies of the high temperature plasma found in solar flares and active regions. For sources other than the Sun, the gas filled proportional counter detector with E/ΔE ~ 5 at a photon energy of 7keV was used almost exclusively until the launch of the Einstein mission. This spacecraft included a non-dispersive solid state detector which offered high quantum efficiency and resolution some three times better than that of the proportional counter. In addition transmission grating and Bragg spectrometers, although of much lower throughput, obtained high resolution spectra of a number of important sources. These results have made it quite clear that, following the exploratory phases of the subject, it is now essential to continue high resolution spectroscopic observations.

Several missions, planned for the next decade, will include a spectroscopic capability. In the following sections, after a brief review of the role of high resolution spectroscopy in X-ray Astronomy, a number of these missions will be discussed. They include the moderate missions Spectrum-X, Astro-D and Spectrosat together with the two major observatories AXAF (NASA) and XMM (ESA). After a largely tabular presentation of the spectroscopic performance of these missions the role of non-dispersive (CCD's and microcalorimeters) and dispersive (grating and Bragg) spectrometers will be discussed.

2. X-RAY SPECTROSCOPY - POSSIBILITIES AND EXAMPLES

In this paper we will deal only with instruments having E/ΔE > 25 at photon energies above 5keV. This will permit the discussion of non-dispersive systems such as solid state detectors or CCD's but will exclude a variety of gas-filled devices. This choice has been made given that with future missions, particularly those of observatory class, the subject will have passed the "discovery" phase and will require that spectroscopic techniques offer a range of diagnostic information about the sources being studied. The justification for this approach will become clear below.

2.1 Systems with E/ΔE < 10

Proportional counters and, very narrowly, Gas Scintillation detectors fall in this category. While many discovery observations have been made with these systems (eg Cluster plasma, Mitchell et al, 1976), they offer the ability to do little more than detect the existence of emission features at E > 2keV. Such detections provide evidence for the presence of hot plasma and permit rough estimates of temperature and element abundance. Crude phase-related emission and absorption spectra have been obtained for the brighter X-ray binaries (Pravdo et al., 1979) while in addition to clusters of galaxies, emission features have been discovered in the spectra of AGN (Hayes et al., 1980) and stellar flares (White et al., 1986).

2.2 Systems with $10 < E/\Delta E < 100$

Solid state spectrometers and CCD's, together with the earlier examples of transmission grating and Bragg spectrometers can be included here. Detection and measurement of emission line intensities become possible for elements ranging from O to Ni. Temperature, emission measure and element abundances may be determined though in a model-dependent manner. Moderate resolution phase-related spectra may be obtained for binaries (McCray et al., 1982; Kahn et al., 1984) together with crude diagnostic data for extended sources such as clusters and supernova remnants (Canizares et al., 1979; Becker et al., 1979). Thus both thermal and photoionised plasma can be studied in emission and, more crudely, in absorption for a wide range of objects.

2.3 Systems with $100 < E/\Delta E < 1000$

In this energy range gratings, Bragg spectrometers and, in the future microcalorimeters, can be used to undertake model independent line ratio diagnostics. In addition the ionisation state of gas in the line of sight may be studied through both emisison and absorption spectroscopy. Thus detailed studies of both the interstellar medium and of a wide range of X-ray sources can be undertaken. A spectrum of Puppis A (Figure 1) obtained with the Einstein FPCS by Canizares et al., 1981, provides an excellent example of what can be achieved.

2.4 Systems with $E/\Delta E > 5000$

At present this kind of capability can only be provided by Bragg spectrometers and, so far, instruments with this performance have not been used to study X-ray sources other than the Sun. While all of the observations described in 2.3 could be better undertaken at this level of resolution, the available throughput will necessarily be much less than that of the microcalorimeter. However high resolution Bragg spectroscopy offers the only possiblity to measure line profiles and so to deduce both directed and turbulent plasma velocities at values less than $100 Kms^{-1}$. An example of solar flare data (Antonucci et al, 1986) is given in Figure 2.

3. FUTURE MISSIONS WITH SPECTROSCOPIC CAPABILITY

The missions may be considered under three headings and in each case a tabular summary of performance is presented. The sensitivities have been calculated for an observing time of either 10^5s or a time which would give rise to a figure five times better than the broad band confusion limited sensitivity - whichever is smaller. In all cases the sensitivity is calculated for a background limited (5σ) or photon limited (25 detected counts) situation.

3.1 Moderate Missions (see Table 1)

These include the USSR's Spectrum-X with West European X-ray telescopes, the Japanese ASTRO-D with US supplied CCD detectors and the German Spectrosat.

TABLE 1 FUTURE MISSIONS WITH SPECTROSCOPIC CAPABILITY - MODERATE MISSIONS

MISSION/ INSTRUMENT	ANGULAR RESOLU-TION (arc sec)	FIELD OF VIEW (arc min)	SPECTRAL RESOLVING POWER ($E/\Delta E$)	EFFECTIVE COLLECTING AREA (CM^2)	COMMENTS ON SENSITIVITY
SPECTRUM-X					
JET-X CCD (2 MODULES) (1993)	20"	20'	**8** (.7 keV) **50** (7 keV)	**250** (.7 keV) **130** (7 keV)	a) BROAD BAND SENS: 3.10^{-15} erg cm^{-2} s^{-1} (6.10^4 s) b) MULTILAYER DIFFRACTOR: 1.10^{-5} hv cm^{-2} s^{-1} (10^5 s, 7 keV, $E/\Delta E = 350$) c) MIN DET LINE FLUX: 2.10^{-6} hv cm^{-2} s^{-1} (10^5 s, 7keV)
ASTRO-D					
CCD (2 MODULES) (1993)	120"	30'	**8** (.7keV) **50** (7keV)	**450** (.7keV) **180** (7keV)	a) BROAD BAND SENS: 7.10^{-14} erg cm^{-2} s^{-1} (2.10^3 s) b) MIN DET LINE FLUX: 2.10^{-5} hv cm^{-2} s^{-1} (10^4 s, 7keV)
SPECTROSAT					
LE TRANS-MISSION GRATING (WITH HRC) (1995)	2"	POINT SOURCES	**50** (1keV) **160** (.28keV) **400** (.1keV)	**15** (1keV) **35** (.28keV) **5** (.1keV)	a) SELECTED MIN DET LINE FLUXES: NE IX (.92keV); FE XVII (.73keV); O VII (.56keV); $1.5.10^{-5}$ hv cm^{-2} s^{-1} (10^5 s) ---- SI XII (.28keV); FE XVI (.25keV); FE XVI (.18keV); 7.10^{-6} hv cm^{-2} s^{-1} (10^5 s) ---- b) ALSO LOWER RESOLUTION SYSTEM WITH PSPC, $E/\Delta E \sim 1 - 100$

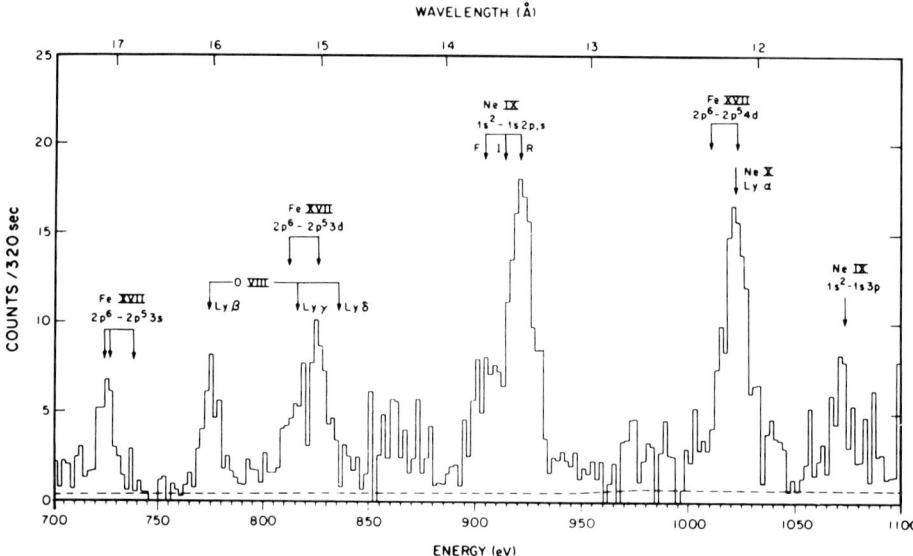

Figure 1. An emission line spectrum of the Puppis-A supernova remnant obtained with the Einstein Focal Plane Crystal Spectrometer

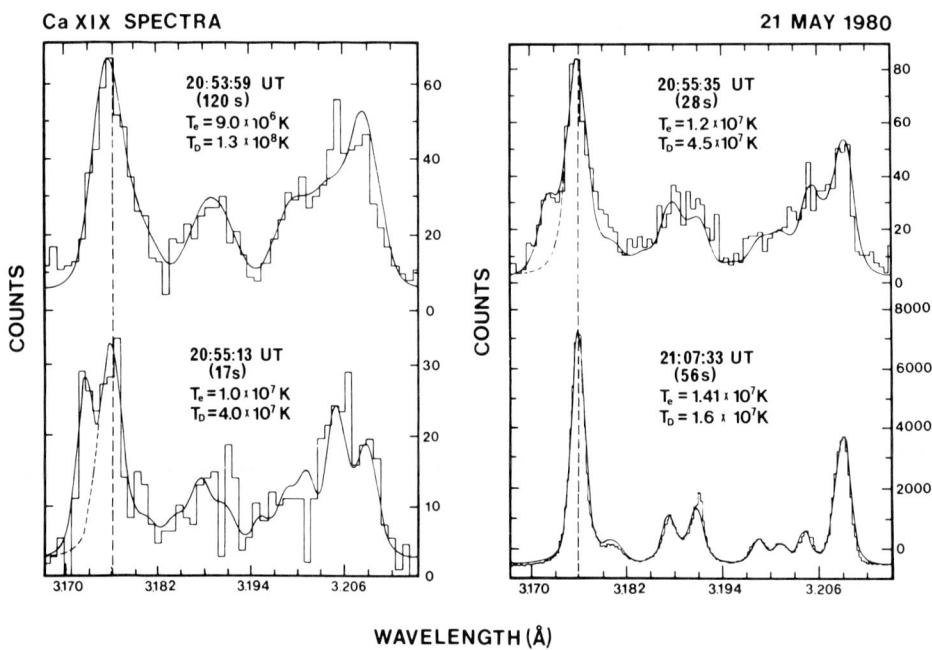

Figure 2. Spectra obtained during the impulsive and cooling phases of a solar flare with the Solar Maximum Mission X-ray Polychromator.

3.2 X-ray Observatories - AXAF (see Table 2)

This major NASA mission is currently scheduled for launch in 1996. It carries a range of powerful spectroscopic instruments into a near earth orbit. Although the single telescope will advance the state of the art in achieving an angular resolution of better than 1``, it will have a relatively modest collecting area of 180cm^2 at 7keV.

3.3 X-ray Observatories - XMM (see Table 3)

This ESA mission, which will include three large aperture X-ray telescopes, is scheduled for launch in 1998. It will be placed in a highly elliptical orbit (apogee - 1000Km; perigee - 70,000Km) of 60° inclination and 24 hour period. With an overall mass of 2400kg, it will be launched on an Ariane IV rocket. Payload electrical power will be 200W out of a total of 875W. The absolute pointing accuracy will be better than 1 arc min while the spacecraft axis will drift at less than 2.5``/min. Attitude reconstruction will be to < 10`` (pitch and yaw) and < 30` (roll). The telescopes will have an angular resolution of better than 30`` (HEW) and a total collecting area of 3000cm^2 at 7keV.

4. CHARGE COUPLED DEVICES (CCD'S)

The use of Lithium drifted Silicon detectors in X-ray spectroscopy led to significant advances when deployed at the focal plane of the Einstein telescope. Since a CCD is effectively an array of such devices which, due to its readout scheme, has the potential for even better spectral resolution than the ordinary Silicon detector, it represents a natural choice for the non-dispersive X-ray spectroscopy role in future missions.

It will be clear from tables 1 - 3 that CCD's can respond over a broad energy range (0.2keV to telescope cut-off) with high quantum efficiency thus offering a high throughput. Given recent developments towards the achievement of noise levels at the 1-2 e$^-$ level RMS (Janesick and Elliott, 1988), the energy resolution figures listed in the tables are somewhat conservative. The small pixels (currently 20μm - 30μm) permit the ultimate angular resolution (0.5``) and sensitivity of the AXAF mission to be realised. Thus moderate resolution spectroscopy may be undertaken for both point and extended sources provided the devices are cooled to a temperature of around 180K.

In order to maintain good quantum efficiency over the whole energy range of the CCD, it is necessary to construct "back-illuminated" devices in high resistivity Silicon. The latter feature is required to permit large (~80μm) depletion depths and thus good high energy stopping power to be achieved. Deep depletion also ensures that background events can be more easily differentiated from X-ray events due to the greater energy deposited in the Silicon. Some further work is required in order to make these two features simultaneously available in flight devices.

The present restrictions on format (~< 10^3 x 10^3 pixels of up to 30μm x 30μm) mean that multiple chip arrays (~< 10 chips) must be employed to achieve adequate coverage of telescope fields of view. In this connection work in progress with fully depleted pn-CCD's (Strüder et al, 1988) offers the prospect of devices with stopping depths of up to 250μm together with pixel sizes of the same order. This latter feature is relevant for telescopes of large aperture but with somewhat poorer angular resolution.

TABLE 2 FUTURE MISSIONS WITH SPECTROSCOPIC CAPABILITY - AXAF

INSTRUMENT	ANGULAR RESOLU-TION (arc sec)	FIELD OF VIEW (arc min)	SPECTRAL RESOLVING POWER (E/ΔE)	EFFECTIVE COLLECTING AREA (CM^2)	COMMENTS ON SENSITIVITY
1) CCD	0.5" (pixel)	12' x 12' 39' x 3'	8 (0.7 keV) 50 (7 keV)	400 (.7 keV) 100 (7 keV)	BROAD BAND SENS: $2 \cdot 10^{-15}$ erg cm^{-2} s^{-1} (10^5 s) MIN DET LINE FLUX: $2 \cdot 10^{-6}$ hv cm^{-2} s^{-1} (10^5 s, 7 keV)
2) MICRO-CALOR-IMETER	~5" - 10"	1' x 1'	50 (.7 keV) 350 (7 keV)	550 (.7 keV) 110 (7 keV)	MIN DET LINE FLUX: $4-20 \cdot 10^{-7}$ hv cm^{-2} s^{-1} in 10^5 s ($20 \cdot 10^{-7}$ hv cm^{-2} s^{-1} in 10^5 s, 7 keV)
3) BRAGG SPECTRO-METER	20" (1-D)	3' (1-D)	200 - 2000 (0.5-8 keV) 50 - 70 (<0.5 keV)	fA ~ 1 - 50 (0.5 - 8 keV) fA ~ 4-40 (<0.5 keV)	MIN DET LINE FLUX: $5-50 \cdot 10^{-6}$ hv cm^{-2} s^{-1} in 10^5 s ($50 \cdot 10^{-6}$ hv cm^{-2} s^{-1} in 10^5 s, 7 keV)
4) LE TRANS-MISSION GRATING (WITH HRC)	< 1"	POINT SOURCES	80 - 1200	10 - 40 (0.1 - 2.5 keV)	MIN DET LINE FLUX: $4-20 \cdot 10^{-6}$ hv cm^{-2} s^{-1} in 10^5 s ($8 \cdot 10^{-6}$ hv cm^{-2} s^{-1} in 10^5 s, 0.56 keV)
5) HE TRANS-MISSION GRATINGS (WITH CCD)					
A) METG	< 1"	POINT SOURCES	100 - 1000 (0.4 - 4 keV)	10 - 200 (0.4 - 4 keV)	MIN DET LINE FLUX: $10-20 \cdot 10^{-6}$ hv cm^{-2} s^{-1} in 10^5 s
B) HETG	< 1"	POINT SOURCES	100 - 1000 (1.2 - 7 keV)	15 - 50 (1.2 - 7 keV)	MIN DET LINE FLUX: $4-10 \cdot 10^{-6}$ hv cm^{-2} s^{-1} in 10^5 s

TABLE 3 FUTURE MISSIONS WITH SPECTROSCOPIC CAPABILITY - XMM

INSTRUMENT	ANGULAR RESOLU-TION (arc sec)	FIELD OF VIEW (arc min)	SPECTRAL RESOLVING POWER (E/ΔE)	EFFECTIVE COLLECTING AREA (CM^2)	COMMENTS ON SENSITIVITY
1) CCD	< 30"	30'	8 (.7 keV) 50 (7 keV)	2500 (.7 keV) 1200 (7 keV)	BROAD BAND SENS: 2.10^{-15} erg cm^{-2} s^{-1} (2.10^4 s) MIN DET LINE FLUX: 2.10^{-7} hv cm^{-2} s^{-1} (10^5 s, 7 keV)
2) REFLECTION GRATINGS	< 30"	POINT SOURCES	400 (.5 keV) 300 (1.5 keV)	150 (.5 keV) 440 (1.5 keV)	MIN DET LINE FLUX: $3\text{-}6.10^{-7}$ hv cm^{-2} s^{-1} (10^5 s) (5.10^{-7} hv cm^{-2} s^{-1} in 10^5 s, 0.56 keV)
3) OBJECTIVE BRAGG SPECTROMETER	< 30"	POINT SOURCES	1000 (7keV)	40 (7keV)	MIN DET LINE FLUX: 3.10^{-5} hv cm^{-2} s^{-1} (10^5 s, 7keV)

	ANGULAR FIELD RESOLUTION OF VIEW		LIMITING MAGNITUDE	WAVELENGTH RANGE	$\lambda/\Delta\lambda$
4) OPTICAL MONITOR	1"	8'	24.5 (10^3 s, B)	1800-6000Å	50-100 (GRISMS FOR 3000-6000Å RANGE) ALSO BROAD-BAND FILTERS

In considering the missions listed in table 1 - 3, all of them employ CCD's with the exception of the German Spectrosat. The two missions for launch in 1993 (Astro-D and Spectrum-X, table 1) have better effective apertures at 7keV than AXAF but inferior angular resolution. Nevertheless they will play a substantial role in providing an immediate spectroscopic follow-up to the ROSAT mission.

Of the two major observatory class missions, AXAF offers the better angular resolution (<1``) but a modest high energy collecting area (180cm$^2$ at 7keV). On the other hand XMM can achieve 3000cm$^2$ at 7keV but with an angular resolution of < 30``. In addition XMM offers the possibility of long continuous observation of variable sources given its highly eccentric orbit. Hence the two major missions provide highly complementary qualities in their use of focal plane CCD's.

5. THE MICROCALORIMETER

This device consists of an X-ray absorber of heat capacity C connected to a heat sink at temperature T by a link of conductivity G. If an X-ray photon of energy E is absorbed, the temperature of the absorber rises by $\Delta T = E/C$ while the subsequent decay to the equilibrium temeprature T takes place in a time $\propto C/G$. A detailed description of the version of this device being developed for AXAF is given elsewhere in these proceedings by Holt and co-workers (see also McCammon et al., 1987).

In an ideal situation, where the temperature transducer contributes only its Johnson noise, the energy resolution scales as $(kT^2C)^{1/2}$. In a practical device, the GSFC group has achieved an energy resolution of 17.3eV (FWHM) for 5.9keV photons and figures below 10eV are anticipated. Thus the microcalorimeter can combine the photon stopping power of a non-dispersive Silicon detector with the resolution of a dispersive spectrometer (see table 2). Response is also provided over a broad energy range and a useful though moderate resolving power of 50 to 70 is still available at 0.7keV.

The requirement to operate the device at T < 100mK in order to achieve the resolving powers mentioned above presents a difficult though by no means insoluble problem in the space environment. However the need to minimise heat capacity leads to a detector size of around 250μm x 250μm and so an array of such devices would be required to provide sufficient coverage of the telescope field of view. The cryogenic systems being considered include Stirling cycle coolers and/or passive radiators paired with liquid Helium dewars and with the final stage of cooling being provided by an adiabatic demagnetisation refrigerator. The liquid Helium systems sets a limit to lifetime which could be around two to three years.

The microcalorimeter has the potential to revolutionise the subject. However the difficulties mentioned above and, in particular, the limited lifetime in orbit have led to the device being considered only for inclusion in AXAF; a mission for which servicing by the shuttle is envisaged. The data in table 2 show that the microcalorimeter provides by far the best combination of energy resolution and sensitivity of any of the instruments listed.

6. DIFFRACTION GRATINGS

Among the dispersive spectroscopic instruments, transmission gratings, which were flown on the Einstein observatory, are proposed for the Spectrosat and AXAF missions while reflection gratings are part of the XMM model payload. The transmission gratings when used with either high spatial resolution microchannel plate or CCD detectors in the telescope focal plane can provide resolving powers of between 100 and 1200 (see tables 1 and 2). However they rely on high angular resolution X-ray telescopes ($\Delta\theta \sim 1``$)

to achieve this performance. Reflection gratings can deliver resolving powers of around 400 when used with telescopes of more modest angular resolution ($\Delta\theta \sim 30"$).

Grating systems have the major advantage that they present a dispersed spectrum simultaneously for registration and analysis. However they tend to work better at lower photon energies although the high energy gratings on AXAF still give some useful performance at 7keV. Transmission gratings only offer their full resolving power for point sources and performance degrades rapidly with increasing source extent. Reflection gratings are more tolerant of source extent and so are particularly suitable for observations of distant clusters or of extragalactic supernova remnants.

From a consideration of the future missions that will use gratings it is clear that Spectrosat, which will combine transmission gratings with another version of the excellent ROSAT X-ray telescope will provide a low energy spectroscopic follow-up to the ROSAT sky survey. Of the major observatories, AXAF offers better spectral resolution while XMM offers better sensitivity.

7. BRAGG SPECTROMETERS

Although much useful data has been obtained from solar observations undertaken with Bragg spectrometers, serious use of these instruments for cosmic observations began with the flight of the Einstein observatory. Bragg spectrometers generally offer the highest resolving power of any currently available instrument but, given that many of the designs proposed operate by scanning the spectrum, they lack a multiplex advantage in addition to having an inherently low throughput.

The set of ten different crystal and multilayer diffractors being developed for the AXAF misison (see table 2) provides an impressive and comprehensive capability for high resolution spectroscopy in the 0.5 keV to 8keV energy range. Although the effective apertures in the range 1-50cm^2 are modest, resolving powers of up to 2000 are available in selected spectral ranges. The AXAF spectrometers also have some spatial resolution in one dimension. A field of view 3` in extent can be covered with a spatial resolution of around 20``.

Given the earlier stage of development, a strong candidate instrument for high resolution spectroscopy has not yet emerged for the XMM. A possible approach using an objective crystal spectrometer design is outlined in table 3. Such a system would use the prime-focus CCD camera as a detector but would have the disadvantage of observing sources in a direction orthogonal to that of the X-ray telescope axes. The use of multilayer diffractors in the focal plane in selected limited spectral ranges for high resolution observations is also a possibility which may prove easier to implement.

8. CONCLUSIONS

It will be clear from the material in the previous sections that the next decade will see the deployment in orbit of a number of very powerful spectroscopic instruments. Following the ROSAT mission, two generations of X-ray sky surveys will have been completed and thus the need for detailed spectral studies will become compelling.

The moderate missions (table 1) will provide a useful and relatively rapid follow up to ROSAT. However major advances for a very wide range of sources will follow the deployment of the two observatory spacecraft AXAF and XMM. AXAF includes a broad selection of spectroscopic instruments and, in particular, will carry the microcalorimeter, a device which for the first time combines high spectral resolution with very good quantum efficiency. However the spacecraft will be placed in near earth orbit and so it will be more difficult to arrange efficient observing strategies particularly in the case of variable sources. In addition, the mission will place a substantial emphasis on positional astronomy, spectrophotometry and on deep surveys of limited regions of the sky. Given a single, although very powerful telescope, the spectroscopic observations will have to compete for focal plane time with the other topics mentioned above.

The XMM is designed to achieve much greater collecting area than AXAF, particularly at energies above 7keV, but at the cost of degraded angular resolution. Thus XMM is more directly targeted at studies of source spectra and variability. In addition the availability of multiple focal planes and the proposed deep orbit will lead to a high degree of operational flexibility.

In fact these two major X-ray observatories are superbly complimentary and together will lead to very considerable advances in our understanding of a very large number of objects and to the solution of many pressing astrophysical problems.

REFERENCES

Antonucci, E., Dennis, B.R., Gabriel, A.H., Simnett, G.M., 1985, Solar Phys., 96, 129.

Becker, R.H. et al., 1979, Astrophys. J., 234, L73.

Canizares, C.R. et al., 1979, Astrophys. J., 234, L33.

Canizares, C.R., Winkler, P.F., 1981, Astrophys. J., 246, L33.

Hayes, M.J.C., Culhane, J.L., Blissett, R.J., Barr, P., Bell-Burnell, S.J., 1980, Mon. Not. R. Astr. Soc., 193, 15.

Janesick, J.R., Elliott, S.T., 1988, Paper presented at SPIE 23rd Annual Symposium (Conference 982), San Diego, 14-19 August.

Kahn, S.M., Seward, F.D., Chlebowski, T., 1984, Astrophys. J., 283, 286.

McCray, R.A., Shull, J.M., Boynton, P.E., Deeter, J.E., Holt, S.S., White, N.E., 1982, Astrophys. J. 262, 301.

McCammon, D., et al., 1987, Japanese Journ. Appl. Phys., 26, (Suppl. 26-3).

Mitchell, R.J., Culhane, J. L., Davison, P. J. N., Ives, J. C., 1976, Mon. Not. R. Astr. Soc., 176, 29.

Pravdo, S.H., et al., 1979, Astrophys. J., 231, 912.

Strüder, L., Lutz, G., Predhl, P., Sterzik, M., 1988, Paper presented at SPIE 23rd Annual Symposium (Conference 982) San Diego, 14-19 August.

White, N.E., et al., 1986, Astrophys. J., 301, 262.

THE ROSAT MISSION

J. Trümper

Max-Planck-Institut für Physik und Astrophysik,
Institut für Extraterrestrische Physik
8046 Garching, W.-Germany

ABSTRACT. The scientific payload of ROSAT consists of a 83 cm X-ray telescope (6 - 100 Å) and a 60 cm XUV-telescope (60 - 300 Å) which are looking parallel. An important objective of the mission is to perform the first all-sky survey with imaging X-ray telescopes providing an improvement in sensitivity by several orders of magnitude compared with previous surveys. A large number of new X-ray sources ($\sim 10^5$) is expected to be discovered and located with an accuracy of 1 arcmin or better, depending on source strength. The sources discovered will represent almost all types of astronomical objects, ranging from nearby normal stars to distant quasars.

After completion of the sky survey which will take half a year, the instruments will be used for detailed investigations of selected sources with respect to spatial structures, spectra and time variability. In this pointing mode, which will be open for guest observers, ROSAT is expected to provide substantial improvements over the imaging instruments of the Einstein observatory.

1. INTRODUCTION

During the last 20 years X-ray astronomy has become one of the major disciplines in the exploration of our universe. A large variety of phenomena have been discovered in this area which comprise almost all kinds of astronomical objects - from the nearby stars to the most distant quasars at the edge of the known universe. In many cases, X-ray emission is the primary and dominating emission process (X-ray binaries with neutron stars or black holes, cooling neutron stars, hot gas in clusters of galaxies etc.), in other objects, the observation of X-rays gives very important complementary information (e.g. stars, degenerate stars, active galaxies and quasars). In X-rays we see the "hot universe", explosive phenomena, large concentrations of relativistic electrons in magnetic or intense radiation fields. It is clear today that these high energy phenomena play a very important role in the evolution of matter in our universe.

2. ROSAT X-RAY TELESCOPE

ROSAT is a free flying satellite to be launched by a Delta II rocket in early 1990. Its anticipated lifetime is 3 years. The scientific payload includes two instruments, a large X-ray telescope (6 - 100 Å) and a smaller XUV-telescope (60 - 300 Å) which are oriented in parallel. In this talk I'll concentrate on the X-ray aspects.

The ROSAT X-ray telescope consists of a fourfold nested mirror system with 83 cm aperture and three focal instruments. Two of them will be position-sensitive proportional counters (PSPC, 6 - 100 Å, 0.1 - 2 keV) having a field of view of $2°$. In the pointing mode the angular resolution will be ~ 30". The positions of point sources may be determined with higher accuracy (to ~ 10 arcsec). The spectral resolution of the PSPC's will be 45 % FWHM at 1 keV. Both PSPC's are being developed and built by MPE Garching. The third focal instrument will be a high resolution imager (HRI, ~ 3") which is an improved version of the corresponding EINSTEIN instrument, provided by SAO/NASA. The X-ray mirror system will have a half power width of ~ 3". It is manufactured by Carl Zeiss Company in Oberkochen near Stuttgart. The calibration of mirrors and the complete X-ray telescope is done in the 130 m X-ray test facility of MPE Garching.

3. THE ALL SKY X-RAY SURVEY

A primary objective of the mission is to perform the first all-sky survey with an imaging X-ray telescope which will lead to an improvement in sensitivity by several orders of magnitude compared with previous "counter" surveys.

The sky survey is performed by continuously scanning great circles perpendicular to the earth-sun-line. This will result in full sky coverage in half a year. The scan of one great circle takes one orbit and thereby avoids earth occultations. At the ecliptic equator a particular source is scanned ~ 30 times during two consecutive days with an integral exposure of ~ 600 sec. The corresponding survey sensitivity (5 σ) is 2×10^{-13} erg/cm^{2}s (0.1-2 keV). The ecliptic pole is crossed during every orbit which results in ~ 60.000 s of total observation time and a sensitivity of 1.5×10^{-14} erg/cm^{2}s. We note that this latter figure is close to that of the Einstein deep surveys.

The sky survey will be done with the PSPC only because of its sensitivity and its relatively large field of view ($2°$ circular). The spectral resolution of the instrument allows to distinguish four "colour" bands. Positions will be accurate to 1 arcmin (90 % confidence radius) for point sources near the sensitivity limit S_{min}. For stronger sources the position determinations will be more accurate due to the better photon statistics. It is expected that for sources at $\sim 10\ S_{min}$ a limiting resolution of 0.5 to 0.1 arcmin (90 % confidence radius) will be reached depending on the final instrument and spacecraft characteristics.

The most important aspect of the ROSAT star survey is that it will yield large unbiased samples of various classes of sources which are detected in a homogeneous and systematic way.

4. THE POINTED OBSERVATIONS

The second main objective of the ROSAT mission is to perform detailed investigations of selected sources. These will be carried out in the pointing mode and will improve our understanding of spatial structures, spectra and time variability of X-ray sources. In this mode the ROSAT X-ray telescope is expected to provide a substantial improvement over the imaging instruments of the EINSTEIN-observatory, particularly in terms of sensitivity (factor $\sim$ 5 to 10), spectral resolution (PSPC versus IPC factor $\sim$ 3) and angular resolution (factor $\sim$ 3 both for the PSPC and the HRI). In addition, the XUV telescope will allow the observations to be extended to long wavelengths. In the pointing mode there will be four "X-ray colours" and three "XUV colours".

5. THE ROSAT SATELLITE AND GROUND OPERATIONS

ROSAT is a three-axis stabilised satellite with CCD-type star trackers (two for the X-ray telescope, one for the XUV telescope) and attitude control by momentum wheels which are desaturated by magnetic torquing. Two tape recorders are used for on-board data storage. The satellite which has a mass of more than 2,5 tons will be launched by a Delta II into an orbit of $\sim$ 580 km height and 57° inclination. Mission control will be performed by the German Space Operations Center (GSOC) at Weilheim near Munich from where the data will be sent to the ROSAT science operation center at MPE Garching which is equipped with a dedicated VAX 8600 facility and a Microvax cluster.

The technical and scientific quicklook activities will be placed at GSOC and MPE as well. Date obtained during the pointed programme will be routed directly to the ROSAT data centers in the US and UK.

6. SCIENTIFIC DATA RIGHTS

The analysis of the ROSAT all sky surveys will be the responsibility of MPE and of the UK WFC consortium for the X-ray data and for the XUV data, respectively.

The pointed programme will be devoted entirely to a guest observer programme. The first call for proposals will be issued one year before launch, viz. in spring 1989, in the US, UK and Germany.

7. CONCLUSIONS

The ROSAT mission will provide the only X-ray telescope in space during the early 1990's. It will be evident from the preceding discussion, that because of its extended waveband coverage, its sensitivity, and its spectral resolution, ROSAT will provide the scope for a wide range of novel investigations on known astrophysical objects as well as an exciting capability for new astrophysical discoveries. This is true for the pointed observations, but also for the all-sky survey which will have a great impact on the work in other regions of the electromagnetic spectrum. Last, but not least, the survey will help to guide the observations of followup X-ray astronomy missions, including in particular the large permanent X-ray observatories to be launched in the late 1990's, AXAF and its European complement XMM.

8. REMARK

A more detailed description of the project including references is given by the present author in Physica Scripta, T7, 1984.

THE SPEKTROSAT MISSION

Peter Predehl

Max-Planck-Institut für Physik und Astrophysik
Institut für Extraterrestrische Physik
8046 Garching, FRG

ABSTRACT

High resolution spectroscopy will be an important diagnostic tool in future X-ray astronomy. SPEKTROSAT, the follow-up mission to ROSAT will be equipped with a transmission grating spectrometer with a spectral resolution of 0.2 Å over the wavelength range between 6 Å and 300 Å. Third order aberations are minimized by a mounting of the grating elements according to the Rowland-torus geometry. The ROSAT mirror system will be slightly changed by a redesign of the optical stops in order to reduce the field of view. The Focal Instrumentation has to be modified for a better match with the spectroscopic requirements. Since SPEKTROSAT is almost identical to ROSAT, this concept offers a large amount of science at relatively low costs. Its limiting line sensitivity will be about 20 times better than that of the EINSTEIN-OGS.

1. SCIENTIFIC RATIONALE

ROSAT will perform the first all-sky survey in X-rays using focussing optics. After the ROSAT survey about 10^5 individual sources are known with a position better than 1 arcmin but low spectral resolution. Detailed investigations on a subset of these sources will be carried out after the sky survey in the pointing phase of the mission. The spectral resolution is limited by the intrinsic energy resolution of the focal instrumentation detectors. A logical next step would be high throughput, high resolution spectroscopy. By using dispersive methods, a mirror system with a quality of at least that of the ROSAT mirror system is required in order to avoid unfeasable grating ruling densities and very large focal plane detectors. According to that mirror, SPEKTROSAT will be a 'soft' instrument dedicated to a wavelength band between 6 and 300 Å.

The soft energy bandpass (below 2 keV) especially is important for the study of stellar coronae. Almost all types from 'normal' (like the sun) to strong variable stars (e.g. cataclysmic variables) are interesting objects. Luminous sources like X-ray binaries and supernova remnants can be observed spatially resolved also in the Magellanic Clouds and even in the Andromeda Nebula. The hot gas in clusters of galaxies can be investigated. The non-thermal character of sources like the Crab

Nebula or the emission from Active Galactic Nuclei can be verified only be means of spectroscopy.

The absorption due to the Interstellar Medium produces prominent edges in the (continous) spectrum of background sources. The investigation of the strength of these edges leads to a knowledge about the abundances of the lighter elements (C, N, O, Ne, Mg and Al).

In this context, the spectral range between 90 and 200 Å is especially interesting, because it contains very intense lines from highly ionized iron (Fe XVIII - Fe XVII) as well as lines produced in a cooler plasma (Fe IX - XII). Therefore temperature and density diagnostic of heterothermal plasmas as likely parts in stellar coronae are possible (Schmitt, 1988).

In addition, the XUV part of the spectra is of particular interest for the photosperic X-ray radiation of hot white dwarfs as well as X-ray and XUV radiation from hot stars whis is - presumably - produced by numerous shocks in their winds.

Summarizing, SPEKTROSAT is the only proposed mission, dedicated purely to high resolution X-ray spectroscopy. It is particularly important for studies of stellar coronae, hot white dwarfs and low N_h AGN.

2. THE SPEKTROSAT CONCEPT

The use of transmission gratings as dispersive elements in X-ray telescopes with an angular resolution in the arcsec range is preferable to the use of reflection gratings, because a transmission grating spectrometer can be incorporated into an existing telescope without major changes of the overall design.

SPEKTROSAT will be an almost identical copy of ROSAT except for the inclusion of a transmission diffraction grating and a modified focal plane instrumentation. As a baseline it carries two high resolution imagers and one proportional counter (the ROSAT focal instrumentation contains two PSPC and one HRI, Pfeffermann, 1986). Minor changes will also be made in the mirror system: in order to avoid spectral confusion by having more than one X-ray source in the direction of dispersion, the field of view is reduced from 2 degrees to 45 arcmin by a redesign of the optical stops. From the ROSAT experience it seems to be feasible to improve the on-axis angular resolution from 3" to 2" (half energy width) thereby increasing the spectral resolving power (Beckstette, 1988).

This concept takes advantage from the fact that most of the S/C-hardware as well the management experience can be carried over from ROSAT thereby reducing costs drastically. Moreover, the extensive ROSAT data analysis system can be used with only minor modifications. The ground operations concept differs from that of ROSAT if SPEKTROSAT is launched into an orbit with 28° inclination (wich is the baseline for Shuttle launches). On the other hand, the scientific performance is enhanced, because the loss due to the belt passages is reduced.

In addition to its main telescope, ROSAT contains also a 'passenger payload', the XUV Wide Field Camera. The possible replacement on SPEKTROSAT by another instrument is investigated currently in industry. Several proposals from the U.S. are already made in order to extend the energy bandpass of the grating spectrometer.

The design of the spectrometer is determined by the environment of the (modified) ROSAT-telescope and by the scientific requirements regarding spectral resolution and the wavelength band to be covered. A resolving power $\lambda/d\lambda = 100$ is needed in order to separate the individual lines of the O VII triplet at 22 Å. The wavelength range should extend at least 200 Å for reasons mentioned above.

According to the grating formula $\sin \alpha = \lambda/d$ (d=grating period) the achievable spectral resolution and the wavelength range are complementary. At a given angular resolution of the telescope, the spectral resolution is determined by the ruling density of the grating. Increasing the ruling density requires a larger focal plane detector (or the wavelength band is reduced).

The environment for the spectrometer is given by the geometry of the telescope (Aschenbach, 1986 and Table 1). The on-axis resolution is affected not only by the mirror-detector combination, but also by other spacecraft subsystems, e.g. the attitude measurement and thermal control (Table 2). The error budged leads to a half energy width of 4.8 arcsec (3.8 arcsec design goal) for the worst case of faint sources. By observing bright sources, where autocorrelation methods within the data analysis can be applied, some of the errors are reduced (or even neglected) leading to only 2.5 arcsec on-axis resolution. This value corresponds to 0.18 Å spectral resolution assuming a grating ruling density of 1000 l/mm and a distance of 1700 mm between the grating and the focal plane.

TABLE 1: Geometry of the (SPEKT-)ROSAT telescope

entrance diameter paraboloids	(mm)	834.9, 695.7, 573.0, 466.1
interface diameter par./hyp.	(mm)	785.5, 653.2, 537.1, 436.4
exit diameter hyperboloids	(mm)	660.0, 548.8, 451.2, 366.6
length of all mirror shells	(mm)	500
total collecting area	(mm)	1141
focal length	(mm)	2400
distance grating focal plane	(mm)	1700
detector size	(mm)	< 120

TABLE 2: Subsystem error budged (68% encircled energy)

Subsystem	ROSAT (verified)	SPEKTROSAT (baseline)	SPEKTROSAT (goal)
Mirror Assembly	1.9	1.28	1.28
Thermal Control Mirr.	0.9	0.9	0.9
Structure Telescope	2.2^1	-	-
Thermal Control Struc.	1.15^2	1.15	1.15
Fiducial Light System	1.5	1.5	1.5
Focal Plane Instrument	2.0	0.64	0.64
Attitude Measurement	1.75	1.75	0.9^3
RMS (half energy width)	8.1	4.8	3.8

[1]: bias only; [2]: random only; [3]: noise only

Since the grating is mounted in the convergent beam behind the mirror, it is exposed to nonparallel light. A plane mounting would produce optical aberrations thereby reducing the spectral resolution. This effect is dominant over the intrinsic resolution given by the mirror-detector combination. Therefore, the grating is required to have an appropriate curvature, the so-called Rowland torus. This geometry is approximated by a

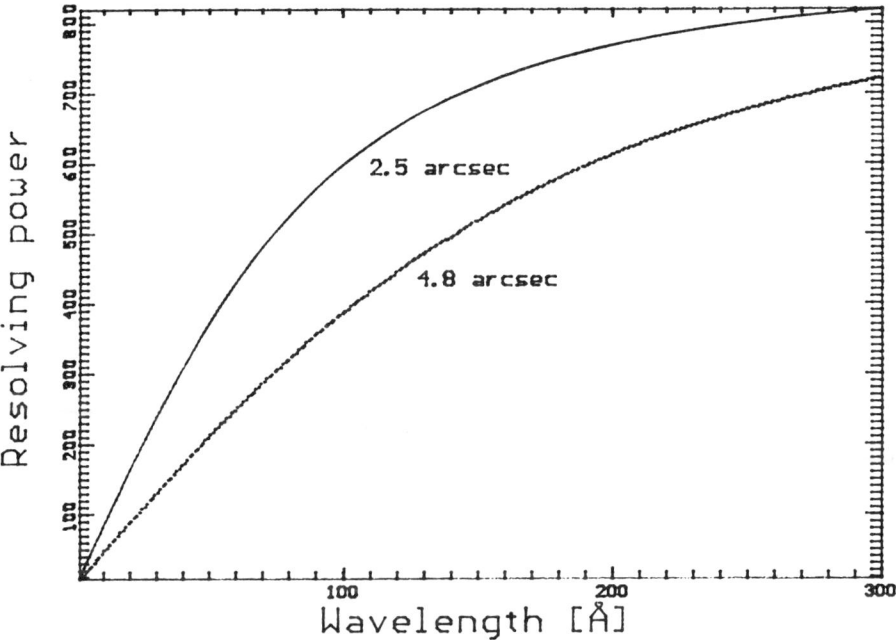

Figure 1: Spectral resolving power of the SPEKTROSAT instrument for faint sources (worst case) and strong sources (best case).

large number of (flat) individual grating facets and also by a curvature of the detector. Residual optical aberrations result from the finite size of the elements, which reduce the resolution at longer wavelengths (Predehl, 1988 and Figure 1). A small extension of the spectrum perpendicular to the direction of dispersion (astigmatism) does not affect the spectral resolution but extends the size of the resolution elements thereby increasing the sensitivity against background (intrinsic detector noise, particles, also diffuse X-ray's).

3. SPECTROMETER PERFORMANCE

The sensitivity of the telescope without grating is quite similar to that of ROSAT (with the HRI in focus). Differences arise from the fact that the wavelength band is extended and therefore modified filters have to be used. For spectroscopy, at least 100 events are required in an individual line. On the other hand, typical observation times will range from several hours to a day. Table 3 contain the sensitivy of SPEKTROSAT for a number of different spectral lines in terms of 100 counts per 10000 seconds. The background is always less than 1 count per resolution element.

TABLE 3: Line sensitivity (100 cts / 10^4 s)

Element	λ (Å)	S (erg/cm2/sec)
C VI	33.7	$2.1 * 10^{-13}$
N VII	24.8	$4.2 * 10^{-13}$
O VII	21.6	$5.8 * 10^{-13}$
O VIII	18.6	$7.2 * 10^{-13}$
Ne IX	13.5	$10. * 10^{-13}$
SI XII	44.1	$1.3 * 10^{-13}$
S XII	39.9	$1.5 * 10^{-13}$
Fe XVI	50.5	$1.2 * 10^{-13}$
	63.7	$1.1 * 10^{-13}$
	66.4	$1.0 * 10^{-13}$
Fe XXII	135	$0.8 * 10^{-13}$
Fe XXIV	192	$0.6 * 10^{-13}$

The quality of a transmission grating can be characterized by its first order efficiency and the degree of higher order supression. For a classical amplitude grating, both depend only on the bar-slit ratio within the grating period. An optimum is achieved at a ratio of 1:1. Then, the first order efficiency reaches its maximum of about 10% (for each of both sides of the symmetric spectrum) and all even orders are cancelled out. At shorter wavelengths, the efficiency depends also on the wire thickness due to transparency effects, which can be used in order to increase the efficiency up to 20%. A long term development-program at MPE lead to gratings having an efficiency very close to this theoretical optimum (Predehl, 1986 and Figure 2).

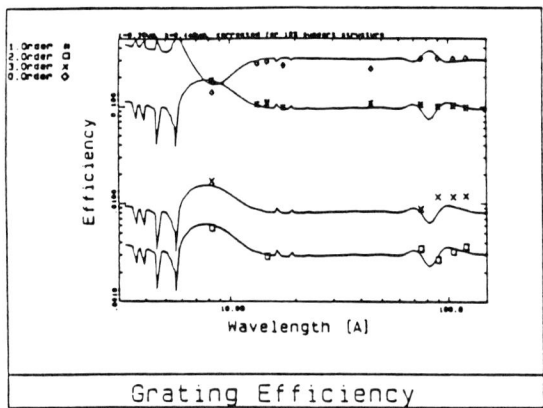

Figure 2: Measured and calculated efficiencies for a grating with 1000 gold wires per mm. These are one-side efficiencies, corrected for the obstruction due to the support grid (18%).

The quality of a grating produces spectrum does not only depend on the geometry of the spectrometer and the bar slit ratio. Any distortion of the grating itself would be reflected in a degradation of the spectrum. Possible sources of distortion may be: broken or missing bars, deviation of the grating foil from the plane, varying grating period, or a 'scatter' of the bars around their nominal positions. These effects reduce the efficiency and (more seriously) the spectral resolution.

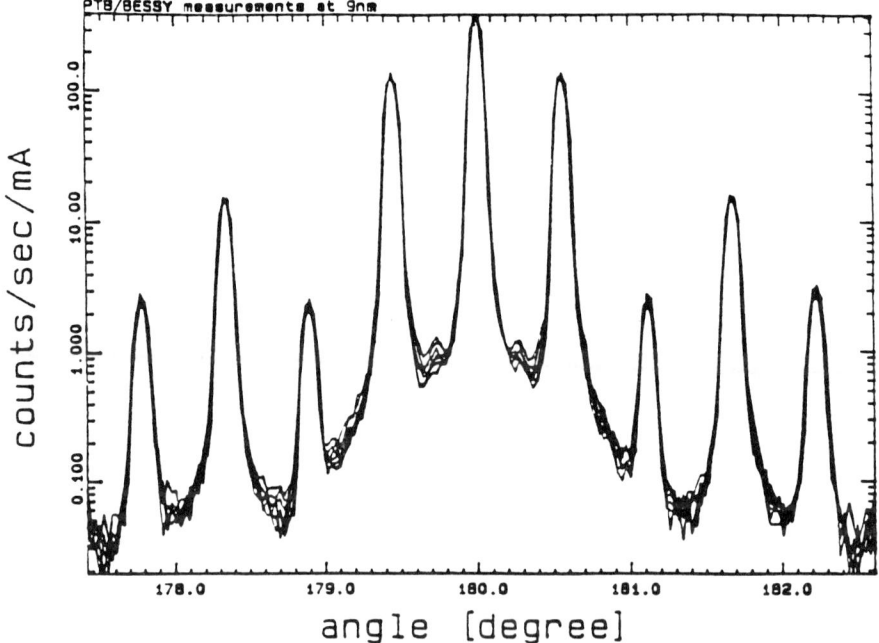

Figure 3: Spectra obtained at different locations on the grating. The wavelength is 90 Å, the intensity is normalized to the synchrotron storage ring current.

Measurements in the visual wavelength range yielded resolving powers $\lambda/d\lambda$ > 6000 for our gratings, limited by the non-flatness of the grating foil. Deviations of the order of μm are sufficient to distort the diffraction pattern. Due to the diffraction geometry, this effect can be neglected in the X-ray region, because a remarkable fraction of a mm is needed in order to produce the same effect. Unfortunately, there is no way to measure the spectral resolution directly in X-rays, because no focussing optics with an angular resolution in the arcsec range is available at the present time.

For SPEKTROSAT, several hundred individual facets will be assembled to the torus. All facets have to be identical within very narrow tolerances. Figure 3 demonstrate the excellent degree of uniformity achieved so far.

ACKNOWLEDGEMENT

We may thank the Bundesministerium für Forschung und Technologie (BMFT) for supporting several grating studies in the context of ROSAT and SPEKTROSAT.

REFERENCES

Aschenbach, B., SPIE **733**, 186, 1986.

Beckstette, K., Aschenbach, B., SPIE **982**, 1988.

Pfeffermann, E., Briel, U., Hippmann, H., Kettenring, G., Metzner, G., Predehl, P., Reger, G., Stephan, K.-H., Zombeck, M., Chappel, J., and Murray, S. SPIE **733**, 519, 1986.

Predehl, P., and Bräuninger, H., SPIE **733**, 203, 1986.

Predehl, P., Bräuninger, H., Burkert, W., Aschenbach, B., Trümper, J., Kühne, M., and Müller, P., SPIE **982**, 1988.

Schmitt, J.H.M.M, this conference, 1988

THE X-RAY ASTRONOMY SATELLITE SAX

R.C.Butler
Piano Spaziale Nazionale, CNR, Rome, Italy

ABSTRACT

The SAX satellite is forseen for launch at the end of 1992 to study the X-ray emission from galactic and extra-galactic sources in the energy range 0.1-200 keV. The payload consists of four concentrator/spectrometer systems (3 units 1-10keV, 1 unit 0.1-10keV), a high pressure gas scintillation proportional counter (3-120keV), a phoswich scintillation counter (15-200keV), and two wide field cameras (2-30keV). Together these instruments will perform the following:-
- Broad band spectroscopy ($E/\Delta E=12$) in the energy range 0.1-10 keV with imaging resolution of 1 arcmin
- Continuum and cyclotron line spectroscopy ($E/\Delta E=5-20$) in the wide energy range 3-200 keV
- Variability studies of bright source energy spectra on time scales from milliseconds to days and months
- Systematic long term source variability studies in selected regions of the sky down to a source intensity of 1 mCrab.

THE SAX MISSION

SAX, 'Satellite for Astronomy in X-rays', is a major program of the Italian Space Plan (PSN) involving a collaboration with the Netherlands Agency for Space Programs (NIVR). SAX is currently finishing its Phase B activities with the satellite forseen for launch at the end of 1992. Its scientific objectives are to carry out systematic and comprehensive observations of celestial X-ray sources in the very broad energy band 0.1-200 keV with special emphasis on spectral and timing measurements.
SAX was originally designed for launch by the Shuttle and insertion into a circular orbit at 550 km and 12 deg inclination, but has been redesigned for launch by the Atlas G-Centaur directly into a 600 km circular orbit at 2 deg inclination. The ground station will be situated near the equator (at the San Marco Base, Malindi, Kenya), while the operations control centre, connected by relay satellite will be in Italy. The satellite will achieve better than a 2 arcmin pointing stability over a series of orbits for a total single observation time of typically upto 100,000 seconds, with a postfacto pointing accuracy of 1 arcmin. During each orbit 450 Mbits of information will be stored onboard and transmitted to the ground during station passage. SAX has a design lifetime of just over two years, but will remain in orbit for upto four years.

THE SAX PAYLOAD INSTRUMENTS AND THEIR CAPABILITIES

The chief characteristics of the SAX payload instruments, described in more detail by Spada (1983) and Perola (1988), are given in Table 1. Fig.1 illustrates the arrangement of the instruments in the payload module, with the concentrator/spectrometer, HPGSPC, and PDS, collectively called the narrow field instruments, all pointing in the same direction, while the WFC point in diametrically opposed directions perpendicularly to them.

	Aperture FWHM	Ang.Res.	Total Eff. Area	Energy Res. FWHM
Concentrator/Spectrometer (3 units, 1-10 keV, MEC/S) (1 unit, 0.1-10 keV, LEC/S)	30'	1'	200 cm^2 7keV 56 cm^2 0.25keV	8% 6keV 39% 0.27keV
High Pressure Gas Scint. Proportional Counter (3-120 keV, HPGSPC)	60'	—	280 cm^2 60keV 300 cm^2 6keV	3% 60keV 10% 6keV
Phoswich Detector System (15-200 keV, PDS)	87'	—	140 cm^2 200keV 680 cm^2 20keV	17% 60keV
Wide Field Cameras (2 units, 2-30 keV, WFC)	20° x 20°	5'	250 cm^2 (/unit through mask)	20% 6keV

Table 1. The Chief Characteristics of the SAX Payload Instruments.

Each of the four mirror assemblies in the concentrator/ spectrometers consists of 30 nested gold coated, double cone approximations to the Woltjer 1 configuration with a focal length of 185 cm. (Citterio et al 1987). Thier design maximises the effective area around 7 keV for iron K-line studies, while the position sensitive GSPC's in their focal planes' will have energy resolutions comparable to that of the solid state spectrometer of the Einstein Observatory and more than a factor of two better than previous proportional counters at 7 keV. The concentrator/spectrometers' sensitivity-confusion limit will be a factor of about three better than that of the ROSAT (E $\leqslant$ 2 keV) all-sky survey so they will be capable of fully exploiting the survey in their choice of representitive samples of faint objects for detailed study upto 10 keV. Their ability to detect emission lines and determine the spectra of different regions of extended sources is shown in Fig.2.

The very good energy resolution of the HPGSPC will be particulary important in the detailed study of narrow cyclotron emission and absorption line features as illustrated by the simulated Her X-1 cyclotron line in absorption in Fig.3. It will also complement the iron K-line studies of the concentrator/spectrometers and the source continuum studies of the PDS.

The high source flux sensitivity over a broad energy range and the energy resolution of the PDS ideally suit the instrument to detailed studies of the continuum in galactic and extra-galactic sources down to its ultimate 5σ sensitivity limit of 1/20 3C273 in 300,000 seconds. It will also study cyclotron line features in known sources and search for these in other binary pulsars extending the measurements of the HPGSPC particularly for broad line features.

The balance in overall broad band sensitivity of the narrow field instruments from 0.1-200 keV is best illustrated by the simulated results on a typical AGN (1 mCrab at 3 keV, energy spectral index =0.6, N =4.10**20 cm**-2) in Fig.3 from which the spectral index can be obtained with about an order of magnitude smaller uncertainty than with previous experiments (Exosat, HEAO-1). Soft excesses or spectral breaks at higher energies should be clearly visible if present.

The payload is completed by the two WFC. Each contains a position sensitive proportional counter filled with two atmospheres of xenon, with a random mask aperture. Their energy range will complement those of the narrow field instruments. They will perform long term monitoring of both galactic and extra-galactic sources, and the detection/localisation of transients on timescales from 2 ms upwards.

SAX SCIENTIFIC OBJECTIVES

The SAX payload has been chosen to cover the energy range 0.1-200 keV in a balanced manner to notably extend the spectroscopic and time variability studies performed to date. In the mission life of atleast two years SAX will perform between 2000 and 3000 seperate observations. The majority will be governed by the narrow field instruments while the WFC will perform long term monitoring of selected sources. Periodically the WFC will survey the galactic plane with particular emphasis on transients. The observations will be based on a core program chiefly devoted to systematic studies of various classes of objects, and a guest observer program allocated about 20% of the time. A selection of the areas, discussed in more detail by Perola (1983), inwhich SAX is expected to make its most significant contributions is given below:-

- Compact galactic sources: the shape and variabilty of their continuum and temporal studies of features such as iron flourescence lines, cyclotron lines and absorption effects as a function of orbital phase and rotation, transient detection and light curve studies
- Supernova remnants: spatially resolved spectra of extended ($\gg 1'$) galactic SNR's, and the spectra of the Magellanic Clouds remnants
- Stars: coronal emission spectra with a sensitivity comparable to that of the Einstein Observatory upto 10 keV
- Active galactic nuclei: spectral and temporal variability studies of their continuum upto 200 keV, the spectra upto 10 keV of very distant sources (z=3.2 for sources equivalent to 3C273) and soft X-ray excess, photoelectric absorption and iron flourescence line studies
- Clusters of galaxies: spatially resolved spectra upto 10 kev of the nearby clusters with iron flourescence line and temperature gradient studies and high energy spectra for $z < 0.1$ and temperature measurements out to $z \sim 1$
- Normal galaxies: spectral studies of their extended emission.

Its launch date at the end of 1992 will give SAX a first opportunity to take advantage in a systematic way of the many new results that should become available from the all-sky imaging survey of ROSAT, and it will preceed the large observatory type missions due for the second half of the 1990's which will concentrate on X-ray astronomy upto 10 keV only.

ACKNOWLEDGEMENTS

The data presented here is the result of the work of many people who are participating in SAX. I wish to acknowledge the contributions of L. Scarsi, G.C. Perola, G. Boella and G. Di Cocco, and in particular O. Citterio and B. Sacco for the concentrator mirrors, A. Smith for the focal plane detector of the LEC/S, G. Manzo for the HPGSPC, F. Frontera and E. Costa for the PDS, and J. Bleeker, B. Brinkman and P. Ubertini for the WFC. Finally I wish to thank the rest of the SAX team at the PSN, G. Manarini, M. Casciola and B. Negri.

REFERENCES

Citterio, O., et al. 1988, Applied Optics, 27, 1470.
Perola, G.C., 1988, Cospar XXVII, Helsinki, in the press.
Perola, G.C., 1983, Proc. of Workshop on Non-thermal and Very High Temperature Phenomena in X-ray Astronomy, eds. Perola, G.C., and Salvati, M., Rome, 19-20 December 1983, p.175.
Spada, G.F., 1983, ibid.

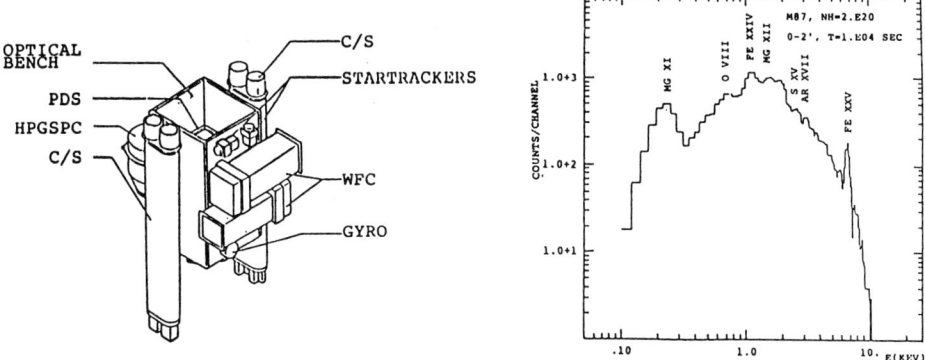

Fig.1 The SAX payload module

Fig.2 Concentrator/spectrometer counts in $10^{**}4$s from the central region of M87

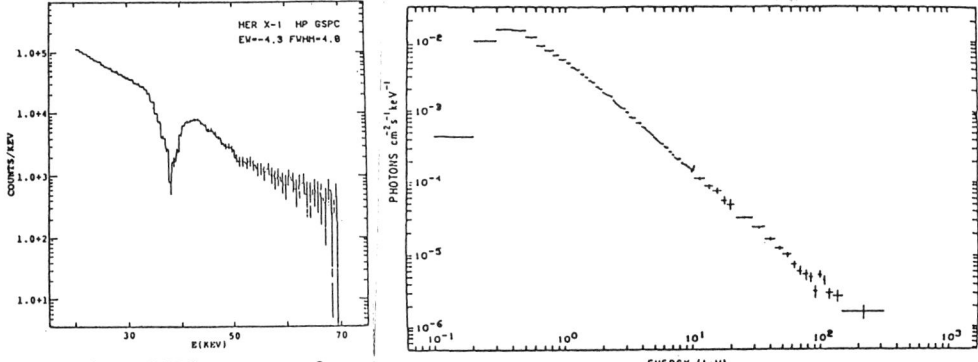

Fig.3 HPGSPC counts from Her X-1 in $10^{**}5$s showing the cyclotron line in absorption

Fig.4 The deconvolved spectrum of a typical AGN observed for $10^{**}5$s

DISCUSSION-R.C. Butler

DISCUSSION

R. Pallavicini: Since the LEGSPC on SAX is sensitive up to 10 keV, what is the reason to have only 1 LEGSPC and 3 MEGSPC, rather than four LEGSPC detectors?

R. Butler: The choice of 3 MEGSPC's and only 1 LEGSPC in the concentrator/ spectrometer comes above all from reliability considerations given that the LEGSPC is now position sensitive and thus capable of very similar results above 1 keV to the MEGSPC's, which was not the case in the original proposal where no imaging capability was foreseen from the LEGSPC. The reliability of a 1.5-2 μm polypropeline window over a mission lifetime of at least two years is very difficult to assess, and so the configuration with 3 MEGSPCs with 25 μ beryllium windows was retained to ensure mission success.

XSPECT: A TELESCOPE/SPECTROMETER SYSTEM ON SPECTRUM RÖNTGEN GAMMA

Herbert W. Schnopper

Danish Space Research Institute, DK-2800 Lyngby, Denmark

ABSTRACT. The SPECTRUM RÖNTGEN-GAMMA mission is being developed by the Babakin Center (BC) together with the Space Research Institute (IKI) of the Academy of Sciences, USSR and is scheduled for launch in 1993. Mission objectives include broad and narrow band imaging spectroscopy over a wide range of energies from the EUV through gamma rays with particular emphasis on the study of extragalactic objects. The Danish Space Research Institute (DSRI) BC and IKI share the responsibility for the preparation of the XSPECT system. Two thin foil telescopes which are conical shell approximations to Wolter 1 geometry, each with an aperture of 60 cm and a focal length of 8 m, are designed to have a half-power width of less than 2 arcmin and will have collecting areas of 1700 and 1200 cm at 2 and 8 keV, respectively. Images and spectra will be recorded with position sensitive proportional counters with good spectral resolution. An objective Bragg crystal panel, placed in front of one of the telescopes, will make high resolution spectroscopic studies ($E/\Delta E \sim 10^3$) of point- and extended sources. Other instruments are under consideration.

1. INTRODUCTION

SPECTRUM RÖNTGEN-GAMMA is expected to be the first of a new series of astronomical missions announced by the Academy of Sciences of the USSR. The payload under study (see Figure 1) consists of instruments intended for a broad range of astrophysical studies but with particular emphasis on extragalactic objects. Concentrating telescopes will be provided for EUV (EUVITA) and X-ray (JET-X and XSPECT) studies and a coded aperture mask telescope will image hard X-rays and soft gamma rays (MART). Supplementary instrumentation will provide all sky coverage for X- and gamma ray bursts. An articulated platform with its own star tracker will perform independent observations with X-ray and EUV telescopes. The satellite will be launched into a deep, highly eccentric, orbit with a period which could be as long as four days and from which long duration observations can be made.

2. XSPECT TELESCOPE SYSTEM

A brief description of the X-ray optics, the focal plane and other instrumentation is contained in the sections which follow.

2.1 X-ray Optics

Specular reflection of X-rays is possible only at small incident angles under conditions of total external reflection. The critical angle, beyond which reflection does not occur, depends upon the density of electrons in the reflector and the energy of the incident X-ray. These requirements have led to the design of X-ray imaging systems which have two reflectors in the form of surfaces of revolution proposed by Wolter

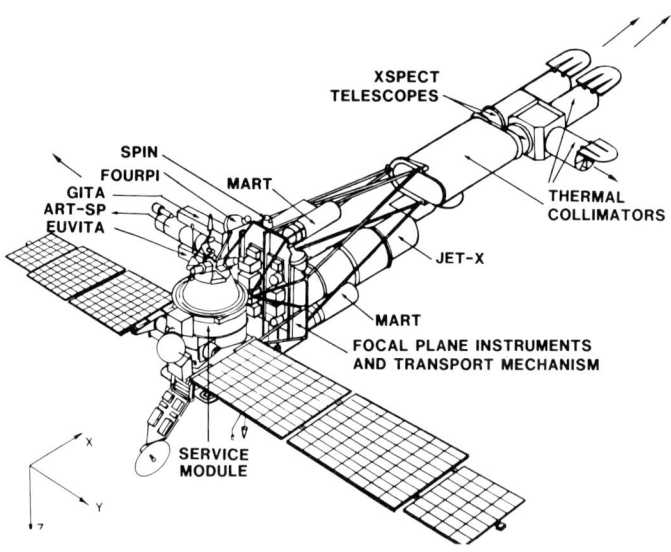

Figure 1. SPECTRUM RÖNTGEN GAMMA satellite. (R.S. Kremnev, Babakin Center)

(1952 a,b). The best use of the available telescope aperture is obtained by nesting as many pairs of reflectors as possible but care must be taken to minimize the deviation fron the condition of exact alignment and distortions from the exact shell figure.

Roughness, both short and long range, on scales as small as atomic sizes will introduce scattering halos around the geometrically perfect image of a point source. The angular diameter of the circle which encloses half of the photons in the image of a point source is defined as the Half Power Width (HPW) and is very useful in determining the performance of the system. It is usually referred to when questions of source confusion arise. For a recent review of these topics see Aschenbach (1985).

If the length of a conical surface is made sufficiently short, then the deviation from the ideal paraboidal and hyperboloidal Wolter 1 surfaces can be made extremely small. Ray tracing studies have shown that the contribution to the HPW from these deviations can be held to <20 arcsec. A quite large fraction of the aperture can be filled by densely nesting many shells of thin foils. This approach not only produces a very large collecting area, but also extends the energy range of the telescope to perhaps 20 keV, or more, since shells with very shallow grazing angles can be placed close to the optical axis.

The design goal for the XSPECT telescope HPW is <2 arcmin and the difference between this value and the contribution from the conical geometry represents the combined error budget which is allowed for all the manufacturing tolerances. Our approach, which follows the original suggestion by Serlemitsos, Petri, Glasser, and Birsa (1984), is to manufacture the telescope in two sections joined together by a flange which also serves as the interface with the optical bench. Each half is divided into independent

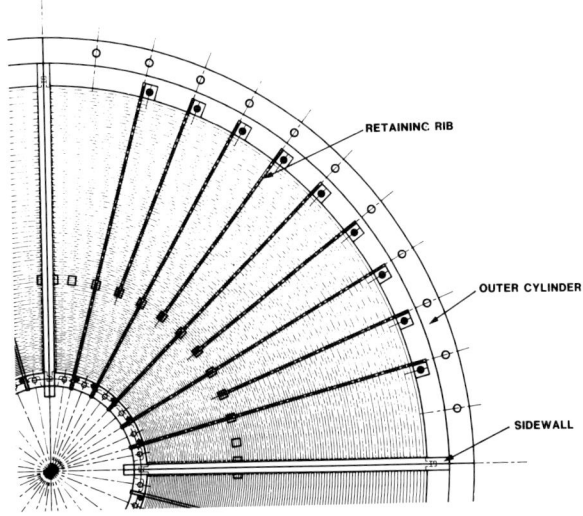

Figure 2. One quadrant of the XSPECT telescope. (J. Polny, DSRI)

quadrants which are keyed to a central reference hub (Figure 2). Foils are fixed into grooves in the quadrant side plates and held in place by adjustable radial spokes. Critical manufacturing steps are carried out in numerically-controlled, spark-cutting machines under tight temperature control.

Commercially available aluminum foils with thicknesses between 0.1 and 0.3 mm can be obtained with relatively smooth surfaces. Samples are evaluated by measuring the scattering of light from a He-Ne laser which samples the long range surface structure in the surface of the foil. Acceptable material is cut to the appropriate shape and rolled to approximately the correct figure. Next, the foils are coated with a thin layer of acrylic resin which tends to smooth out the short range residual graininess in the surface. Finally, a layer of gold (or perhaps other materials under study) is evaporated to form the high density reflecting surface. Surfaces prepared in this way can have a surface roughness on the order of 3 Å RMS which is smooth enough to remove totally the effects of small angle X-ray scattering. The long rang structure inherent in rolled aluminum foils is not removed and can contribute ~ 1 arcmin or more to the HPW (Kunieda and Serlemitsos, 1988).

Materials other than aluminum foils are being investigated by various groups. Tanaka and Makino (1988) report on plastic foils which are cut, rolled and glued to form the complete conical surface and are then assembled into the structure after gold evaporation. Laboratory tests are very impressive, but in flight thermal effects which could cause figure altering gradients may pose practical limitations. More promising is the possibility of using galvano-plastically deposited nickel foils (Hudec and Valnicek, 1985) either as full cones or as rolled foils. Laser scattering tests at DSRI of foils produced on ordinary window glass indicate very smooth surfaces since almost no scattering is observed. The foils are 20 x 30 cm in size and a large area in the central portion has a thickness variation of no greater than 1.6% (Hudec, 1988). The foils lie flat, an indication of relatively small internal stresses and a property which becomes important in insuring the proper figure of the rolled foil. The long range waves in the glass surface are, however, faithfully reproduced in the foil. Super-polished glass substrates are being prepared for the next production run.

Foils are mounted into the quadrant under the control of a well collimated optical beam 30 cm in diameter. The quadrant is mounted to the reference hub whose axis is made accurately parallel to the axis of the beam. This proceess allows individual foils to be evaluated as they are mounted and also will measure, apart from optical diffraction effects, the total contribution to the HPW from mechanical defects in the foils and the structure. The completed quadrant can be evaluated in a facility, now under construction, which will provide a pencil beam of extremely well collimated and monochromatic X-rays. The quadrant is mounted on an x-y table which can be used to make a raster scan of the beam. The reflected beam is registered by an imaging detector and the synthesized full beam image can be constructed by computer processing. Similar measurements can be made for each of the full four quadrant sections and for the fully assembled telescope. These tests will be repeated for several energies. It is expected that the final calibration will take place in orbit. The laboratory measurements are used only to control the manufacturing processes.

The parameters being considered for the baseline design are listed in Table 1.

Table 1. XSPECT Conical Approximation to Wolter 1 Optics

Focal length	8	m
Number of shells	ca. 118	
Outer shell diameter	60	cm
Inner shell diameter	16	cm
Shell length	20	cm
Shell thickness	0.3	mm
Minimum shell separation	0.8	mm
Total reflecting surface	65	m^2
Mass per telescope	50	kg
Field of view (FOV)	60	arcmin
Half power width (HPW)	<2	arcmin
Shell material		aluminum (nickel)
Reflecting material		gold (iridium)
Effective area (one module, without detector efficiency)	1700 1200 60	cm^2 @ 2 keV 8 20

Figure 3 compares the collecting area vs. energy which can be expected for a single XSPECT telescope with that of several other past or future missions. Measurements of reflection efficiencies for gold at several energies have been made at DSRI and have been found to agree well with the theoretical values which have been used to construct the XSPECT curve.

2.2 X-ray Detectors

A detector with a minimum diameter of 15 cm will match the focal plane scale of 2.3 mm/arcmin and a field of view (FOV) of 1 degree. The resolution of the detector should be fine enough to adequately sample the 4.6 mm diameter pixel which is equivalent

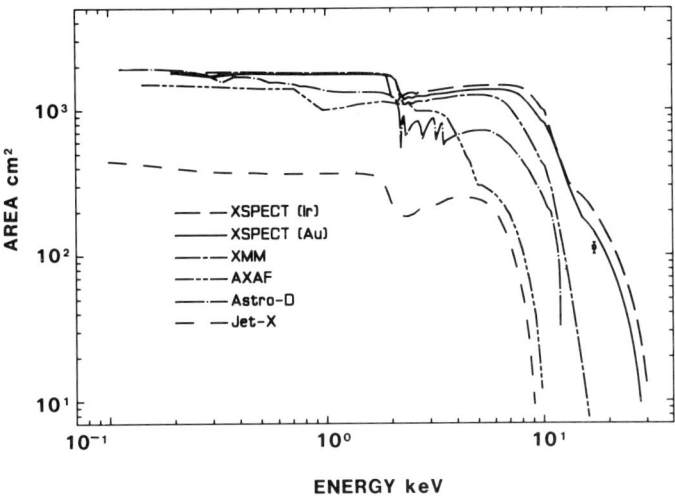

Figure 3. On axis effective area v.s. energy for various telescopes (single module, without detector efficiency). (N.-J. Westergaard, DSRI)

to the HPW (and allow for some improvement). The final requirement is that the detector span the entire range of energies reflected from the telescope with an energy resolution which will allow significant astrophysical interpretation of the data. From the wide variety of gas-filled and solid-state detectors, a large area position sensitive proportional counter and a small array of silicon solid state detectors have been chosen to meet the scientific requirements of the mission. Other detectors are under consideration but they are not discussed here.

2.2.1 <u>Position sensitive proportional counter (PSPC)</u>. Detectors of this type have been flown successfully on EINSTEIN, unsuccessfully on EXOSAT and are to be flown on ROSAT. They offer the advantage of well established technology and can be made to cover large areas. All gas filled detectors must have an entrance window and the strong variation of X-ray attenuation in typical window materials, particularly at energies <2 keV, makes it difficult to cover the very broad XSPECT energy range in a single detector. The baseline detector system consists of two independent detectors, one for low energy which has a thin stretched plastic window and a gas flow system to replenish leakage and the other for high energy which has a sealed beryllium window. The characteristics of these detectors are summarized in Table 2.

Table 2. XSPECT Detector Specifications

	Low Energy	High Energy
Diameter of active area	15 cm	15 cm
Energy range	0.5 - 3.0 keV	2 - 20 keV
Energy resolution	25% at 2 keV	13% at 6 keV
Position resolution (x and y)	2 mm	1 mm
Quantum efficiency	>0.7; E>0.8 keV	>0.7; 2<E<10 keV

At present, multiwire (MWPC), microstrip (MSPC) and parallel gap (PGPC) proportional counters are under development at DSRI (Budtz-Jørgensen, et al, 1989). Of these, the first has been chosen as the baseline even though it does not meet the (self-imposed) energy resolution requirements specified in Table 1. The details of this detector have been given elsewhere (Madsen, et al, 1985). Only two layers of wires are used; an anode plane with 12 μm NiCr wires placed 2 mm apart and a field-controlling cathode plane with wires also spaced 2 mm apart located 1.5 mm above and parallel with the anode plane. The wires are supported at each end on insulating bars which positions each wire to within ±2 μm.

A cathode plane divided into four segments is located at a suitable distance below the anode plane. Each segment is divided into strips which are interleaved with those of the other segments. Primary electrons created in the initial X-ray absorption event drift towards the anode wires in the relatively weak field between the entrance window and the cathode wires. An avalanche occurs in the strong field surrounding the anode and the movement of ions and electrons induces charges of opposite polarity on the anodes and cathodes. The geometry allows more than 50% of the charge produced in the avalanche to be induced on the cathode plane. The variation in strip geometry allows the centroid position of the charge cloud to be calculated from the charge signals collected on the four segments.

The dominant contribution to the detector resolution is noise in the cathode plane preamplifier and it is, therefore, important to reduce the interelectrode capacitance on the cathode plane and to use low noise preamplifiers (less than 300 rms electrons). Results obtained from a breadboard detector filled with a mixture of Ar + 10% CH (P10) gave position resolutions of 0.42 and 0.34 mm (FWHM) and energy resolutions of 40% and 22% (FWHM)for 1.5 and 5.9 keV X-rays, respectively. Pulse height and rise time discrimination together with inter-electrode voltage comparison yield a combined background rejection efficiency of >99% and an acceptance efficiency of 71% at 5.9 keV (Madsen, et al, 1985).

Limitations imposed by having suspended wire anode and cathode planes restrict the achievable energy resolution. It is difficult to insure that the wires stay parallel in the applied electric field and non-uniform gas amplification is a commonly seen, particularly in large area detectors. A novel approach, in which the wires are replaced by very narrowly spaced conducting "microstrips (MS)" accurately deposited (±2 μm) on an insulating substrate, has made it possible to circumvent this problem (Oed,1988). The wire planes in the DSRI breadboard detector were replaced by a MS which was supplied by Oed. A gas gain of 10^3 was obtained at a voltage of 600 V and energy resolutions of 27.4% and 13.4% were obtained with P10 gas for 1.5 and 5.9 keV X-rays, respectively. Similar resolutions were obtained with propane gas (Budtz-Jørgensen , et al, 1989). Tests with new thin large area MS plates which are designed to induce strong signals on the cathode segments are in progress. Various readout schemes are also under study.

Ultra-thin, strong window materials which are nearly leak-proof have recently become available (Sipilä, 1988). At present they are being produced with relativly small areas but it is expected that the requirements for the XSPECT detectors can be met in the near future. The window transmission is equivalent to 0.5 - 1.0 μm of stretched polypropylene. The ideal solution for XSPECT is to combine both the high- and low energy detectors into a single unit which has an ultra thin window and a MS plate and, thereby, fulfills the requirements specified in Table 2.

2.2.2 Solid state spectrometer array (SSSA). The significant improvement in energy resolution obtained from these devices more than compensates for the requirement that the detector be cooled. Solid state X-ray detectors are usually made from lithium-drifted-silicon [Si(Li)]. The primary difference from a gas detector is that, although much less energy is required to liberate an electron within the silicon, there is no equivalent to gas amplification. The primary charge must be detected by extremely low noise, cooled preamplifiers which can contribute ~ 100 eV to the FWHM and this is the limiting resolution for X-rays with energies below ~ 2 keV. At 6 keV, a resolution of 150 eV ($E/\Delta E=40$) is common, but recent preamplifier improvements have lowered this value to 130 eV ($E/\Delta E=46$) with expectations of 50 eV ($E/\Delta E=10$) at 500eV.

The combination of layers of gold which acts as the charge collector and of non-active ("dead") silicon on the front surface of the detector limits the low energy response of the detector while the thichness (usually 2-4 mm) limits the high energy response. Filters to reduce infrared, visible and ultraviolet background also limit the low energy response. Imaging detectors can be either arrays of discrete elements or segmented blocks within a large single crystal. In either case it would be difficult to match the pixel size requirements of a high resolution telescope such as AXAF, but the needs of XSPECT do not pose a practical limitation.

The deep orbit makes it possible to use a radiative cooler to bring the detector system to the proper operating temperature. The number of pixels will be limited be the minimum power required by the preamplifier and the heat leaks between the detector assembly and the heat radiator. The minimum design is 7-9 pixels with 4-6 mm diameter, but the number could increase when ongoing development work is evaluated.

2.3 Objective Crystal Spectrometer (OXS)

Bragg crystal instruments can provide very high energy resolution but with low efficiency since the multiplex advantage inherent in broad band and grating spectrometers is lost because of the restrictive nature of the X-ray diffraction process. In addition, their high dispersion insures that the effects of telescope blur do not dominate the energy resolution and it is, therefore, possible to design instruments which will operate successfully in conjunction with low resolution telescopes. Substructure within the helium like emission lines including the diagnostically important satellite lines can be resolved for all of the cosmically abundant species and, in some cases, the structure of individual lines can be studied. Many important spectral lines are well separated in energy and bright sources can be imaged at each energy without the confusion of overlapping common for grating instruments.

The approach chosen for the XSPECT spectrometer is to separate the spectroscopic and imaging functions of the instrument. A large, flat, free-standing crystal would be a fine spectrometer for a bright point source or, with suitable mechanical collimation, for an extended source. Excellent solar spectra have been obtained this way. For the relatively weak cosmic X-ray sources the background introduced from the necessarily large active detector volume overwhelms the signal. With a telescope following the crystal (Figure 4) the background attributable to the detecting pixel volume shrinks dramatically and extended sources can be imaged as well (Schnopper and Byrnak, 1987). The angular range for which a crystal reflects a monochromatic beam is typically 0.5-1.0 arcmin and, therefore, this concept is ideally suited for XSPECT.

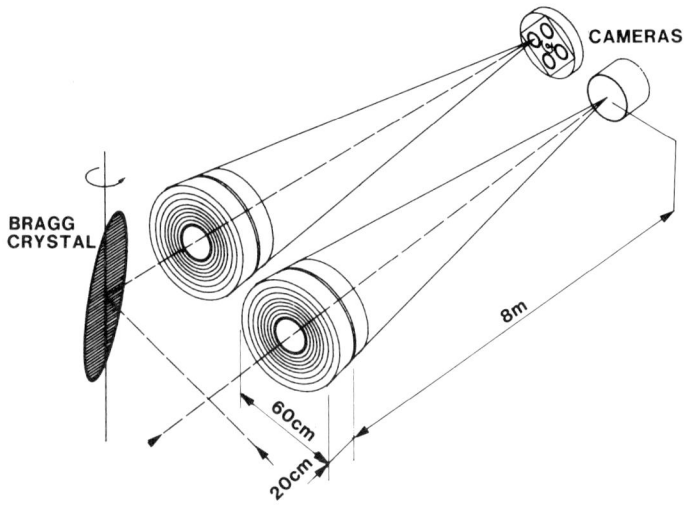

Figure 4. XSPECT spectrometer, telescope and detectors.

2.4 The XSPECT Telescope Assembly

XSPECT is illustrated in Figures 1 and 4 and consists of the telescope modules; the Bragg crystal panel; the complete complement of focal plane instruments; the structural, thermal and focal plane assemblies; and the star tracker. The telescope is launched in a folded configuration which allows the satellite to fit within the shroud of the PROTON launcher and is deployed when orbital altitude is reached. Babakin Center is responsible for the structure and deployment mechanism, the thermal collimators (which keep the telescopes within the allowed temperature range and the focal plane transport mechanism. The Institute of Electromechanics of the USSR Academy of Sciences is responsible for the thermal radiator.

3 SCIENTIFIC PROGRAM

The large collecting area and broad energy range provided by the telescope and the deep orbit of the satellite will lead to a comprehensive observing program. With low resolution telescopes such as XSPECT, source confusion can become a limiting factor in the performance of the telescope. The slope of the relationship between the number of sources N(>S) with flux greater than S together with the spectral shape of a typical source determine when a particular telescope system becomes confused. Sky surveys already carried out are consistent with a slope of -1.5 for the logN(>S)-logS curve, but recent work by Hamilton and Helfand (1987) shows that fluctuations in the pixel to pixel background data indicate a flattening of the curve for flux levels below the EINSTEIN deep survey limit [$S_0 = 2.6 \times 10^{-14}$ erg cm^{-2}s^{-1}(1-3 keV) and 4.0×10^{-14} erg cm^{-2}s^{-1} (3-10 kev)]. The implications are that for $S<S_0$ there can be many more faint objects than expected from extrapolation of surveys for which $S>S_0$ and that these faint sources will not be detected with the most sensitive telescopes now planned.

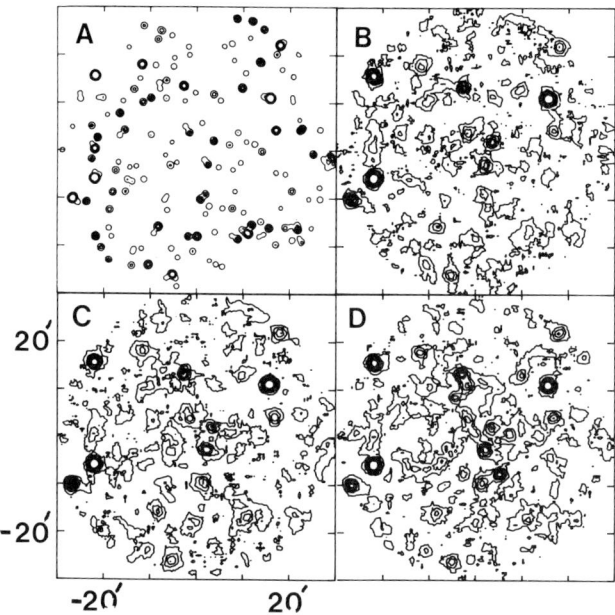

Figure 5. Simulated response of the PSPC for 10^5s exposures for sources distributed according to LogN(>S)-LogS with various slopes. a) The input catalog; b), c) and d) Slopes of 1.05, 1.20 and 1.50. (N.-J. Westergaard, DSRI)

Simulations for observations of 10^5 s have been made under various assumptions for the slope of the log(N>S)-logS curve. These are shown in Figure 5. The point response function of the telescope has been modeled to included construction errors and scattering up to a HPW of 2 arcmin^2. Sources with spectra similar to the X-ray background are assumed. The results indicate that sources with fluxes well below S_0 can be detected if the distribution is truely flattened. Knowledge of precise source positions to be obtained from the ROSAT catalog will be invaluable in sorting out the problems arising from confused regions. Assuming a 4 arcmin^2 resolution and requiring that there be 40 beam areas surrounding each source there are 22.5 resolved bins/deg^2 or 10^6 on the whole sky @ $S > 8 \times 10^{-14}$ erg cm^{-2}s^{-1}.

A few examples illustrate the sensitivity of the instrument. The Seyfert galaxy NGC4151 (d=20 Mpc, z=0.004) with $S \simeq 10^{-10}$ erg cm^{-2} s^{-1} (2-10 keV) could be detected at a distance of 750 Mpc (z=0.15, $S=7 \times 10^{-14}$ erg cm^{-2} s^{-1}). The quasar 3C273 (d=950 Mpc, z=0.2) with $S=6 \times 10^{-11}$ erg cm^{-2} s^{-1} could be detected at a redshift for which evolution is an important factor. The Coma Cluster (d = 113 Mpc, z = 0.02) with $S = 3.2 \times 10^{-10}$ erg cm^{-2} s^{-1} could be detected at a z=0.5 with 1 count s^{-1}. Cooling flows could be detected at z=0.3. These detections include physically significant measurements of spectral index or temperature.

The large collecting area combined with the broad energy range will allow studies of the surface brightness distribution of clusters at large distances from the center and, if they exist, measurements of the surface brightness distribution of superclusters. Dark matter would contribute to both of these distributions.

A 10^5 s observation binned in 10 arcmin^2 bins will include ~140 and <50 counts of signal and noise, respectively. The gain uniformity in the gas detector will permit precise studies of the bin-to-bin fluctuations to be made which will reveal the spectrum and luminosity function of the unresolved portion of the X-ray background. These studies should also discover 1 quasar every 10^4 s or 10^3 sources/year with 25% observing efficiency.

Provided that they are not confused, bright x-ray sources in nearby galaxies, i.e. a CYG X-1 in M31 and SN1987A in the LMC, are easily detected. M31 is a good example since all the EINSTEIN sources with the exception of those at the very center of the galaxy could be studied in great detail.

Within the Galaxy there are many topics to chose from where the XSPECT system can be expected to provide excellent data. Of particular importance are sub-millisecond studies of time variability, spectra from all classes of stars, plasma diagnostics of the interstellar medium, diagnostics of accretion discs, and high- and low resolution temperature and density mapping of supernova remnants. Emission lines from compact objects which can be broadened by various scattering processes can be detected over the broad XSPECT energy range as can low energy cyclotron emission- and/or absorption features.

4. ACKNOWLEDGEMENTS

The staff at DSRI including: B.P. Byrnak, F.E. Christensen, A. Hornstrup, P. Jonasson, M.M. Madsen, H.U. Nørgaard-Nielsen and N.-J. Westergaard have made significant contributions to the work reported here. I thank them warmly for their assistance in preparing the manuscript.

5. REFERENCES

Aschenbach, B. 1985, *Rep. Prog. Phys.*, **48**, 579-629.
Budtz-Jørgensen C., Madsen, M.M., Jonasson,P., Schnopper, H.W., and Oed, A. 1989, *SPIE Proc.*, **982**, in press.
Hamilton, T.T., and Helfand, D.J. 1987, *Ap. J.*, **318**, 93-102.
Hudec, R., and Valnicek, B. 1985, *SPIE Proc.*, **597**, 111-118.
Hudec, R. 1988, *private communication*.
Kunieda., and Serlemitsos, P.J. 1988, *SPIE Proc.*, **830**, 12-15.
Madsen, M.M., Jonasson, P., Jensen, P.L., Rasmussen, H.E., Ørup, P., and Schnopper, H.W. 1985, *SPIE Proc.*, **597**, 199-205.
Oed, A. 1988, *Nucl. Instr. Meth.*, **A263**, 351-359.
Schnopper, H.W., and Brynak, B.P. 1987, *Appl. Opt.*, **26**, 2871-2876.
Serlemitsos, P.J., Petre, R., Glasser, C., and Birsa, F. 1984, *IEEE Trans. Nuc. Sci.*, NS-31, 786-790.
Sipilä, H. 1988, *private communication*.
Tanaka, Y.,and Makino, F., 1988, *SPIE Proc.*, **830**, 242-244.
Wolter, H. 1952a, *Ann. Phys.*, **10**, 94.
    1952b, *Ann. Phys.*, **10**, 286.

DISCUSSION-H. Schnopper

S. Kahn: Why are you using a single crystal plane rather than a venetian blind approach, which would take less room in front of the mirror?

H. Schnopper: A venetian blind would only be effective for shallow incident angles, i.e. for reflection gratings. At $\sim 45°$ incident angle the blind elements would occult each other; a single plane is the best geometry.

B. Smith: Did you mention the wavelengths your crystals will cover?

H. Schnopper: We plan to cover the oxygen, silicon, sulfur and iron K energy bands.

P. DeKorte: What will be the spectral resolution of the Bragg Spectrometer taking into account the system resolution and the spacecraft attitude uncertainties.

H. Schnopper: We expect an $E/\Delta E$ in the range 5×10^2 - 2×10^3, with the best performance at iron.

JET-X
A JOINT EUROPEAN X-RAY TELESCOPE FOR SPECTRUM-X

A. Wells[1], D.H. Lumb[1], K.A. Pounds[1], G.C. Stewart[1], B. Aschenbach[2],
H. Brauninger[2], G. Hasinger[2], J. Trumper[2], O. Citterio[3], L. Scarsi[4],
A. Peacock[5] & B. Taylor[5].

[1]Physics Department, Leicester University, Leicester, U.K.
[2]Max Planck Institute fur Extraterrestriche Physik, Garching, W. Germany.
[3]Osservatorio Astronomica di Brera, Merate, Italy.
[4]Instituto Fisica Cosmica e Informatica del CNR, Palermo, Italy.
[5]Space Science Department, ESTEC, Netherlands.

Abstract

The Joint European X-Ray Telescope, JET-X, is one of the core instruments of the scientific payload of the USSR SPECTRUM-X astrophysics mission due for launch in 1993. The JET-X instrument concept is described and its scientific performance and capability discussed.

1 INTRODUCTION

X-ray emission constitutes an important fraction, often most, of the energy emitted from whole classes of astronomical objects ranging from stars to distant clusters of galaxies and the cosmic X-ray background. The Joint European X-ray Telescope, JET-X, is designed to study the X-ray emission from these sources in the band 0.15-10 keV; particularly to meet primary scientific goals in cosmology and extragalactic astronomy. JET-X has been selected for the core payload of the USSR's SPECTRUM-X Project whose combined response extends over the range 20eV-100keV.

JET-X is being developed by a consortium from the UK groups from the Universities of Leicester and Birmingham; the Rutherford Appleton Laboratory; the Mullard Space Science Laboratory; the Max Planck Institut for Extraterrestrial Physics, Garching, W. Germany; groups from the CNR and Universities in Milan, Rome and Palermo, Italy and the Space Science Department of ESTEC, ESA.

JET-X consists of two identical, coaligned X-ray imaging telescopes, each with a spatial resolution of 30 arcsecond or better. Focal plane imaging is provided by a cooled X-ray sensitive CCD detector which will combine high spatial resolution with good spectral resolution, with particular emphasis on high sensitivity and spectral resolution ($\frac{E}{\Delta E} \geq 50$) around the 7 keV Fe-line complex. An optical monitor is co-mounted with the X-ray telescopes to permit simultaneous observation and identification of the optical counterparts of X-ray target sources.

JET-X is conceived as a stand-alone instrument with simple interfaces to the SPECTRUM-X spacecraft. Development risks have been minimised by selecting mature technology for the mirror and detector systems; the mirrors will be produced using the replication techniques already under development for the Italian SAX mission, and the detectors will be based on development from current Leicester and ESTEC CCD programmes. JET-X will generate its own aspect solution, through pointing measurements with an integrally mounted attitude monitor and incorporates on-board data handling electronics for controlling the instrument with mass memory sufficient to store 24 hours-worth of scientific data. Data will be read out, on command, by the spacecraft on-board computer and transmitted to the ground receiving station via the spacecraft telemetry system during ground contacts.

2 INSTRUMENT DESCRIPTION

2.1 Design Objectives

The primary objectives of JET-X are:

1. Imaging with ≤ 30 arcsec resolution with a limiting sensitivity at 1 keV of ~ 0.5 nJy

2. Medium resolution spectroscopy ($\frac{E}{\Delta E} \geq 10$) in the 1-10 keV band with emphasis on high sensitivity and spectral resolution ($\frac{E}{\Delta E} \geq 50$) around the 7 keV Fe-line complex.

3. Time variability of X-ray spectra on timescales ranging from milliseconds to months.

4. Simultaneous (and continuous) optical monitoring of the X-ray sources to a limiting magnitude of $m_v \sim 22$.

The limiting x-ray sensitivity will be ~ 3 orders of magnitude better than previous sky-survey instruments, better than that of the EINSTEIN Observatory (by 1 order of magnitude) and comparable to that of the ROSAT XRT but with a wider energy range. Observations with JET-X will be of significant importance for cosmological studies. The instrument will be able to study the emission from large, statistically significant, samples of all the known classes of X-ray source. JET-X provides one of the first opportunities in X-ray astronomy to use a CCD detector at the focus of a high resolution imaging telescope. The energy resolution is sufficiently high to use the diagnostic emission and absorption features in the source spectra to study the emission mechanisms, chemical composition and the environment of cosmic sources in more detail than has hitherto been possible.

Simultaneous and complementary observations can be carried out using any of the other instruments in the SPECTRUM-X payload. For example, low energy (0.1-1.0 keV) imaging photometry with moderate spectral resolution using JET-X can be combined with the imaging continuum spectrophotometry offered by proposed extreme ultraviolet imaging instruments. Alternatively, simultaneous spectrophotometry and imaging with JET-X can be combined with time-resolved spectroscopic studies with the large area, lower resolution foil telescope X-SPECT instrument provided by IKI and the Danish Space Research Institute. Naturally, response overlaps between the different instruments allow for simultaneous cross-calibration of the responses in the overlapping wavebands.

2.2 Instrument Concept

Two identical telescopes, each with a nested array of 12 mirrors, aperture of 0.3m and a focal length of 3.5m, have a total effective area of 360 cm^2 at 1.5keV, 140 cm^2 at 8keV. The angular resolution will be better than 30 arcseconds with a design potential of 10 arcseconds. Figure 1 shows the effective collecting area over the energy range.

The focal plane detector provides a spatial resolution compatible with the optics ($<$ 100 microns), together with broad band spectroscopy with a resolution in the 0.15-10keV energy band in the range 5 to 60. A CCD instrument has been chosen for its potential to provide the necessary energy resolution, simultaneously with good quantum detection efficiency and background rejection, together with a spatial resolution which suitably oversamples the telescope angular response function.

Front-illuminated (FI) three-phase CCDs covering the energy band 0.8-10 keV have already been developed to a sufficiently advanced stage to meet the requirements for the JET-X instrument (Chowanietz, Lumb and Wells, 1986). These devices are fabricated on high resistivity (1000 ohm-cm) silicon to obtain good quantum detection efficiency at energies above 3 keV. A back-illuminated (BI) version is now under development using higher resistivity epitaxial silicon, to further enhance the higher energy response, and allow the energy response to be extended down to 0.2 keV.

A 3 x 3 array of CCDs covers a field of view of 40 arcmin x 30 arcmin in the field of view of each telescope but dead spaces between the individual CCDs limit the field coverage to 75 %. By mounting the arrays at right angles to each other in the two telescopes, an overall field coverage of 95% is achieved with 20% of the field covered at reduced efficiency. With the CCDs cooled to around 170-180 K, energy resolution of 150 eV FWHM at 8 keV can be achieved over the whole field of view. The quantum detection efficiency and other device parameters are summarised in Section 2.4.

An option under consideration is to mount a retractable grazing incidence synthetic multi-layer structure in one telescope to provide high dispersion in the vicinity of the iron K- line ($\frac{E}{\Delta E} \geq 350$), with the CCD acting as the readout for the dispersed spectrum.

The two telescopes will be co-aligned in a common structure, together with an attitude monitor which will provide an independent attitude solution to 5 arcsec accuracy, or alternatively to provide spacecraft attitude correction data for JET-X pointings. An optical monitor with sensitivity down to $m_v = 22$ to allow simultaneous optical observation on the chosen X-ray targets is currently under study. Electronics units drive the CCD detectors, digitize and process the x-ray images and deliver digital science data to a JET-X central data management unit which monitors the JET-X instrument status and housekeeping and controls the optical and attitude monitors.

The SPECTRUM-X data handling system consists of a spacecraft on-board computer (OBC) which can combine data from up to 16 separate experiments. The OBC will control data handling for downlink and provide instrument control through up-linked commands. Telemetry is provided via a low speed link (up to 65 kb/s) for command verification and housekeeping and a high speed link (up to 18 Mb/s) for science data. Telemetry and telecommand links are for up to 1.5 hours daily during normal mission operations. These operational constraints mean the instrument requires its own on-board mass memory with capacity to store 24-hours worth of science data. The JET-X mass memory will be part of the central data handling unit and will be read-out on command by the spacecraft OBC during ground contacts.

2.3 Telescope Mirrors and Mirror Mount

Arrays of nested electroformed mirror shells, using Wolter-I geometry, are used to form the x-ray imaging optics for JET-X. The construction technique is based on replication in which a mandrel is machined to the required paraboloid-hyperboloid figure and then super-polished to a surface finish of around 5 Å rms. A gold layer is deposited (by sputtering or evaporation) on to the mandrel surface, and a nickel layer is electroformed on top of this. Next the nickel shell is separated from the mandrel and the gold layer adheres to the nickel.

This technique has already reached an advanced state of development for the Italian SAX programme, where nickel replica mirrors 1 mm thick and with diameters of 165 mm and 65 mm have been manufactured and shown to have an angular resolution (HPW) of 60 arcsec at 8 keV (limited mainly by the conical approximation) when mounted in a nested pair (Citterio et al., 1987). For JET-X, two modifications to the SAX technique are being pursued which will lead to the attainment of better intrinsic angular resolution whilst enabling larger diameter mirrors to be produced. The conical approximation of SAX will be replaced by a true Wolter I configuration. This will ensure that on-axis resolution is dominated by figure errors and surface finish only. The target specification per mirror is 5 arcseconds. Secondly, the nickel layers will be made thicker (2 mm) so as to ensure better stability, especially with the larger mirrors.

The mirror parameters are summarised in Table 1. Figure 2 shows the mirror mounting system.

Table 1 Mirror Characteristics

Field of View	20 arcmin (50% vignetting)
Angular resolution	10-30arcsec (HPW)
Focal length	3.5 m
Reflecting surface	Gold
Configuration	Wolter-I
Mirror length	2 x 300 mm
Outer mirror diameter	300 mm
Inner mirror diameter	87.5 mm
Shell thickness	2 mm
Number of shells	12
Distance between shells	3-5.8 mm
Surface finish (microroughness)	5 Å rms

2.4 Focal Instrument Assembly

Development of a 3x3 array of fully depleted back-illuminated CCDs is currently in progress. Performance data has so far been obtained on front-illuminated fully depleted devices (Lumb and Holland, 1989). The imaging capability was demonstrated by resolution tests using a mask with an array of 150 micron pinholes spaced 0.5 mm apartand and illuminated with 1.5 keV radiation. Resolution down to an equivalent telescope HPW of 10" was achieved.

Figure 3 compares the predicted efficiency for a 30 micron depletion depth with measured data with total detection efficiency (including split events) and detection efficiency for isolated pixels, both being recorded. The good spectroscopic response of the CCD is demonstrated by the similarity

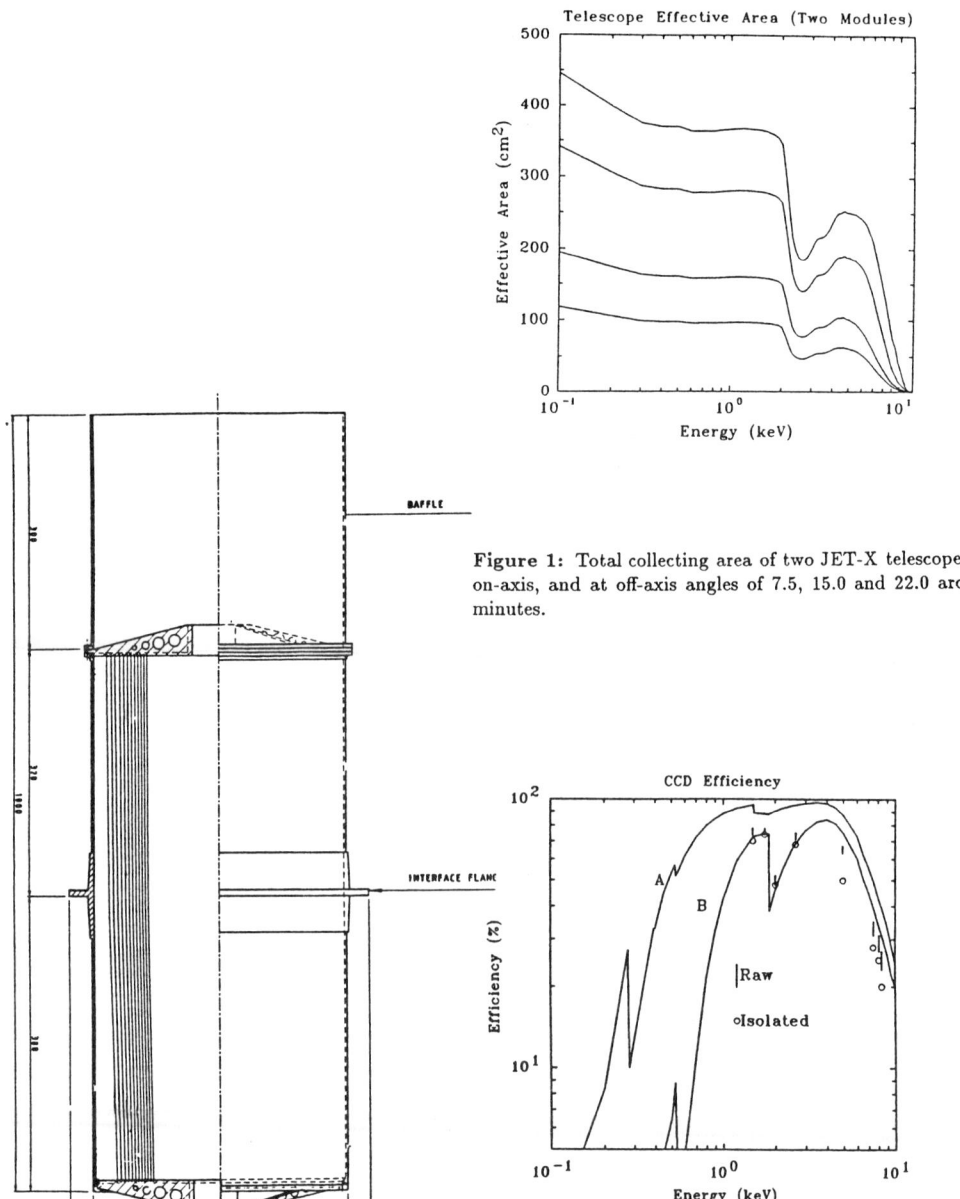

Figure 1: Total collecting area of two JET-X telescopes on-axis, and at off-axis angles of 7.5, 15.0 and 22.0 arc-minutes.

Figure 2: Design for the mirror mounting system.

Figure 3: Current efficiency measurements for a 30 micron CCD. Both raw data (bars) and isolated pixel data with full spectral resolution (circles) are shown. For comparison curve B shows the theoretical prediction for such a device. Curve A shows the predicted efficiency for the JET-X baseline detector - a 40 micron thick, back-illuminated CCD including the attenuation of an aluminised light filter.

between these two data sets. Figure 3 also shows the target response of the back-thinned device.

Figure 4 shows the energy resolution resulting from isolated event data with the pulse height distributions obtained by illuminating the CCD with characteristic K- emission lines of various metallic elements, with approximately 2000 photons incident in each line. This illustrates the effects of both energy resolution and efficiency. Neighbouring elements are well resolved down to an energy of ~1keV, whilst H- and He-like ion lines are resolved down to an energy of ~2keV. Figure 5 shows that measured spectral resolutions agree precisely with the prediction from equation 1 for the measured CCD noise level of 10 electrons rms and the expected photoelectron Schott noise contribution for a Fano factor of 0.12.

$$\Delta E(FWHM) = 8.58(N^2 + FE/3.65)^{0.5} \tag{1}$$

where N = rms electron noise, F = Fano factor for silicon (0.12), and E = Photon Energy (keV).

Figure 6 shows the pulse height response (both before and after pixel anti-coincidence rejection is applied) to Co-60 induced photo-electrons which are close to minimum ionising and therefore representative of the in-orbit background response of the CCD to protons and cosmic ray ions. A combination of anti-coincidence and energy veto allows an overall rejection efficiency of > 98%.

Thinning and surface treatment processes are being developed to enhance the soft X-ray and short wavelength visible light responses. The CCD substrate can be etched back to the edge of the depletion region, so as to ensure high detection efficiency below ~1keV, with this surface treated to ensure that photo-electrons generated near the surface are pushed towards the depletion layer. A high concentration p-type layer is formed by ion implantation, and activated by annealing. The latest data show a dead surface layer of only ~500 Angstroms, which is confirmed by optical efficiency data at a wavelength of 400nm where an efficiency of 60% is achieved. The absorption lengths of silicon at 0.5keV and 400nm are comparable, indicating that excellent soft X-ray efficiencies will be attainable. These techniques are now being applied to the deep depletion CCDs, and it is expected that the instrument detection efficiencies below 1keV will be determined largely by the aluminised lexan light block rather than the CCD itself. The CCD characteristics are summarised in Table 2

Table 2 CCD Detector Characteristics

Energy Range	0.2-10 keV	(> 20% efficiency)
Geometric Area:		
(single detector)	12.7 × 8.6 mm²	
(3*3 CCD array)	42 × 30 mm²	
Spatial Resolution	< 0.1 mm	
Background rejection	> 98.5%	
Spectral Resolution	See Figures 4,5	
Quantum Efficiency	See Figure 3	
Time Resolution:		
(imaging mode)	20 s	
(timing mode)	1 ms	
Operating Temperature	170 – 180 K	

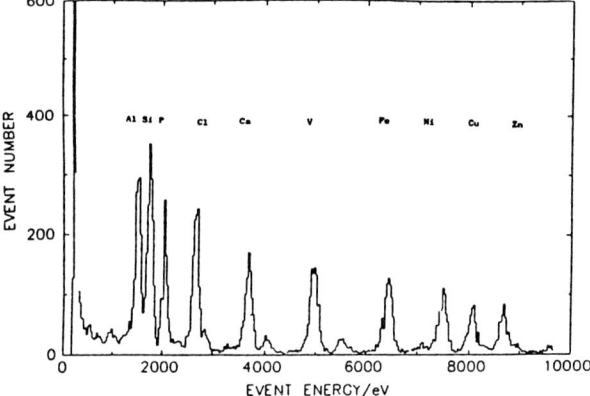

Figure 4: Measured pulse height response to a range of K emission lines of different elements.

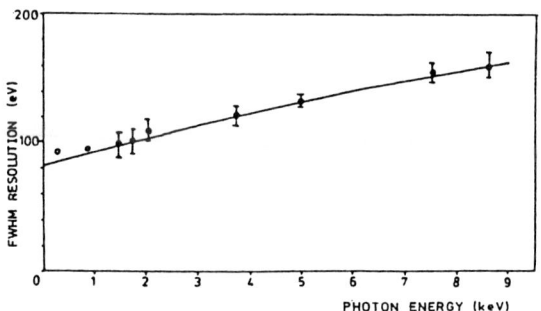

Figure 5: Comparison of measured and predicted energy resolution. Open circles represent predictions for the sub-keV performance in a back-illuminated CCD with same noise.

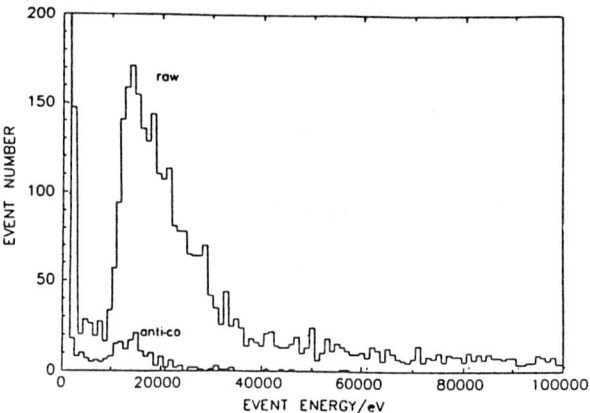

Figure 6: Pulse height response to charged particle background. Upper data - total energy deposition; lower data - residual data after pixel-pixel anti-coincidence.

2.5 Optical Monitor

In addition to providing *post facto* attitude reconstruction, the optical monitor can simultaneously observe the optical counterparts of the X-ray sources and serendipitously study stellar microvariability. The first objective is directly related to the astrophysical understanding of X-ray sources while the second allows JET-X to undertake a completely novel form of optical astronomy, because of the ability of JET-X to monitor continuously for up to 10^5 s.

The optical system consists of an F/5.7 Ritchey-Chretien telescope with a field corrector and an aperture of 30cm. The detector system is an optical CCD system which would have a high quantum efficiency leading to a photometric accuracy in a 1s integration of 0.04 mag for a star with $m_v \sim 15$. Read-out noise in the CCD would however lead to a limiting magnitude of $m_v \sim 20$, given the constraint on readout time imposed by the requirement that the monitor can also give attitude reconstruction measurements.

2.6 Background Monitor

The CCD detectors must be switched off for perigee passage and for occasions such as solar flares, when the background particle rates may be excessive. Whilst the former may be accommodated automatically within the mission timeline, the provision of a background monitor is required to allow the detection of a high background environment.

2.7 Layered Synthetic Mono-structure

A retractable synthetic multi-layer crystal may be included in one telescope module to give JET-X the additional capability of high resolution Bragg spectroscopy. The baseline design is a crystal constructed of 420 Ni/C layers, each 10 Å thick. The crystal is shaped to intersect all X-rays with a constant Bragg angle in the light-path from the mirrors. The Bragg angle is chosen such that the same focal plane detector as used for imaging studies can be used. An annular dispersed spectrum is then focussed on the CCD detector. The operating energy is $\sim$ 7keV with fine tuning provided by a small stepping motor. A resolution of $\frac{E}{\Delta E} \sim 350$ is currently possible, sufficient to resolve the individual components of the Fe XXV He-like triplet. It is believed that a resolution of 500 will be possible.

Preliminary measurements show that the throughput of this system is high, reflectivities of $\sim$ 30% having been attained. Further work will be undertaken to further improve this figure, the goal being a reflectivity of 50%.

3 SCIENTIFIC CAPABILITY

3.1 Instrument Performance

The derived instrument performance assumes the telescope and detector data given in tables 1 and 2, a spatial resolution of 20 arcseconds, and a back-illuminated CCD. The anticipated background has three components :

a) The cosmic X-ray background: $N(E)dE = 8E^{-1.4}$ph cm^{-2}s^{-1}keV^{-1}sr^{-1}

b) A galactic diffuse X-ray background: $N(E)dE = 120E^{-1.0}\exp(\frac{-E}{0.3})$ph cm^{-2}s^{-1}keV^{-1}sr^{-1}

c) Particle background for altitudes > 40000 km: 1.5×10^{-2}cts^{-1}cm^{-2} in the 0.1 - 10keV band after rejection.

3.1.1 Point Source Sensitivity

In Figure 7 the time taken to achieve a 5σ detection of the source count rate integrated over the instrument passband is shown as a function of the 0.5-10 keV flux for a source with a power-law spectrum of photon index 1.7, with low energy absorption corresponding to intervening cold matter with a neutral hydrogen column density of 10^{20}cm^{-2}.

Also shown in the figure is the confusion limit (calculated using the 1 source per 30 beamwidth criterion) for a detector with 20 arcsec beamwidth. The limit has been extrapolated from the observed source number - flux relation measured for extra-galactic X-ray sources with the Einstein Observatory by assuming a continuing Euclidean population of sources. (The correction for the differing passbands of the instruments has been made assuming a power-law spectrum with photon index of 1.7 as is measured for the majority of local AGN).

The time taken to reach this confusion limit, roughly 10^5s, is close to the expected orbital period of the SPECTRUM-X satellite, which therefore might be a typical exposure time, showing that the instrument throughput is well matched to its spatial resolution.

A by-product of the high sensitivity is that flux variability of approximately 10-20% on timescales of a few tens of seconds will be detectable in sources at flux levels of $\sim 1\mu$Jy. The simultaneous background measurement available with an imaging detector means that systematic errors will not substantially affect this type of measurement.

3.1.2 Spectral Capability

The energy resolution afforded by a CCD is well matched to the spectral properties of thermal plasmas with temperatures in the range $10^6 - 10^8$K. In addition to the strong iron lines at around 7 keV, other strong lines from various ionisation stages of iron (at $\sim$ 1keV), silicon, sulphur, oxygen, calcium and argon fall within the passband of JET-X. JET-X is capable of resolving the resonance lines from the He- and H- like ionisation stages of these elements.

The time taken to measure lines at 7 kev with equivalent widths of 2 keV, 1keV and 100eV, with respect to a power law continuum with a photon index of 1.7, is shown in Figure 8. A line with an equivalent width of 1keV can be detected at the 5 sigma level in 3×10^5s for sources ten times brighter than the confusion limit (e.g. high redshift clusters). A value of 750eV is typical for cosmic plasmas with temperatures in the 2-8 keV range. (At this observation length the line detection is still essentially photon limited and the limiting line intensity is $\sim 4 \times 10^{-6}$ ph cm^{-2}s^{-1}). For sources down to $\sim 100\mu$Crab strength lines of 100eV equivalent width can be detected.

In addition to the medium resolution spectroscopy provided by the CCD detectors the inclusion of the layered synthetic monostructure (LSM) would enable studies at much higher resolution ($\frac{E}{\Delta E} \sim 350 - 500$) in the region of the Fe K line at $\sim$ 7keV to be made. For the current 30% reflective efficiency the limiting line sensitivity in a 10^5s observation would be $\sim 10^{-5}$ phcm^{-2}s^{-1}. For sources brighter than 10^{-12} erg cm^{-2} s^{-1} (1-10 keV), 60 times the EINSTEIN deep survey limit, lines with equivalent widths down to 1keV can be studied in detail.

3.1.3 Spatial Resolution

Objects larger than 500 Kpc, such as clusters of galaxies, will be resolved whatever their distance. Cooling flows, with typical extents of 100-200 Kpc will also be spatially resolved at all distances, while individual galaxies would be resolved to redshifts of 0.5. This is a significant advantage over, for example, a 2 arcminute resolution telescope.

3.2 Astrophysical Impact

The sensitivity and spectroscopic capability of JET-X are such that studies of astrophysical importance for almost all classes of astronomical object can be made. Sufficiently large samples for each class of object will be observable that a major impact in all areas of astronomy, from planetary studies to cosmology, in addition to the studies of 'classical' X-ray sources will be possible. JET-X, however, has been designed primarily for the study of faint, distant X-ray sources of cosmological importance. Here a small sample of the key problems which can be addressed with JET-X are outlined.

3.2.1 Faint Source Studies and the X-ray Background

The high sensitivity of JET-X to point sources (for example, luminous AGNs will be detected to redshifts $\sim$ 4) will enable source counts to be studied directly at flux levels $\sim$ 10 times fainter than achieved with the EINSTEIN observatory, allowing a cross-comparison with the ROSAT X-ray telescope deep survey results in a complementary energy region.

If the $\log N - \log S$ is Euclidean to this flux level then JET-X will have imaged at least 50% of the diffuse X-ray background. On the other hand fluctuations analyses (Fabian 1981, Hamilton and Helfand 1987) of EINSTEIN data suggest that the source counts flatten at flux levels just below the EINSTEIN limits. This is at the flux level directly addressed by JET-X. Flattening of the source counts also implies that a new class of faint X-ray sources (e.g. young galaxies) may be required to explain the background. If these sources have a luminosity function similar to that suggested one should be visible in each 10^5s JET-X exposure.

JET-X can also measure the spectrum of the residual X-ray background over the 0.1-10keV band on scales down to less than 1 arc-minute. Comparison of this residual background with the spectra measured for individual populations of X-ray emitters will further constrain possible models for the origin of the background.

3.2.2 Clusters

The high spatial resolution, coupled with the continuum spectroscopic capability of JET-X, will enable detailed studies of the temperature and density profiles of the gas in clusters of galaxies to be determined. The gravitational potential to which the gas is responding may thus be mapped, allowing the determination of the distribution of both the visible and dark matter in clusters. Measurements of cluster gas parameters when coupled with measurements of the microwave decrement through the Sunyaev-Z'eldovitch effect will give model independent estimates of the cosmological parameters H_0 and q_0. Measurements of the angular diameters of cooling flows and clusters as a function of redshift will also lead to estimates of these parameters.

JET-X can test models of the evolution of clusters and the enrichment of the intra-cluster gas by measuring the abundance of heavy elements as a function of radius in clusters and as a function of redshift.

JET-X will strongly constrain the cooling-flow rates in nearby clusters, the distribution of flow rate with radius and will also determine the magnitude of flows in clusters at high redshift.

3.2.3 Galaxies

The study of X-ray emission from normal and star-burst galaxies has to-date been very limited. The EINSTEIN observatory results suggested that most elliptical galaxies contain large amounts of hot gas. JET-X will be able to detect the most luminous ellipticals out to redshifts of order 0.25 and obtain spectra for typical galaxies out to 100 Mpc (e.g. the Coma cluster) in exposures of $< 10^5$s. If, as is suggested, this gas is in hydrostatic equilibrium, measurements of the radial temperature and density distribution of the gas will allow the mass distribution of these galaxies to be determined.

Studies of cooling flows in nearby elliptical galaxies will also be possible. The 'canonical' wisdom that elliptical galaxies contained little or no gas and that consequently no star-formation was taking place is now untenable. X-ray observations of the cooling gas in these galaxies, particularly for a range of galaxy environments, will be essential in understanding the true picture.

The dominant emission from spiral galaxies, which in general have lower luminosities than ellipticals, is not from hot gas but from discrete sources, in particular X-ray binaries. Spectroscopy of these discrete sources will be possible throughout the local group. Comparative studies of the source populations will provide new data for evolutionary calculations through studies of their luminosity functions, distribution of orbital periods, location in the galaxies. The relationship of the X-ray source properties to their metallicities can also be investigated.

The spectral difference between the discrete sources and the diffuse ISM (if our galaxy is representative of spirals) will mean that studies of the ISM in galaxies at distances of up to $\sim$ 100Mpc. will be possible. If these galaxies have gravitationally bound haloes with gas temperatures of $\sim 10^6$K, such as our galaxy, then it will be possible to derive the mass distribution within spirals as well as elliptical galaxies.

Star-burst galaxies have a much higher ratio of $L_x : L_{opt}$ than average and thus will be detectable out to distances of order 500 Mpc. The correlation between X-ray properties and the star-formation rate will mean that for an unbiased selection of star-burst galaxies the influences of effects such as inter-galaxy interaction can be studied. The relationship of X-ray emission to the star-burst phenomenon in nearby galaxies will be important in undertanding the potential contribution of young galaxies to the X-ray background.

3.2.4 Active Galaxies

The recent discovery (Nandra *et al.*, 1989, Pounds *et al.*, 1989) that AGN X-ray spectra are not simply featureless power-laws but that a substantial fraction show evidence for emission and absorption by cold iron imprinted on their spectra suggests another scientific area where the capabilities of JET-X are well-matched to the astrophysical problem. The good spectral resolution of the instrument below 1keV when coupled with the ability to precisely determine the intrinsic continuum will mean that detailed information about the physical condition and location of the absorbing material intrinsic to the AGN will be obtainable for a significant sample of active galaxies.

Some AGN such as MR2251-178 show evidence for absorption by partially ionised gas. As the intrinsic luminosity of the source changes so does the ionisation of this gas. The sensitivity of JET-X to the absorption edges, and hence ionisation states, of intermediate Z elements will be important in determining the location, density and abundance of this material.

A substantial fraction of AGN show excess emission at energies below 1 keV. These soft excesses are believed to be caused by emission from the accretion disc surrounding the central source. The low energy response of JET-X is sufficient to detect and investigate the temporal and spectroscopic behaviours of these soft excesses.

4 Mission Profile

In a lifetime of 5 years JET-X will be capable of making $\sim 10^3 - 10^4$ observations of durations between $10^4 - 10^5$s. In addition to the detailed information obtained on the few thousand prime targets JET-X will serendipitously obtain information on $\sim 10^5$ X-ray sources. For at least 10% of these sources good quality spectra (sufficient, for example, to determine the power-law index of a serendipitous AGN to ~ 0.1) will be obtained. Multi-colour broad-band photometry will be possible for most of the remainder. For comparison, previous imaging X-ray observatory instruments, provided little or no spectral information for most of the few thousand sources observed serendipitously and were not sensitive in the spectral band of most interest.

The ROSAT X-Ray Telescope has a sensitivity similar to JET-X but only at energies < 2 keV and with limited spectral resolution. The ROSAT all-sky survey will provide invaluable information for optimising the scientific return from JET-X observations.

Scheduled for early 1993, well before either of the world class missions, AXAF and XMM, Spectrum-X will be one of the first of several international missions to exploit the rich astrophysical potential of X-ray spectroscopy. In comparison with contemporary instruments, JET-X will have a lower throughput, but a much better limiting sensitivity, spectral and imaging resolution than X-SPECT (also included in the core payload of SPECTRUM-X). It will have comparable throughput but much better spatial resolution than the Japanese/US ASTRO-D and larger throughput and higher resolution than the imaging spectrometer on the Italian SAX missions likely to be in orbit on the same time scale.

5 References

Chowanietz, E., Lumb, D.H. and Wells, A. 1986. *Proc. SPIE.,* **597** 381.
Citterio, O., Bonelli, G., Conti, G., Mattaini, E., Santambrogio, E., Sacco, B., Lanzara, E., Brauninger, H. & Burkert, W. 1987. *Proc. SPIE.***830**,139.
Fabian, A.C. 1981. *Ann. N.Y. Acad. Sci.***375**, 235.
Hamilton, T.T. & Helfand, D.J. 1987. *Ap. J.***318**, 93.
Lumb, D.H. & Holland A.D. 1989. *Proc. SPIE.***830**,116.
Nandra, K., Pounds, K.A., Stewart, G.C., Fabian, A.C. & Rees, M.J. 1989. *MNRAS***236**,39P.
Pounds, K.A., Nandra, K., Stewart, G.C. & Leighly, K. 1989. *MNRAS Submitted.*

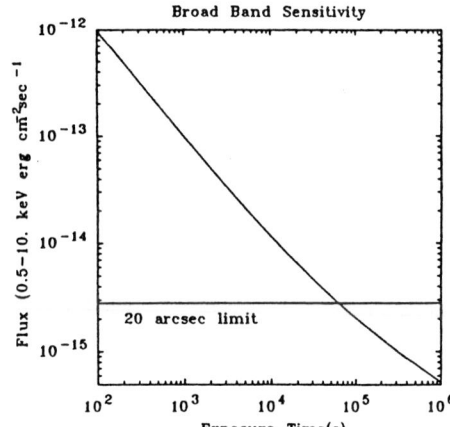

Figure 7: Time required for a 5σ detection of a point source for JET-X. Also shown are the confusion limits for 20 arc-seconds and 2 arc-minutes assuming a Euclidean extrapolation of the EINSTEIN log N - log S.

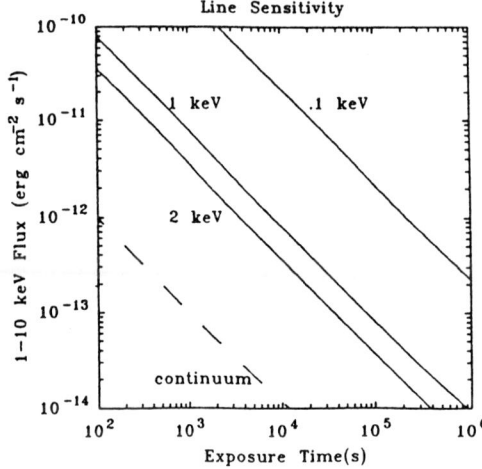

Figure 8: Time required by JET-X for a 5σ detection of lines of varying equivalent widths as a function of the continuum intensity.

9. Future X-ray Observatories, Detectors and Instrumentation

A HIGH RESOLUTION ECHELLE SPECTROMETER FOR SOFT X-RAY AND EUV ASTRONOMY

James Green and Stuart Bowyer
Space Sciences Laboratory, University of California, Berkeley, CA, USA

ABSTRACT. We present a new design for high resolution spectroscopy from 80 to 400 Å. This design employs grazing incidence optics and variable line-spaced gratings to achieve high resolution. Unlike some previously proposed EUV echelles, this design employs straight groove planar gratings, which are a well-proven, easily manufactured design. The instrument delivers a peak resolution of $\lambda/\Delta\lambda = 7500$ and a peak effective area of 3 cm^2.

I. INTRODUCTION

High resolution spectroscopy shortward of the IUE cutoff is a high priority for the next generation of space instruments. The Far Ultraviolet Spectroscopic Explorer will give high resolution down to 900 Å. The Orbiting Retrievable Far and Extreme Ultraviolet Spectrometer will give resolution in excess of 5000 down to 400 Å. Therefore, we have designed this instrument to provide resolution on the order of 5000 with a wavelength coverage for 80–400 Å. The instrument has been designed with a 1-m diameter optic, although the concept is also valid for smaller systems. We discuss the tradeoffs resulting from a reduction in aperture.

II. MIRROR DESIGN

For high efficiency at the wavelengths of operation, it is necessary to employ a grazing incidence telescope. Additionally, optical systems in this wavelength region are highly susceptible to "contamination" from diffuse emission. In this particular application, problems could arise from diffuse emission at 304 and 1216 Å. Even though 1216 Å is outside the instrument bandpass, the diffuse emissions are so intense that the scattered intensity from the grating can be larger than the signal if the instrument is not properly baffled. Fortunately, one grazing incidence optical design allows strict control of off-axis light (Green and Bowyer 1986). This is the Hettrick-Bowyer Type I design, which is a solution to the coma-free Schwarzschild equations. The primary mirror is a slightly modified paraboloid, and the secondary mirror is a slightly modified ellipsoid. The light comes to a focus between the mirrors. This design represents the grazing incidence analogue to a Gregorian telescope, just as a Wolter-Schwarzschild Type II represents the grazing incidence analogue to a Cassegrain. By placing an aperture at this primary focus, the field of view and, therefore, the diffuse emission are strictly controlled. A few additional baffles are all that is needed to eliminate problems from diffuse emission. The mirror chosen for this experiment is an f/10 optic with a 3-m front-to-focus length. The mirror parameters are presented in Table 1.

III. THE GRATINGS

Three sets of two gratings intercept the beam directly after the secondary mirror. The first grating operates in a high order and acts as the echelle; the second grating operates in first order and acts as a cross disperser. Each echelle grating intercepts 1/8 of the total beam. Since we have chosen to use straight groove variable line-spaced gratings, the resolution is inherently limited by the speed of the beam accepted by the grating in the direction perpendicular to the dispersion (Hettrick 1984). In order to achieve our resolution goal, the grating could not be any wider than one which intercepted 1/8 of the beam. If the grating width were increased, resolution would decrease, but effective area would increase. The echelle line

Table 1. Instrument Parameters

Mirror	Hettrick-Bowyer Type I
Diameter	1 meter
Focal Length	3 meters
Gratings (6)	Varied line-spaced, planar
Detectors	Microchannel plate, Wedge and strip anode
Resolution	40µ
Photocathode	KBr
Instrument Performance	
Coverage	80–400 Å
Resolution	> 5000
Average effective area	~ 1 cm^2

spacing is optimized at one $m\lambda$, where m is the order number and λ the wavelength.

The cross disperser is oriented so that the beam intersects the grating normal to the ruled direction (standard in-plane mount). This allows the line spacing to be optimized again for one wavelength. The gratings designed in this manner are quite straightforward to manufacture. Their parameters are listed in Table 1.

IV. THE DETECTORS

The detectors employed are microchannel plate detectors with wedge and strip anodes (Martin et al. 1981). The system has been designed for detectors with a 25-mm active area. In all performance calculations we have assumed that a potassium bromide photocathode will be applied (Siegmund et al. 1988). To achieve the spectral resolution stated, it is necessary that the detector have a resolution of 40 microns. This is a reasonable expectation for a 25-mm detector.

V. SYSTEM PERFORMANCE

An assembly drawing of the instrument is provided in Fig. 1. For clarity, only one of the three echelle/cross disperser grating sets is shown. A direct imaging detector is placed at the telescope focus, allowing tracking of the target and compensation for drift for maximum resolution. The system's effective area was calculated by determining the reflectivity of each optical component, assuming the use of a gold coating and incorporating the effects of polarization. The grating efficiency was calculated with a standard diffraction efficiency code that includes shadowing effects. Since the gratings operate near the blazed $m\lambda$ at nearly all λ, the diffraction efficiency remains high, and groove shadowing becomes the dominant limiting factor. We assumed the detector quantum efficiency matched the experimental values for potassium bromide as measured in our laboratory.

The resolution of the instrument was verified by raytrace analyses of the entire system. Randomly placed parallel rays were raytraced through the mirror and then given a random three-arcsecond blur simulating the intrinsic blur in the mirror caused by small figure errors. The rays were then traced through both gratings to the detector plane. The resulting images are displayed in Figs. 2 and 3. It can be seen that the absolute resolution of the system (dispersion/image width) is in excess of 7500 and that the FWHM resolution is even higher, ~ 10,000. These instrument performance characteristics are also included in Table 1.

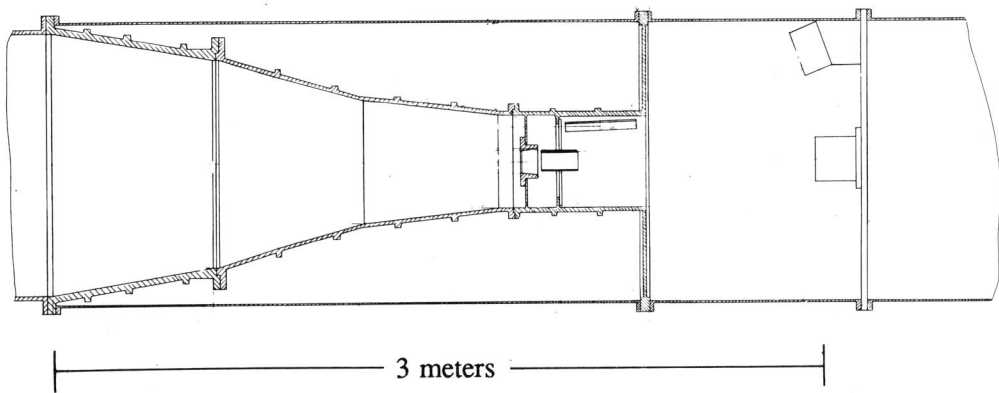

Figure 1. A schematic drawing of the instrument assembly. The primary and secondary mirrors are joined by an aluminum metering structure that is precision-machined to maintain the correct mirror alignment. For clarity, only two of the six gratings are shown, along with the detector. The detector at the mirror focus performs direct imaging and drift correction.

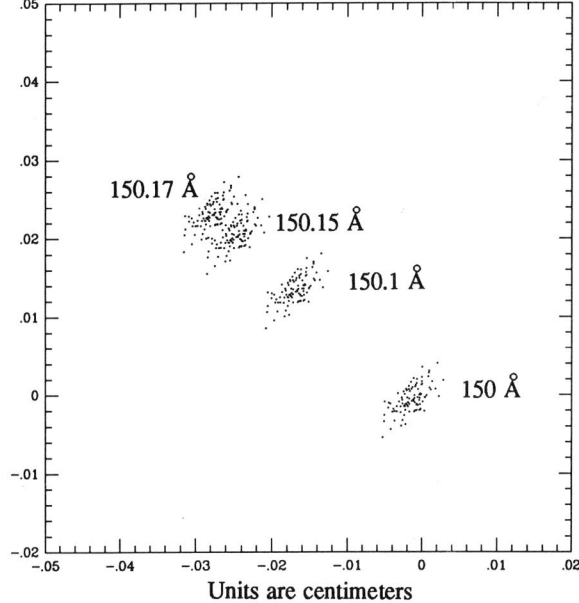

Figure 2. A raytrace result demonstrating the resolution achieved by this instrument. The raytraces include a three-arcsecond blur from the optics. Clearly, with 40 μ detector resolution, 0.02 Å at 150 Å can be resolved for a resolution of 7500. With better detectors, a FWHM resolution in excess of 10,000 can be achieved.

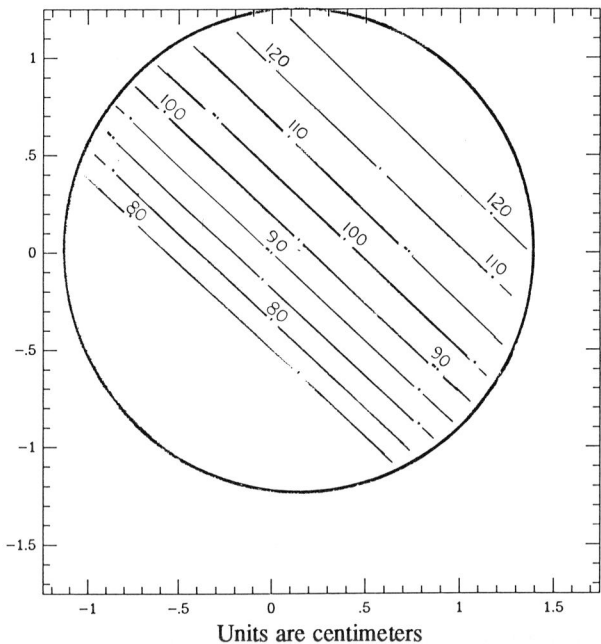

Figure 3. This raytrace result demonstrates coverage of the echelles from 80 to 120 Å. The lines drawn show the various orders, and the small dots are the actual image sizes. Several wavelengths are labeled for identification. Note that each wavelength appears in at least two orders. The circle drawn represents the active area of the detector.

When a similar design is incorporated into a smaller package, instrument performance is not degraded significantly. Since each grating sees only one section of the mirror aperture, the effective area goes linearly with mirror radius, not as the square. Also, as the mirror radius decreases, if the length is maintained, the graze angles on the mirror decrease, and the reflectivity increases. This is especially true at shorter wavelengths. For example, if the instrument diameter is reduced to 50 cm, fully 75% of the 1-m diameter instrument average effective area can be achieved.

ACKNOWLEDGMENTS

This work was supported by NASA grant NGR-05-003-450.

REFERENCES

Green, J., and Bowyer, S. 1986, *Appl. Opt.*, **25**, 1991.

Hettrick, M. C. 1984, *Appl. Opt.*, **23**, 3221.

Martin, C., Jelinsky, P., Lampton, M., Malina, R. F., and Anger, H. O. 1981, *Rev. Sci. Instrum.*, **52**, 1967.

Siegmund, O. H. W., Everman, E., Vallerga, J. V., and Lampton, M. 1988, *Appl. Opt.*, **27**, 1568.

EXPECTED SCIENTIFIC PERFORMANCE OF THE THREE SPECTROMETERS ON THE EXTREME ULTRAVIOLET EXPLORER

J. V. Vallerga, P. Jelinsky, P. W. Vedder, and R. F. Malina
Space Sciences Laboratory, University of California, Berkeley, CA, USA

ABSTRACT. The expected in-orbit performance of the three spectrometers included on the Extreme Ultraviolet Explorer (EUVE) astronomical satellite is presented. Recent calibrations of the gratings, mirrors and detectors using monochromatic and continuum EUV light sources allow the calculation of the spectral resolution and throughput of the instrument. An effective area range of 0.2 to 2.8 cm^2 is achieved over the wavelength range 70–600Å with a peak spectral resolution $\lambda/\Delta\lambda$ (FWHM) of ~ 360 assuming a spacecraft pointing knowledge of 10 arc seconds (FWHM). For a 40,000 sec observation, the average 3σ sensitivity to a monochromatic line source is 3×10^{-3} photons cm^{-2} sec^{-1}. Simulated observations of known classes of EUV sources such as hot white dwarfs and cataclysmic variables are also presented.

I. INTRODUCTION

During the first six-month phase of its mission, EUVE will carry out an all-sky survey in the EUV (70Å to 750Å) in four photometric bandpasses defined by thin film filters, as well as a deep survey along the ecliptic (Bowyer 1983). The second phase of the mission is devoted to the spectrometer with long pointings (typically 40,000 sec) at targets selected by guest observers. The calculations made in this paper are based on calibration results of flight hardware components (Jelinsky et al. 1988). These components are to be integrated in 1989, and launch is set for September, 1991, on a Delta II rocket.

II. INSTRUMENT DESCRIPTION

The EUVE spectrometers consist of three varied line-spaced plane reflection gratings mounted in the converging beam behind a single grazing incidence Wolter-Schwarzschild type II telescope (Hettrick et al. 1985). Each grating has a dedicated microchannel plate (MCP) detector (Siegmund et al. 1986), thus providing simultaneous coverage of an EUV source in three overlapping spectroscopic bands: 70–190Å ("short"); 140-380Å ("medium"); and 280-760Å ("long"). Each detector has a set of thin film filters, which act as order filters and attenuate any scattered Ly α, the dominant component of nighttime airglow with an intensity of 3500 Rayleighs (Chakrabarti 1984). The other major airglow lines at 304Å and 584Å, caused by resonance scattering of solar flux by He II and He I, fall withing the spectroscopic bandpasses and require the inclusion of a wire grid collimator in the medium and long wavelength channels to limit this diffuse flux to a small part of the spectrum.

The effective area of the three spectrometer channels is shown in Fig 1. This includes the geometrical area (1/6 of telescope area), mirror reflectivity, grating efficiency, order filter transmission, and detector quantum efficiency. Also shown is the spectral resolution, $\lambda/\Delta\lambda$ (FWHM), of each channel, which is limited primarily by the aberrations of the gratings. Given these characteristics, a diffuse background model (Chakrabarti, Kimble, and Bowyer 1984), and an integration time, we can calculate a minimum detectable line flux (photons cm^{-2} sec^{-1}) vs. wavelength, which is shown in Fig 2. Note the decrease in sensitivity around 304Å and 584Å due to the diffuse airglow. Observations in the Earth's shadow are more sensitive near 304Å than those shown in Fig. 2, but the 584Å sensitivity remains the same since most of the 584Å flux is interplanetary in origin.

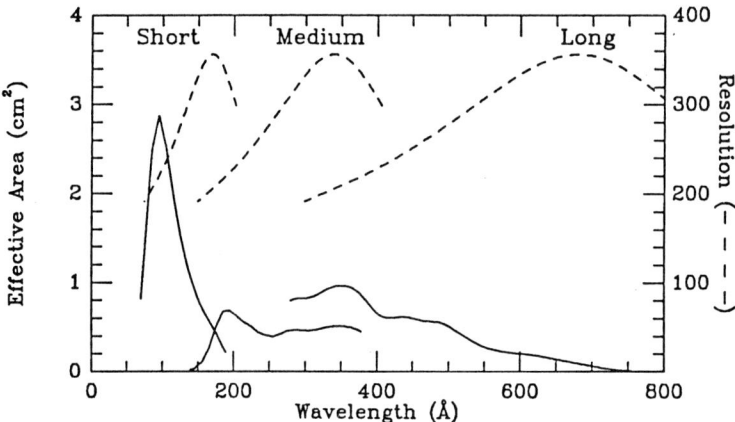

Figure 1. EUVE Spectrometer Characteristics. The plotted effective area (solid line, left scale) for the three spectrometer channels combines the geometric area with the measured throughput of the individual components in each channel (mirror, grating, filter, detector). The calculated resolution R (= $\lambda/\Delta\lambda$ FWHM -- dashed line, right scale) for the three channels assumes spacecraft pointing knowledge of 10"(3σ) and detector spatial resolution of 70 microns FWHM.

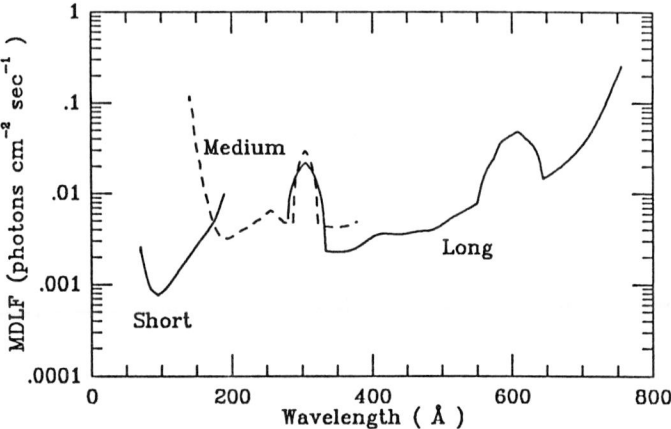

Figure 2. Minimum Detectable Line Flux. Plotted is the minimum flux from an unresolved line in order to be detected at the 5σ level above background in a 20,000 second observation for the three spectrometer channels. The wide features centered at 304Å and 584Å are due to the partially collimated diffuse background lines for observations away from the Earth's shadow. The MDLF would be lower by a factor of ≈6 at 304Å in the Earth's shadow, but the sensitivity at 584Å would remain unchanged since the neutral helium which scatters the solar line is interplanetary.

III. SPECTROMETER SIMULATION METHOD

To simulate the count rate spectra of known classes of EUV sources, we take modelled spectra of EUV sources, calculate the flux for some assumed distance, and attenuate the flux assuming a column density of hydrogen (N_{HI}) and a model of the interstellar medium, or ISM (Cruddace et al. 1974). The spectra are then weighted by the effective area (including second and third order diffraction) and convolved with the mirror point spread function, telescope aspect, grating resolution, and detector spatial resolution. Monochromatic diffuse background (airglow) is added convolved with the collimator response, and a uniform detector background count rate is also added. The resultant counts/bin are then randomly distributed based on Poisson statistics. The resultant "raw" count rate spectrum does not include the effects of nonuniform detector response (flat field), image distortion, and scattering, all of which are expected to affect the counts in any bin by no more than 5%.

IV. SIMULATION RESULTS OF EUV SOURCE SPECTRA

We have simulated two classes of known EUV sources, hot white dwarfs and cataclysmic variables, to show the capabilities of the EUVE spectrometer in observing continuum sources and spectral line sources, respectively. For the hot white dwarf we have taken a model atmosphere (Malina et al. 1982) at 55,000 K with a small fraction of helium (1×10^{-5}) and placed it at the distance (49 pc) of the known EUV source G191-B2B with a neutral hydrogen column of $N_{HI} = 8 \times 10^{17}$ cm^{-2} (Green, Jelinsky, and Bowyer 1988). Figure 3 shows the resulting raw count spectrum in all three channels. Note the edge due to the stellar He II at 228Å and at 456Å (228Å in second order). The strong edge at 504Å is due to neutral helium in the ISM with an assumed column density equal to 10% that of neutral hydrogen. White dwarf spectra will serve as sensitive probes of the ISM, as well as provide vital information about the last stages of stellar evolution.

EXOSAT observations of the SU Uma cataclysmic variables VW Hyi and OY Car in superoutburst imply the presence of an extended hot, optically thin corona with temperatures between 10^6 and 10^7 K (Van Der Woerd 1987). Spectra of these hot corona in the EUV are very strong functions of temperature, and the EUVE spectrometers can do much to pin down the physics of the emitting region. Figure 4 shows a simulation in the medium wavelength

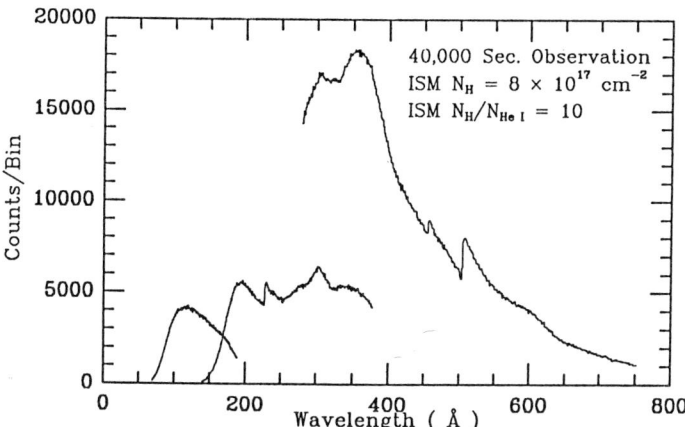

Figure 3. Hot White Dwarf G191-B2B. The simulated raw count spectrum of known EUV source G191-B2B in the three spectrometer channels plotted together. Note the strong detection of the interstellar medium absorption edge of He I at 504Å and the white dwarf He II edge at 228Å.

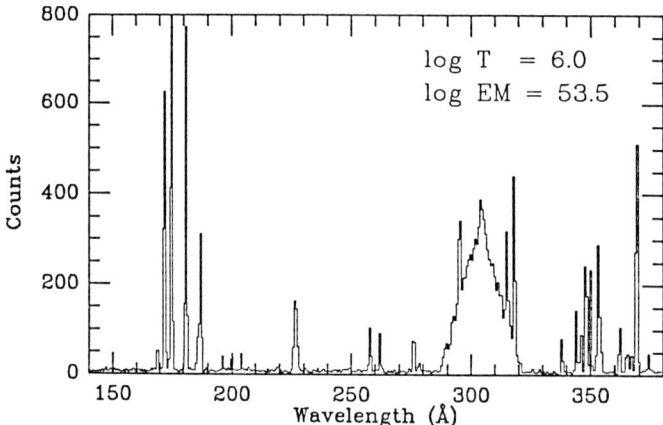

Figure 4. Cataclysmic Variables. Plotted is a medium wavelength channel simulation of a 20,000 second raw count spectrum of a cataclysmic variable in superoutburst with a hot, optically thin corona. The CV was placed at 50 pc with a hydrogen column of 10^{19} cm^{-2}. The 304Å diffuse background line shows the triangular response of the wire grid collimators.

spectrometer of a coronal spectrum based on the Raymond and Smith code (Raymond and Smith 1977) at 10^6 K with an emission measure of $10^{53.5}$ (derived from the EXOSAT data, Van Der Woerd 1987) placed at 50 pc and an ISM column of $N_{HI} = 10^{19}$ cm^{-2} and $N_{HI}/N_{HeI} = 10$. The large triangular-shaped bump centered at 304Å is diffuse airglow through the collimator. The strong emission near 175Å is a complex of Fe X, Fe IX, and Fe XI lines. Similar spectra can be generated using the coronal model for less exotic but much more abundant and closer late type stars such as M dwarfs.

ACKNOWLEDGEMENTS

This work was supported by NASA contract NAS5-29298.

REFERENCES

Bowyer, S. 1983, *Adv. Space Res.*, **2**, 157.

Cruddace, R., Paresce, F., Bowyer, S., and Lampton, M. 1974, *Ap. J.*, **187**, 497.

Chakrabarti, S., Kimble, R., and Bowyer, S. 1984, *J. Geophys. Res.*, **89**, A7.

Green, J., Jelinsky, P., and Bowyer S. 1988, in *Proc. Symposium on Cosmic Abundances of Matter*, in press.

Hettrick, M. C., Bowyer S., Malina, R. F., Martin, C., and Mrowka S. 1985, *Appl. Opt.*, **24**, 1737.

Jelinsky, P., Jelinsky, S. R., Miller, A., Vallerga, J., and Malina, R. F. 1988, *Proc. SPIE*, **628**, in press.

Malina, R. F., Bowyer, S., and Basri, G. 1982, *Ap. J.*, **262**, 717.

Raymond, J., and Smith, B. 1977, *Ap. J. Suppl.*, **35**, 419.

Siegmund, O. H. W., Lampton, M. L., Chakrabarti, S., Vallerga, J. V., Bowyer S., and Malina, R. F. 1986, *Proc. SPIE*, **627**, 660.

Van Der Woerd, H. 1987, *Astrop. Space Sci.*, **130**, 225.

DISPERSIVE SPECTROSCOPY ON AXAF

T.H. Markert

Massachusetts Institute of Technology, Cambridge, MA, USA

ABSTRACT. There are two transmission grating spectrometers and one Bragg crystal spectrometer being developed for the Advanced X-ray Astrophysics Facility (MIT is building the crystal spectrometer and one of the grating spectrometers; the Laboratory for Space Research in Utrecht is responsible for the other grating spectrometer). The gratings divide the AXAF energy band (80 eV - 10 keV) into three regions (the MIT instrument contains gratings with two different periods) and attain resolving powers for point sources between 100 and 1800. The gratings are composed of arrays of small facets mounted on plates which can be inserted immediately behind the AXAF telescope. The dispersed spectra from the grating arrays are read out by one of the AXAF imaging instruments.

The Bragg Crystal Spectrometer (BCS) is a focal plane instrument. One of eight selectable curved diffractors intercepts the AXAF X-ray beam as it diverges beyond the focal point. X-rays that satisfy Bragg's law are reflected from the crystal which, because of its curvature, re-focuses the beam onto an imaging detector. Narrow spectral regions are scanned by rocking the crystal over a range ~0.1 to 1°. Nearly the entire AXAF energy range can be studied by selecting the appropriate crystal and rotating it to the proper Bragg angle. The BCS achieves the highest spectral resolutions of the AXAF spectrometers: for 500 eV < E < 1600 eV, the FWHM of a narrow line (ΔE) is $\leq$ 1 eV.

1. INTRODUCTION

Three of the six instruments which are being developed for NASA's Advance X-ray Astrophysics Facility (AXAF) are dispersive spectrometers, i.e., the X-radiation is diffracted at a particular angle depending on the wavelength. There are two transmission grating spectrometers and one crystal spectrometer. The Bragg Crystal Spectrometer (BCS) and the High Energy Transmission Grating spectrometer (HETG) are being developed at MIT as part of the High Resolution X-ray Spectroscopy Investigation (Claude Canizares is Principal Investigator). The Laboratory for Space Research in Utrecht, the Netherlands, is designing the Low Energy Transmission Grating spectrometer (LETG). Albert Brinkman is the Principal Investigator.

2. TRANSMISSION GRATING SPECTROMETERS

Figure 1 shows, schematically, how the transmission gratings are situated in AXAF (for a detailed description of the HETG and LETG see Canizares, Schattenburg and Smith 1985, Brinkman *et al.* 1985 and Schattenburg *et al.* 1988). The gratings are made of small (approximately 1 inch square) elements which are used to cover the slots in a grating assembly plate (there is one plate for the HETG and an independent plate for the LETG). Either grating assembly can be rotated into position immediately behind the

AXAF telescope so that the slots line up with the telescope annuli (the AXAF mirror assembly consists of 6 Wolter I grazing incidence telescopes; each mirror produces an annulus of X-rays with a width of ≤ 2 cm). Each grating element (about 500 are required per plate) must be co-aligned to within about 1 arc minute so that the dispersed spectra from each element will superpose correctly.

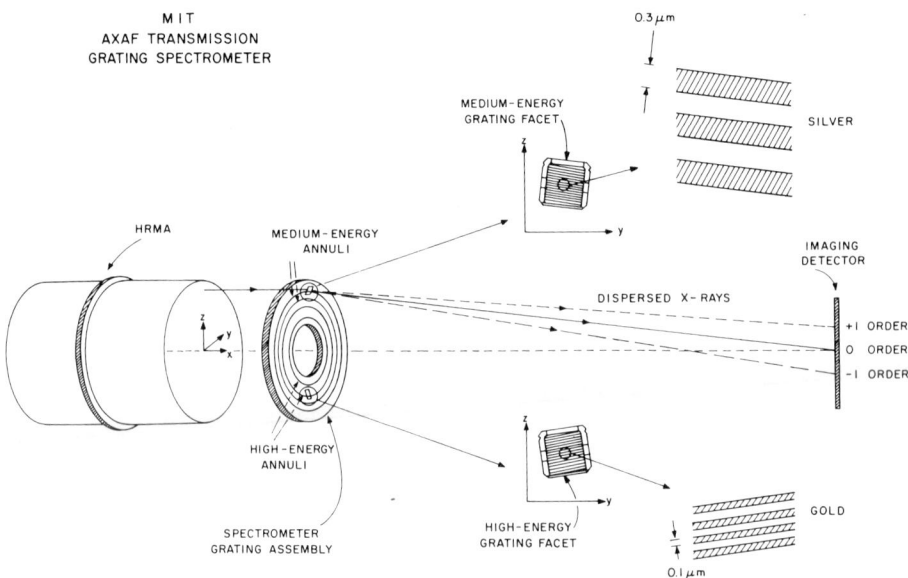

Figure 1 - The MIT Transmission Grating Spectrometer on AXAF (the HETG). 2000 Angstrom period gratings cover the inner three AXAF mirrors (the HEG); 6000 Angstrom period gratings cover the outer three mirrors (the MEG). The Utrecht spectrometer (the LETG) has the same general appearance, except that all of the grating elements are identical (10,000 Angstrom period).

Figure 1 is an illustration of the MIT instrument (the HETG). The HETG is composed of two kinds of gratings elements, 2000 Å period gold gratings (the High Energy Gratings - HEGs) and 6000 Å period silver gratings (the Medium Energy Gratings - the MEGs). The HEGs are situated over the inner three AXAF telescopes (where virtually all of the high-energy response of the AXAF is) and the outer gratings (the MEGs) over the outer three mirrors (where about 70% of the low-energy response is found). The LETG has a single grating type (10,000 Å period gold) which covers all 6 of the mirror annuli.

The gratings diffract the X-rays as they emerge from the mirror assembly, dispersing them over the roughly 8.4 meter distance between the grating assembly and the focal plane. The dispersed spectrum is read out by either of the AXAF imaging instruments (the AXAF CCD Imaging Spectrometer [ACIS] or the microchannel plate instrument [the High Resolution Camera, HRC]). The HEG and MEG grating elements (which operate simultaneously as part of the HETG) are tilted slightly with respect to one another. The dispersed spectra from the two grating types form a shallow "x" on the imaging detector, so there is no confusion between the two spectra.

Figure 2 shows the resolving powers expected for the three grating types. As the figure shows, the three kinds of gratings have high resolution over most of the AXAF energy band. At most energies, E/ΔE is determined by the spatial resolving power (the pixel size) of the imaging instrument and the telescope blur (E/ΔE ~ x/Δx). In order to remove geometric aberrations, the grating assembly and the detector array are designed to approximate a Rowland circle. The residual geometric effects limit the maximum resolving power attainable (e.g. Beuermann, Brauninger and Trumper 1978).

The AXAF gratings have high resolving powers (at most energies) and high sensitivities. Because the resolving power is a function of the image size, the grating spectrometers are most effective for observations of point sources. They can also be used as "slitless spectrometers" to image extended sources in the light of bright lines. Because of the high sensitivity of the transmission gratings, there are a great many candidate sources. For example, we estimate that there are 35,000 active galactic nuclei which can be studied in a reasonable amount of time (i.e. that will provide at least 25 counts in a 200 eV equivalent width line in a 20,000 second observation using the HETG and the ACIS detector).

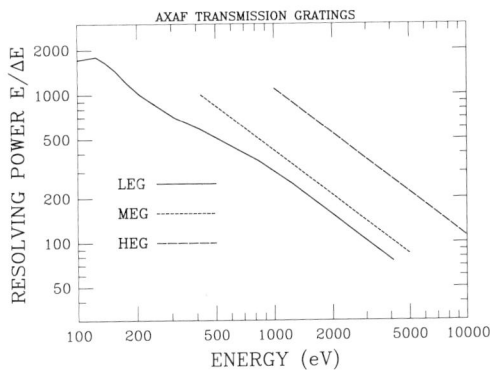

Figure 2 - Resolving powers of the three AXAF Transmission Grating types.

3. BRAGG CRYSTAL SPECTROMETER

3.1 General Description

The Bragg Crystal Spectrometer (BCS) is an expanded and improved version of the Focal Plane Crystal Spectrometer (FPCS) that operated successfully on the HEAO-2 satellite (for a more detailed discussion of the BCS design, see Canizares, Markert and Clark 1985, and Markert *et al.* 1988). The BCS is illustrated, schematically, in Figure 3. X-rays from the AXAF telescope come to the focus, where they intercept an assembly containing 10 apertures of different sizes and shapes, a number of filters, and on which sit two sealed monitor proportional counters (for monitoring the time behavior of point sources). If the monitor counters do not intercede, the X-rays continue beyond the focal plane, the beam diverges and strikes one of eight, selectable curved crystals. Those X-rays that satisfy Bragg's law [1] are reflected efficiently and are refocused onto an imaging proportional counter (there are two for redundancy). The crystal curvature accomplishes two thing: 1) it refocuses the beam so that the X-rays are concentrated onto a very small area (typically ~1 cm^2). For such a sharp focus the non-X-ray background makes a very small contribution to the signal; 2) the diverging X-ray beam strikes the crystal at nearly a

---

1. $n \times \lambda = 2d \times \sin\theta$, where λ is the wavelength, n is an integer (the order of the diffraction, usually 1), d is the lattice spacing of the crystal, and θ is the angle between the ray and the tangent to the crystal surface (the Bragg angle)

constant Bragg angle; therefore the crystal satisfies Bragg's law for only a very narrow range of wavelengths for any one setting of the central angle.

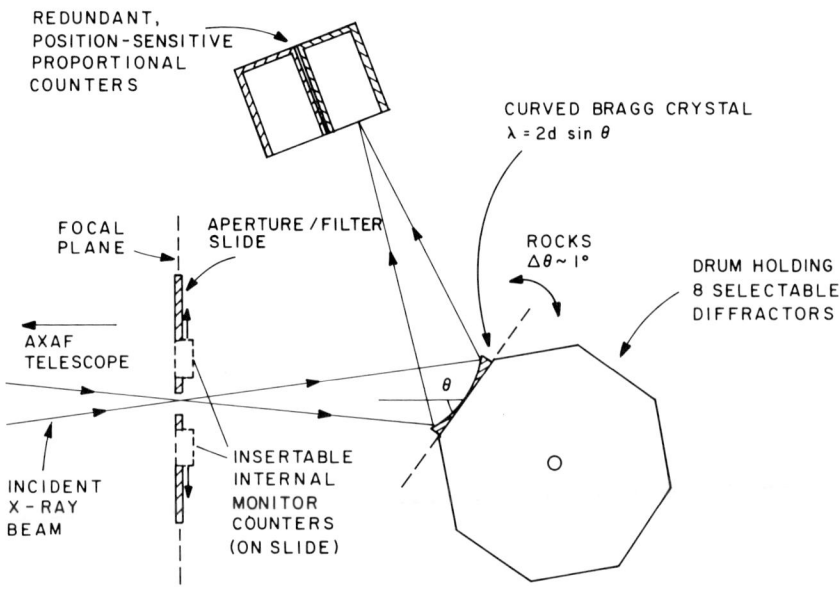

Figure 3 - Schematic of the Bragg Crystal Spectrometer on AXAF.

In order for the crystal curvature to accomplish these two things, it is essential that the three elements of the spectrometer (the X-ray focus, the crystal center and the detector) lie on a circle (a Rowland circle) with diameter equal to the radius of curvature of the crystal itself. This means, for example, that the distance from the crystal to either the detector or the focus must equal $R_{crystal} \times \sin\theta$. Clearly the crystal must be able to rotate over a large range in order to be able to select a particular wavelength, and the detector must also rotate in order to follow the reflected beam. In total, the Rowland circle geometry requires 5 motor motions. Translating the aperture slide and rotating the drum to select a new crystal add two more motions.

To operate the BCS one selects the appropriate crystal (geometry limits the range of angles accessible to a crystal, so that the wavelength range is also limited; covering the complete AXAF range requires 8 diffractors) and then sets up the Rowland circle for a particular wavelength. The crystal is rotated back and forth over an angle of ~0.1 to 1.0°. For rocking motions this small, deviations from the ideal Rowland configuration are small. During data reduction, the detector count rate can be correlated with the crystal angle and the spectrum reconstructed.

3.2 Techniques for Achieving Higher Resolving Power

As one instrument among 6 on AXAF, the particular strength of the BCS lies in the extremely high spectral resolving powers that it can achieve. One of our goals during the AXAF definition phase has been to optimize the BCS to function as a part of a complement of X-ray spectrometers, each of which has certain strengths and weaknesses. Consequently, at MIT we have expended much of our efforts in maximizing the BCS resolving power. By careful selection of diffractor materials and geometries, and by working to conform the crystal shapes to the ideal geometries, we are able to achieve resolving powers significantly better than those of the FPCS (at some energies our goal is $E/\Delta E > 2000$). For example, while all of the FPCS crystals had shapes that were circular in cross-section (so-called Johann geometry) we have baselined a somewhat different shape (the exponential spiral) for two of our crystals. This choice of geometry leads to an extremely high resolving power over a limited range of wavelengths.

We have also been experimenting with another technique for obtaining high resolving power (as suggested, for example, by Schnopper et al. (1976)) which we call the High-Res Setting. This is illustrated in Figure 4. For the standard Johann (diffractor with circular cross-section, Rowland circle configuration) geometry, the diverging X-rays strike the crystal at slightly different Bragg angles. In fact, a ray that makes an angle α with the central ray strikes the crystal at a Bragg angle $\theta_0 + \alpha^2/2 \times \cot(\theta_0)$. Therefore, any one setting of the crystal encompasses a range of wavelengths. This can amount to a broadening of the rocking curve by several arc minutes which, in many cases, can be the dominant factor in limiting the resolution of the crystal. If the detector is moved off the Rowland circle, however, then each X-ray will strike at a different point. Since the detector can image ($\delta x \sim 0.5$mm), one can trace each ray back to the crystal and thus determine the value of α. The Bragg angle can, therefore, be corrected for each ray and the resulting spectrum made considerably sharper.

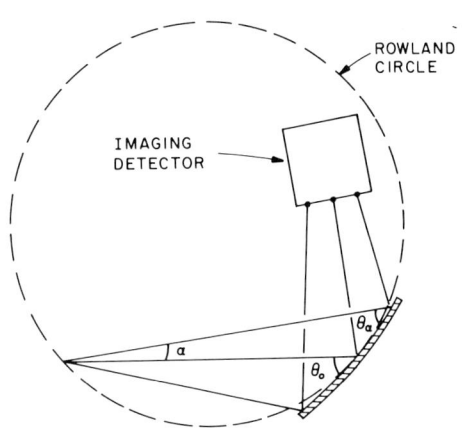

Figure 4 - BCS High-Res Setting. Here the detector is moved off the Rowland circle. The observer uses its imaging capability to determine the source of the ray (the angle α) and corrects the Bragg angle by $\alpha^2/2 \times \cot\theta_0$.

We have explored the High-Res setting extensively via computer ray-tracing and in the lab. Figure 5 shows two spectra taken at the MIT crystal evaluation facility. The dashed line is a spectrum of a Ti Kα line taken with the standard setting (i.e, the imaging capabilities of the detector were not used). The solid line is the same data (renormalized) in which the location of the X-ray on the detector was used to correct the Bragg angle (and thus the photon energy). A significant decrease in ΔE is seen for the High-Res case.

Figure 6 shows the resolving powers we hope to attain with the BCS in the Standard (detector on the Rowland circle) and the High-Res Settings. Each short line is the resolving power of one of the eight diffractors. The High-Res setting is only useful for observations of point sources, since there is an ambiguity in the image due to location on the sky and location on the crystal. In the Standard Setting, however, the resolving power achieved for observations of extended objects should be as shown in the figure.

Note that for much of the AXAF energy range, the resolution is better than 1 eV and is ≤ 10 eV at all energies.

Although achieving high resolving powers is the primary goal of BCS design, we note that, for some kinds of observations, the BCS sensitivity is comparable to that of the other AXAF instruments. In fact, we estimate that there are ~2000 X-ray sources in the sky that are suitable targets for detailed spectral analysis with the BCS (i.e., for which we can detect at least 25 photons in at least one X-ray emission line in 20,000 seconds).

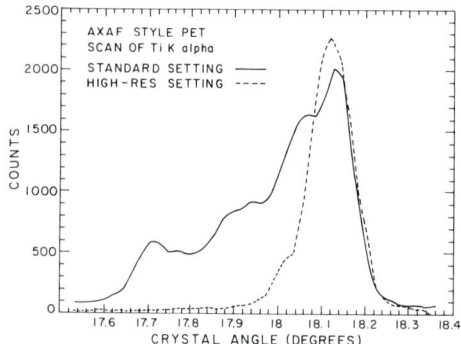

Figure 5 - Spectra of Ti K α made with a Bragg Spectrometer at the MIT crystal evaluation facility. Both the Standard (Rowland circle) and the High-Res Settings (detector off the Rowland circle) are shown.

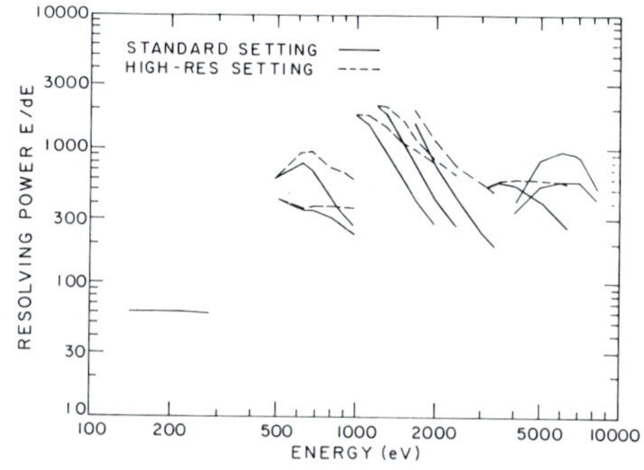

Figure 6 - Resolving powers (E/ΔE) expected for the BCS on AXAF. Both the Standard Setting (solid lines) and the High-Res Setting (dashed lines) are shown.

ACKNOWLEDGEMENTS. I thank the many workers at MIT and Utrecht who have contributed to the design of the transmission gratings and the BCS. In particular, I am grateful to Mark Schattenburg (HETG Instrument Scientist) for his assistance. This work was supported in part by NASA Contract NAS8-36748.

REFERENCES

Beuermann, K.P, Brauninger, H., and Trumper, J. 1978, *Appl. Optics*, **17**, 2304.
Brinkman, A.C., van Rooijen, J.J., Bleeker, J.A.M., Dijkstra, J.H., Heise, J., de Korte, P.A.J., Mewe, R., and Paerels, F. 1985, *Proc. SPIE*, **597**, 232.
Canizares, C.R., Markert, T.H., and Clark, G.W. 1985, *Proc. SPIE*, **597**, 241.
Canizares, C.R., Schattenburg, M.L., and Smith, H.I. 1985, *Proc. SPIE*, **597**, 253.
Markert, T.H., Powers, T.R., Levine, A.M., McCullum, C.B., Mohr, J.J. and Canizares, C.R. 1988, *Proc. SPIE*, **982**, in press.
Schattenburg, M.L., Canizares, C.R., Dewey, D., Levine, A.M., Markert, T.H. and Smith, H.I. 1988, *Proc. SPIE*, **982**, in press.
Schnopper, H.W., Delvaille, J.P., Epstein, A., Kalata, K. and Sohval, R. 1976, *Space Science Instrumentation*, **2**, 243.

DISCUSSION-T. Markert

J. Vallerga: Do you have any concerns about the stability of the silver transmission gratings on the ground and in orbit?

T. Markert: We are aware of a potential problem with oxygen contamination. On the ground, silver can tarnish (oxidize). We will test tarnished gratings on the ground to see if oxidation affects their performance. In space, oxygen atoms bombard satellites at high velocities causing damage to plastics and silver. This should not be a problem for us except when the telescope is pointing in the direction of orbital motion. If necessary we may deposit a thin coating of non-reactive material on the gratings (e.g. aluminum) or make the gratings out of a non-reactive metal such as ruthenium or rhodium.

J. Linsky: Your high resolution mode produces a narrow, symmetric rocking curve compared to the asymmetric rocking curve for the standard mode. Could you comment on whether or not the high resolution mode also produces more symmetric line profiles.

T. Markert: The High-Resolution Setting removes the systematic geometric effects on the rocking curve shape. The remaining contributors to the rocking curve shape are departures from the nominal geometry that occur during the bending process and the inherent shape of the curve which is a characteristic of the particular crystal. These contributors are reasonably symmetric, particularly for an imperfect crystal, where the mosaic orientations are random. Therefore we expect the profile of a line scanned in the High-Res Setting to be fairly symmetric.

H. Schnopper: When the instrument is in the high-resolution setting what is the penalty in sensitivity which is caused by the larger area on the detector that is required to detect the line?

C. Canizares: Because the background rate of the BCS is so low, most observations are still signal-limited. For very weak sources the increased background makes a difference (it increases by approximately a factor of 10), but only a few objects in the BCS target population are affected by the higher background.

S. Labov: What are the advantages of a Bragg crystal spectrometer if an X-ray calorimeter is built which achieves 1 eV resolution?

T. Markert: A 1 eV X-ray calorimeter would be a formidable detector. The BCS would have some advantages, such as a longer lifetime (the gas supply to the proportional counter would last longer than the liquid helium cryogen), a larger field of view, and a somewhat higher resolution at some energies (we hope to achieve 0.5 eV with the BCS). I think it is important to point out, however, that the best resolution thus far demonstrated with a calorimeter is 17 eV (see Holt's paper in this volume), so that the technology has to advance significantly before crystal spectrometers become obsolete. At the same time that calorimeter technology is being improved, of course, we (and others) will be striving to improve crystal techniques.

THERMAL DETECTORS FOR X-RAY ASTRONOMY

Stephen S. Holt

Laboratory for High Energy Astrophysics, NASA/GSFC, Greenbelt, MD, USA

ABSTRACT. Spectroscopy is traditionally characterized by the sacrifice of quantum efficiency for high spectral resolution. Since X-ray astronomy is a photon-limited discipline, the choice between high resolution for very few sources versus much lower resolution for many more has not always been an easy one. The development of new thermal detectors offers the opportunity to "have one's cake and eat it, too."

CHARACTERISITICS OF AN "IDEAL" SPECTROMETER

The most important attributes of a spectrometer depend upon the specific scientific objectives of a particular investigation, but there are some which are so generally useful that they can be safely assumed to be characteristic of an ideal spectrometer. These attributes include energy resolution sufficient to address the most important scientific objectives for all classes of sources, and simultaneous sensitivity over a wide bandpass with near-unit efficiency and negligible background. In particular, the following discussion will consider such a spectrometer for the AXAF (Advanced X-Ray Astrophysics Facility).

The primary characteristic of any spectrometer is its energy resolution, or equivalently, its resolving power:

Resolving power: $R(E) = E/FWHM(E)$

> where $FWHM(E)$ is the full-width-half-maximum energy resolution at energy E, so that $R(E)$ is the X-ray energy measured in units of the resolution

Clearly, higher resolving power is "better," but it is worth examining whether or not a case can be made for a practical upper bound to the resolving power based upon purely astrophysical considerations.

A good starting place for the determination of the required resolving power derives from an examination of standard coronal equilibrium spectra in the temperature range 10^{6-8} K. Figure 1 displays such spectra for solar abundances as viewed by a detector at the focus of the AXAF telexcope with perfect efficiency over the entire bandpass. The three traces in each panel are for FWHM = 1, 10 or 100 eV.

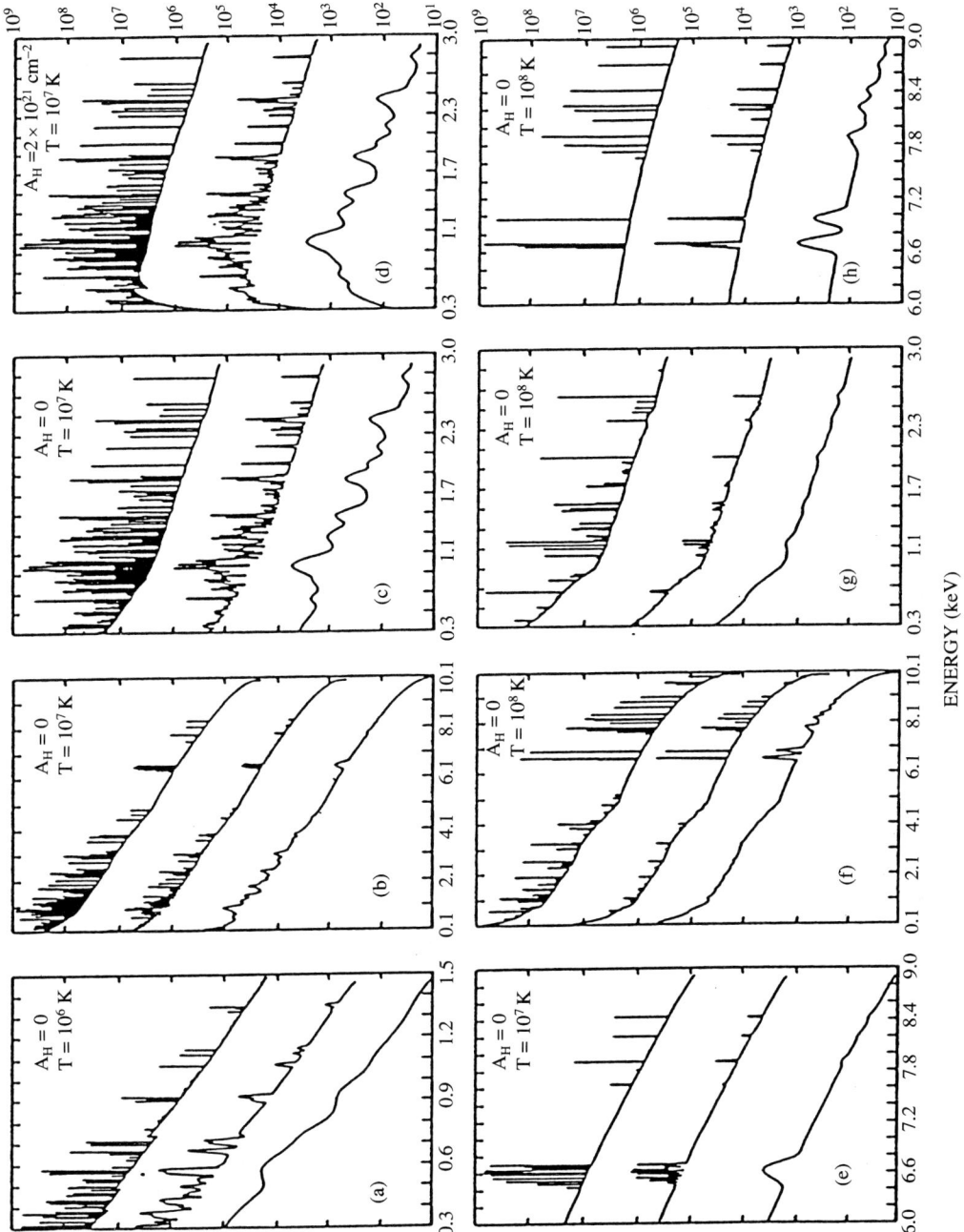

Figure 1. Coronal equilibrium spectra of plasmas with solar abundances, as viewed with detectors having FWHM resolutions of 1, 10 and 100 eV (upper, middle and lower trace of each panel).

In general, the identification and interpretation of features that are superposed on the continuum requires increasingly better spectral resolution as the features become weaker, or as they become confused with other features from which they cannot be resolved. At temperatures near 10^8 K, an equilibrium thermal spectrum is dominated by bremsstrahlung continuum, and exhibits strong isolated Fe K-emission at energies >6.6 keV, as well as Fe L-lines near 1.2 keV. Note, however, that the "completely ionized" lower-Z elements exhibit strong recombination lines: H-like Si XIV (at 2.0 keV) and S XVI (at 2.6 keV), each with about 30 eV equivalent continuum widths, are barely discernible with FWHM 100 eV, but FWHM 10 eV can reveal O VIII at 0.65 keV and even separate Ne X and Mg XII from the Fe L-band emission. At temperatures near 10^7 K, where the power emitted in the lines and the continuum are comparable, FWHM 100 eV can separate the K-emission complexes of the abundant elements (here dominated by He-like rather than H-like analogs for Z>10), while 10 eV uncovers most of the detail that 1 eV can fully resolve. At 10^6 K, the equilibrium spectra are line-dominated; here O and Ne may be recognized with 100 eV, but FWHM 10 eV is necessary to extract quantitative information.

Atomic lines provide diagnostics for electron temperature, ionization temperature, density, mass motion and elemental abundances. The model-dependent interpretation of these diagnostics can also be used to distinguish among thermalization, shock heating or photoionization, as well as the extent to which the plasma is in the ionizing or recombining phase, so that even its history may be gleaned from its current status. The required minimum sensitivities depend, of course, on the details of the scientific problems which the observations are intended to address. The utilization of these diagnostics can be found in many references (e.g. Bahcall and Sarazin 1978; Pradhan and Shull, 1981).

For L-shell to K-shell transitions, which dominate the E > 1.3 keV spectrum of hot plasmas, the most important lines are those from:
- o the fluorescence of cold (neutral) material,
- o the analog of Lya from hydrogen-like material, and
- o the resonance (r), forbidden (f) and intercombination (i) transitions of helium-like material.

Fluorescence and Lya-analog recombination are clearly the low and high temperature limiting cases, and the helium-analog lines are important because there is a wide range of plasma conditions and temperatures for which there is substantial population of this ionization state. Here the ratio (i+f)/r can be a useful diagnostic for electron temperature, for example, while f/i and f/(i+r) are more sensitive to density. Table 1 gives very simple analytic approximations (that are good to < 10%) for the separation energies of these five lines for elements 7 < Z < 27. The energy of Lya is, of course, Z^2 times the 10.2 eV for hydrogen, so that demanding equivalent performance of a single spectrometer for all Z would require its resolution FWHM(E) to scale the same way. For a spectrometer with constant FWHM, the required value would be that for the lowest Z that is expected to be generally useful.

TABLE 1: LINE SEPARATIONS

Line pairs	Approximate energy separation (eV)
Lyα - Resonance	10 Z
Resonance - Intercombination	Z
Resonance - Forbidden	2 Z
Resonance - Neutral	10 Z

Table 1 demonstrates that while about 1 eV may be required to completely resolve all the lines of potential interest from all elements, 10 eV is sufficient to separate the most important lines from oxygen, and can totally resolve them for iron. The strong dielectronic satellite lines that solar studies have identified as useful diagnostics are also typically somewhat more than 10 eV apart from the Fe XXVI and Fe XXV lines, although there is blending of fainter companions. High velocity mass motion is easily discernible with 10 eV, e.g. the broadening expected to match that for optical lines in AGN (active galactic nuclei), which corresponds to velocities of about 5000 km s^{-1}, is $4v_{1000}Z_{10}^2$ eV (v_{1000} in units of 1000 km s^{-1}). The natural width of a resonance line is about $10^{-2}Z_{10}^4$ eV, however, so that resonance and even narrower forbidden lines have widths that cannot be measured even with 1 eV resolution. Thermal broadening at $2(T_8 Z_{10}^3)^{1/2}$ eV would also require sub-eV resolution.

It then follows that most of the important K-shell line diagnostics from O to Fe require no better than 10 eV resolution. To obtain a general capability for the measurement of thermal or natural broadening, the resolution would have to be improved by orders of magnitude. All of the above would suggest, therefore, that about 10 keV represents an important threshold for an X-ray spectrometer.

There are at least three "efficiency" parameters that each deserve separate consideration in a detailed study of scientific objectives, but which can be combined into a single parameter that has general utility.

Quantum efficiency: $\epsilon(E)$ is the efficiency with which photons of energy E incident on the detector surface are measured with resolving power R(E)

Instantaneous bandpass: $\Delta_{ins}E$ is the energy range over which the spectrometer can operate simultaneously

Effective overall bandpass: $\Delta_{tot}E$ is the energy range over which the quantum efficiency $\epsilon(E)$ is greater than (say) 10% of its maximum value -- for a well-matched detector-telescope combination, this would be the effective band-pass of the telescope $\Delta_{tel}E$

More is better for all three parameters, but a variety of subjective arguments can be mustered to justify "required" values. It is especially important to emphasize the scientific value of having <u>both</u> $\Delta_{ins}E$ and $\Delta_{tot}E$ approach $\Delta_{tel}E$ in terms of discovery potential. Any spectroscopic observation performed with such a spectrometer will not just test for a narrow spectral feature at one energy that might be predicted by a model that is currently fashionable; it will simultaneously measure the spectrum throughout the entire telescope bandpass. An additional advantage of having maximally large $\Delta_{tot}E$ is in measuring the continuum: while the sensitivities to many spectral features will scale as the square root of exposure parameters, the precision with which continuum slopes can be determined scales linearly with the bandpass.

The combining "grasp" parameter Φ is meant to be the most useful possible definition of the average efficiency. It can also be considered a measure of the "speed" of the spectrometer. Φ is a function of the source spectrum and mirror response as well as the detector, i.e. if a source spectrum integrated over the total AXAF bandpass (here taken to be 0.1-10 keV) is S ($cm^{-2}s^{-1}$), then the rate at which photons can be detected with a perfect detector at the focus of the telescope is the convolution of the telescope area as a function of energy $A(E)dE$ (cm^2) with the source spectrum over the telescope bandpass:

$$C\ (s^{-1}) = \int_{\Delta_{tel}E} S(E')\ A(E')\ dE'$$

We define Φ as the average quantum efficiency taken <u>simultaneously</u> over the entire telescope bandpass:

$$\Phi = \langle\epsilon\rangle = \frac{1}{C} \int_{\Delta_{tel}E} \epsilon(E')\ S(E')\ A(E')\ dE'$$

in order that it be clearly distinguishable from the average instantaneous efficiency $\langle\epsilon(E)\rangle$ of a spectrometer over a limited bandpass $\Delta_{ins}E$ about E. for such a spectrometer:

$$\Phi = \langle\epsilon(E)\rangle \frac{\Delta_{ins}E}{\Delta_{tel}E}$$

General requirements for the spectroscopy of a large number of sources require the ability to investigate a reasonable portion of the telescope bandpass with a respectable level of sensitivity. Simulations can be used to demonstrate that a minumum of 10^{3-4} detected photons are generally required for detailed spectral analysis. Figure 2 displays the "logN-logS" relation for extragalactic sources with slope -1.5 (e.g. Maccacaro et al. 1982) in AXAF units, i.e., the source intensity is given in units of C s^{-1} as defined in the previous paragraph for a detector with PHI=1 at the focus of the AXAF telescope. The flatter trace (with slope -0.7) is that for sources in the galaxy, including binary accretors and supernova remnants.

The "weak source" limit of a spectrometer will be determined by the detector background B(E) in addition to Φ. As a practical matter, the acceptable background is a function of Φ; if the total detector background is a small fraction of the expected source counts, then it is effectively trivial. For a detector with Φ approaching unity, for example, there will be thousands of sources for which C > 1 ct s^{-1}, so that B(E) as high as 10^{-2} might be acceptable.

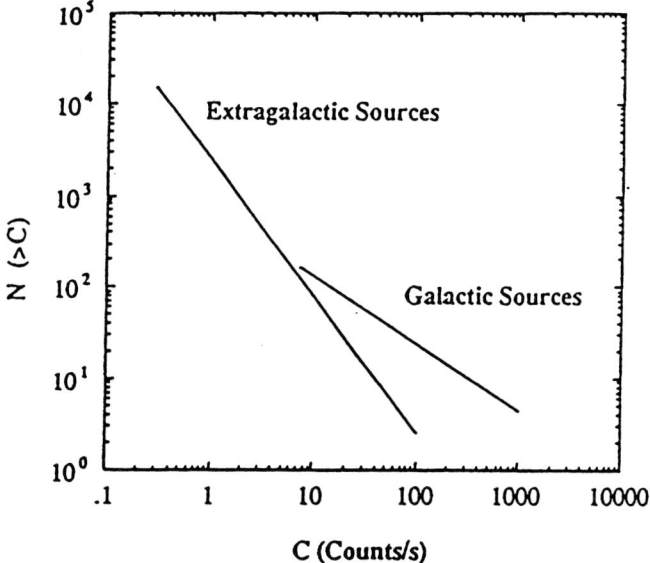

Figure 2. The number of observable sources as a function of the count rate of a perfect detector in the focal plane of AXAF.

A totally objective (albeit incomplete) measure of the capability of a spectrometer is its sensitivity to the equivalent continuum width (EW) of a single narrow spectral feature. If the feature is narrower than FWHM(E), the detectable EW will scale as the inverse square root of the following parameters for any spectrometer:
- the source strength,
- the exposure time, and
- the detector parameters $\epsilon(E)$ and $R(E)$.

For a scanning spectrometer, the same formalism may be used if $\epsilon(E)$ is the mean efficiency to the feature over the entire energy range that must be scanned during the exposure time. Additionally, the effect of background (either non-X-ray or X-rays multiplexed from other energies) on the detectable EW can be approximated with a multiplicative factor containing the ratio of the background counts B' to the source counts S' recorded within FWHM: $(1 + B'/S')^{1/2}$.

Two other parameters that can be important in determining the utility of instruments for astronomical spectroscopy are associated with angular sizes: the total instantaneous field-of-view FOV and the differential pixel size $d\Omega$ within FOV. For maximum generality, it would be nice to have FOV as large as the 30 arc min that characterizes the AXAF telescope, itself, and $d\Omega$ similarly matched to the AXAF sub-arc second spatial resolution. For purposes of estimating minimum requirements, however, the great majority of the sources in Figure 2 are point sources (i.e. not resolved by the telescope at a level < 1 arc sec), but there are important exceptions. Both supernova remnants (SNR) and clusters of galaxies are extended sources. Young SNR in our galaxy have typical sizes of arc minutes, and those in nearby galaxies are (of course) even smaller. Similarly, the core radii of bright clusters of galaxies have typical values of the order of an arc minute. It is clear, therefore, that about 1 arc minute is the minimum FOV necessary to obtain exposures that simultaneously cover large portions of the extended areas of X-ray emission in most SNR and clusters of galaxies.

From Figure 2, it is clear that there is no point source confusion problem for FOV of order 1 arc minute at 10^4 s, even for a detector with $\Phi=1$. The important issue for $d\Omega$ is, therefore, the extent to which gradients can be measured across the same extended sources discussed in the last paragraph. There is important differential spectral-spatial information to be obtained from both SNR and clusters, e.g. cooling flows near the center in the latter, and non-equilibrium effects inside the blast wave in the former. Since the timescales for variability in these objects are longer than the lifetime of AXAF, spectral-spatial mapping with a "fast" detector could be performed in a raster mode, but the observatory would be utilized more efficiently if the whole spectal-spatial correlation was performed in a single exposure.

A practical limit for dΩ can be obtained from the source catalog itself. For a given dΩ and FOV, we can estimate how strong an extended source would have to be in order to meet the two conditions that there were enough counts accumulated in each pixel for independent spectroscopy, and that the accumulation time was short enough to allow many different such exposures in the observing timeline. For dΩ of about an arc sec over an arc min FOV, for example, it would take a source as intense as the Crab nebula to allow a perfect focal plane detector to accumulate 10^3 counts per pixel in 10^4 s. This means that there are very few areas of extended X-ray emission for which a dΩ as small as 1 arc sec can be justified with the AXAF telescope.

CHARACTERISTICS OF AN X-RAY CALORIMETER

An X-ray calorimeter works by converting the energy of a single X-ray photon entirely into heat. For it to function as a spectrometer for individual X-rays, the effective noise in the temperature measurement must be small compared to that temperature rise.

A practical device consists of an X-ray absorber and a thermometer mounted on a substrate with small heat capacity C_0, connected by a weak thermal link of conductance G_0 to a heat sink at temperature T_0 (see Figure 3). When an X-ray of energy E is absorbed, the temperature of the substrate is increased by an amount $\Delta T = E/C_0$ in a time short compared to the time constant $\tau = C_0/G_0$ associated with recovery of the substrate back to the bath temperature T_0. The random transfer of energy between the substrate and heat sink produces fluctuations in the energy content of the substrate.

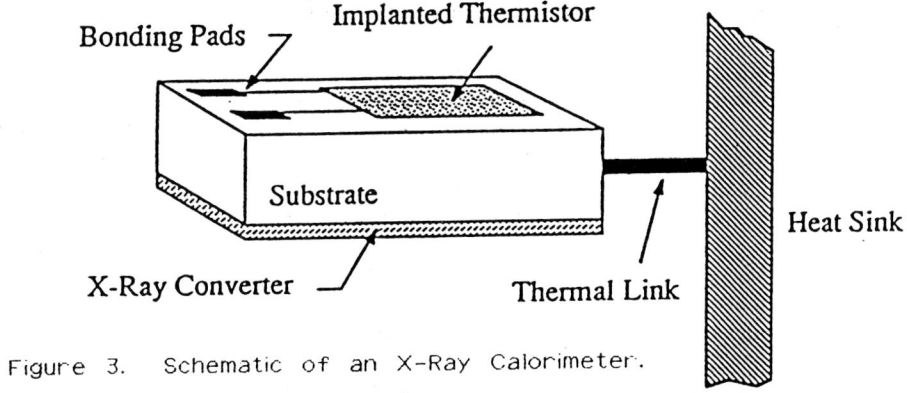

Figure 3. Schematic of an X-Ray Calorimeter.

An elementary statistical calculation gives the mean square magnitude of these fluctuations as:

$$\langle \Delta E \rangle_{rms} \text{ (FWHM)} = 2.35[kT_0^2 C_0]^{1/2}$$

where k is Boltzmann's constant. Note that this is independent of G_0 and any details of the thermal link. These fluctuations can be thought of as square-root N variations in the number of phonons contained in the substrate, and represent the background noise level against which the increase in energy due to an absorbed signal event must be measured. The advantages of low temperature operation are obvious, particularly when consideration is given to the fact that C_0 scales as T_0^3 with the proper choice of materials. As an example, a piece of silicon 0.5 mm on a side and 25 μm thick with heat capacity 4×10^{-15} J K^{-1} at 100 mK would have a limiting fluctuation noise of about $\Delta E_{rms} = 0.2$ eV.

The most important additional noise sources are those which might arise from the conversion of X-ray energy to heat (e.g. if some energy is trapped in states that are long-lived compared to the rise time of the heat pulse), and from the measurement of the temperature. Conversion noise may be made arbitrarily small, but the Johnson noise in ion-implanted thermistors provides an unavoidable increase in the minimum system noise achievable for this thermometry technique, as demonstrated in the seminal work of Moseley, Mather and McCammon (1984) on this subject, based on analyses of noise sources in infra-red bolometers (Mather, 1982; Mather, 1984). These authors demonstrate that in the case where only Johnson noise is an important addition, the expression for achievable noise can be approximated with a multiplicative factor ξ (with a value of about 2 for well-chosen thermistor parameters) to that for the basic fluctuations:

$$\Delta E_{rms} \text{ (FWHM)} = 2.35\xi[kT_0^2 C_0]^{1/2}$$

An equivalent circuit for the calorimeter consists of a voltage source V_{bias} in series with both a fixed load resistor R_L and a thermistor R (across which the signal equivalent to the heat depostition Q is measured), so that $V_R = V_{bias} R/(R+R_L)$. The temperature dependence of the thermistor is given by a $\alpha = -d(\log R)/d(\log T)$, which has a typical value of 6. From the definitions of α and C_0:

$$\frac{dR}{R} = \frac{-\alpha dT}{T} = \frac{-\alpha dQ}{C_0 T} \quad \text{so} \quad \frac{dV}{dQ} = \frac{\alpha V_{bias} R R_L}{C_0 T (R+R_L)^2}$$

The responsivity S (volts/watt), essentially dV/dQ multiplied by the effective recovery time τ, should increase linearly with V_{bias}, but is found instead to reach a maximum value and then <u>decrease</u>. One reason for this limitation is the self-heating of the thermistor increasing its heat capacity, but there are a variety of other non-ideal effects that limit the achievable responsivity.

In order to keep the heat capacity down, especially when considering the additional components arising from the X-ray absorber and the electical connections to the thermistor, the detector sizes are limited to < 1 mm^2. For reasons originally associated with *a priori* pointing limitations that prevented AXAF from guaranteeing that a point source image could be located on such a small detector, we have baselined a 6×6 array of calorimeters in the focal plane that each operate as independent spectrometers. In addition to enlarging its FOV, this arrangment also provides some imaging capability to the spectrometer.

The small detector volumes assure that even without active anticoincidence the non-X-ray background will be small. Scaling by the relative volumes from solid state detectors that have been flown before, the backgound summed over the entire array of detector should be of order 10^{-3} s^{-1} over the entire 10 keV bandpass. Recalling from Figure 2 that there are thousands of sources that will provide count rates in excess of 1 s^{-1} for a detector as efficient as a calorimeter, this background should be negligible.

The detector efficiency is limited at low energies by the windows that must be placed between the cold detector and the thermal noise that direct radiation from warmer surroundings would produce. The current AXAF design, based upon windows that have already been tested under launch conditions, results in efficiency that rises rapidly above about 10% at 0.3 keV to a value that is prevented from reaching unity if meshes are required to maintain the integrity of the windows.

SUMMARY

The X-ray calorimeter for AXAF is still under development, but test results suggest asymptotic values for relevant spectroscopic parameters. Recent development work, including practical limitations of thermistor-implanted X-ray calorimeters and associated cryogenic systems, can be found in McCammon et al (1987), Moseley, et al (1988) and Kelley et al. (1988), and will not be discussed here. With respect to energy resolution, the current "best" value obtained in a single test is 17 eV FWHM at 100 mK, but the values for each of the contributing noise components have each been separately measured in other tests to be less than 10 eV -- this suggests that the limiting value for AXAF will certainly be better than 10 eV.

TABLE 2: SUMMARY OF EXPECTED AXAF CALORIMETER PARAMETERS

FWHM	< 10 eV
R_{max}	10^3
$\Delta_{tot}E$	0.3-10 keV
$\Delta_{ins}E$	0.3-10 keV
$\Phi = \langle \epsilon(E) \rangle$	> .6
Background	effectively zero
FOV	1 arc min
$d\Omega$	10 arc sec

A rough comparison of the expected AXAF calorimeter values for two of these parameters (R_{max} and Φ) with those from spectrometers that were flown onboard the Einstein observatory is displayed in Figure 4. Note that the effects of the larger area and increased bandpass of the AXAF telescope are not represented in this figure -- one that could properly include them would make this comparison even more striking.

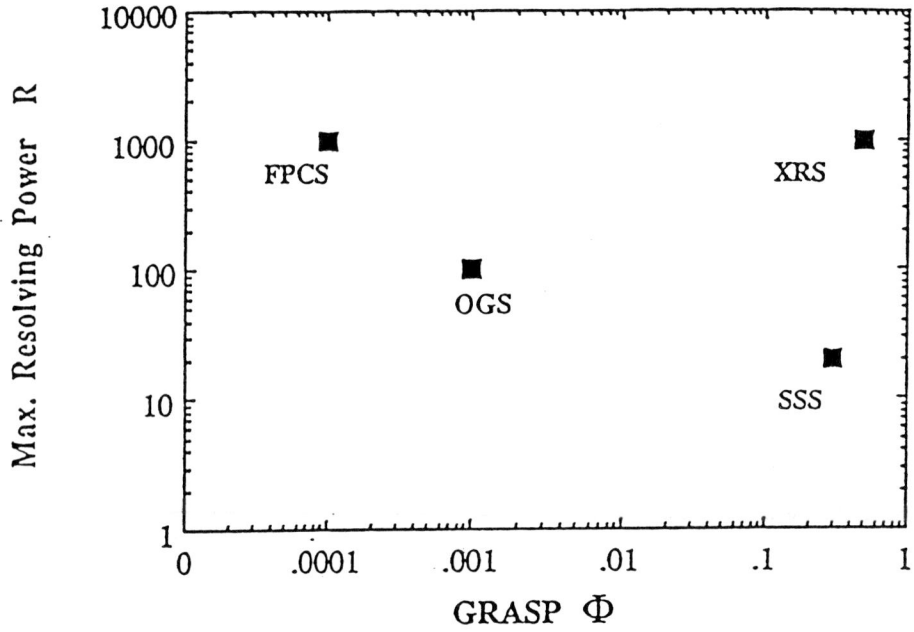

Figure 4. Comparison of maximum resolving power and grasp of the Einstein Observatory spectrometers with the AXAF calorimeter (XRS).

Observing a point source with milli-Crab intensity and an AGN-type spectrum (alpha=0.7, $N_H=10^{21}$ H-atoms/cm^2) for 10^4 s, the calorimeter described here can detect features at "interesting" energies with the following equivalent continuum widths (EW) at 99% confidence:

TABLE 3: LINE DETECTABILITY IN 10^4 S FOR A MILLICRAB SOURCE

E = 0.6 keV	Oxygen	EW = 0.5 eV
2	Silicon	2
7	Iron	21

Because the scaling laws go like the square roots of parameters, the formal sensitivities are not all that much better for slightly better $\epsilon(E)$ or FHWM; it is only in comparison with dispersive spectrometers that the sensitivity differences are truly dramatic.

INNOVATIVE TECHNIQUES FOR X-RAY CALORIMETRY

S. Labov[1], E. Silver[1], D. Landis[2], N. Madden[2], F. Goulding[2], J. Beeman[2], E. Haller[2], J. Rutledge[3], G. Bernstein[4] and P. Timbie[4]

[1]Laboratory for Experimental Astrophysics,
 Lawrence Livermore National Laboratory, Livermore, CA, USA

[2]Lawrence Berkeley Laboratory, Berkeley, CA, USA

[3]University of California, Irvine, CA, USA

[4]University of California, Berkeley, CA, USA

ABSTRACT. In our x-ray calorimetry effort, we have developed several techniques which may be helpful to other groups working in this field. We are studying several different monolithic and composite calorimeter designs. In our readout configuration, the preamplifier circuit employs negative voltage feedback which allows us to accurately measure the temporal profile of the thermal pulse produced by an x-ray absorbed in a micro-calorimeter. Rise times of less than two microseconds have been observed in monolithic devices operating at .3 K. Furthermore, the feedback preamplifier can be configured for either positive or negative electro-thermal feedback. This preamplifier system is followed by an analog pulse shaping amplifier with a frequency response that can be adjusted to yield the maximum signal to noise ratio for a given thermal response of the calorimeter. In addition, we have developed several diagnostic procedures which have been useful in determining the operating and noise characteristics of our devices. These include an infrared light-emitting diode which flashes a discrete amount of energy on to the calorimeter, and a capacitively coupled test input to the preamplifier which allows us to directly determine the total noise in the thermal detection system. Finally, we are developing an adiabatic demagnetization refrigerator with a temperature control system that is designed to stabilize the 0.1 K cold stage to better than 8 μK. This is required for a resistive thermal detector with resolving power of 1000.

1. INTRODUCTION

Cryogenic x-ray calorimeters are of great interest to x-ray astronomy because they offer high resolving power with nearly 100% quantum efficiency below 10 keV. X-ray spectroscopy with microcalorimeters was first discussed by Moseley et al. (1984) of Goddard Space Flight Center (GSFC) and tested by McCammon et al. (1984) of Wisconsin. The collaboration between GFSC and Wisconsin has been steadily progressing toward the goal of practical high resolution x-ray calorimeters. They have measured a resolution of 17 eV (Moseley et al. 1988), and have plans to use such a device in a upcoming sounding rocket experiment (Zhang and McCammon, 1988) Additional research has been reported by Coron et al. (1985) and Fraser (1987). In October 1986, the Laboratory for Experimental Astrophysics initiated a program to develop x-ray microcalorimeters. This paper describes some of the techniques we have developed with emphasis on those which may be of use to other groups working in this field.

2. ELECTRONICS AND DIAGNOSTICS

With practicality and high count rate capability in mind we have conceived of a new approach to low noise pulse counting electronics for microcalorimeters. The circuit configuration we have adopted is based upon JFET preamplifiers with negative voltage feedback that Goulding, Landis, and Pehl (1969) developed for use with silicon and germanium charge collecting detectors. Negative voltage feedback significantly improves the temporal response of the calorimeter over previous techniques, without adding any additional noise (Silver et al., 1989). It has enabled us to measure the thermalization time of high resistance germanium calorimeters to be less than 1 μs. We hope to take advantage of this swift thermalization time and develop detectors that operate at high count rates (Silver et al., 1988).

The negative voltage feedback amplifier also allows the calorimeter to be operated with either a constant voltage (V) across or a constant current (I) through the calorimeter. In the constant voltage mode, the calorimeter is placed at the input to the amplifier and a bypass capacitor holds the voltage constant while the calorimeter's resistance (R_C) undergoes quick excursions. This configuration is shown in Figure 1. When a photon strikes the calorimeter, its resistance drops. Since the voltage is held constant during this time, the biased induced heating (V^2/R_C) increases and the duration of the pulse is increased. The result is positive electro-thermal feedback. In contrast, the constant current configuration results in negative electro-thermal feedback since the bias heating ($I^2 R_C$) drops when the calorimeter temperature rises. In the constant current case the calorimeter is placed within the feedback loop of the amplifier, as shown in Figure 2.

Our pre-amplifier configuration provides flexibility since it allows us to switch from constant voltage to constant current mode at any time during an experiment, without opening the cryostat. When quicker pulses are required, we switch to constant current mode with negative electro-thermal feedback. If longer pulses are preferred, then we can return to constant voltage mode.

The JFET we are currently using is the Interfet 2N6453 which has been preselected for noise less than 1 nV/$\sqrt{Hz}$ at 300 K. This JFET is ensconced in plastic which helps reduce microphonics which originate when the drain lead inside the JFET moves relative to the gate lead. The total gate input capacitance is 15 pF and the drain voltage is set to 2.5 V. The drain current is held at 4 mA, which maintains the device near its 130 K optimal operating temperature. A small copper wire provides a thermal link between the JFET and the liquid nitrogen bath. This link has two beneficial effects. First, the 9 mW of power dissipated in the JFET flows into the nitrogen bath, thus extending the hold time of the cryostat. Secondly, the link to the liquid nitrogen bath prevents the JFET temperature from dropping to such a low temperature that it will not operate.

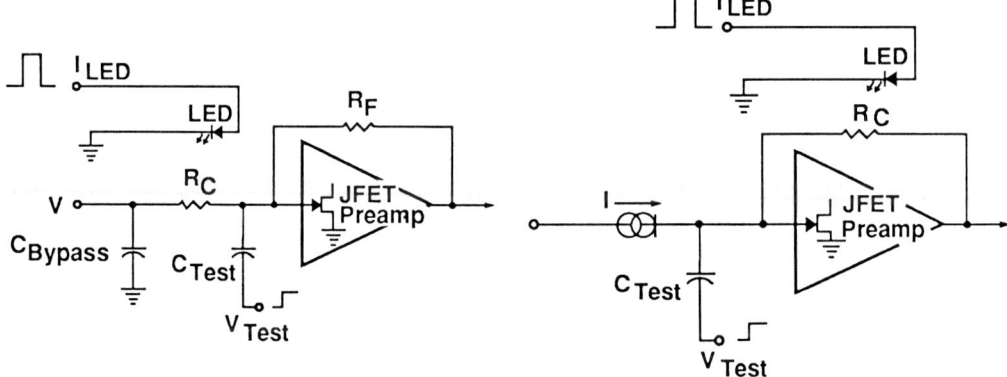

Fig. 1. The constant voltage preamplifier configuration is shown with test capacitor C_{TEST}, and infrared LED.

Fig. 2. The constant current preamplifier configuration.

The preamplifier output is filtered and amplified by our analog pulse shaping post-amplifier. This filter system uses one differentiator and five integrators to limit the bandpass of the signal and noise. The time constant of this filter system can be adjusted to maximize the signal to noise ratio. The post-amplifier output is fed to a pulse height analyzer to obtain spectra. The post-amplifier is described in detail by Goulding and Landis (1982).

We have implemented several new techniques for diagnosing calorimeter performance. First, a test input capacitor (C_{TEST}) enables us to exercise the entire electronic chain without irradiating the calorimeter. This provides a combined measure of the electronic, Johnson, and phonon noise, independently of any "conversion noise" in the detector. This test input also allows us to measure the important time constants in the system. With the calorimeter in the feedback loop (constant current), the time constant $R_C C_C$ can be measured, where C_C is the intrinsic capacitance of the calorimeter. With the calorimeter out of the feedback loop (constant voltage), the time constant $R_F C_F$ can be measured, where C_F is the intrinsic capacitance of the feedback resistor R_F. Second, a controllable infrared LED is used to independently test the operation of the calorimeter. This LED produces short bursts of infrared light at regular intervals, which are absorbed by the calorimeter. These heat pulses are extremely useful for measuring the thermalization time, thermal decay time, and in conjunction with the thermal conductance obtained from the calorimeter load curve, the heat capacity. Both the infrared LED and the test input capacitor are shown in Figures 1 and 2.

One other useful laboratory technique we have been using is to locate our x-ray sources outside of the cryostat. This allows us to make spectra with a number of different sources, and at various count rates. It also allows us to remove the x-ray source and measure the total noise of the system in the absence of pulses. The photons enter the cryostat through a thin beryllium window which blocks the 300 K radiation, maintains the vacuum integrity of the cryostat, and passes x-rays with energies above 4 keV with nearly 100% efficiency. The photons then pass through thin aluminum foils located on the 77 K nitrogen shield, and the 1.5 K helium shield. The total distance from the source to the detector is only 5 cm.

3. ADIABATIC DEMAGNETIZATION REFRIGERATOR (ADR)

We have invested considerable effort in a parallel program to construct and test ADRs for use in the laboratory and eventually in space. Our system uses ferric ammonium sulfate as its paramagnetic salt. Two parts of the salt (by weight) are dissolved in one part 7% sulfuric acid solution held at 37°C. The crystal is then grown directly on a brush of gold wires which have been silver soldered into a copper post. Both the copper and the silver solder are plated with gold to prevent corrosion by the salt. The 2 cm diameter, 6 cm long salt crystal is then sealed with Stycast epoxy into a stainless steel cylinder which is supported in the magnet bore by twelve lengths of nylon monofilament each 4 cm long and .28 mm diameter. The measured heat leak of this system is about 0.2 µW when operated at 0.1 K. A temperature of less than 50 mK has been achieved and the hold time at 100 mK is more than 12 hours.

High resolution x-ray spectroscopy using microcalorimeters sets stringent bounds on the thermal stability of the refrigeration system. For example, if S is the voltage step produced by a photon of energy E absorbed by a calorimeter with a total heat capacity of C_V, then

$$S = I \Delta R = I \frac{dR}{dT} \frac{E}{C_V},$$

where I is the current (assumed to be constant in this example), ΔR is the change in calorimeter resistance and T is the temperature. If the dominant contribution to the calorimeter heat capacity is crystalline in origin, the Debye law predicts $C_V \propto T^3$. At the temperatures of interest, semiconductor resistors tend to vary as $R \propto e^{\sqrt{\Delta/T}}$. In this case

$$\frac{d \log|S|}{d \log T} = -\left[\frac{9}{2} + \frac{1}{2} \frac{\sqrt{\Delta}}{\sqrt{T}}\right] \cong \frac{T}{|S|} \frac{\delta S}{\delta T}.$$

So for $S/\delta S > 1000$, a stability of $\delta T < 8$ μK is required for $\Delta = 25$ and $T = .1$ K.

To achieve this temperature stability, we have begun developing a microprocessor-based servo-control system that will adjust the superconducting magnet current to the desired precision. By controlling the temperature directly with the magnetic current, we avoid adding extra heat into the system that is characteristic of heater driven control systems. This extends the total hold time of the system, and reduces the mechanical complexity. A low resolution prototype of this controller has been tested, and a stability of +/− 20 μK has been already been achieved. We are confident that the new control system will surpass the required stability.

4. CONCLUSIONS

With high energy resolution, high quantum efficiency and a broad energy bandwidth, a practical x-ray calorimeter system would offer many advantages over solid state detectors and crystal spectrometers. The techniques described here have been useful in the development of x-ray calorimeters, and may be important in making practical devices. The negative voltage feedback amplifier allows fast signals to be processed which is essential if these detectors need to be used in moderate count rate applications. This electronic technique has already enabled us to study the intrinsic time response at temperatures below 1 K. With the constant current to constant voltage switching ability, a calorimeter can be designed so that resolution can be traded for speed at any time. Finally, the analog pulse processing is operationally straightforward, and the test pulse diagnostics provide direct ways of determining the system parameters.

We wish to thank Paul Richards and Yolanda Wai for their assistance. This work was performed under the auspices of the U.S. Department of Energy by Lawrence Livermore National Laboratory under contract No. W-7405-ENG-48.

5. REFERENCES

Coron, N., Artzner, G., Dambier, G., Jegoudez, G., Leblanc, J., Lepeltier, J. P., Deschamps, J. Y., Rocchia, R., Tarrius, A., Testard, O., Hansen, P. G., Jonson, B., Ravn, H. L., Stroke, H. H., and Turlot, E. 1985, *Proc. of SPIE,* **567**, 389.

Fraser, G. W. 1987, *Nucl. Inst. Meth.,* **A256**, 553.

Goulding, F. S., Landis, D. A., and Pehl, R. H. 1969, in *Semiconductor Nuclear-Particle Detectors and Circuits*, ed. W. Brown, W. Higinbotham, G. Miller, and R. Chase (Washington, D.C.: National Academy of Sciences), p. 455.

Goulding, F. S. and Landis, D. A. 1982, *IEEE Trans. Nucl. Sci.,* **NS- 29**, 1125.

McCammon, D., Moseley, S. H., Mather, J. C., Mushotzky, R. F. 1984, *J. Appl. Phys.,* **56**, 1263.

Moseley, S. H., Mather, J. C., and McCammon, D. 1984, *J. Appl. Phys.,* **56**, 1257.

Moseley, S. H., Kelly, R. L., Schoelkopf, R. J., Szymkowiak, A. E., McCammon, D., and Zhang, J. 1988, *IEEE Trans. Nucl. Sci.,* **35**, 59.

Silver, E., Labov, S., Landis, D., Madden, N., Goulding, F., Beeman, J., Haller, E., Rutledge, J., Bernstein, G., and Timbie, P. 1988, *Proc. IAU 115*, this volume.

Silver, E., Labov, S., Goulding, F., Madden, N., Landis, D., and Beeman, J. 1989, *Nucl. Instr. Meth.*, submitted.

Zhang, J. and McCammon, D. 1988, *Proc. IAU 115*, this volume.

OBSERVING SOFT X-RAY LINE EMISSION FROM THE INTERSTELLAR MEDIUM WITH X-RAY CALORIMETER ON A SOUNDING ROCKET

J. Zhang[1], B. Edwards[1], M. Juda[1], R. Kelley[2], G. Madejski[2], D. McCammon[1], H. Moseley[2], M. Skinner[1], R. Schoelkopf[2], and A. Szymkowiak[2]

1 Physics Department, University - Madison, Madison, WI, USA
2 NASA/Doddard Space Flight Center, Greenbelt, MD, USA

ABSTRACT. We have been developing X-ray calorimeters that have high spectral resolution and high quantum efficiency. For an X-ray calorimeter working at 0.1 K, the energy resolution ideally can be as good as one eV for a practical detector. A detector with a resolution of 17 eV FWHM at 6 keV has been constructed. We expect to be able to improve this by a factor of two or more. With X-ray calorimeters flown on a sounding rocket, we should be able to observe soft X-ray line emission from the interstellar medium over the energy range 0.07 to 1 keV. Here we present a preliminary design for an X-ray calorimeter rocket experiment and the spectrum which might be observed from an equilibrium plasma. For later X-ray calorimeter sounding rocket experiments, we plan to add an aluminum foil mirror with collecting area of about 400 cm^2 to observe line features from bright supernova remnants.

INTRODUCTION

The Soft X-ray Background (SXRB, E < 1.0 keV) has been studied for about twenty years, primarily with proportional counters. All the SXRB observations suggest that the major source of the SXRB is thermal emission from a hot, thin interstellar plasma of temperature ~10^6 K (Williamson et al.1974, McCammon et al. 1983, Marshall and Clark 1984). Observations of O VI and OVII (Jenkins 1974, Inoue 1979, Schnopper 1982, Rocchia 1984) also support a thermal model. This hot plasma should be rich in line emssion. Getting high resolution X-ray spectra of such a diffuse source is very difficult, but could add much information about its physical state and possibly its past history. X-ray calorimeters have adequate resolution for this task, and a modest instrument can be constructed with enough throughput to make useful observations on a sounding rocket flight.

INSTRUMENT

The way in which the X-ray calorimeter works is that a photon is absorbed and its energy is converted to heat; the resulting temperature rise can be measured by detecting the resistance change of an ion-implanted silicon thermistor (Fig. 1.). More detailed descriptions are in McCammon et al. (1987) and Holt (1988). Since the energy resolution is limited by the heat capacity of an individual calorimeter , the collecting area can be increased by using multiple detectors. This also permits imaging operation if the detector array is used with a telescope.

We have been developing X-ray calorimeters at GSFC and at the University of Wisconsin. A detector with 17 eV FWHM resolution at 6 keV has been constructed. We expect to improve the energy resolution to 5 eV for low energy X-rays in the near future. Figures 1 and 2 show different views of a single detector element. Figure 3 is the electrical circuit used to run the calorimeters. Figure 4 is a sketch of the basic strucuture of a sounding rocket cryostat for X-ray calorimeters. The required 0.1 K operating temperature is provided by an adiabatic demagnetization refrigerator (ADR) . The salt pill (ferric ammonium alum) is fixed in a G-10 tube which is suspended by Kevlar fibers. A set of blocking filters is necessary to limit the shot noise in the detectors caused by low-energy photons from thermal infrared radiation and ultraviolet air glow. We plan to use four filters in series. Each one has 200 Å aluminum evaporated on to a 1000 Å parylene-N substrate. The calculated net IR transmission of four filters is $<10^{-10}$. The dashed line in Figure 5 is the net X-ray transmission based on measurements of the individual filters. The X-ray calorimeter has an intrinsic quantum efficiency of almost 100%. The X-ray throughput is limited by the filter transmission and the detector array area. We are planning to observe emission from the interstellar medium (ISM) over about 1 steradian of the sky with 0.2 cm^2 total detector area.

OBSERVATION OF INTERSTELLAR MEDIUM EMISSION

The solid curve in figure 5(a) is a calculated emission spectrum convolved with the expected 5 eV resolution of the calorimeters. We have used the equilibruim plasma models of Raymond and Smith (1987), with solar abundances, and with the temperature and emission measure of each component chosen so that total B-band, C-band, and M-band count rates agree with the Wisconsin sky survey (McCammon 1983). Figure 5(b) shows a Monte Carlo simulation of a 300 second observation with the sounding rocket instrument described above. Even with such short observing time, the statistics are sufficient to permit a considerable amount of useful scientific analysis. For later X-ray calorimeter sounding rocket experiments, we plan to add an aluminum foil mirror with 400 cm^2 collecting area (Serlemitsos 1981) to observe bright supernova remnants.

REFERENCES

Holt, S. S. 1988, invited talk in this colloquium.
Inoue, H. Koyama, K., Matsuoka, M., Ohashi, T., Tanaka, Y., and Tsuremi, H. 1979, *Ap. J., (Letters)*, **227**, L85.
Jenkins, E. B. and Meloy, D. A. 1974, *Ap. J., (Letters)*, **293**, L115.
Marshall, F. J. and Clark, G. W. 1984, *Ap. J.*, **287**, 633.
McCammon, D. Burrows, D. N., Sanders, W. T., and Kraushaar, W. L. 1983, *Ap. J.*, **269**, 107.
McCammon, D., Juda, M., Zhang, J., Holt, S. S., Kelley, R. C., Moseley, S. H., and Szymkowiak, A. E. 1987, *Proc. 18th Int. Conf. on Low Temperature Physics*, Kyoto, 1987. Japanese Journal of Applied Physics, Supplement, 26,3
Raymond, J. C. and Smith, B. W. 1977, *Ap. J. Suppl.*, **35**, 419
_____, 1987, *private communication* (update to Raymond and Smith 1977)
Rocchia, R.,Araud, M., Blondel, C., Cheron, C., Christy, J. C., Rothenflug, R., Schnopper, H. W., and Delvaille, J. P. 1984, *Astron. Astrophys.*, **130**, 53
Schnopper, H. W., Delvaille, J. P., Rocchia, R., Blondel, C., Cheron, C., Christy, J. C., Ducros, R., Koch, L., And Rothenflug, R 1982, *Ap. J.*, **253**, 131
Serlemitsos, P. J. et al., 1981, *X-ray Astronomy in the 1980's*, ed. S.Holt, NASA Technical Memorandum **83848**, 441
Williamson, F. O., Sanders, W. T., Kraushaar, W. L., McCammon, D., Borken, R., and Burner, A. N. 1974, *Ap. J., (Letters)*, **193**, L133.

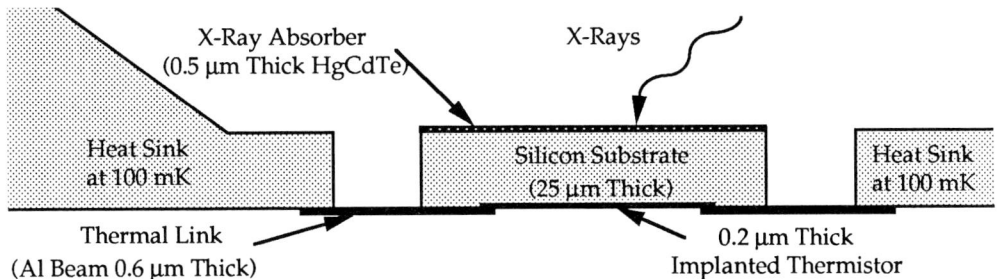

Figure 1. A side view of an X-ray calorimeter (not to scale). When an X-ray hits the absorber (HgCdTe), the photon energy is converted to thermal energy. The resulting temperature rise can be measured by the ion-implanted thermistor. Aluminum beams provide thermal and electrical links.

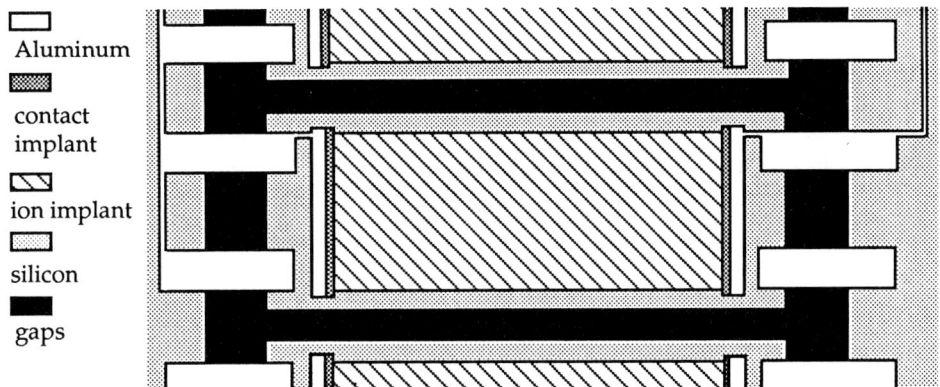

Figure 2. A single detector elemnt of a testing arraye

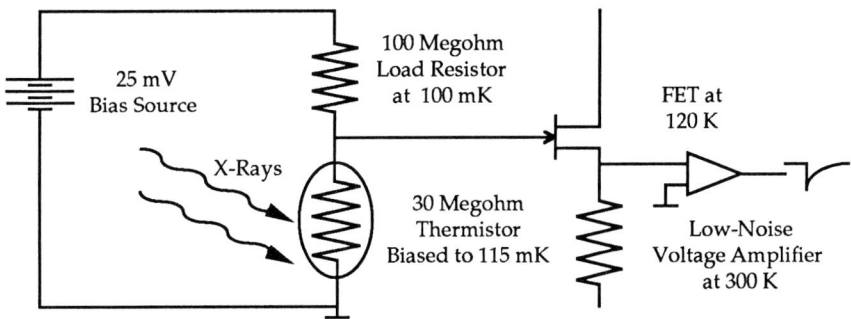

Figure 3. An electrical circuit used to run a calorimeter.

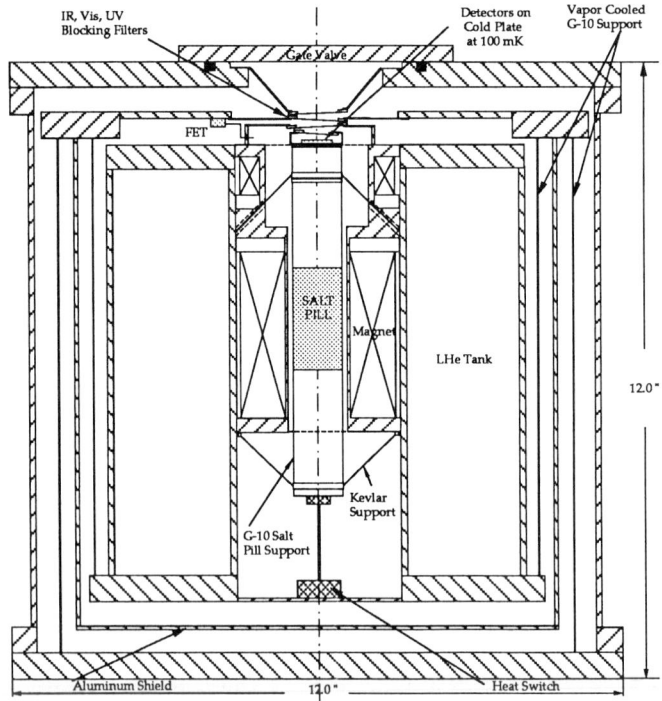

X-Ray Calorimeter Sounding Rocket Cryostat

Figure 4. The required 100 mK operating temperature of a X-ray calorimeter is provided by an adiabatic demagnetization refrigerator. This shows the basic structure of the cryostat. The salt pill (ferric ammonium alum) is fixed in a G-10 tube which is suspended by kevlar fibers. A set of blocking filters is used to limit the shot noise in the detectors caused by low-energy photons from thermal infrared radiation and ultraviolet air glow. The G-10 shell supports for the liquid helium dewar are vapor cooled.

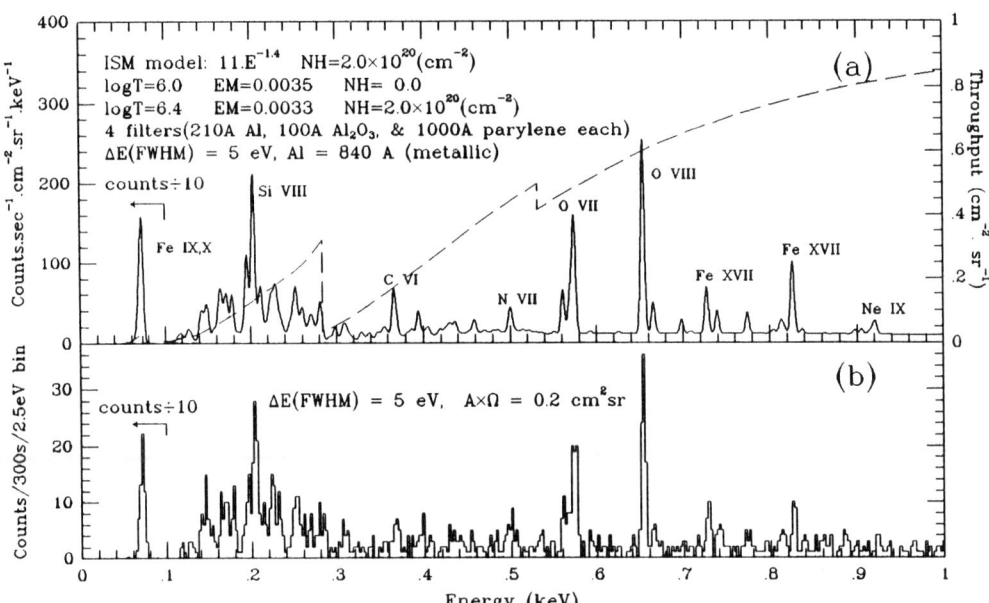

Figure 5. (a) A calculated emission spectrum convolved with the expected 5 eV resolution of the calorimeters. We have used equilibrium plasma models of Raymond and Smith (1987), with solar abundance, and with the temperature and emission measure of each component chosen so that total B-band, C-band, and M-band count rates agree with the Wisconsin sky survey (McCammon 1983). The dashed line is the instrument response (net X-ray transmission of four filters). (b) A monte Carlo simulation of 300 second observation with sounding rocket instrument described in the text.

HIGH THROUGHPUT SOFT X-RAY SPECTROSCOPY WITH REFLECTION GRATINGS

Steven M. Kahn[1]

[1] Department of Physics and Space Sciences Laboratory, University of California, Berkeley, CA 94720 USA

ABSTRACT: As dispersing elements, grazing incidence reflection gratings offer the unique combination of high dispersion and wide spectral coverage at high efficiency. They can therefore be coupled with large area, low-resolution mirrors and high quantum efficiency detectors to yield moderate resolution spectroscopy of faint X-ray sources. Various design options are presented and compared, including both objective and convergent-beam configurations and both in-plane and off-plane grating mountings. A specific reflection grating payload design for ESA's X-Ray Multi-Mirror Mission (XMM) is reviewed in more detail. Predicted performance curves derived from ray trace studies are presented along with preliminary X-ray reflectivity measurements of prototype grating samples.

1. INTRODUCTION.

In many astrophysical discussions of the scientific potential of X-ray spectroscopy, primary emphasis has been given to the Fe K-line region between 6 and 7 keV. The Fe K lines have been detected in a wide variety of sources, and have certainly provided useful physical constraints for many cosmic systems. However, the softer regions of the spectrum, particularly near 1 keV, may in the long run prove even more interesting. The soft X-ray band (0.2 - 2 keV) is densely permeated with important features, including the K-shell transitions of C, N, and O, and the L-shell transitions of Fe. The Fe L-shell lines are especially promising. They cover a very wide range of ionization and are quite well-separated.

Unfortunately, it is the high density of expected transitions that makes it difficult to work in this band. Rather high spectral resolving power, $E/\Delta E \geq 200$, is required even to correctly identify the elemental species associated with observed features. Non-dispersive detectors, including the state-of-the-art solid state devices, are not up to the task. Their resolution, (ΔE), is either fixed or increases slowly with energy, so that ($E/\Delta E$) decreases at lower energies. Even the new cryogenic detectors (Holt 1989) are not sufficient below 1 keV unless their resolution can be improved by at least a factor of ten. If spectral features cannot be uniquely identified in the raw data, then the power of high resolution spectroscopy to provide model-independent constraints becomes severely reduced.

For dispersive devices, on the other hand, the resolution in wavelength, ($\Delta \lambda$), is approximately fixed across the band. Hence, for these cases, ($E/\Delta E$) increases at lower energies. The transmission grating spectrometers which were flown on *Einstein* and EXOSAT (Seward et al. 1982; Brinkman et al. 1980) exhibited this property; they achieved their highest resolving power at the lowest accessible energies. Transmission grating experiments are also planned for the upcoming SPEKTROSAT (Predehl et al. 1988) and AXAF (Schattenberg et al. 1988) missions.

In a dispersive spectrometer, an imaging optic must be incorporated somewhere in the system to provide the concentration. The limiting achievable spectral resolution scales essentially linearly with the angular resolution of the focussing mirror. A high resolution mirror is thus required to obtain the desired resolving power. Both AXAF and SPEKTROSAT will be ~ arc-second facilities.

The next really substantial increase in sensitivity for soft X-ray spectroscopy, however, is more likely to be associated with the so-called "high throughput" missions like SPECTRUM X, XMM, and LAMAR (Peacock and Ellwood 1988; Gorenstein 1978). These experiments utilize an array of telescopes in order to obtain very large collecting area. In order to reduce costs, significant compromise in angular resolution is required. The design specifications for these facilities are more on the order of 0.5 to several arc-minutes.

With reduced angular resolution, very high dispersion is required to maintain the spectral resolution. It turns out that this is not possible with transmission gratings. The necessary groove densities are far in excess of current fabrication limits. Reflection gratings, on the other hand, do not suffer from this problem. A reflection grating used at grazing incidence can offer much higher dispersion than a transmission grating for a given line spacing. As is shown below, reflection gratings also yield fairly high diffraction efficiency in the soft X-ray band. Hence, when coupled to a high throughput mirror system, they provide a very sensitive, high resolution, soft X-ray spectrometer.

In this paper, we review several important design considerations relevant to the incorporation of reflection gratings on high throughput X-ray spectroscopy missions. We begin with the basic optics of reflection gratings in Section 2, and then move on to discuss grating orientation and various optical design options in Sections 3 and 4 respectively. We concentrate primarily on the motivations which have led to a specific design we are proposing for the XMM mission in collaboration with the Laboratory for Space Research at Utrecht. The optimization of this design and a brief summary of its expected performance and scientific capabilities is provided in Section 5. Finally in Section 6, we discuss grating fabrication issues and present some experimental results obtained for prototype grating samples.

2. THE BASIC OPTICS OF REFLECTION GRATINGS.

A schematic diagram of a simple reflection grating is given in Figure 1. We have defined a coordinate system in which the Z-axis is parallel to the grating grooves and the X-axis is oriented along the normal to the grating plane. An arbitrary incident ray makes an angle θ with the Z-axis and its projection in the XY-plane makes an azimuthal angle α with the Y-axis. If the grating is large compared to all relevant scales, then the outgoing ray reflected off the grating will also have polar angle θ. This is the so-called "conical diffraction" condition, since all outgoing rays emerge in a cone. The outgoing azimuthal angle, β, is related to the incoming angle, α, by the dispersion equation:

$$m \lambda = d \sin\theta (\cos\beta - \cos\alpha). \tag{1}$$

where d is the groove spacing, and m is an integer, i.e. the spectral order. Note that the zero order, $\alpha = \beta$, corresponds to a pure reflection off the grating surface. The m < 0 cases are referred to as the "inside orders" since the outgoing ray falls between the zero order and the normal on the cone. The m > 0 cases are referred to as the "outside orders".

Most applications which require high diffraction efficiency from the grating invoke "blazed" gratings in which the facets exhibit the staircase or sawtooth pattern illustrated in Figure 2. The tilt angle of the facets, indicated by δ in the Figure, is called the blaze angle. Referring to the geometry above, we can calculate the incoming and outgoing graze angles made with the facet surface in terms of previous quantities:

$$\sin\gamma_{in} = \sin\theta \sin(\alpha+\delta) \equiv \sin\theta \sin\mu_{in}. \tag{2}$$

$$\sin\gamma_{out} = \sin\theta \sin(\beta-\delta) \equiv \sin\theta \sin\mu_{out}.$$

These definitions apply to the inside order in particular. When γ_{in} is equal to γ_{out}, then each of the facets behaves like a tiny mirror, and they all add coherently to give maximum diffraction efficiency. This is the "blaze condition". A simple manipulation of the dispersion equation shows that at blaze:

$$|m|\lambda = 2d\ \sin\gamma\ \sin\delta. \qquad (3)$$

The wavelength which satisfies this condition for a given spectral order and for a given graze angle is called the blaze wavelength. Note that this expression holds for all polar angles, θ.

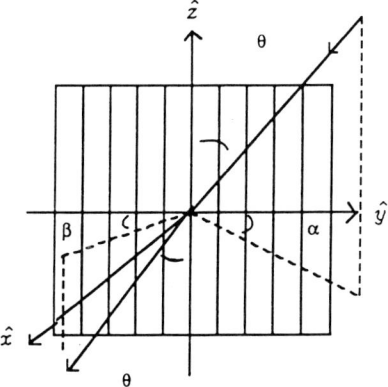

Figure 1: A schematic of a reflection grating showing the orientation of arbitrary incoming and outgoing rays.

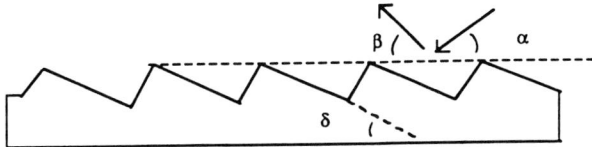

Figure 2: A blow-up of a blazed reflection grating illustrating the triangular shape of the grooves. The blaze angle δ is defined in this figure.

For an ideal blazed grating, it is possible to apply Fraunhofer scalar diffraction theory to derive the diffraction efficiency of the grating in a given orientation for each of the spectral orders (Madden and Strong 1958). A straightforward calculation gives:

$$Eff_m = \frac{g^2}{4\sin^2\theta \sin\alpha \sin\beta_m} P_m^2 [\sin Q_m/Q_m]^2. \tag{4}$$

where:

$$g \equiv \sin\alpha/\sin(\alpha+\delta) = \frac{\sin\alpha}{\sin\mu_{in}}.$$

$$P_m \equiv \sin\theta [\sin(\alpha+\delta) + \sin(\beta_m-\delta)].$$

$$Q_m \equiv (\pi g d/\lambda) \sin\theta [\cos(\alpha+\delta) - \cos(\beta_m-\delta)].$$

Here β_m is the outgoing angle for the mth spectral order. These expressions assume that the grating surface is perfectly reflecting. Of course, in the X-ray band, no surface is even close to perfectly reflecting and one must normalize by a reflection efficiency evaluated at some "effective graze angle". There is some debate in the literature over what is the correct normalization to use (Hutley 1982). The one we prefer involves the geometric mean of the reflectivities evaluated at the incoming and outgoing angles:

$$[R(\gamma_{in}) R(\gamma_{out})]^{1/2},$$

which at least preserves the time reversal symmetry of the final expression.

Not surprisingly, even with the normalization above, the scalar diffraction calculations are only approximate. Because the incoming and outgoing graze angles are different, the reflection and diffraction aspects of the problem simply cannot be separated. The only correct approach is to self-consistently solve Maxwell's Equations in free space subject to the material boundary conditions imposed at the grating surface. This is the "exact electromagnetic calculation" (Petit 1980). Comparisons show the largest discrepancy between the two calculations for the in-plane configuration, $\theta = 90°$, because the difference between γ_{in} and γ_{out} is largest in that orientation. Despite its inaccuracy, the simple diffraction theory does give the right qualitative behavior, and can be very useful for analytical optimization studies. The exact electromagnetic calculation is, of course, required for eventual instrument calibrations.

3. GRATING ORIENTATION.

In this Section, we address the question of the grating orientation, i.e. how should the grating be mounted with respect to the incoming beam? This issue has received a lot of attention in the literature in recent years, which has led to some confusion in the field. In order to compare the options fairly, it is necessary to parametrize the problem in an appropriate way and evaluate competing geometries which are individually optimized for the problem at hand. For a given wavelength band of interest, one must first select the first order blaze wavelength, λ_B, which can crudely be viewed as the wavelength of maximum sensitivity. The wavelength band also determines the graze angle, γ, which provides reasonable reflection efficiency. Hence, λ_B and γ can be regarded as fixed by scientific considerations. The other parameters are then coupled by the various relations. Specifically, d and δ are related by the blaze equation (Equation 3) and θ and μ are coupled by the graze angle definition (Equation 2). Hence, in addition to λ_B and γ, there are at most two additional parameters which need to be specified. For reasons that will be clear later, our preferences for the parametrization are: μ and δ/μ. Since $\mu \equiv \alpha + \delta$ (for the inside orders) and $\alpha \geq 0$, δ/μ must be ≤ 1.

There are, in fact, only two configurations which have been discussed extensively in the literature: the "in-plane" or "classical" configuration, in which the rays come in perpendicular to the grooves; and the "off-plane" or "conical" configuration, in which the rays come in nearly parallel to the grooves. In our notation, the in-plane case is characterized by $\theta = 90°$, $\mu = \gamma$, and the off-plane case by $\theta = \gamma$ and $\mu = 90°$. There is a nice symmetry between them.

How does one choose between these two cases, or between any intermediary cases? Initially, we must look at the relative performance, i.e. the sensitivity and the resolution. This is easy to do analytically at the blaze wavelength itself. For example, the complicated Fraunhofer expression we had earlier (Equation 4) reduces to the simple form:

$$Eff_B = R(\gamma) \frac{\sin(\mu-\delta)}{\sin(\mu+\delta)} \qquad (5)$$

when evaluated at the blaze. If the resolution is dominated by the blur introduced by the mirrors, ε, (as is true for most of the high throughput missions), and if this blur is essentially isotropic as seen by the grating, then another straightforward calculation gives the resolving power at blaze:

$$(\lambda/\Delta\lambda) = \frac{\sin\gamma}{\varepsilon} [\sin^2\mu - \sin^2\gamma + \frac{\sin^2(\mu-\delta)}{4\sin^2\delta}]^{-1/2}. \qquad (6)$$

For simplicity, we shall refer to the ratio of sines which appears on the right hand side of Equation 5 as the "sensitivity factor", η, and the term involving square brackets on the right hand side of Equation 6 as the "resolution factor", ρ. Clearly the optimal design will have the highest possible values of η and ρ. In Tables 1 and 2 respectively, we list η and ρ as a function of the parameters we chose earlier: μ and δ/μ. These values are calculated for a prototype soft X-ray design: $\lambda_B = 15$ Å and $\gamma = 2°$.

| Table 1 - Sensitivity Factor η Versus μ and δ/μ ||||||
| Calculated for λ_B = 15 Angstroms and $\gamma = 2°$ ||||||
δ/μ \ μ	2°	5°	10°	20°	60°	90°
0.1	0.82	0.82	0.82	0.82	0.89	1.0
0.3	0.54	0.54	0.54	0.55	0.68	1.0
0.7	0.18	0.18	0.18	0.19	0.32	1.0
0.9	0.053	0.053	0.054	0.057	0.11	1.0
0.95	0.026	0.026	0.026	0.028	0.059	1.0
1.0	0	0	0	0	0	1.0

Table 2 - Resolution Factor ρ Versus μ and δ/μ						
Calculated for λ_B = 15 Angstroms and $\gamma = 2°$						
δ/μ \ μ	2°	5°	10°	20°	60°	90°
0.1	0.22	0.22	0.22	0.23	0.25	0.30
0.3	0.86	0.86	0.85	0.83	0.72	0.71
0.7	4.7	4.4	3.7	2.5	1.1	0.97
0.9	18	10	5.6	2.9	1.2	1.0
0.95	38	12	5.8	2.9	1.2	1.0
1.0	∞	13	5.9	2.9	1.2	1.0

There are two conclusions which can be drawn from these tables: (1) First, there is an obvious complementarity between resolution and sensitivity. Parameter combinations that give high resolution give low sensitivity and visa versa. (2) Second, there is not actually that much variation in either parameter; for mid-range values of δ/μ, all values of μ give acceptable resolution and sensitivity. The in-plane design (left-most column) gives the highest resolution, whereas the off-plane design (right-most column) gives the highest sensitivity, but not by more than a factor of a few in either case. Interestingly, if we define a "figure of merit" as the product $\eta \times \rho$, it peaks at the two extreme cases, in-plane and off-plane, so there is, in fact, no real reason to consider the intermediate designs.

Evidently, there is not a strong case to be made for in-plane versus off-plane geometries based on theoretical performance alone. However, the two alternatives are distinguished in terms of practical considerations. In the high throughput soft X-ray application, each requires pushing the fabrication state-of-the-art in a particular direction. For the in-plane mount, this involves fabricating low blaze angles. Since $\mu = \gamma = 2°$ in this case, and since δ is a fraction of μ, δ is typically required to be less than ~ 1°. 1.5° blaze angles are easily fabricated, but < 1° is conventionally viewed as a technical challenge. For the off-plane mount, the problem involves very high groove density. Since $\mu = 90°$ in this case, $\delta \cong$ tens of degrees, and d must be much smaller. In fact, virtually all of the off-plane designs shown in the tables require groove densities > 40,000 l/mm. That is far beyond the state-of-the-art for reflection gratings by at least a factor four. As it turns out, the low blaze angles required for the in-plane designs are much less of a concern (see Section 6), so, by this analysis, the in-plane configuration appears to be preferred for the high throughput X-ray application.

3. OPTICAL DESIGN.

We now turn to questions of the optical design. In particular, where should the grating be placed with respect to the X-ray optical path? At longer wavelengths, a reflection grating is typically located behind a slit in the telescope focal plane, as in the classical Rowland circle design. However, since the grating must be used at grazing incidence in the soft X-ray band, this configuration introduces a number of technical problems, the most important of which is that it requires an extremely long (~ 2 m) instrument bay behind the focal plane. There are some tricks which can be invoked to partially get around this difficulty, but not enough so to make it easy to incorporate this kind of design.

Another possibility is to mount the grating in front of the telescope itself (Cash 1988). This has the advantage that it does not disturb the telescope focal plane assembly. The grating is tilted and the telescope is offset pointed from the source so that the desired wavelength band is diffracted down the optical axis. In principle, one can convert between direct imaging and spectroscopic modes by simply rotating the grating in and out of the field of view of the telescope.

The principal disadvantage of this configuration is that the telescope is used as a camera to image the spectrum, i.e. one relies on both its on-axis and off-axis properties. Conventional X-ray telescopes have, by optical standards, very poor off-axis performance; aberrations grow quadratically with off-axis angle. In addition, vignetting by the mirror shells seriously limits throughput off-axis. Hence, the spectrum is degraded near the edges of the band.

As an alternative, the grating can be placed behind the mirror. The telescope is then used on-axis at all wavelengths. However, in this configuration, the grating does not see parallel light. That introduces aberrations if not properly corrected. As first shown by Hettrick and Bowyer (1983), the aberrations can be removed if the groove spacing is varied across the grating plane. Hettrick's varied line-space solution has the additional advantage that it yields a focal plane which lies at near normal incidence to the beam. A potential drawback to placing the grating behind the telescope is that the detector must then be offset from the telescope focal plane. Thus, this configuration is not easily convertible between imaging and spectroscopic applications.

Since the grating is used at exteme grazing incidence in the soft X-ray band, a very large grating would be required to completely cover the beam exiting a typical high throughput telescope. A single large grating can be assembled from a set of smaller gratings arrayed in a co-planar configuration, however this solution is prohibitively expensive. Since the required groove spacing depends on the distance from telescope focus, the individual smaller gratings cannot be identical. In addition, this approach leads to reduced angular resolution, since the angular blur introduced by the mirrors increases nearly linearly with distance down the optic axis of the telescope.

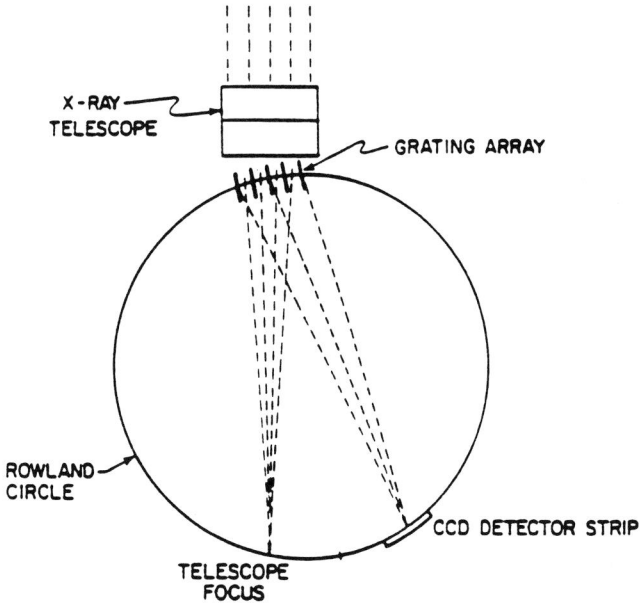

Figure 3: The reflection grating optical design we are proposing for incorporation on XMM. The gratings sit on a Rowland circle in a near-parallel stack at the exit from the X-ray mirror.

Alternatively, a set of gratings can be arrayed alongside each other in a near parallel stack (see Figure 3). In this case, the gratings can be identical replicas of a single master since they all lie at roughly the same distance from the telescope focus. A truly parallel array will generate aberrations. However, these can be cured by an appropriate choice of geometry. The array is arranged so that all gratings are oriented at the same angle with respect to the incident ray at grating center. In addition, the gratings are aligned along a Rowland circle, as shown in Figure 3, which also includes both the telescope and spectroscopic foci. If the spectroscopic detectors are aligned along this circle, comatic aberrations associated with the array are eliminated for all wavelengths in the spectrum.

A complication arises with this design due to vignetting of outgoing rays by neighboring gratings in the stack. Since the outgoing rays leave the grating at larger angles than the incident rays, this can only be corrected by spacing the gratings at larger separations, i.e. by not intercepting the entire beam exiting the telescope. The unintercepted rays can be effectively utilized by a complementary instrument in the telescope focal plane. This yields a kind of "built-in redundancy" in the use of the telescope, with no moving parts.

5. OPTIMIZATION AND SCIENTIFIC PERFORMANCE OF THE XMM DESIGN.

Motivated by the considerations outlined above, we are proposing a reflection grating spectrometer for XMM which consists of an array of thin reflection gratings placed in the Rowland circle configuration at the exit from the X-ray telescope. The gratings are mounted in the in-plane configuration and are oriented so that the first, second, and third inside spectral orders are diffracted at moderate to high efficiency. The diffracted light is imaged by an array of charge-coupled device (CCD) detectors offset from the telescope focal plane. The separation of the spectral orders is accomplished using the energy resolution of the CCDs.

The complete specification of the optical design for this configuration requires a choice of grating parameters, specifically the average line spacing on the gratings, d, the mean incidence angle, α, and the blaze angle, δ. These in turn determine the length of the detector array and its position with respect to the telescope focus. The optimization begins with a choice of first order blaze wavelength. Given the prevalence of important spectral features between 0.5 and 1 keV, we choose $\lambda_B = 15$ Å. d, δ, and the graze angle, γ are then coupled by the blaze equation (Equation 3). Hence, only two of the three parameters are sufficient to specify the design. It is convenient to choose γ and $\eta = \sin(\gamma-\delta)/\sin(\gamma+\delta)$ as the relevant parametrization. The resolution of the design clearly increases as we increase the detector length. Hence the number of CCD chips required becomes the limiting consideration. In terms of γ and η, the detector length is approximately given by:

$$l_D \approx \frac{2\gamma L}{(1+\eta)} [\{\eta^2(1 - \frac{\lambda_{max}}{\lambda_B}) + \frac{\lambda_{max}}{\lambda_B}\}^{1/2} - \{\eta^2(1 - \frac{\lambda_{min}}{\lambda_B}) + \frac{\lambda_{min}}{\lambda_B}\}^{1/2}], \qquad (7)$$

where L is the distance between the center of the grating box and the telescope focal plane, and λ_{max} and λ_{min} are the maximum and minimum wavelengths in the spectral band respectively. For a maximum allowable detector length, there is a coupling between η and γ, with γ an increasing function of η.

The total diffraction efficiency of the system at blaze also depends on these two parameters. From the scalar Frauhofer theory (which is a reasonable approximation near the blaze), we get:

$$Effic = \eta^2 R(\gamma, \lambda_B) \qquad (8)$$

where $R(\gamma, \lambda_B)$ is the reflection efficiency of the surface at angle γ and wavelength λ_B. Since γ increases with η to maintain the detector length, and $R(\gamma, \lambda_B)$ is a sharply decreasing function of γ, the efficiency is maximized at a well-defined optimal set of η, γ values.

For XMM, we take λ_{max} = 35 Å, λ_{min} = 5 Å, and L = 6.7 m. We assume a maximum detector length of 235.8 mm. We find the maxium mix of first and second order efficiencies at an optimal value of η of 0.53. The implied blaze angle is 0.70° and the mean groove density is 646 l/mm. The mean incident angle on the gratings is 1.58°.

The effective area for this optimized design incorporated behind two of the three XMM telescopes is illustrated in Figure 4. Separate curves are given for the first, second, and third spectral orders diffracted by the gratings. As can be seen, the effective area of the system is extremely high, greater than 100 cm^2 over most of the instrument bandpass. The resolving power is shown as a function of wavelength in Figure 5 for first and second spectral orders. As can be seen, it exceeds the requirement $\lambda/\Delta\lambda > 200$ over the entire band. This resolution is more than adequate to yield unique identifications for nearly all discrete features expected to be prominent in soft X-ray spectra of cosmic sources.

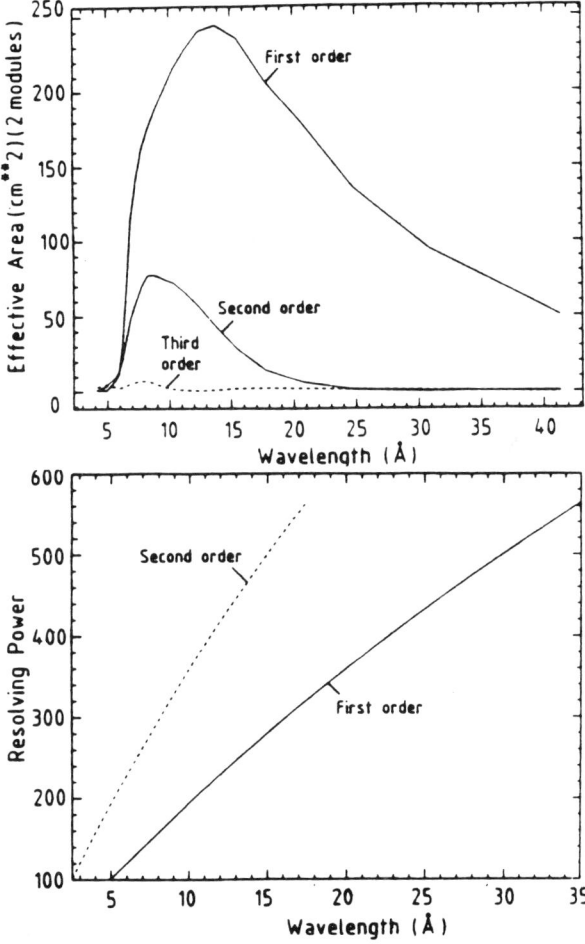

Figures 4 (top) and 5 (bottom): The effective area and resolving power respectively as a function of wavelength for the proposed XMM design.

6. FABRICATION ISSUES.

We now turn to the question of fabrication, can the required reflection gratings be made with current technology? As mentioned earlier, the principal challenge for the in-plane design involves the low blaze angle, $\delta = 0.70°$. However, since the required line spacing is rather coarse, $1/d = 646$ l/mm, the actual groove depth is not much less than has been previously fabricated.

There are two methods of producing gratings: by mechanical and holographic means (Michette 1986). Mechanical ruling is a burnishing process, a diamond stylus is dragged across the surface displacing the metal coating. At high pressures, the metal is fluid; the process is more akin to plowing a path through snow rather than cutting into metal. The groove shape is not well-defined by a single pass. It only begins to take its final shape when the neighboring grooves ae ruled as well. Hence, it is difficult to control accurately. However, the varied spacing required of our design does not present a problem for this technique. Since each groove is independently ruled, its position can be precisely controlled.

The holographic technique uses a photoresist exposed to a laser interference pattern to create the periodic structure. The result is then developed and ion-etched in a controlled manner to produce the desired triangular groove profile. Since this is not a mechanical process, it yields more predictable results. However, there is some question as to whether the required continuous variation in groove spacing can be produced with this technique.

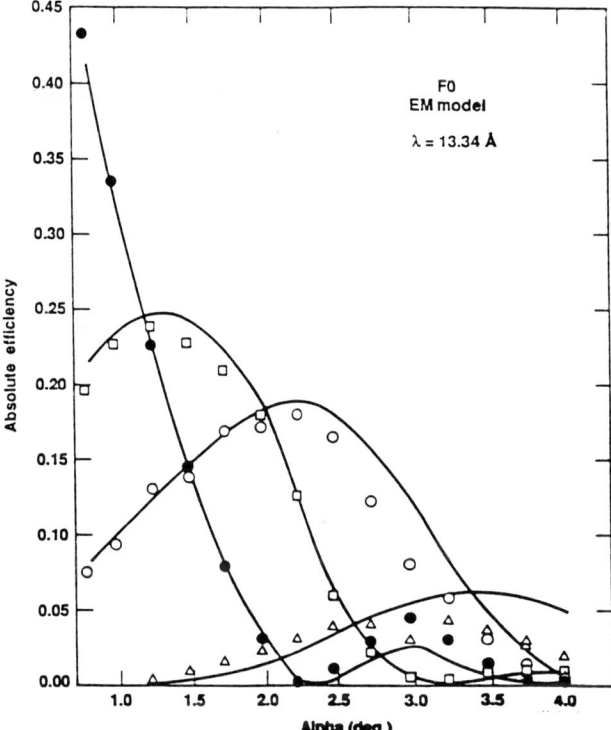

Figure 6: Measured absolute reflectivities in 0th (filled circles), 1st (open squares), 2nd (open circles), and 3rd (open triangles) spectral orders at 13.34 Å, plotted as a function of incident angle for a holographic prototype XMM grating. The solid lines give the predicted reflectivities from the exact electromagnetic theory assuming a perfect blazed groove shape. (From den Boggende et al. 1988).

In connection with the XMM proposal effort, we have acquired prototype grating samples of both varieties and tested them in dedicated facilities at both Utrecht and Livermore (den Boggende et al. 1988). The high diffraction efficiency expected has, in fact, been verified for both cases. The results for the holographic sample are shown in Figure 6. Here 0th, 1st, 2nd, and 3rd order absolute reflectivities at 13 Å are plotted as a function of incident angle. The solid lines give the theoretical predictions using the exact electromagnetic treatment. As can be seen, the agreement with the data is excellent, indicating that the groove shape exhibits a nearly perfect blaze profile. Further tests with varied space samples are currently in progress.

7. SUMMARY.

Reflection gratings offer special promise for incorporation on upcoming high throughput X-ray spectroscopy missions where limited angular resolution in the X-ray telescope prevents use of conventional transmission gratings. A variety of general considerations suggest that an optical design incorporating an array of gratings mounted in the in-plane configuration behind the X-ray mirror is particularly attractive for this application. A reflection grating spectrometer of this design is currently being proposed for the XMM mission. Preliminary measurements of prototype grating samples have verified that it is possible to achieve near theoretical reflectivity at the appropriate soft X-ray wavelengths.

8. REFERENCES.

Brinkman, A.C., Dijkstra, J.H., Geerlings, W.F.P.A.L., van Rooijen, F.A., Timmermann, C., and de Korte, P.A.J., 1980, *Appl. Opt.*, **19**, 1601.

Cash, W., 1988, *Proc. SPIE*, **830**, 204.

den Boggende, A.J.F., de Korte, P.A.J., Videler, P.H., Brinkman, A.C., Kahn, S.M., Craig, W.W., Hailey, C.J., and Neviere, M., *Proc. SPIE*, **982**, 283.

Gorenstein, P., 1978, in *New Instrumentation for Space Astronomy*, ed. K. van den Hucht and G.S. Vaiana, Oxford, Pergamon Press, p. 232.

Hettrick, M.C., and Bowyer, S., 1983, *Appl. Opt.*, **22**, 3921.

Holt, S.S., 1989, this volume.

Hutley, M.C., 1982, *Diffraction Gratings*, Academic Press, London.

Madden, R.P., and Strong, J., 1958, *Diffraction Gratings*, Appendix P, in *Concepts in Classical Optics*, Freeman and Co., San Francisco.

Michette, A.G., 1986, *Optical Systems for Soft X-Rays*, Chapter 7, Plenum Press, New York.

Peacock, A., and Ellwood, J., 1988, *Proc. SPIE*, **982**, 277.

Petit, R., 1980, *Electromagnetic Theory of Gratings*, in *Topics in Current Physics*, **22**, Springer Verlag, Berlin.

Predehl, P., Brauninger, H., Burkert, W., Aschenbach, B., Trumper, J., Kuhne, M., and Muller, P., 1988, *Proc. SPIE*, **982**, 265.

Schattenburg, M.L., Canizares, C.R., Dewey, D., Levine, A.M., Markert, T.H., and Smith, H.I., 1988, *Proc. SPIE*, **982**, 210.

Seward, F.D., Chlebowski, T., Delvaille, J.P., Henry, J.P., Kahn, S.M., Van Speybroeck, L., Dijkstra, J., Brinkman, A.C., Heise, J., Mewe, R., and Schrijver, J., 1982, *Appl. Opt.*, **21**, 2012.

THE SHEAL DIFFUSE X-RAY SPECTROMETER EXPERIMENT

W. T. Sanders, S. L. Snowden, and R. J. Edgar

University of Wisconsin-Madison, Madison, WI, USA

ABSTRACT. The Diffuse X-ray Spectrometer (DXS) experiment is part of NASA's SHEAL 2 mission, scheduled to be flown as an attached Shuttle payload in 1992. The DXS is designed to measure the spectrum of the low energy (0.15 to 0.28 keV) diffuse x-ray background with energy resolution better than 0.01 keV. This paper describes the DXS experiment and presents the results of calculations of the anticipated data.

1. INTRODUCTION

At energies less than 0.28 keV the diffuse x-ray background is thought to originate mostly in a local interstellar plasma with a temperature near 10^6 K, if collisional equilibrium is assumed (McCammon et al. 1983; Bloch et al. 1986). X-ray emission in this energy range from such a plasma is believed to be primarily in the form of collisionally excited lines of such ions as Si VIII, Mg IX, S VIII, and Ne IX, and the purpose of the Diffuse X-Ray Spectrometer (DXS) experiment is to search for and study those emission lines. It is important to obtain spectra of this diffuse emission, both to verify that it arises from a hot plasma, and to study the properties of that plasma. Such properties include the temperature, the relative abundances of the heavy elements responsible for the emission lines, and the ionization state of the gas. The plasma may not yet have reached ionization equilibrium if, for example, the local component of the emission is due to a blast wave with an age of order 10^5 yr (Cox and Anderson 1982; Edgar 1986; Rothenflug and Arnaud, this volume).

2. THE DXS INSTRUMENT

Figure 1 illustrates DXS in the Shuttle orbiter in the SHEAL 1 configuration and Table 1 lists some instrument parameters. The DXS employs a curved lead stearate Bragg reflector to disperse the x-rays across the face of a one-dimensional position-sensitive proportional counter. Figure 2 presents a cross section of a DXS detector assembly and Table 2 lists some detector parameters. By having two detectors facing in complimentary directions and scanning back-and-forth across the same arc of the sky, complete wavelength coverage over the entire scan arc can be obtained. The effective DXS area - solid angle product is greater than 0.01 cm^2 sr from 0.15 keV to 0.27 keV, peaking near 0.019 cm^2 sr around 0.18 keV. In a mission with four days devoted to the collection of data by DXS, it is possible to obtain 5000 seconds of prime low-background data in each detector at all wavelengths for each of ten 15° x 15° regions along a 150° arc. This should result in about 10,000 counts in each spectrum, which will allow useful spectral fitting for the individual regions.

Figure 1. (below) DXS in the SHEAL 1 configuration. The DXS experiment consists of two instruments, one attached to each side of the Shuttle's cargo bay. Each DXS instrument is mounted on a SPOC (Shuttle Payload of Opportunity Carrier) plate that is attached to the Shuttle. The SPOC avionics delivers electrical power and ground commands to DXS and receives scientific and engineering data from DXS for transmission to the Operations Control Center. The arrows labeled "Diffuse X-ray Spectrometer" point to the rotating detector assemblies.

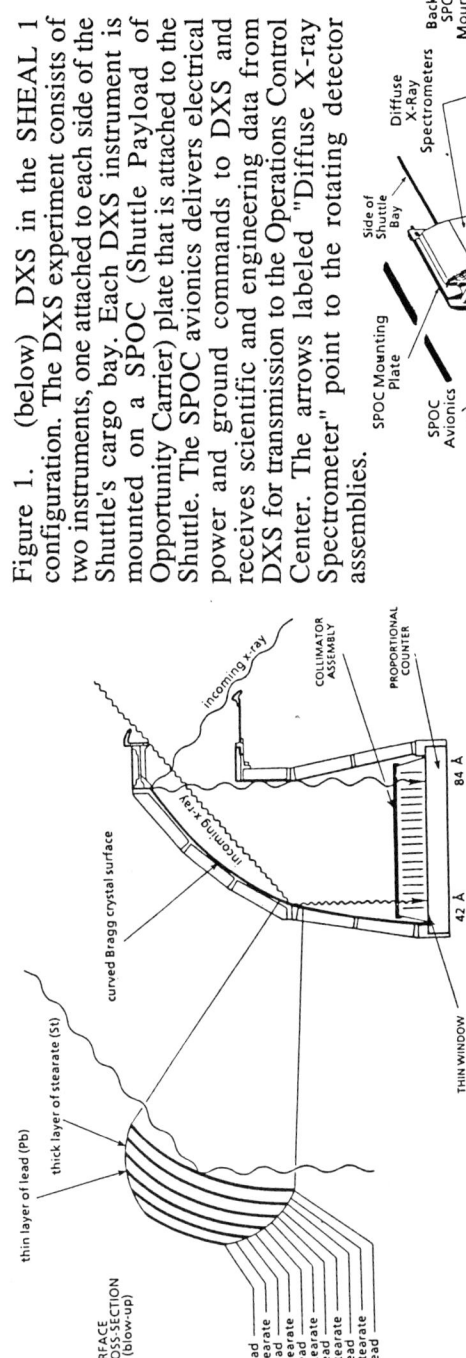

Figure 2. (above) A cross section of a DXS detector assembly. Incident x-rays are Bragg-reflected from a cylindrical lead stearate crystal panel and enter a collimated proportional counter that records the x-ray position along the direction of dispersion. Because a particular element of a counter sees only a limited piece of the crystal panel, it receives only those x-rays that are Bragg-reflected at a particular angle. At any given time, different elements of the counter along the dispersion direction receive different energy x-rays from different parts of the sky. Rotation of a detector extends the spectral coverage of sky directions in the middle of the scanned arc in addition to simply observing more of the sky. Adding a second detector that mirrors the scanning motions of the first greatly increases the fraction of the scan path that receives full spectral coverage.

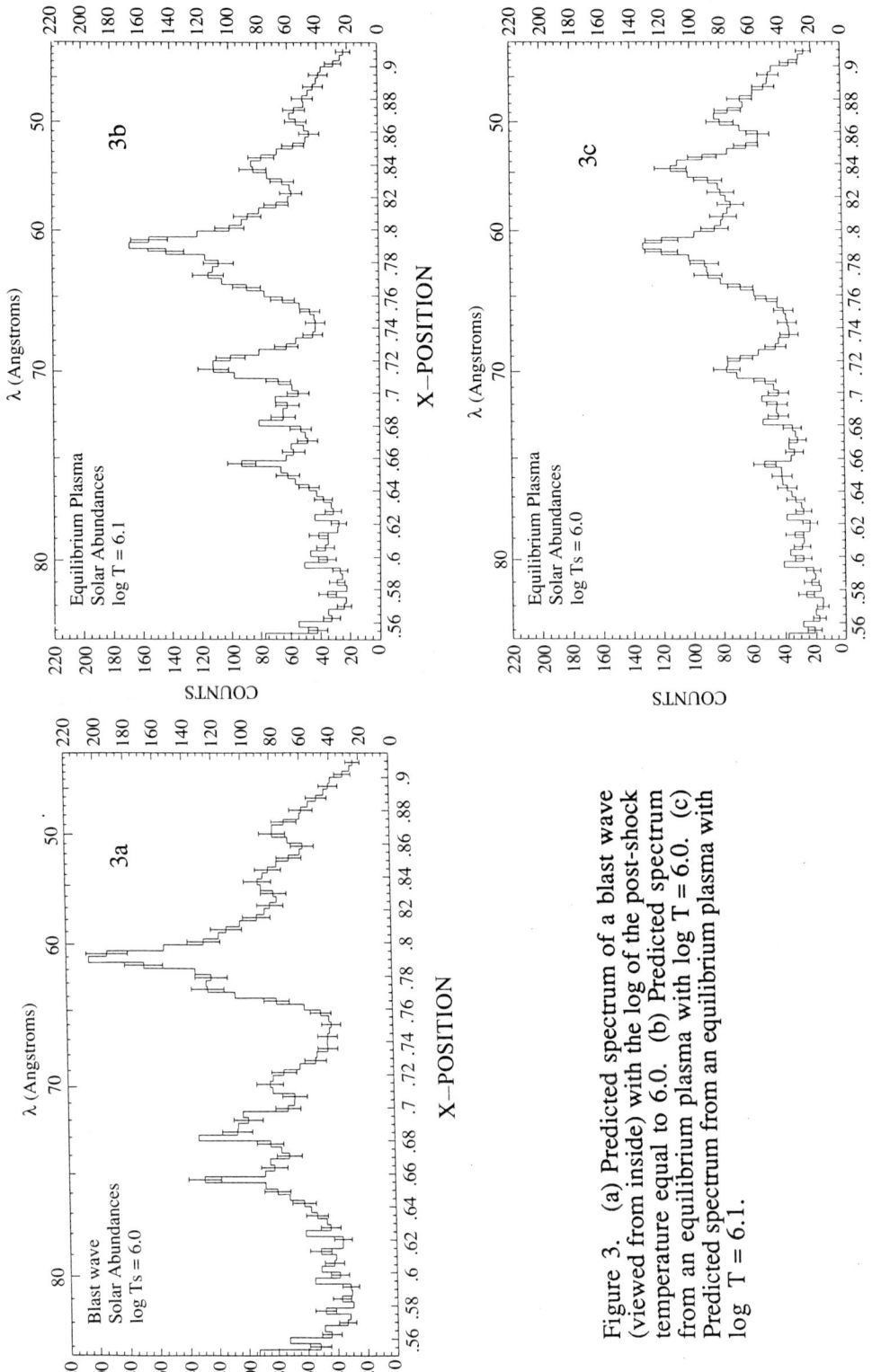

Figure 3. (a) Predicted spectrum of a blast wave (viewed from inside) with the log of the post-shock temperature equal to 6.0. (b) Predicted spectrum from an equilibrium plasma with log T = 6.0. (c) Predicted spectrum from an equilibrium plasma with log T = 6.1.

TABLE 1. Experiment Parameters

Number of Instruments:	2
Weight per instrument:	700 lbs
DC Power Consumption:	900 Watts
Size of SPOC Plate:	50" x 60"
Region Scanned:	15° x 150°
Observing Time per Resolution Element:	5000 s
Total Good Observing Time:	50,000 s
Mission Operational Time:	60 orbits

TABLE 2. Detector Parameters

Crystal Panel Radius of Curvature:	63 cm
Crystal Panel Dimensions:	14" x 24"
2d Spacing of Lead Stearate Crystals:	101 Å
Integrated Reflectivity of Crystals:	0.6 mr
Field of view:	15° x 15° (FWHM)
Energy Resolution:	$\Delta E \approx 10$ eV
Proportional Counter Open Area:	9" x 21"
Mesh Transmission:	68%
Thin Window Composition:	
Formvar:	60 µg cm^{-2}
UV 24:	30 µg cm^{-2}

3. THE MODELS

The blast wave spectrum shown in Figure 3a is that of a Sedov model with negligible external pressure. Ionization and recombination rates, as well as emission line parameters, are taken from the Raymond and Smith (1977, 1987) code. This allows for the gas to be in a non-equilibrium ion state, which can have profound effects on the resulting x-ray spectrum. Temperature effects can be seen by comparing Figures 3b and 3c, both of which assume an equilibrium plasma, with atomic physics again given by Raymond and Smith. Solar abundances are assumed, and the integration time in each case is 10,000 seconds (5000 seconds for each of two detectors). Each of the spectra also includes two absorbed components: one with log T = 6.5 absorbed by a hydrogen column $N_H = 1.0 \times 10^{20}$ cm^{-2}, and a 11E$^{-1.4}$ photons cm^{-2} s^{-1} sr^{-1} spectrum, absorbed by $N_H = 5.0 \times 10^{20}$ cm^{-2}. These two absorbed components contribute approximately 10% of the predicted counts.

It is a pleasure to acknowledge the guidance of the Principle Investigator of the DXS experiment, W. L. Kraushaar, and the work of D. McCammon, who designed and built the DXS detectors. The engineering staff of the University of Wisconsin's Space Science and Engineering Center deserves particular mention for design, contruction and testing of the instrument. This work was supported by NASA contract NAS 5-26078.

REFERENCES

Bloch, J. J., Jahoda, K., Juda, M., McCammon, D., Sanders, W. T., and Snowden, S. L. 1986, *Ap. J. (Letters)*, **308**, L59.
Cox, D. P. and Anderson, P. R. 1982, *Ap. J.*, **253**, 268.
Edgar, R. J. 1986, *Ap. J.*, **308**, 389.
McCammon, D., Burrows, D. N., Sanders, W. T., and Kraushaar, W. L. 1983, *Ap. J.*, **269**, 107.
Raymond, J. C. and Smith, B. W. 1977, *Ap. J. Supp.*, **35**, 419.
———. 1987, private communication.
Rothenflug, R. and Arnaud, M. 1988, this volume.

STIGMATIC SPECTROSCOPIC INSTRUMENTS FOR THE WAVELENGTH RANGE 30-300 Å WITH A HIGH ANGULAR AND SPECTRAL RESOLUTION USING MULTILAYER MIRRORS

E.N.Ragozin

P.N.Lebedev Physics Institute, Moscow, USSR

A concept of stigmatic spectroscopic instruments for the range $\lambda \sim 30\text{-}300$ Å is presented. A parallel beam is dispersed by means of a plane diffraction grating at grazing incidence, whereas focusing of the XUV radiation is performed by a concave multilayer mirror at normal incidence. A spectroheliograph of the new type may have dispersion an order of magnitude higher than traditional Wadsworth-type instruments. The theoretical resolving power of a spectrograph of the new type is limited by apertures of the multilayer mirrors or by the total number of grooves of the grating and many reach $(\lambda/\delta\lambda)_{theor} \sim mN \sim 3 \cdot 10^5$ using presently available optical elements.

1. INTRODUCTION

In 1882 Prof. Rowland conceived the idea of combining dispersive properties of a plane diffraction grating (DG) with the focusing performance of a concave mirror (Sampson, 1967). Presently, spectroscopic instruments with a concave DG are still the main tools of research in the vacuum UV. For normal incidence schemes with moderate astigmatism, the practical short wavelength limit due to a decrease of reflectivity lies in the range 200-300 Å. Grazing incidence instruments can be efficient down to ~ 10 Å, but the spherical DG at grazing incidence does not focus the beam in the direction perpendicular to the Rowland circle. Beside a decrease in brightness of spectral lines, this causes a loss of spacial information about the source and partly depreciates the spectral information. The aberrations of a concave DG at grazing incidence confine its width to the so-called optimum width W_{opt} (Eldén et al. 1932) providing the maximum brightness in the spectrum without significant loss of resolution. The theoretical resolving power in this case equals $(\lambda/\delta\lambda)_{theor} \sim mpW_{opt}$ (p is the groove density, m - the spectral order). We want to point out the prospect of developing a generation of stigmatic XUV spectroscopic instruments (spectroheliographs, spectrographs, dispersive microscopes etc.) with a high spacial and spectral resolution using multilayer mirrors (MM) (e.g., Gaponov et al. 1987). The elimination of astigmatism is bound with a separation of functions of dispersing and focusing: a plane DG should decompose a parallel beam into a spectrum, whereas normal incidence concave MMs should focus the radiation. The new scheme is void of limitations of principle so far as the width of the DG is concerned; this allows to increase the theoretical resolving power by an order of magnitude.

2. OPTICAL SCHEME OF A SPECTROHELIOGRAPH

Fig. 1 presents an optical scheme of a spectroheliograph. A plane DG at grazing incidence disperses a parallel beam into a spectrum; a concave MM at "nearly" normal incidence forms stigmatic monochromatic images of a remote source on its focal surface. Let the normal to the MM lie in the plane of dispersion, while the width of the diffracted beam be limited by the aperture of the MM. Then the focal curve is a circle of r/2-diameter tangent to the MM at its aperture center $f=(r/2)\cos\gamma(\lambda)$ (r - the radius of curvature of the MM, $\gamma(\lambda)$ - the angle of incidence for the axial ray). Let α and β denote grazing angles of incidence and diffraction, respectively. Firstly, we shall notify that the angular magnification of the instrument in the plane of dispersion differs from unity, namely $d\beta/d\alpha = \sin\alpha/\sin\beta$. For a source with equal angular dimensions in the two mutually orthogonal directions, the length of any image (except for the zero-order image) in the direction of dispersion will be either smaller (for $m>0$, $\beta>\alpha$), or larger (for $m<0$, $\beta<\alpha$) than its height in the orthogonal direction. An angular dispersion $d\beta/d\lambda = 10^{-8} mp/\sin\beta$ and a linear dispersion $dl/d\lambda = (r/2)d\beta/d\lambda$ can assume higher values (by a factor of $1/\sin\beta$) than for a normal-incidence mounted DG. The spectral interval $\Delta\lambda$ sufficient for separating two monochromatic images of the source with the angular dimension θ (rad) is

$$\Delta\lambda = \theta \cdot (d\beta/d\alpha)/(d\beta/d\lambda) = \theta \cdot 10^8 \sin\alpha/mp \quad (1)$$

Note that $(m\Delta\lambda)$ is constant over the spectrum and can be made sufficiently small by mounting the DG at small grazing angles $\alpha \ll 1$. The practical limit is imposed by the source, for the value of α cannot be notably smaller than the source size. For instance, for a DG with p=1200 grooves/mm at $\alpha = 8°$ and $2°$, the ratio $m\Delta\lambda/\theta_{min.arc}$ equals 0.34 Å/min and 0.085 Å/min, respectively. For $\theta = 32'$, the relevant values of $m\Delta\lambda$ in these cases are 10.8 Å and 2.71 Å. Relation (1) can be interpreted alternatively: it yieldes an angular resolution in the plane of dispersion corresponding to the spectral linewidth $\Delta\lambda$. Table 1 presents values of

β, $d\beta/d\alpha$, $d\beta/d\lambda$ and an inverse linear dispersion $d\lambda/dl$ (at r=4 m) versus λ. The angular dimension of a slightly astigmatic image of a point source $\gamma^2(\lambda)D/f$ characterizes the angular resolution and may be brought down to 1-10 sec of arc. It can be inferred that the scheme allows to achieve excellent parameters and imagery along with extending the operation range down to $\lambda \sim 30\text{Å}$.

3. OPTICAL SCHEME OF A STIGMATIC SPECTROGRAPH

This scheme comprises two MMs. The beam passing through the entrance slit is collimated by the normal incidence MM_1 and successively encounters the DG and the MM_2; the latter two are mounted, with respect to the beam, similarly to an instrument of Fig.1. For small angles γ and moderate apertures $D \sim 2(\lambda f^2/\gamma)^{1/3}$, a MM does not induce beam aberrations that deteriorate the spectral resolution of the instrument, so that $\delta\lambda \sim \delta\varphi \cdot d\lambda/d\beta \sim (\lambda/D_2)\sin\beta/mp$, or $(\lambda/\delta\lambda)_{theor} \sim mpD_2/\sin\beta$, where $\delta\varphi$ is the angular size of the diffraction pattern due to the aperture of MM_2. Since $L=D_2/\sin\beta$ is the illiminated width of the DG, the ultimate resolution is bound with manufacturing excellent plane DGs with a high overall number of grooves Lp. Assuming available values L=300 mm and p=1200 grooves/mm, we obtain a theoretical resolving power $(\lambda/\delta\lambda)_{theor} \sim 3.6 \cdot 10^5$, which exceeds that obtained using traditional grazing incidence spectrographs ($\lambda/\delta\lambda \approx 2 \cdot 10^4$) in the range $\lambda \sim 100\text{Å}$. For sufficient values of $dl/d\lambda$ to overcome the finite spacial resolution of the detector, outside spectral orders and small grazing angles $\beta \longrightarrow 0$ may be used at moderate values of f_2.

4. SPECTRAL RANGE

A stigmatic imagery, a higher resolving power and efficiency are achieved at the price of reducing the spectral range due to selective reflection from MMs. Vinogradov and Kozhevnicov (1986) calculated the integral reflectivity coefficient $\mathcal{R}(\lambda_0) = \int R(\lambda)d\lambda \sim R(\lambda_0)\Delta\lambda_{1/2}$ for multi-layer periodic structures optimized to yield the maximum value of $\mathcal{R}$ (λ_0 - resonance wavelength in Angstroms where the reflectivity coefficient R assumes its maximum value, $\Delta\lambda_{1/2}$ - the width (FWHM) of the reflectance peak). The computed ratio $\mathcal{R}/\lambda_0$ increases in the range 30 to 100Å; at a wavelength $\lambda_0 \sim 100$ Å it is largest for the pair Ru-B and equals $\mathcal{R}/\lambda_0 \sim 2 \cdot 10^{-2}$. It was shown that with an increase of the number of layers $\mathcal{R}$ reaches its maximum value considerably earlier than $R(\lambda_0)$, and that allows to vary $\Delta\lambda_{1/2}$ and $R(\lambda_0)$ maintaining the constant value of $\mathcal{R}$. Thus, the spectral range of the instrument will be a compromise with efficiency (transmission) within this range. Assuming, for instance, a peak reflectivity of 10% at $\lambda_0 \sim 100$ Å, we can reckon upon a bandwidth $\Delta\lambda_{1/2} \sim 20$ Å.

5. REFERENCES

Gaponov, S.V. et al. 1987, JTP Letters (Sov.Phys.), 13, 214.
Mack, J.E., Stehn, J.R., Edlen, B. 1932, J.Opt.Soc.Amer. 22, 245.
Sampson, J.A.R. 1967, Techniques of Vacuum Ultraviolet Spectroscopy (N.Y.: John Wiley & Sons).
Vinogradov, A.V., Kozhevnicov, I.V., 1986, Moscow, P.N. Lebedev Physics Institute, Preprint N° 103.

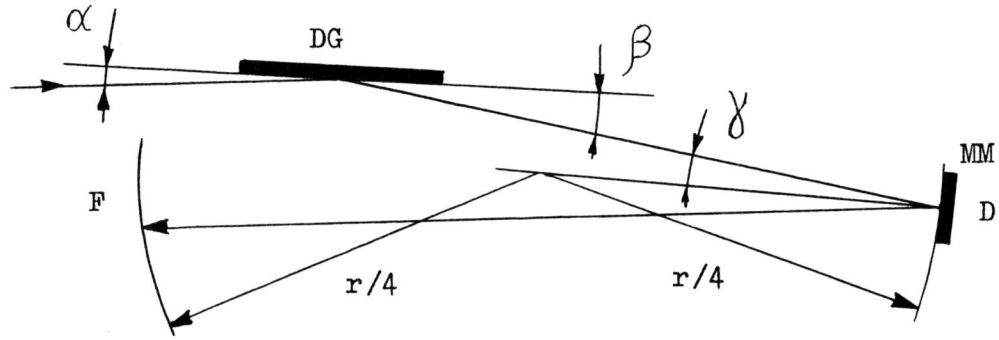

Fig. 1. Optical scheme of a spectroheliograph. DG - a plane diffraction grating; α and β - grazing angles of incidence and diffraction, respectively; D and r - an aperture diameter and radius of curvature of the multilayer mirror MM; $\gamma(\lambda)$ - angle of incidence for the diffracted ray, F - the focal curve. The grooves are orthogonal to the plane of the drawing.

Table 1. Parameters of a spectroheliograph (p=1200 grooves/mm, r=4 m) for two grating mountings.

	$m\lambda$, Å	β, rad	$d\beta/d\alpha$	$m^{-1} d\beta_0/d\lambda$ rad/Å	$md_0\lambda/dl$ Å/mm
$\alpha=8°$	-75	3.83-2	3.64	3.14-3	0.16
	-50	8.64-2	1.61	1.39-3	0.36
	50	1.78-1	0.79	6.79-4	0.74
	75	1.94-1	0.72	6.23-4	0.80
	100	2.09-1	0.67	5.79-4	0.86
	200	2.60-1	0.54	4.66-4	1.07
	300	3.04-1	0.47	4.01-4	1.25
$\alpha=2°$	25	8.50-2	0.41	1.41-3	0.35
	50	1.15-1	0.30	1.05-3	0.48
	100	1.59-1	0.22	7.58-4	0.66
	200	2.22-1	0.16	5.44-4	0.92
	300	2.71-1	0.13	4.48-4	1.12

Subject Index

ALEXIS	160
ASTRO-D	229, 247, 329
AR Lac	105, 177
AXAF	8, 11, 49, 53, 71, 94, 120, 171, 185, 229, 247, 263, 275, 276, 279, 329, 339, 351, 364
B band	146
BBXRT	229
Be band	154, 160
blazar	269
Bragg Crystal Spectrometers	136, 313, 349, 376
Broad Line Region	267
Burgess formula	3
C I	260
C II	37, 103, 132
C IV	20, 132
C V	102
C VI	299
Calorimeter	353
C band	146, 160
cellular structure of space	92
Ca XIX	20, 100, 102, 116, 127
Centaurus	224
collisional excitation	13
Coma Cluster	209, 224
coronal holes	11, 40
cyclotron lines	76, 78, 303
dielectronic recombination	5
diffraction grating	380
Distorted Wave cross section	5
double layer	92
double radio source	92
Elwert factor	32
Einstein Observatory	1, 17, 75, 111, 124, 167, 172, 192, 199, 209, 220, 271, 282, 285, 291, 295, 311, 319, 327

Subject index

EUVE	161, 335
EXITE	279
EXOSAT	1, 15, 16, 36, 110, 111, 122, 132, 157, 197, 202, 205, 210, 220, 311, 338, 365
Fe K-line	263, 346, 348, 365
Fe K-shell (Helium-like)	193, 200
Fe L-line	346, 348, 365
Fe VII	172
Fe VIII	160
Fe IX	149, 160, 296, 338
Fe X	148, 160, 338
Fe XI	338
Fe XII	291
Fe XVI	4, 299
Fe XVI-XIX	49
Fe XVII	4, 10, 49, 99, 137, 144, 211, 296
Fe XVIII	21, 117, 118, 141, 296
Fe XIX	118
Fe XX	90, 118, 141
Fe XXI	15, 102, 118
Fe XXII	102, 111, 120, 299
Fe XXIII	195, 268
Fe XXIV	3, 71, 97, 117, 118, 211, 299
Fe XXV	102, 116, 210, 211, 308
Fe XXVI	15, 193, 211, 263
Fe XIX	20
field aligned currents	91
Field criterion	44, 188
Gaunt factor	32
Ginga	70, 110, 140
HEAO-1	192
He II	38
Hinotori	15, 97
Hyades	176

Hubble Space Telescope	71, 263
HUT	71
ionization rate	2
interstellar medium	160, 362, 376
IRAS	168
IUE	103
JANUS	55
Konus experiment	64, 77
LAMAR	366
Leidenfrost layers	92
magnetosphere	90
M band	147
Mg VIII	40
Mg IX	376
Mg XII	348
MR2251-178	329
M-type stars	115
Multi-layer mirrors	320, 380
neutron stars	63, 70, 78, 85
Nd-YAG laser	55
non-equilibrium ionization	137, 155, 166, 170, 173
N II 6583	261
N VII	299
Ne III	141
Ne IX	18, 146, 172, 299, 376
Ne X	172, 348
OB star	201
OSO-5	15, 16, 119
O I	141
O II	141, 228, 260
O III	141, 228, 260

O VI-line	134
O VI	361
O VII	5, 8, 132, 138, 172, 299, 361
O VIII	99, 132, 134, 142, 172, 195, 210, 269, 299
P Cygni profile	267
photo-electric absorption	201
post T-Tauri stars	125
Rayleigh - Taylor instabilities	157
Raymond and Smith model	4
recombination rate	2
reflection gratings	333, 365
ROSAT	71, 110, 120, 161, 228, 262, 281, 291, 295, 311, 319, 329
RS CVn binaries	99, 105, 115, 122, 128, 176
SAX	302, 329
Schwarzschild criterion	44
Sedov model	144
S II	227
S II 6717/30	261
S VIII	376
S XII	299
S XVI	348
Si III	103
Si IV	37
Si VIII	376
Si X	40
Si XII	299
Si XIII	116
Si XIV	348
SIRTF	171
Skylab	11
SMM	79, 97, 116, 127
solar active regions	51
solar corona	11, 21, 40, 101, 111

solar flare	1, 18, 49, 91, 105, 108, 126
solar maximum mission (SMM)	11
SOLEX	8, 33
Spacelab II	11
spectroheliograph	375
Spectrometer	334
SPECTRUM-X-GAMMA	133, 302, 313, 361
SPEKTROSAT	120, 290, 310
star-burst galaxies	323
Stellar Flares	127, 132
Sun	38, 40, 97
Sunyaev-Zel'dovich effect	216, 269
Tenma	86, 74, 140, 157, 192, 210
Ti^{+20}	57
transmission grating	360
transmission grating spectrometer	290, 334
white dwarf	70, 78, 156, 336
XMM	8, 49, 53, 94, 132, 185, 247, 280, 324
X-ray Calorimeter	356
X-ray polarimetry	74, 360

Object Index

1E0839	224
1E0839+2938	259
22 Vul	103
3C196	228
3C 273	300
3C 295	224, 237, 261
4U0115+63	73, 83
4U1543-47	205
4U1700-37	202
4U1700-37/HD153919 (binary system)	201
A85	237
A133	237
A262	237
A278	237
A376	237
A401	237
A426 Perseus	237
A478	237
A496	237
A539	237
A576	237
A592	237
A644	237
A665	237
A1060	237
A1644	237
A1689	237
A1767	237
A1795	222, 237, 261
A1877	237
A1983	237
A1991	237

Object index

A2029	227, 237
A2052	237
A2063	237
A2199	237
A2244	237
A2415	237
A2597	222
AB Doradus (HD 36705)	132
AD Leo	111
AWM4	253
AWM7	237
α CrB	110
α TrA	103
Algol	90, 124
Andromeda Nebula	290
AR Lac	105, 124
AT Mic	123
λ And	36, 124
BL Lac objects	263
β Dra	106
black hole	265, 275
Capella	15, 16, 97, 99, 110, 124
Cas A	138
Castor	125
Centaurus A	224, 237
Coma Cluster	209, 224, 328
Crab Nebula	295
Crab pulsar	69
σ^2 CrB	99, 110, 124, 128
Cygnus Loop	136
Cyg A	237
Cyg X-1	205, 316
Cyg X-2	195, 197

Eradinus	146
EQ Peg	123
Feige 24	111
Fornax	224
G191-B2B	337
GRB 030579	74
HD 560	124
HD 27130	116
(HD 36705)	132
Her A	237
Her X-1	73, 83
Hercules Hole	148
HR 1099	124
Hyades	176
Hydra A	219
HZ 43	111
II Peg	105, 124
III ZW 2	125
λ And	36, 124
LMC X-3	205
M31,	311
M87/Virgo	211, 225, 247
Magellanic Clouds	141, 168, 295, 304
M dwarf	123
MKW3	237
MKW4	237, 251
MKW9	251
MR2251-178	329
MXB 1636-536	85
ND132D	136
NGC1275	47, 227
NGC4151	315
NGC4696	252
North Polar Spur	146

OB star	201
Ophiuchus cluster	210, 224
OY Car	337
II Peg	105, 124
Perseus cluster	209, 237
Π^1 Uma	111
PKS 0745	222, 237
PKS 2155-304	269
Procyon	111
Prox Cen	126
Puppis A	17, 136, 172, 281
SC184	237
SC0004	237
SN1006	165
SN1572, Tycho,	156
SN1987A	316
σ^2 CrB	99, 181, 124
Sirius B	111
Sun	112, 119, 124
Tycho, SN1572,	156
Π^1 Uma	111
UX Ari	97, 111, 124
VW Hyi	337
V711 Tau (=HR 1099)	97
white dwarfs	70, 78, 156, 335
Wolf 630 AB	99, 111, 123
YY Gem	123

AUG 1 4 1990